Liaison Functions 2017

Core Programming Area at the 2017 AIChE Spring Meeting and 13th Global Congress on Process Safety

San Antonio, Texas, USA
26-30 March 2017

ISBN: 978-1-5108-4148-2

Printed from e-media with permission by:

Curran Associates, Inc.
57 Morehouse Lane
Red Hook, NY 12571

Some format issues inherent in the e-media version may also appear in this print version.

Printed by Curran Associates, Inc. (2017)

For permission requests, please contact AIChE
at the address below.

AIChE
120 Wall Street, FL 23
New York, NY 10005-4020

Phone: (800) 242-4363
Fax: (203) 775-5177

www.aiche.org

Additional copies of this publication are available from:

Curran Associates, Inc.
57 Morehouse Lane
Red Hook, NY 12571 USA
Phone: 845-758-0400
Fax: 845-758-2633
Email: curran@proceedings.com
Web: www.proceedings.com

TABLE OF CONTENTS

482030 AGILE Award Keynote Address

Abstract:

Factors in Selecting AGILE Honoree

1. Overall sustained excellent industry leadership in businesses that employ chemical engineers

2. Recognized for efforts in innovation and creating strategies that ensure success

3. Value of the keynote to the meeting attendees and the members of the Institute Schedule

Event	Time	Location
Continental Breakfast	7:00AM - 8:00AM	Henry Gonzalez Convention Center, Lila Cockrell Theatre
AGILE Keynote Address	8:00AM - 9:15AM	Henry Gonzalez Convention Center, Lila Cockrell Theatre

The 2017 AIChE Spring Meeting and 13th GCPS AGILE Keynote Address will be presented by AGILE award recipient John Y. Televantos, Partner, Arsenal Capital Partners.

John Televantos is a Partner at Arsenal Capital Partners, a middle market private equity firm focused on the specialty industrial, healthcare and specialty financial service sectors. Before joining Arsenal in 2006, Dr. Televantos was President of Aqualon Company, owned by Hercules, Inc.; President and Founder of Helios Chemical Company; and Chief Executive Officer for Foamex International.

Dr. Televantos began his career with Union Carbide in 1977 in silicones research. There he developed two key processes for the manufacture of Methyl Chloride and Siloxane polymers that were commercialized. These processes not only improved efficiency and reduced cost but also dramatically reduced waste. He moved through a variety of management positions there and later at ARCO Chemical, where he became vice president of strategy and development.

At ARCO John co-invented and led a team that developed and commercialized a breakthrough technology in polyol manufacture (a key raw material for polyurethanes) eliminating a whole process step along with waste and emissions, while at the same time opening the door to the creation of new higher-performance polyurethanes. The technology has since been licensed by most of the major global polyol manufacturers with significant waste reduction worldwide.

At Arsenal Capital, Dr. Televantos serves as co-head of the specialty industrial practice. During his tenure, he has been part of several acquisitions and successful exits in the chemicals and materials sectors. In addition to serving on the boards of several Arsenal-owned companies, he is lead director on the board of Balchem Corp.

479518 Tray Jet Flooding – Separating Facts from Fiction

Abstract:

One of the mechanisms limiting the capacity of cross-flow trays is jet flooding. This flooding mechanism is normally associated with high levels of entrainment by either the spray or froth hitting the tray above. Over the years several solutions have been proposed to extend the capacity of cross-flow trays by mitigating this flooding mechanism. Through careful experimentation it was found that some of these solutions do not deliver what is promised. In this presentation data from large scale pilot plants will be used to analyze the underlying fundamentals of jet flooding and possible ways of mitigating it. Feedback from the industry on the efficacy of these solutions will be presented.

477025 Dry Tray Pressure Drop of Sieve Trays Revisted

Abstract:

The sieve tray has been in the market place as a distillation device for many decades. It has been used extensively on distillation tray equipment worldwide as a highly efficient vapor/liquid-contacting device. Many people have examined dry tray pressure drop data for sieve tray decks including both authors. The primary author has published a recent paper correlating the data into an equation. However, he had left the "door open" for additional refinement of this correlation for those applications that might have high tray thickness to hole diameter ratios (e.g. tiny hole applications in aluminum as had been practiced in air separations). The secondary author made two more recent presentations on this very subject employing both simulator data and Computational Fluid Dynamics (CFD) for very thick trays. This current paper will combine all three into a single definitive correlation for the dry tray pressure drop of sieve trays. A correlation for sieve tray pressure drop is presented that incorporates the effects of tray thickness, sieve hole diameter and open area and is compared to data.

Can 2-Pass Moving Valve Trays Experience Multiple Hydraulic Steady States?

By

Matthew Olsson, Fluor, Houston, Texas, USA

and

Henry Z. Kister, Fluor, Aliso Viejo, CA, USA

Presented at the Distillation Topical Conference, AIChE Spring Meeting, San Antonio, Texas, March 26-30, 2017

Summary

Our study reveals that multiple steady state vapor/liquid distributions may be experienced in 2-pass moving valve trays at turndown. Due to the symmetry of 2-pass trays, a perfect split of both vapor and liquid between the passes is always one possible distribution. The others involve having uneven vapor distribution. Some of these steady states effect wide variations in L/V ratios from panel to panel. This maldistribution is likely to lead to major efficiency reductions.

The maldistribution described in this paper is not due to poor distribution of vapor or liquid to the trayed section from an inlet, and it is also not caused by differences in tray geometry between one side and the other, though we have the capability to evaluate these situations.

The cause of the maldistribution is that at turndown, the valves close, and while they are closing, the dry pressure drop remains constant regardless of the vapor rate. This makes it easy for the vapor to swing from one pass to another, forming a number of alternative steady states.

Using our model, we investigated the effect of uniform weeping on the steady state multiplicity. While shifting the steady state some, uniform weeping does not largely impact the steady states. We also looked at the effect of weir loads on the steady state multiplicity. The multiplicity appears to exist both at the lower and higher weir loads. Our findings are that this is a vapor-driven phenomenon with little effect on or influence by liquid unless the L/V ratio is extremely high.

One strategy that appears to alleviate the effects of the multiplicity issue is using valves of different weight. This is a common practice used by valve tray manufacturers. Our study shows that this practice can significantly lessen the severity of maldistribution at turndown, but can also cause a wider range of loadings to experience some maldistribution.

Background

Moving valve trays are known to provide a turndown of 4-5:1 from the maximum operating loads (1,2). However there have been reports (3) and experiences in which turndown of multi-pass trays, including 2-pass trays, fell short of the expected turndown, with indication that maldistribution played a role (3).

In an attempt to gain understanding of the loss in turndown, we applied the Fluor Multi-pass Maldistribution Model (MMM) (4) to some hypothetical tray designs to explore the mechanisms detected. This paper describes our investigation.

Model Tray Design

The tower modeled has 8 ft internal diameter, and contains six 2-pass trays at 24" tray spacing. The top tray, Tray 6, is fed by an internal feed pipe and center false downcomer. The bottom tray, Tray 1, has a sealed center downcomer.

Even trays have side downcomers, and odd trays have center downcomers. The downcomers are straight, with the same width at the top and bottom. The side downcomer width is 11 inches, and the center downcomer width is 10 inches. The clearance between the bottom of each downcomer and the tray deck is 1 inch for both side and center downcomers. 1.5-inch inlet weirs are used to seal the downcomers, and the outlet weirs are 2 inches tall.

The tray deck is 14-gauge. Round moving valves are used, all with 16-gauge thickness and 7/16" leg length. The open slot area is 11% of the bubbling area.

To link with a system that has been studied extensively and for which abundant data and reliable flood and efficiency data and predictions are available, we chose to perform the calculations for the separation of cyclohexane from normal heptane at 24 psia and total reflux. Abundant data for this system in single-pass sieve and valve trays have been published and are available in the literature (5-7).

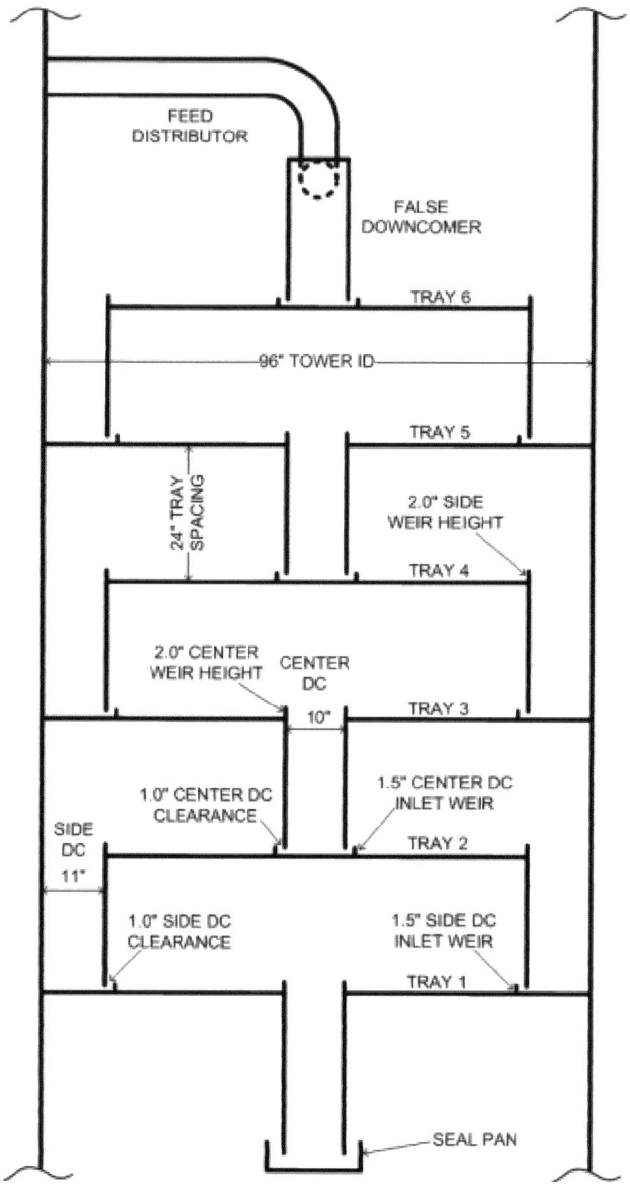

Figure 1. Key Tray Dimensions for Modeled Column.

Modeling Using the Multi-pass Maldistribution Model (MMM)

Our multi-pass maldistribution model is used to evaluate vapor-liquid distributions on multi-pass trays and to find hydraulically balanced distributions. Two equalities are required for a hydraulically balanced distribution for 2-pass trays, and both have to do with shared downcomers. For 2-pass trays, only the center downcomer is shared, and in the modeled column, Trays 1, 3, and 5 have center downcomers as illustrated in Figure 2.

The first equality is pressure drop. On trays for which liquid flows from the center downcomer to the side downcomers, the center downcomer divides the vapor space, and the pressure on either side does not need to be equal. On trays for which liquid flows from the side downcomers to the center downcomer, vapor is continuous above the tray, and the pressure equalizes in the shared space i.e. every two trays. For a balanced solution, the total pressure drop (through both trays) on either side of the shared downcomer must be equal.

The second equality is downcomer backup. As the center downcomer is shared, the calculated downcomer backup for either side must be equal.

The model used for this paper is an extension of our model to calculate maldistribution on 4-pass trays, presented previously (4). The 4-pass model was intended for tray design and rating, dividing the tower in half and hydraulically balancing the asymmetrical passes around the off-center downcomer to find a single solution. The tray design could be refined to match the L/V ratio on either side of the off-center downcomer.

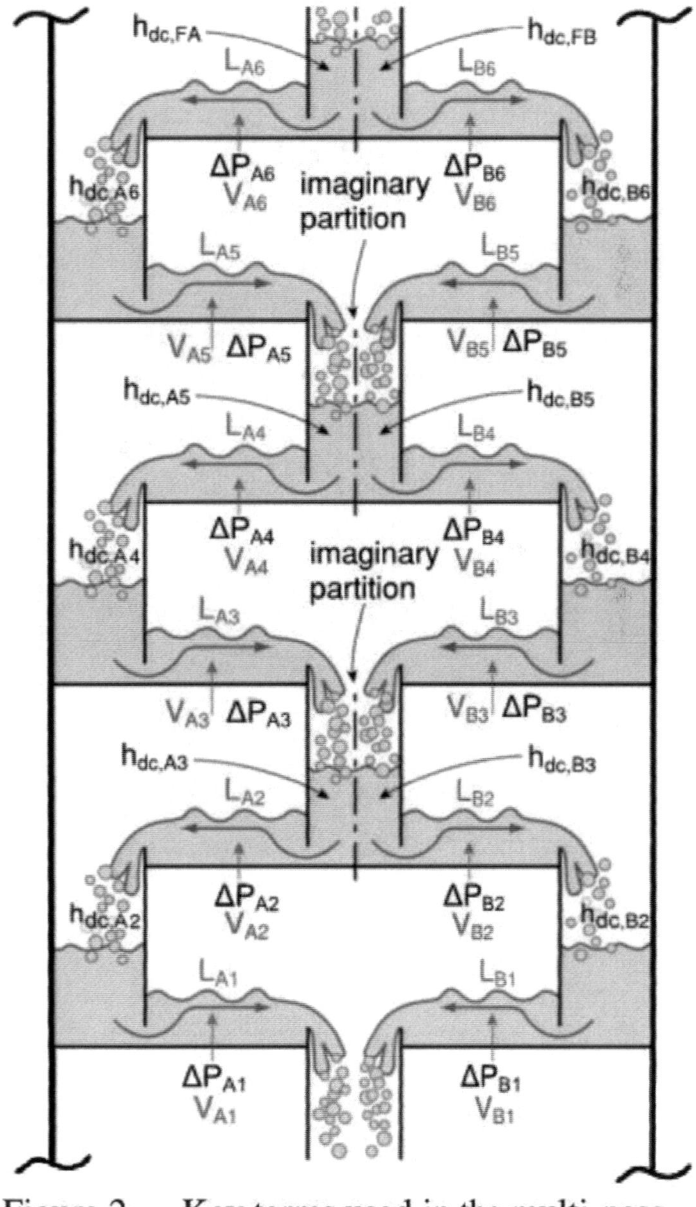

Figure 2. Key terms used in the multi-pass maldistribution model (MMM) equations.

The 4-pass model has only two

trays which are hydraulically linked, and the liquid rate to and from the center or side downcomer did not change. To obtain thorough insight into valve tray maldistribution, we developed a 2-pass maldistribution model that has 6 trays, hydraulically linked top to bottom, with a liquid feed (false downcomer) to the top tray and a vapor feed to the bottom tray. Total vapor and liquid loadings are constant throughout the tray section for both the 4-pass model and the new 2-pass maldistribution model.

With the 4-pass model, when the vapor loadings were such that one or more tray panels were operating in the region where valves open and close, the maldistribution would be much worse than if all valves were open or all were closed. We recommended avoiding this transition operating region throughout the tray section at design rates.

With 2-pass moving valve trays at turndown, the trays are operating in this transition region. The 2-pass maldistribution tool was designed to search for multiple hydraulically balanced solutions in the transition region. It was also designed to study the effects of feed maldistribution on the tray section, and it has features to calculate weeping and use multiple valve weights on a tray deck.

For the hydraulic calculations, we used the hydraulic equations described in our earlier work (4). For pressure drop calculation, we used the Klein method (8) and the Glitsch method (9), as we have had a lot of success with both. For downcomer backup, we used the classic downcomer backup equation in Perry's Handbook with the Klein method, and the Glitsch downcomer backup equation with the Glitsch method.

The vapor and liquid distributions were closely analyzed at 18%, 22%, 29%, 36% and 40% of the estimated jet flood point.

Steady State Multiplicity

2-pass trays are symmetrical, so a perfectly balanced solution with a 50%/50% split of vapor and the same for liquid is always a hydraulically balanced solution.

However, our MMM analysis found that for 2-pass moving valve trays, at some low loadings there is possibly more than one hydraulically balanced vapor and liquid distribution, a phenomenon we refer to as steady state multiplicity. Such multiplicity shows up when using the Klein pressure drop method (8), but not when using the Glitsch pressure drop method (9). The difference in the behavior of these methods at turndown is due to the combination of the valve tray dry pressure drop characteristic and the aeration factor.

For a moving valve tray, as the gas flow rate increases, the dry pressure drop increases to the point that the valves begin to open, the closed balance point (Figure 3). Then, the dry pressure drop stays constant (for the Klein method) from the gas flow rate when the first valve opens to the rate when all valves are open, the open balance point. Dry pressure drop again increases with the gas flow rate once all valves are open.

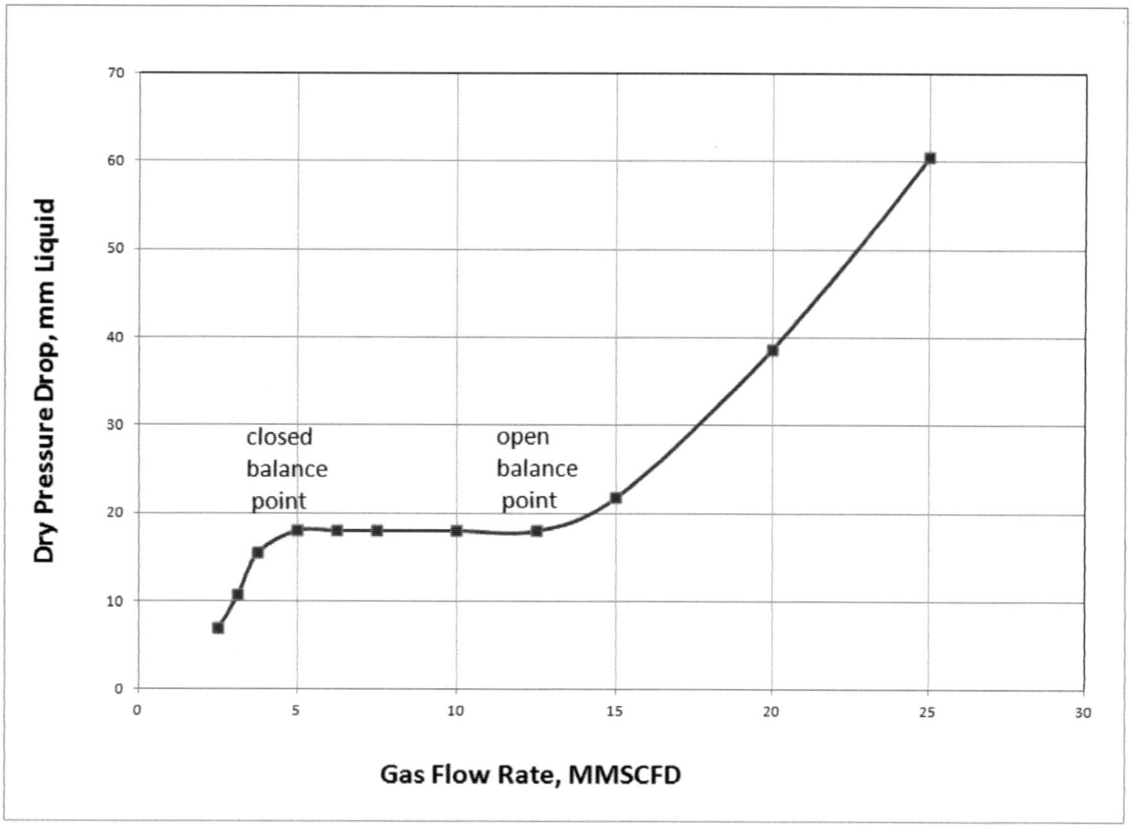

Figure 3. Typical Valve Tray Pressure Drop Characteristic

Multiple Solutions with the Klein Pressure Drop Method

Figure 4 shows the dry pressure drop, pressure drop through liquid, and total pressure drop through Trays 4 and 5 on either side (A and B) of the center downcomer from Tray 5. The loading is 29% of the estimated jet flood point, and the liquid is evenly distributed between passes A and B for both trays. The x-axis is the fraction of vapor to pass A, with the remainder to pass B.

The dashed lines represent the calculated dry pressure drop through both trays for either pass. The pass A dry pressure drop resembles the dry pressure drop characteristic in Figure 3, and pass B is a mirror image. The dotted lines represent the pressure drop through the tray liquid for either pass. The liquid rate remains the same, but per the Klein method, the pressure drop decreases with increasing vapor load as the liquid becomes more aerated.

The solid lines at the top represent the total pressure drop through the trays, the sum of the dry pressure drop and the pressure drop through the liquid. As with the dry pressure drop, the total pressure drop increases with increasing vapor flow rate below the closed balance point and above the open balance point. However, at the rates when the valves open and close, the total pressure drop decreases with increasing vapor flow rate. This is because the dry pressure drop is constant for this region while the pressure drop through the liquid decreases with increasing aeration.

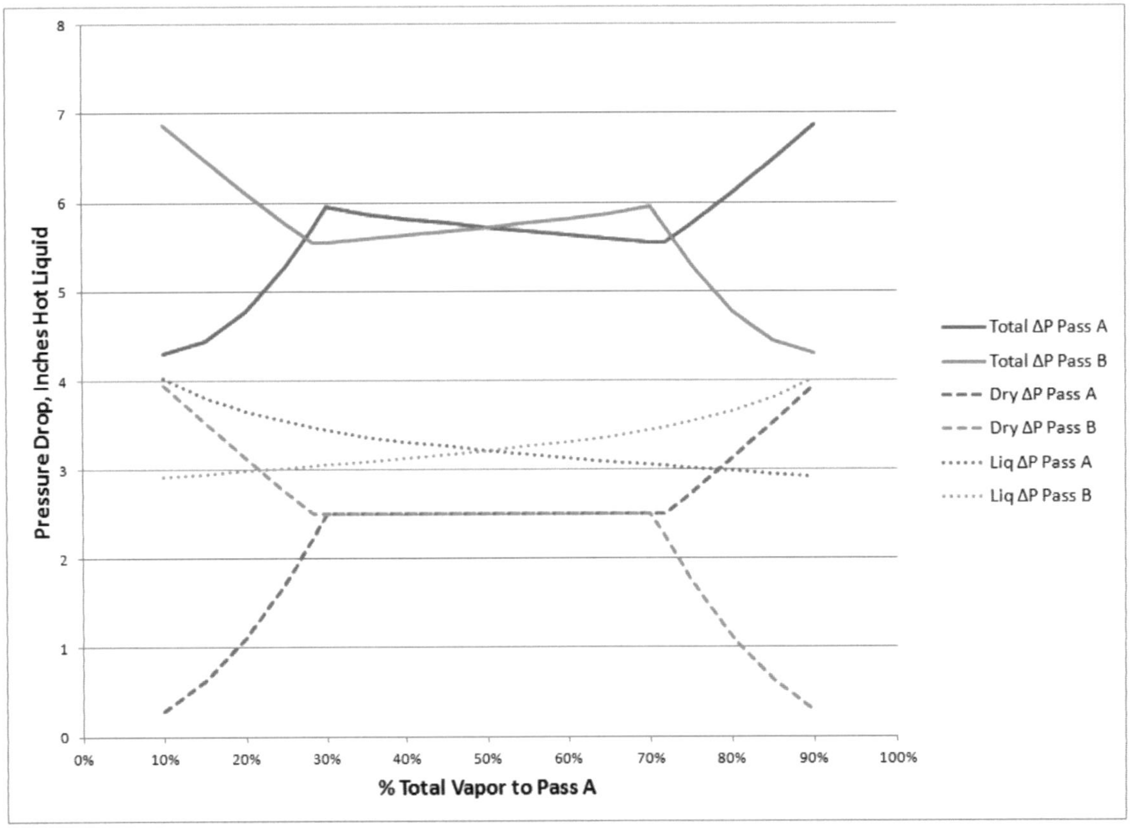

Figure 4. Dry, liquid, and total pressure drop through panels A and B of trays 4 and 5. Klein pressure drop equation, 29% jet flood.

The resulting shape of the total pressure drop curve includes a local maximum at the closed balance point and a local minimum at the open balance point. The total pressure drop curve of pass A intersects that of pass B three times. Each intersection constitutes a potential hydraulically balanced solution, where the pressure is equalized below Tray 4 and again above Tray 5, and the total pressure drop on either side of the center downcomer is equal.

To be a full solution, the downcomer backup in the center downcomer from Tray 5 also needs to be equal for each side. Figure 5 plots the downcomer backup from Tray 5 along with the total pressure drop through Trays 4 and 5 from Figure 4. Their shapes are similar, because the total pressure drop through Tray 5 is a major component of the downcomer backup. The other components (weir height, head over weir, head loss under downcomer, hydraulic gradient) are equal, because the liquid is split evenly between passes A and B for this diagram. The closeness of values of the total pressure drops to the downcomer backups is deceiving, as the plotted pressure drop is the total for two trays (4 and 5), while the downcomer backup is for only one downcomer (from tray 5), so the values of the downcomer backup per tray are of the order of about twice the total pressure drop per tray.

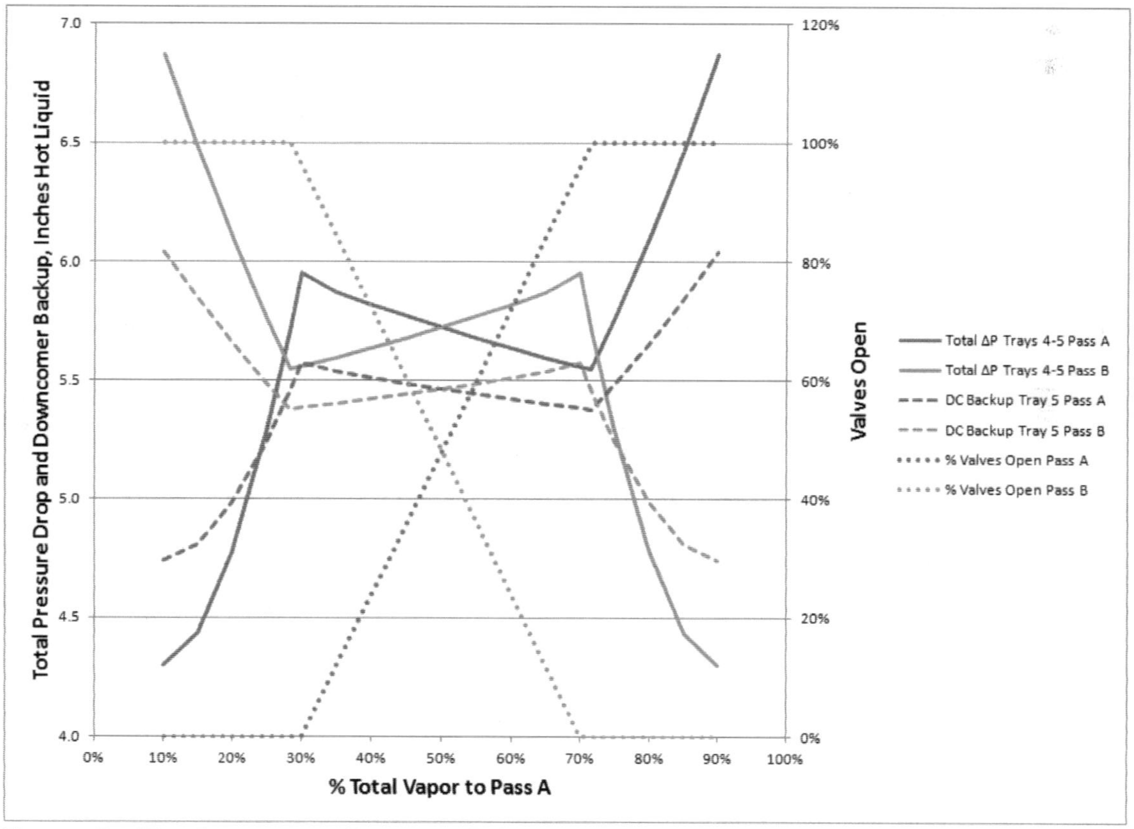

Figure 5. Total pressure drops and downcomer backups through panels A and B of trays 4 and 5. Klein pressure drop equation, 29% jet flood. At this load, multiple solutions exist.

Of course, the solution with both liquid and vapor split evenly between the passes (50% vapor to panel A, 50% liquid to panel A) is a solution on this diagram, as both the pressure drop curves and downcomer backup curves intersect at this point. At about 27% vapor to panel A, and for its mirror image at about 73%, the pressure drop curves and downcomer backup curves each intersect, at nearly the same vapor rate. To line these points up, only a slight liquid maldistribution is needed. For the 27% vapor to panel A, the solution is found with about 50.5% liquid to panel A. For 73% vapor to panel A, the solution is found with about 49.5% liquid to panel A. Each solution other than the even split has this slight liquid maldistribution. Based on our model, the only ways to significantly maldistribute the liquid are either to have a physical difference between the two sides or to have uneven weeping.

Variation of the Multiple Solutions with Percent Jet Flood

Figure 6 is similar to Figure 5, but based on 18% of the estimated jet flood. When the vapor and liquid are evenly distributed, less than 1% of the valves on each panel are open. The only regions present on Figure 6 are the increase in pressure drop with vapor rate when the valves are closed, the local maximum at the closed balance point (at about 50% of the vapor to Panel A or B), and a portion of the region where the valves are opening. Again, multiple solutions are found at the points where all valves are closed on one panel or the other, but, because of the closeness of the evenly split intersection to the closed balance point, the multiple solutions are not much different from each other, so the maldistribution is low. For only slightly less vapor than 18% flood, the evenly split intersection would be the only hydraulically balanced solution because it is in the region where pressure drop increases with vapor rate as all valves are closed.

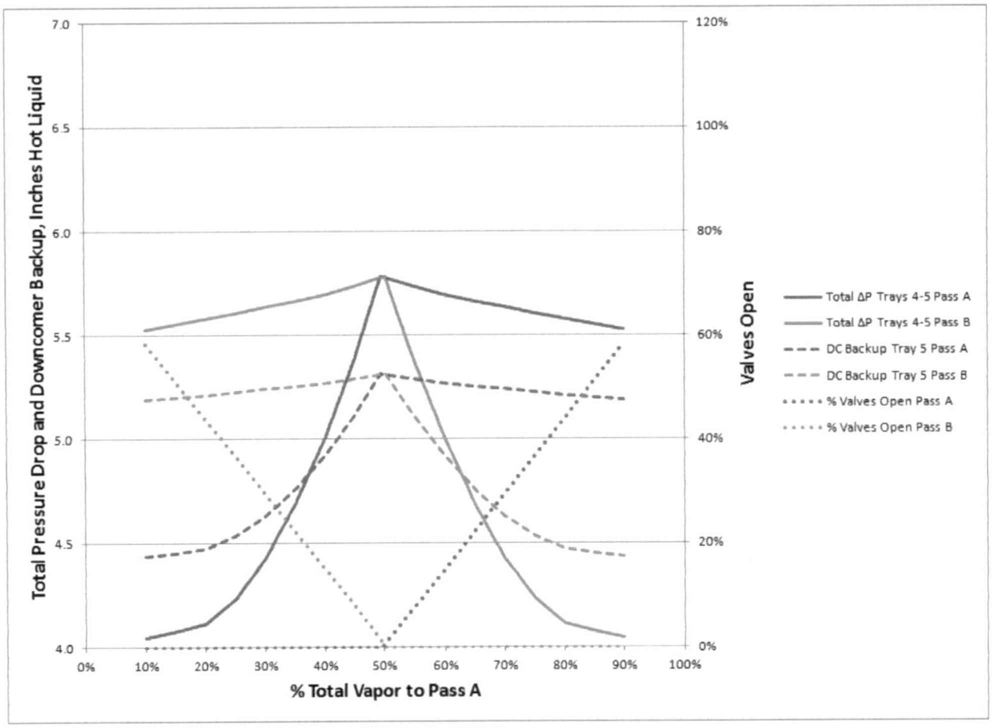

Figure 6. Total pressure drops and downcomer backups through panels A and B of trays 4 and 5. Klein pressure drop equation, 18% jet flood. At this load, multiple solutions still exist, but very close to each other, hardly visible.

For 22% jet flood (Figure 7), over 20% of the valves on each panel open at the evenly split intersection. As with 18% of jet flood, the other solutions are near the points where all valves are closed on one panel or the other. Because of the increased vapor load, these points are farther apart than in Figure 6, so the potential maldistribution at these loadings is greater than for 18% of jet flood.

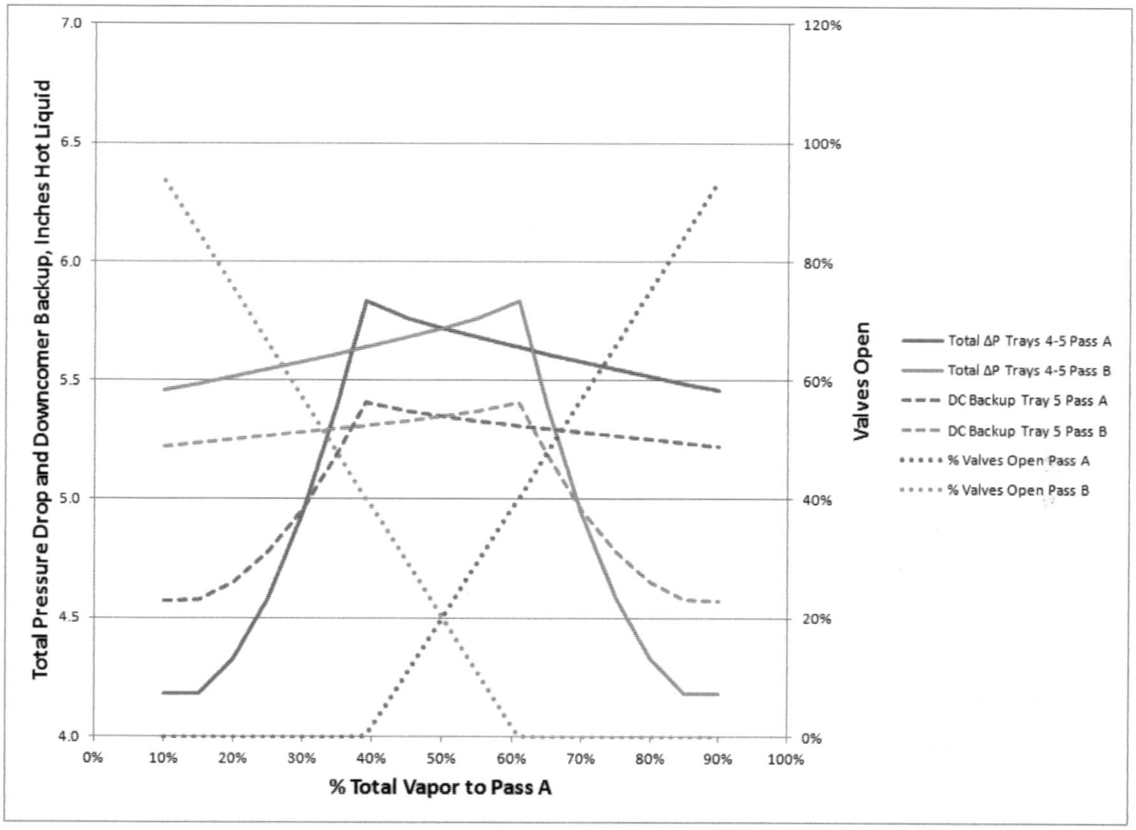

Figure 7. Total pressure drops, and downcomer backups, through panels A and B of trays 4 and 5. Klein pressure drop equation, 22% jet flood. At this load, multiple solutions exist, but closer together than in Figure 5.

For 29% jet flood (Figure 5), nearly 50% of the valves are open at the evenly split intersection, so this is in the center of the pressure drop curve. The other solutions are found at points where all valves are open on one panel and closed on the other. These solutions at 29% of jet flood (27% and 73%) are farther apart than for 22% of jet flood (37% and 63%), so this case has the highest worst-case maldistribution.

For 36% jet flood (Figure 8), about 75% of the valves are open at the evenly split intersection, so this point is closer to the open balance point. The other solutions are at points where all valves are open on one panel and some on the other. Because of the proximity to the open balance point, the multiple solutions are closer together than for 29% jet flood, so it has lower worst-case maldistribution. If the vapor loadings were increased further to the open balance point, the evenly split intersection would be the only hydraulically balanced solution.

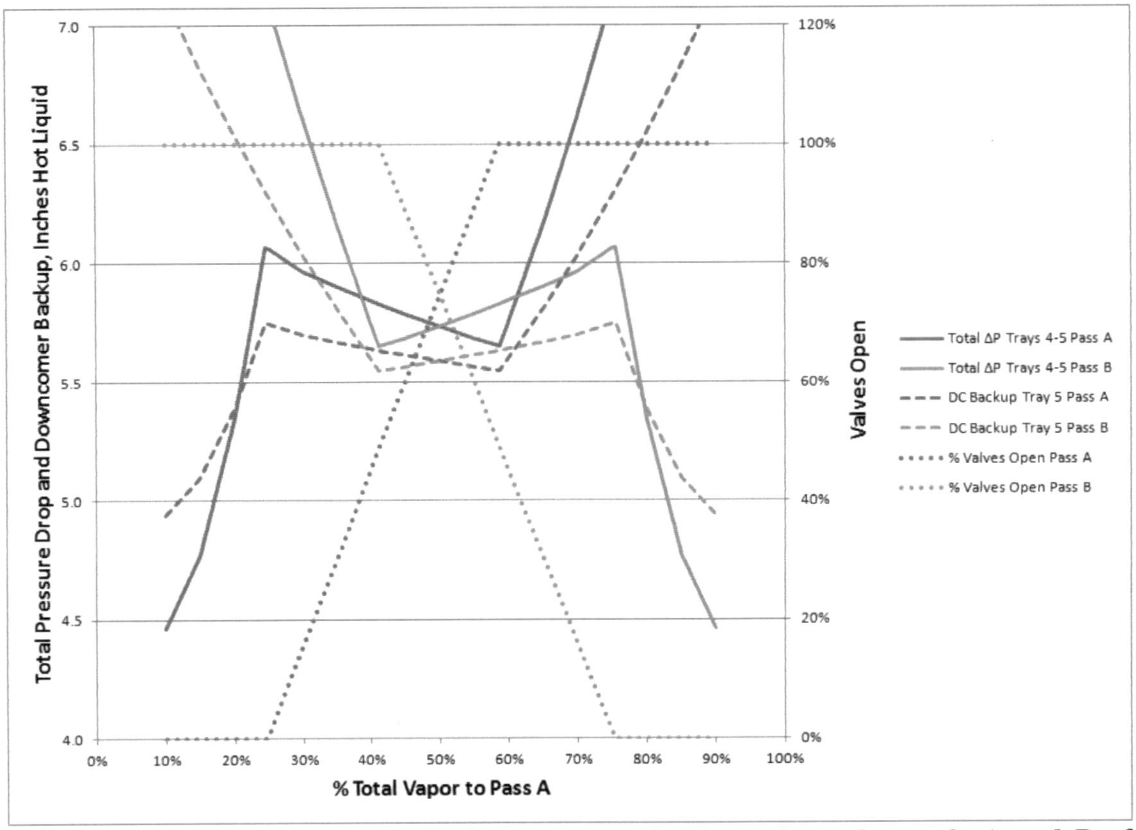

Figure 8. Total pressure drops and downcomer backups through panels A and B of trays 4 and 5. Klein pressure drop equation, 36% jet flood. At this load, multiple solutions exist, and again, closer together than in Figure 5.

The common thread for each of these loads is that at the evenly split intersection, both panels had some valves open but not all, and this is necessary for multiple solutions to exist. Also, the further away from either the open or closed balance point the evenly split intersection falls, the greater the potential maldistribution of the other solutions.

One difference between these loads is how the total pressure drop and downcomer backup for the alternate solutions compare with those for the evenly split solution. For 18% and 22% of jet flood, the maldistributed solutions actually have lower total pressure drop and downcomer backup than the evenly split solution. For 29% jet flood, the pressure drop and downcomer backup for the alternate solutions are slightly lower than for the evenly split solution, as well. For loadings only slightly higher than 29% jet flood, the three solutions would be equal, so switching between balanced solutions may be favored. For 36% jet flood, pressure drop and downcomer backup for the alternate solutions are higher than at the evenly split solution, and this should favor the evenly split solution. Comparing these findings, the implication may be that when greater than 50% of valves are open, even distribution is favored, but when less than 50% of valves are open, maldistribution is favored.

No Multiple Solutions for the Glitsch Pressure Drop Method

The Glitsch pressure drop method does not show multiplicity (Figure 9) because of two differences with the Klein method. The main difference is that for the Glitsch method the aeration factor for the tray liquid is constant, 0.4, so the pressure drop through the liquid does not decrease with increasing vapor load (at constant liquid) as it does with the Klein method. The other difference is that according to the Glitsch method, the dry pressure drop slightly increases with vapor load in the region where valves open and close, as opposed to the Klein method, where the dry pressure drop stays flat in that region (as in Figure 3).

Combining these two factors, the total pressure drop as calculated by the Glitsch method always increases with increasing vapor load. This results in only one intersection between the pressure drop curves of pass A and pass B, which is always at the evenly split distribution.

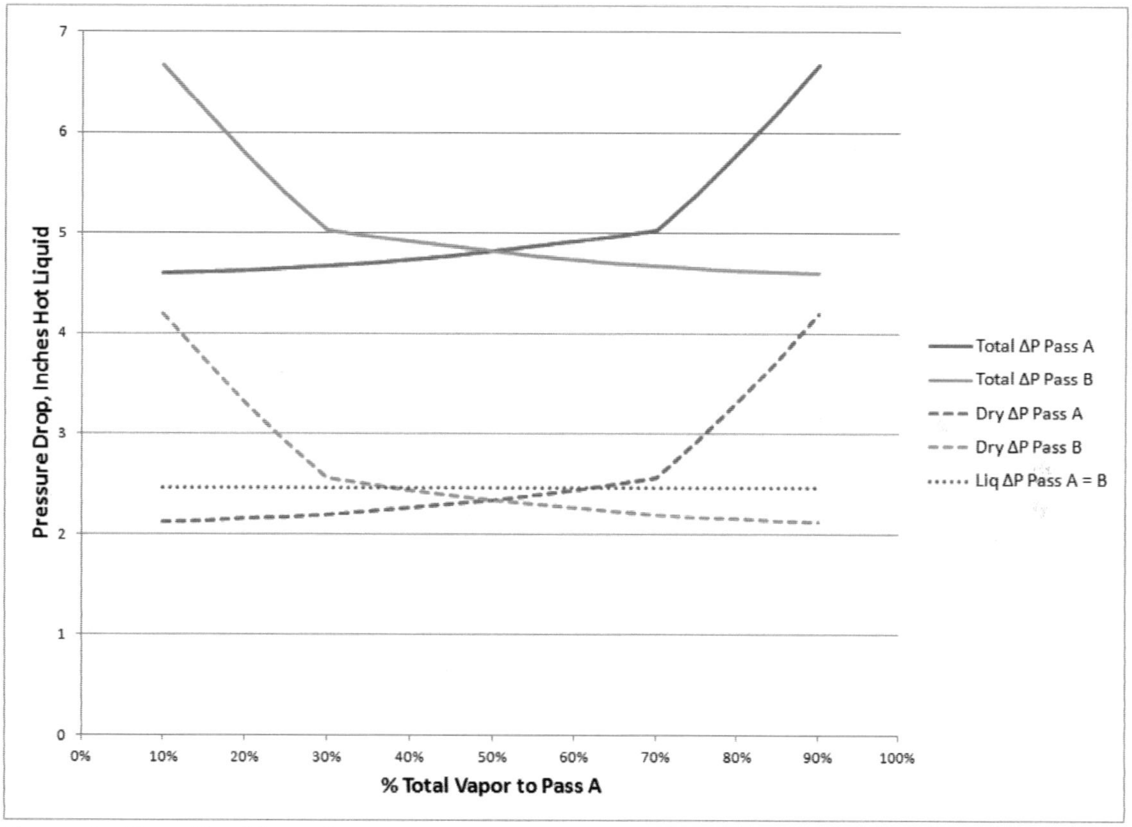

Figure 9. Dry, liquid, and total pressure drop through panels A and B of trays 4 and 5. Glitsch pressure drop equation, 29% jet flood.

Which Correlation is Better Here: Klein or Glitsch?

This brings the question: Which correlation do we believe? The key to the answer is to appreciate that both correlations focus on the region in most important to designers, i.e., when the valves are open and pressure drop is high. This is the region where dry pressure drop dominates. In this region both correlations predict much the same and their trends are robust and reliable.

Neither correlation was intended to address distribution in multi-pass trays at turndown. At turndown, measured pressure drops are low and subject to greater errors, so it is likely that the pressure drop trends are inaccurate. Further, due to the flatness of the pressure drop curve in this region, instability is likely, and gas can flip-flop from one panel to another, similar to the instability described by Brierley et al. (10).

An argument supporting the suitability of the Klein correlation for this analysis is that there is considerable evidence to suggest that due to increased aeration the liquid pressure drop on a tray diminishes as the vapor load is raised. This is featured in the Bolles and Fair correlation for sieve trays (11), the Colwell clear liquid height and froth density correlation (12), and even by a control issue reported by Buckley et al. (13). The Klein method's approximation of a liquid head that diminishes with an increasing vapor flow rate appears much more realistic than the Glitsch method's approximation that the aeration factor for the tray liquid is constant, 0.4, regardless of the vapor load. We therefore proceed on the basis of the Klein pressure drop correlation.

Distribution Criteria

Tray Stability

For low vapor loads, a criterion termed *tray stability factor*, η, was proposed by Summers et al. (3). It is a variation of the Froude number, comparing the upward inertial force of vapor to the downward gravitational force of liquid.

$\eta = (\Delta P_{DRY}/H_{CL})^{0.5}$

ΔP_{DRY} = calculated dry tray pressure drop, inches of liquid

H_{CL} = hydrostatic head, inches of liquid = total pressure drop – ΔP_{DRY}

Summers found that stability factors exceeding 0.6 limit weeping to below 30% and render trays stable. Note that Summers' method (3) derived the hydrostatic head H_{CL} values from total pressure drop measurements by subtracting from them calculated dry pressure drops as defined above. In the absence of measured total pressure drop data, Summers (14) recommends using the Colwell clear liquid height correlation (12) for calculating stability factors, and reports excellent experience with it. The values for the stability factors for our model runs are plotted on Figure 10 below. Figure 10 shows that the stability factors for the modeled trays were above 1.05 down to 20% of jet flood, suggesting no instability, less than 30% weeping, and minimal adverse effects due to weeping.

Summers et al. (3) also proposed an additional crude stability criterion for 2- and multi-pass trays to determine whether the total vapor loading is sufficient to make it likely that vapor will flow through each tray pass. The criterion is whether:

$\Delta P_{ACTIVE} > \Delta P_{INACTIVE}$

For this test, liquid would be distributed evenly among the tray passes. The inactive pass would have no vapor loading, and the full vapor loading would be divided among the remaining passes. In the case of 2-pass trays, all vapor would travel through one pass, with the liquid split evenly. The pressure drop through the inactive side is the sum of the weir height and the head over the weir, using the Francis Weir Formula (2). The pressure drop through the active side is the sum of the dry pressure drop and hydrostatic head calculated for the aerated liquid from Colwell (12). An equivalent expression for the multi-pass test is:

$\Delta P_{ACTIVE}/\Delta P_{INACTIVE} > 1$

We applied this test to the loadings used in our model and added the results to Figure 10. As long as the $\Delta P_{ACTIVE}/\Delta P_{INACTIVE}$ ratio is >1, this Summers et al. criterion expects no additional issues due to maldistribution between the passes. Figure 10 shows that this ratio exceeds unity for loadings above 25% of jet flood and is below unity at lower loadings.

Overall, of the two Summers et al. stability criteria, one (multi-pass instability) is consistent with our findings of 2-pass valve tray maldistribution at turndown, but only below 25% jet flood. The other (tray instability) does not anticipate this maldistribution. Neither anticipates the modeled maldistribution in the range of 25% to 40% of flood as is discussed below.

Bolles Distribution Ratios

Maldistribution varies the liquid to vapor (L/V) ratio from pass to pass. Pinching, and therefore poor separation, occurs in the low L/V regions, while the high L/V ratios in other regions only slightly offset the separation loss. Overall, observed efficiencies diminish, sometimes grossly. Bolles (15) defined the distribution ratio Φ as the ratio of the maximum pass L/V to the minimum pass L/V, and recommended keeping this distribution ratio below 1.2 for good tray efficiency. Summers (16) tightened Bolles' criterion to 1.1.

Distribution ratios can readily be calculated from our MMM model, since our model calculates the L and V for each pass for a given hydraulically balanced solution. Figure 10 shows the distribution ratios for the "worst-case maldistribution" from the MMM for each percent of jet flood based on the Klein pressure drop method. The term "the worst-case maldistribution" is used because we found up to three hydraulically balanced solutions loadings up to the vapor rate where all valves are open. One of the three is an evenly split distribution (distribution ratio of 1). The other two (mirror images in Figures 5 to 8) both have the same high "worst-case maldistribution" distribution ratios.

Figure 10 shows that for the tests between 20% of flood and 40% of flood, the distribution ratios were well above the maximum of 1.2 specified by Bolles, reaching a huge ratio of about 2.6 at 30% of flood. We have seen significant losses in tray efficiency in multi-pass trays when the distribution ratios exceeded 1.3, so 2.6 is huge.

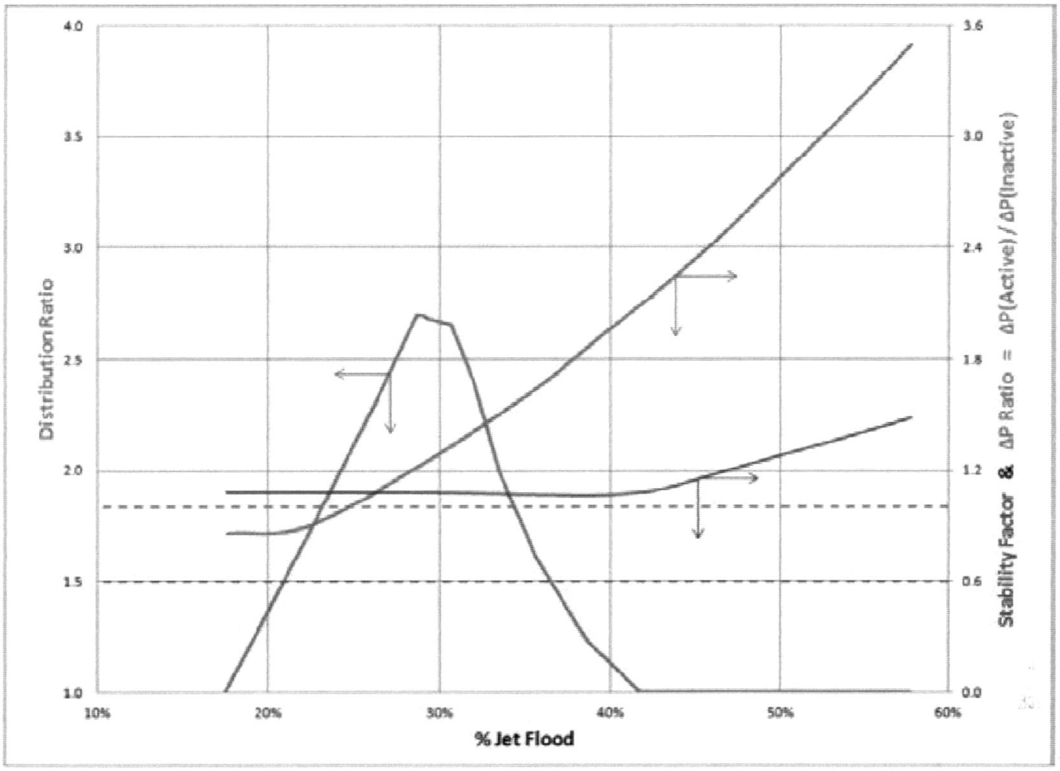

Figure 10. Applying the Summers et al. hydraulic metrics to explore instability: Stability Factor (red curve) and ΔP Ratio (green curve), compared with Bolles Distribution Ratio for Worst-Case Maldistribution calculated from our MMM model (blue curve).

What affects the Steady State Multiplicity?

Multiple Valve Weights

Tray manufacturers to often use two different valve weights on each tray panel, typically with 50% light gauge valves and 50% heavy gauge valves (17), to improve efficiency at turndown. The different weights are evenly spread across the tray (typically one row of lights, one row of heavies, rows perpendicular to the liquid flow) with the light valves forming the front row facing the incoming liquid. We explored the efficacy of this strategy to alleviate steady state multiplicity.

The dry pressure drop characteristic of a two-weight valve tray is similar to that of a single weight valve tray, except that there are two flat regions instead of the one in Figure 3, one for each valve weight. On Figure 11, for 29% flood, as the gas flow rate increases, the dry pressure drop increases to the point that the light valves (16-gauge) begin to open, the *light closed balance point*. Then, the dry pressure drop stays constant to the rate when all light valves are open, the *light open balance point*. Dry pressure drop again increases with the gas flow rate once all light valves are open, up to the point that the heavy valves (14-gauge) begin to open, the *heavy closed balance point*. Then, there is a second region in which the dry pressure drop is constant, up to the gas rate that all valves are open, the *heavy open balance point*. Above this point, dry pressure drop increases with increasing gas flow rate.

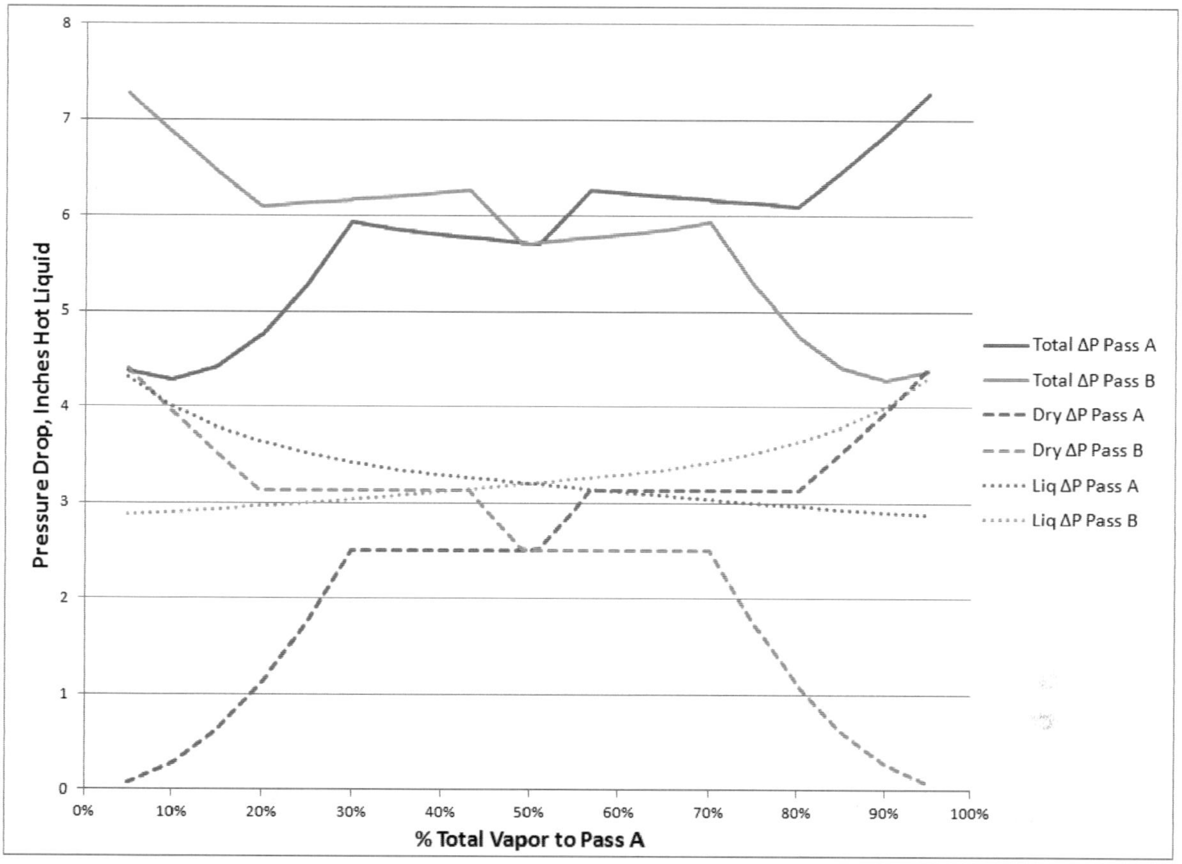

Figure 11. Dry, liquid, and total pressure drop through panels A and B of trays similar to trays 4 and 5, but with two weights of valves, 14-gauge and 16-gauge. Klein pressure drop equation, 29% jet flood.

Figure 11 shows that for 29% jet flood, a two-weight design would nearly eliminate the potential maldistribution seen in Figure 4. However, Figure 13 shows that at 40% of jet flood loadings, which had very little potential maldistribution with a single valve weight (Figure 12a), a two-weight valve design could experience significantly more maldistribution (Figure 12b).

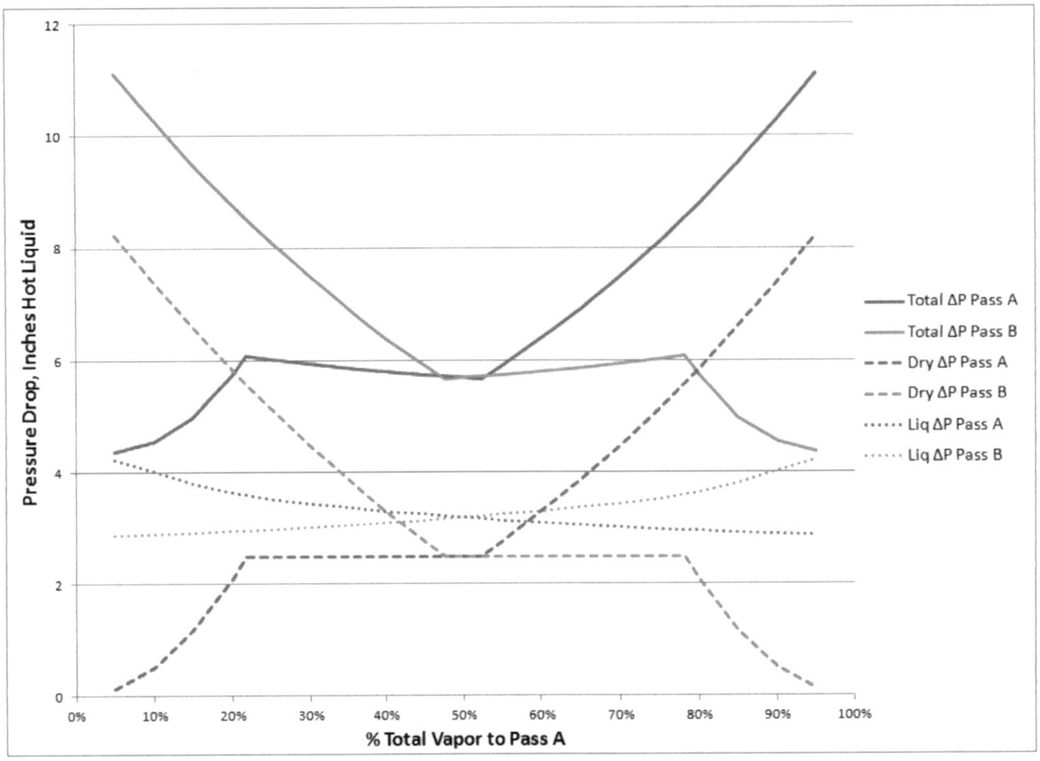

Figure 12a. Dry, liquid, and total pressure drop through panels A and B of trays 4 and 5. Klein pressure drop equation, 40% jet flood.

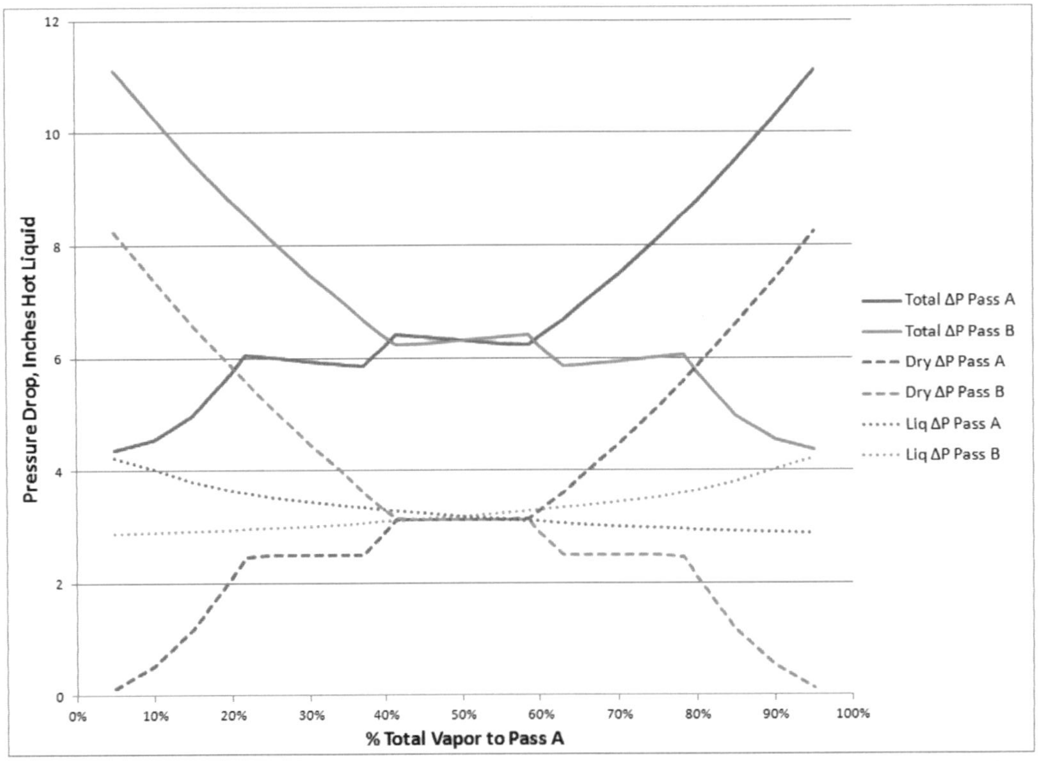

Figure 12b. Dry, liquid, and total pressure drop through panels A and B of trays similar to trays 4 and 5, but with two weights of valves, 14-gauge and 16-gauge. Klein pressure drop equation, 40% jet flood.

Figure 13 compares the distribution ratios for the worst-case maldistribution for the single valve weight to those of similar trays containing 50% light (16-gauge) and 50% heavy (14-gauge) valves. For the two valve weights the distribution ratios are much smaller, indicating a tremendous improvement. The turndown also improved, with the major maldistribution peak moving from about 30% to 23% of jet flood. On the debit side, the opening of the heavy valves generated a new maldistribution peak centered around 40% of jet flood. However, at this peak the maximum distribution ratio was 1.4, much less than the maximum in the single weight valve case, and in half of that peak the distribution ratios were within the Bolles criterion of 1.2 maximum. So the efficiency loss due to that peak should be much less than with the single-weight valves.

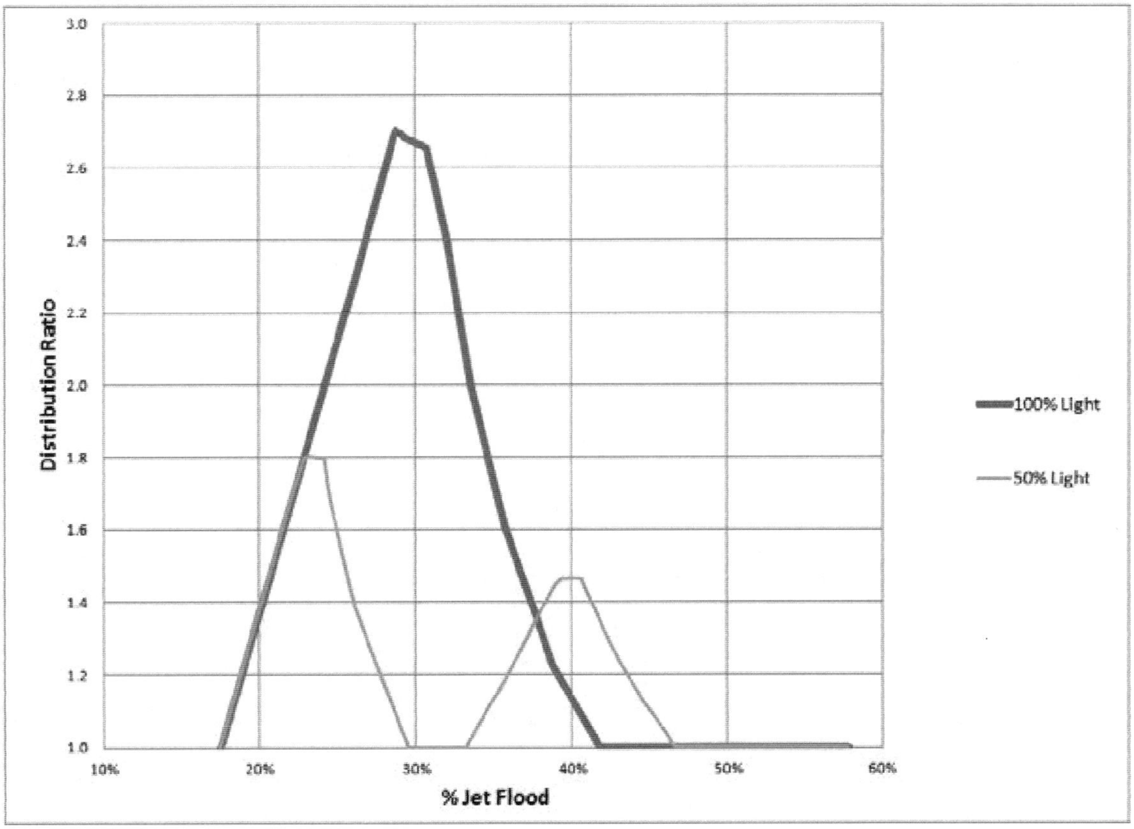

Figure 13. Comparing calculated worst-case distribution ratios of trays 4 and 5 with uniform valve weights (blue) to the calculated worst-case distribution ratios with two weights of valves, 14-gauge and 16-gauge (green).

Heavy Valve Gauge

To explore the effect of valve thickness, we ran the model again, pairing the 16-gauge valves with 12-gauge valves (Figure 14).

Comparing Figure 13 to 14, the distribution ratios are quite similar, and both are much lower than for the single weight valves as described earlier. Compared to the 14-gauge/16-gauge valves, the 12-gauge valves open at higher vapor rates than the 14-gauge valves. The maldistribution peaks associated with the heavy valves closing are shifted away from those of the light gauge valves. This adversely affects distribution closer to design rates, but it increases the range of good distribution between the peaks, where light valves are open and heavy valves are closed. There is no difference between the figures at vapor rates where only light valves are open.

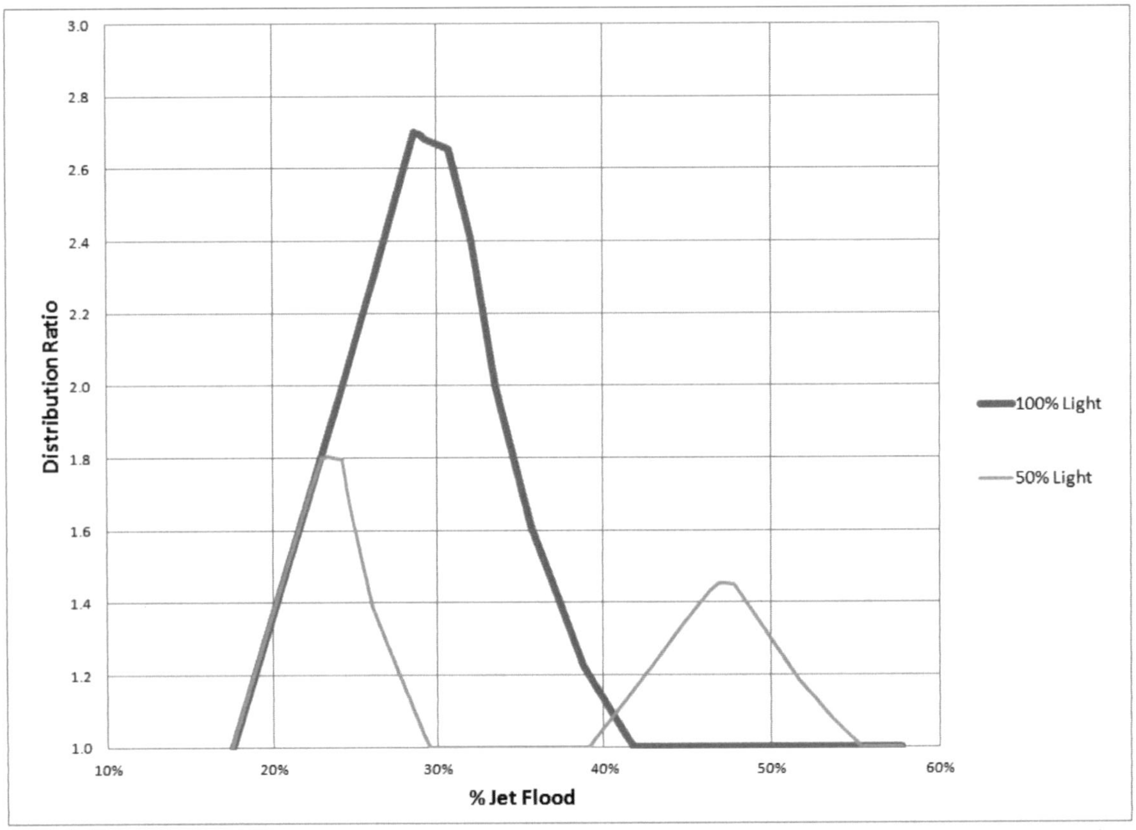

Figure 14. Comparing calculated worst-case distribution ratios of trays 4 and 5 with uniform valve weights (blue) to the calculated worst-case distribution ratios with two weights of valves, 12-gauge and 16-gauge (green).

Multiple Valve Weight Strategy

Figures 13 to 14, show that employing a multiple valve weight strategy tremendously reduces the distribution ratios for the worst-case maldistribution, and therefore has a beneficial effect at turndown. However, some loss of efficiency may be encountered at higher loads due to the opening of the heavier valves which is encountered at higher percent of jet flood value. Since it is also common that loadings vary significantly throughout a trayed section, while the tray design does not, at any point in time different trays in the tower may operate at different points on the same diagram. Whatever the case, the valve weight strategy should be reviewed for consistency with the predicted tower operation.

Vapor Tunnels

One strategy that is sometimes employed for multi-pass trays to allow for vapor to redistribute based on the pressure drop through each tray is to use vapor tunnels (4, 16). Vapor tunnels are passages through a shared downcomer which allow vapor to pass based on pressure differential between the tray passes on either side of the downcomer. These may also equalize the clear liquid height on the tray being fed from the shared downcomer, depending on their location vertically in the downcomer.

For this 2-pass trays case, vapor tunnels through the center downcomer would equalize the pressure on either side of the center downcomer. This would decouple the vapor distribution of the tray with the center downcomer from the distribution to the tray below. This would not address the cause of maldistribution at turndown, which is that the vapor rate is insufficient to open all the valves on the tray decks, resulting in no dry pressure drop difference between the evenly split solution and the maldistributed one. The difference would be that with vapor tunnels, every tray would have its own vapor distribution, rather than every two trays, and by this reasoning vapor tunnels should not significantly change the performance of a trayed section.

Weir Loads

Since our analysis was based on the cyclohexane-normal heptane system at 24 psia and total reflux, the weir loads were low. For any of the loadings between the closed balance point and the open balance point, the weir loading was less than about 2 gpm/inch for the side downcomers and less than about 1.5 gpm/inch for the center downcomers.

To determine if the multiplicity would persist at higher liquid loadings, we evaluated the turndown behavior of an alternate case. The alternate case has the same internal diameter, downcomer dimensions, and valve type as those in our model. The difference is that the weir loading was raised to 8 gpm/inch at 80% of flood. The L/V ratio for this case was 2.4, much higher than the L/V ratio of 1.0 for the total reflux case. An L/V ratio of 2.4 is of the order used in a stripping section of a fractionator, where 2-pass trays are frequently used.

The increase in liquid rate and L/V ratio did not increase the influence of the liquid on maldistribution. The maldistribution chart for the alternate case, Figure 15, looks the same as the one for the lower weir loads in Figure 13. The peaks appear to have shifted to the right. This is because for a given vapor rate, the increased liquid rate increased the percent jet flood. The vapor rates at which valves opened and closed did not change from Figure 13.

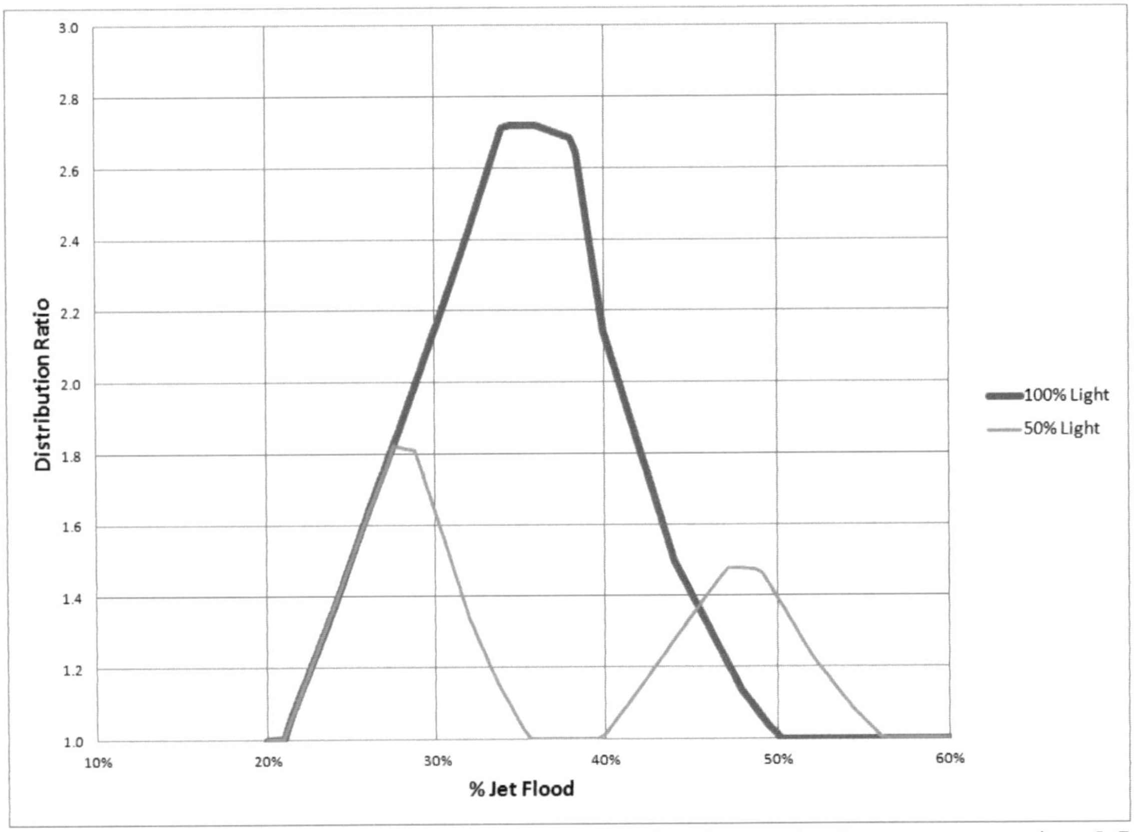

Figure 15. Comparing calculated worst-case distribution ratios between trays 4 and 5 for alternate loadings with higher weir loading than the total reflux, single weight 16-gauge valves (blue) vs. two weights of valves, 14-gauge and 16-gauge (green).

Unique Multiplicity at Extremely High L/V Ratios

For the alternate case in Figure 15, the L/V ratio was increased such that the weir loading would be 8 gpm/in at 80% of flood, and there was not a significant change in behavior from the total reflux case. Indeed, this is true for even higher weir loadings as well. For these cases, using multiple valve weights (16-gauge and 14-gauge) would be a good solution to limit maldistribution at turndown.

However, in some cases, the L/V ratio can be increased further, well above the normal range, such that the limiting factors in tray design are downcomer choke flood and downcomer backup flood rather than jet flood. We evaluated a second alternate case with a much higher L/V ratio to determine whether such liquid-dominated loadings would experience any unique behavior.

For the second alternate case, we used an L/V ratio of 8.6, much higher than the first alternate case (2.4) or the total reflux case (1.0). Unlike for the previous alternate case, the tray geometry needed altering for this high L/V case. The downcomers employed for the lower L/V ratios were far too small for the 8.6 ratio. To achieve a comparable design, the diameter of the tower was increased from 96" to 108", with all the additional area added to the downcomers, maintaining straight downcomers. The active and the valve open slot areas were unchanged. The side downcomer width was increased from 11" to 17.3", and the side downcomer clearance was increased from 1" to 4". The center downcomer width was increased from 10" to 17.75", and the center downcomer clearance was increased from 1" to 3".

With this design and 8.6 L/V ratio, at the vapor rate for which about half the valves are open (29% jet flood for total reflux loadings and modeled column trays), the weir loading is 9.3 gpm/in, and the trays are at 39.3% jet flood. As the loadings are increased to 80% jet flood, the weir loading is 19 gpm/in, the downcomer inlet velocity is just below 0.51 ft/s, and the center downcomer backup is 12.9" clear liquid. These values push against the maximum good practice limits (2), but do not violate them, giving at least a reasonable design for much of the operating range.

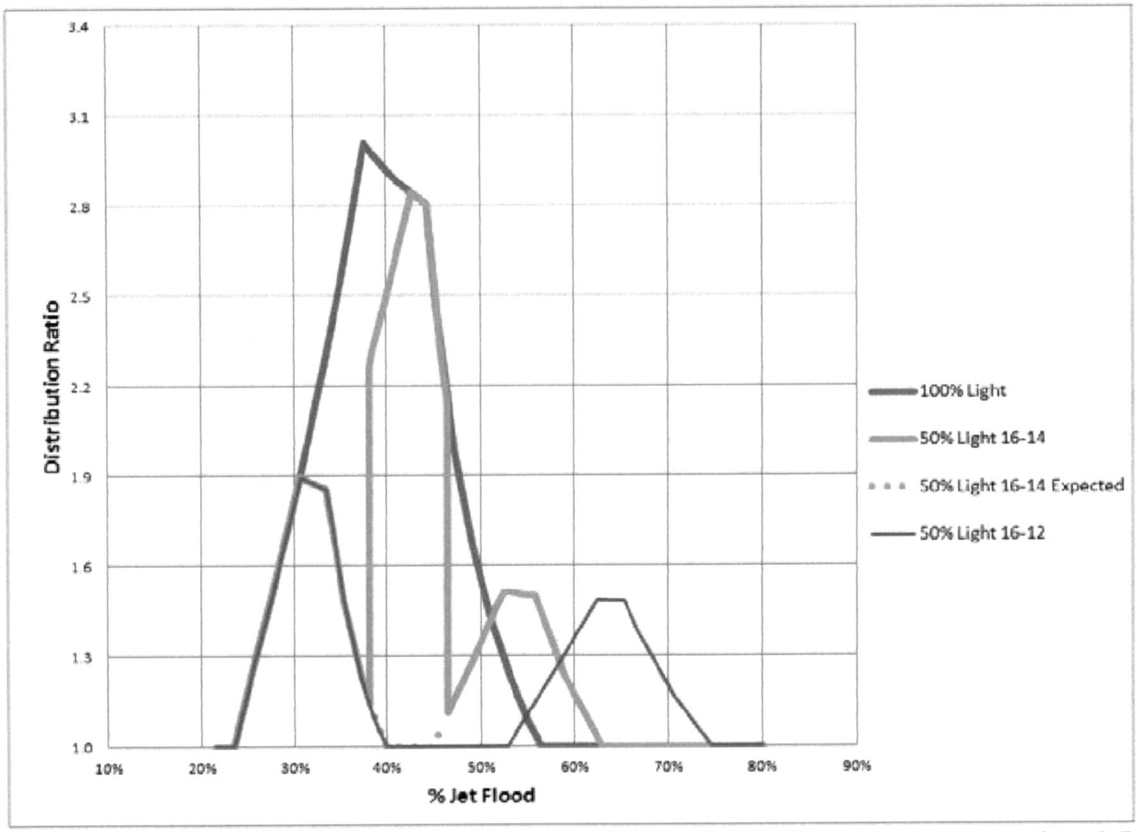

Figure 16. Comparing calculated worst-case distribution ratios between trays 4 and 5 of the larger diameter tower with high L/V loadings, single weight 16-gauge valves (blue) vs. two weights of valves, 14-gauge and 16-gauge (green) or 12-gauge and 16-gauge (red).

The distribution ratios for the 16-gauge single valve weight design and for two multiple valve weight designs (16-gauge/14-gauge & 16-gauge/12-gauge) are shown in Figure 16. For the single valve weight design, as well as for the 16-gauge/12-gauge design, the distribution ratio curves are similar to those for the lower L/V ratios (compare to Figure 14). In contrast, the design using 16-gauge and 14-gauge valves appears to no longer reduce peak maldistribution as effectively as it does for the lower L/V cases.

The portion of the distribution ratio curve for the 16-gauge/14-gauge design which deviates from the other two-weight curves is between 38% flood and 46% flood. The conventional solutions are still possible and shown as the dotted line, but for this range there are more maldistributed solutions. For the portion that the curve shares with the peak of the single-weight curve, all light and heavy valves are open on panel A, and all valves are closed on panel B.

For each of the designs discussed in this paper, the liquid is not maldistributed more than 0.5-1%, and that is also the case for the 8.6 L/V ratio. The difference in pressure drop through liquid for panel A and panel B is then due to the difference in the aeration factor, based on the amount of vapor maldistribution. For lower L/V ratios, the difference in pressure drop through liquid at turndown is very small, as on Figure 13, and is only significant enough to lead to an alternate solution for a single-weight design. With an 8.6 L/V ratio, the difference in pressure drop through liquid becomes much more pronounced. With the relatively small difference between the dry pressure drop at which the 16-gauge and the 14-gauge valves open, the more pronounced difference in pressure drop through liquid can exceed the dry pressure drop difference between the opening of the 16-gauge and 14-gauge valve weights. This allows for the behavior where heavy valves can be open on panel A, while light valves are closed on panel B. This is unique behavior for very high L/V ratios.

In order to eliminate this behavior, it is possible to increase the difference in dry pressure drop between the valve weights such that it is greater than the difference in pressure drop through liquid (due to aeration). This can be done by using 16-gauge and 12-gauge valves, shown in Figure 16.

Weeping

We estimated for each pass the amount of liquid that weeps through the valves on the tray deck and bypasses the downcomer. To do this, we used the Colwell and O'Bara modification (18) of the Lockett and Banik correlation for sieve trays (19), found in *Distillation Design* (20). The correlation uses the Colwell clear liquid height (12), which is a function of the weir load. The weeping liquid was removed from the weir load, and an iterative solution needed to be performed for each tray pass.

As this is a sieve tray correlation, it uses the full open area. This produced results with over 80% of tray liquid weeping for 18% flood and over 40% of tray liquid weeping for 22% flood. These results are unrealistic for moving valve trays because the valves can close, lowering the open area and increasing the vapor velocity, two factors which reduce weeping.

To correct for this, the open area was adjusted based on the number of valves open on the tray. For the Klein correlation, the closed balance point and open balance point are calculated in terms of the vapor velocity through the fully open hole area, with the closed balance point at a lower velocity and the open balance point at a higher velocity. The number of open valves is found by interpolating between these two points. The open area for weep is then adjusted based on the number of valves open. Closed valves are not completely closed, however. As an approximation, the velocity ratio between the open balance point and the closed balance point was used as the area ratio between an open valve and a closed valve.

With the area correction in place, the weeping was limited to less than 5% of tray liquid for all cases, making it mostly irrelevant to the distribution behavior discussed in this paper. It is possible that this is an overcorrection and that more weeping is possible at turndown. This would especially be true in the region where valves are opening and closing with no effect on the dry pressure drop, as some valves could be open for vapor while others are open but mostly weeping. This path was not explored as our model does not have a method to find solutions with this behavior, other than arbitrarily.

Uneven Weeping

We also explored the effects of uneven weeping on vapor and liquid distribution. We specified that for one side of the tower, 20% of the tray liquid would weep, while for the other side, 0% would weep. As the same amount of liquid weeps on each tray, the weeping liquid hydraulically does not exist on the tray, over the weir, in the downcomer, or under the downcomer escape, so it has no hydraulic influence anywhere. With this in mind, the conclusion was not surprising. This 1/5 of the tray liquid which weeps was 1/9 of the overall liquid. 4/9 of the overall liquid flowed across the tray and into the downcomer, with the other 4/9 on the opposite side. The vapor distribution was unaffected. The calculated clear liquid height, total pressure drop, and downcomer backup were all slightly lowered, but there was no hydraulic difference between the weeping side and the non-weeping side.

Final Thoughts

For 2-pass trays, the maldistribution occurs at turndown due to the multiplicity mechanism described in this paper that produces stable but maldistributed hydraulic steady states in which the vapor becomes highly maldistributed in the regions where valves open and close. We expect similar steady state multiplicities at turndown in multi-pass trays (with 3, 4, 6, 8 passes). Worse, these multi-pass trays do not have pressure equalization across the entire tray, as each vapor space is shared by a maximum of two passes. This could cause vapor maldistribution from one tray to propagate several trays upward.

Maldistribution varies the liquid to vapor (L/V) ratio from pass to pass. Lessons from packed tower maldistribution (20-22) extend to multi-pass trays (15, 20). Pinching, and therefore very poor separation, occurs in the low L/V regions, while the excessive L/V ratios in other regions only slightly improve separation. Overall, separation suffers and observed efficiencies diminish, sometimes grossly. Lockett and Billingham (22) show that the efficiency loss depends on the degree of L/V unevenness, as well as the proximity of the pinch.

Tools for reliably predicting multi-pass tray maldistribution are scarce. To our knowledge, no commonly available tray hydraulics rating tool will predict maldistribution in 2-pass trays. It is our hope that the findings and tools in our paper will promote understanding and good design to circumvent the turndown efficiency loss.

Conclusions

This study is the first to report the multiplicity of steady states in 2-pass valve trays at turndown, in the region where some valves open and others are closed. In this region the dry pressure drop becomes constant while the liquid head slowly declines with an increase in vapor load. There are generally three steady states, one with vapor split equally between the passes, while the other two produce stable but maldistributed hydraulic steady states. In the maldistributed steady states, the number of open valves differs from panel to panel, with a corresponding variation in vapor loads. This may cause the vapor to become highly maldistributed between the panels. This multiplicity can lead to unexpectedly reduced turndown in 2-pass valve trays. This multiplicity is found at loadings for which other predictive hydraulic metrics are satisfied.

According to our model, maldistribution due to multiplicity occurs without being induced by maldistribution of adjacent trays or feeds. It is possible whenever the vapor load is such that some valves are open while others are closed. It occurs for pairs of trays which are hydraulically linked by a shared center downcomer.

The distribution ratios (the ratio of the maximum pass L/V to the minimum pass L/V) produced by the maldistributed steady states can exceed 2.5 (a maximum of 1.2 is considered good). An even split of vapor between the two panels (with a distribution ratio of 1), is always one of the steady states. The distribution ratios for the two other maldistributed steady states vary strongly with the vapor rates, approaching unity near the open and closed balance points, and peaking where about half the valves are open.

The multiplicity and distribution ratios are insensitive to weir loads and to weeping. The multiplicity appears to be a vapor-driven phenomenon with little effect on or influence by the liquid.

The maldistribution can be largely alleviated and the distribution ratios reduced by using two valve weights, with 16-gauge and 14-gauge being the most common. For all but the extremely high L/V ratios, this practice would largely reduce the distribution ratios in the regions where valves open and close. However, this practice can also cause a wider range of loadings to experience some maldistribution.

We expect the steady state multiplicity reported here to extend to multi-pass trays. When the distribution ratios are high, we expect a possible significant efficiency loss upon turndown.

References

1. Kister, H. Z., "Distillation Operation", McGraw-Hill, NY, 1990.
2. Kister, H. Z., P. Mathias, D. E. Steinmeyer, W. R. Penney, and J. R. Fair, “Equipment for Distillation, Gas Absorption, Phase Dispersion, and Phase Separation”, Section 14, in R. H. Perry and D. Green “Chemical Engineers’ Handbook”, 8th Ed., 2008.
3. Summers, D. R., L. Spiegel, and E. Kolesnoikov, “Tray Stability at Low Vapor Loads”, Proceedings of “Distillation and Absorption 2010”, Eindhoven University of Technology, The Netherlands, p.611, 2010.
4. Kister, H. Z., R. W. Dionne, W. J. Stupin, and M. Olsson, ”Preventing Maldistribution in Multi-Pass Trays“, Chem. Eng. Prog., p.32, April 2010.
5. Yanagi, T., and M. Sakata, “Performance of a Commercial Scale 14% Hole Area Sieve Tray”, Ind. Eng. Chem. Proc. Des. Dev. 21, p.712, 1982
6. Glitsch, Inc. “Glitsch Ballast Trays A-1, V-1, V-0, Experimental test Data”, Bulletin Nos. 159/160, Dallas, Texas, 1983 (currently obtainable from Koch-Glitsch LP, Wichita, Kansas)
7. Kister, H. Z., “Effect of Design on Tray efficiency in Commercial Towers”, Chem. Eng. Progr., June, p. 39, 2008.
8. Klein, G. F., “Simplified Model Calculates Valve-Tray Pressure Drop”, Chem. Eng., p. 81, May 3, 1982
9. Glitsch, Inc., “Ballast tray Design Manual”, Bulletin 4900, 6th Ed., Dallas, Texas 1993
10. Brierley, R. J. P., P. J. M. Whyman, and J. B. Erskine, “Flow Induced Vibration of Distillation and Absorption Columns”, IChemE. Symp. Ser. 56, p. 2.4/45, 1979.
11. Bolles, W., and J. R. Fair, in J. M. McKetta (Ed.) “Encyclopedia of Chemical Processing and Design”, vol. 16, p. 86, 1982.
12. Colwell, C. J., Ind. Eng. Chem. Proc. Des. Dev. 20 (2), 298, 1981.
13. Buckley, P. S., R. K. Cox, and D. L. Rollins, "Inverse Response in a Distillation Column," Chem. Eng. Prog. 71(6), 1975, p. 83.
14. Summers, D. R., Private Communication, January 3, 2017
15. Bolles, W. “Multipass Flow Distribution and Mass Transfer Efficiency for Distillation Plates”, AIChEJ, 22 (1), 153, 1976
16. Summers, D. R., “Designing Four Pass Trays” Chem. Eng. Prog., p.26, April 2010
17. Bolles, W. L., "Estimating Valve Tray Performance", Chemical Engineering Progress, 72 (9), p. 43, 1976.
18. Colwell, C. J., and J. T. O’Bara, “Estimate Sieve Tray Efficiency in the Weeping Region”, Paper presented at the AIChE Spring Meeting, April 1989.
19. Lockett M. J., S. Banik, Ind. Eng. Chem. Proc. Des. Dev. 25, p.561, 1986.
20. Kister, H. Z., “Distillation Design”, McGraw-Hill, NY, 1992.
21. Pilling, M. “Ensure Proper Design and Operation of Multi-pass Trays”, Chem. Eng. Prog., p.22, June 2005.
22. Lockett, M. J., and Billingham, Trans. IChemE., 80, Part A, p. 373, May 2002; Trans. IChemE., 81, Part A, p. 134, January 2003.

480103 Troubleshooting EB/SM Splitters: How a Maldistribution Analysis Can Help?

Abstract:

Troubleshooting EB/SM Splitters: How a maldistribution analysis can help?

Senthil Krishnamoorthy, Sulzer Chemtech USA Inc.

Lee Ming Yang, Sulzer Singapore Pte. Ltd.

The ethylbenzene (EB) / styrene (SM) splitter (a.k.a. the EB recovery column) is the biggest column in the styrene production process. It can be considered a mega column: column diameters are often larger than 5 m, column heights are taller than 60 m, and 6 beds of structured packing are usually employed.

Generally, packed columns with large diameters are susceptible to large-scale maldistribution. For the EB/SM splitter, the low relative volatility (α-value) of the binary system, and the stringent product specifications, make the column extremely sensitive to maldistribution. Therefore, if an EB/SM splitter is exhibiting poor separation efficiency, a maldistribution evaluation is strongly recommended as part of the troubleshooting evaluation.

This paper will review the various sources of maldistribution susceptibility, which can stem from design, installation, and operational issues. Sometimes, the problem source(s) can only be exactly determined after opening up the column during plant shutdown. As the shutdown time frame is usually limited, there is only a narrow window period to both identify and resolve the source(s) of column underperformance.

Fortunately, it is possible to narrow down the possibilities prior to plant shutdown, via methodologies that encompass process, hydraulic and mechanical considerations. Some important steps include: simulation matchback to operating data, hydraulic evaluations of the column internals, and maldistribution sensitivity analysis of the column. After opening up the column, further methodologies can help to pinpoint the problem source(s). Some important steps include: visual inspection of the column internals, and liquid distributor water tests.

The troubleshooting process will be illustrated via several troubleshooting case studies. They will describe the methodologies used to identify the source(s) of column underperformance, and the subsequent corrective actions to return the column to satisfactory operations. These experiences should help engineers to perform troubleshooting in a structured manner, and allow plant owners to optimize their resources in a troubleshooting activity.

480703 Big Data Analytics Skills - a Revolution Lacking in Revolutionaries

Abstract:

Since the advent of the common use of the term "Big Data[i]" in 2010 to describe large, *unruly* sets of data, and the artful analysis thereof, industry has felt besieged by opportunity. The chemical and refining industry has been faced with "Big Data" long before it was cachet. Yet within industry there is still a struggle to get hearts and minds around what the Big Data Journey means to an organization and how to take advantage of the data opportunities that lie within. The added complication to the chemical industry is that it must add "contextualization" to the data analysis because it's not enough just to know there is a signal, we need to know what the signal means[ii]. Yet during this data revolution there is an underlying challenge that Academia, Industry and Governmental Agencies must tackle together to ensure the US Chemical industry retains its competitive position in the world. Together we need to provide education for the higher level skills and abilities required across roles in manufacturing to assure that no one gets left behind in the new Big Data age. Indeed, the next great technological "step" made may not be in engineering design or IoT connectedness (reference Industry 4.0), it may be in designing and creating a better human capability in "data acuity" out of our university system; operators, engineers and scientists alike who are better prepared for the data age than their predecessors. This talk will take a look at what Industry requires in skills sets and perspective to be "data ready", and what Academia and Government might do to improve the output of data-ready skills.

---

[i] Big Data, What is it? White, Chemical Engineering Progress, AIChE March 2016.

[ii] Big Data, Getting Started on the Journey. Colegrove, Seasholtz, Khare, Chemical Engineering Progress, AIChE March 2016.

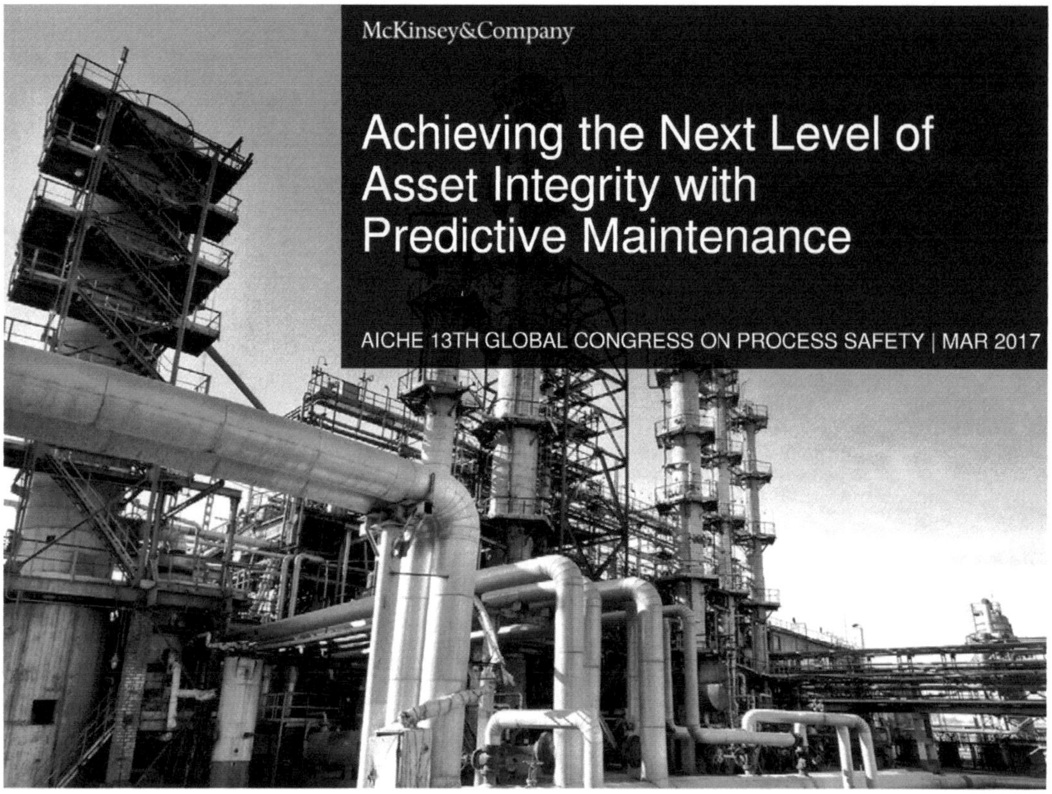

Abstract

Achieving the next level of asset integrity with predictive maintenance

Exceptional organizations manage to deliver superior operational performance without compromising on safety and integrity. On the contrary, high performing players are at the forefront of equipment reliability for critical equipment and safety performance.

In this context, introducing Predictive Maintenance can dramatically help achieve the next level of asset integrity by leveraging data - structured and unstructured, machine and non-machine based – in order to generate insights with advanced analytics methods that would not be possible with conventional techniques
Implementing Predictive Maintenance generates three benefits that reinforce one another:
Predict when equipment will fail and help reduce the safety and integrity threats
Avoid failures and positively contribute to reduce the scale of the value at risk
Enhance specific asset knowledge and learn more about the behavior of critical equipment

In a recent client project for an upstream oil and gas facility, the main gas compressors have been identified as **critical equipment**: they account for many failures and a large portion of production losses. A Weibull analysis showed that out of several known failure modes, some show a random failure pattern, leading to sudden breakdowns releasing high energy and thus representing a safety issue. As per the random failure behavior, classical time- or usage-based maintenance strategies are not fruitful.

Combining 100's of gigabytes of loss data, maintenance work order data and sensor data from 1,200 sensors around the equipment, a data driven model has been created showing that out of the 1,200 sensor sources, 43 explanatory variables predict the time to failure with ~80% accuracy. The model has been tweaked further in order to reduce the number of false negatives while accepting by this a certain number of false positives (i.e. false alarms) due to safety reasons.

As a result of this model, an early warning of an imminent failure means repairs can be planned, spares can be gathered and the equipment can now be shut down in a safe and orderly manner. Overall, this represents an average reduction from 14 to 4 days of downtime per breakdown event.

Outline

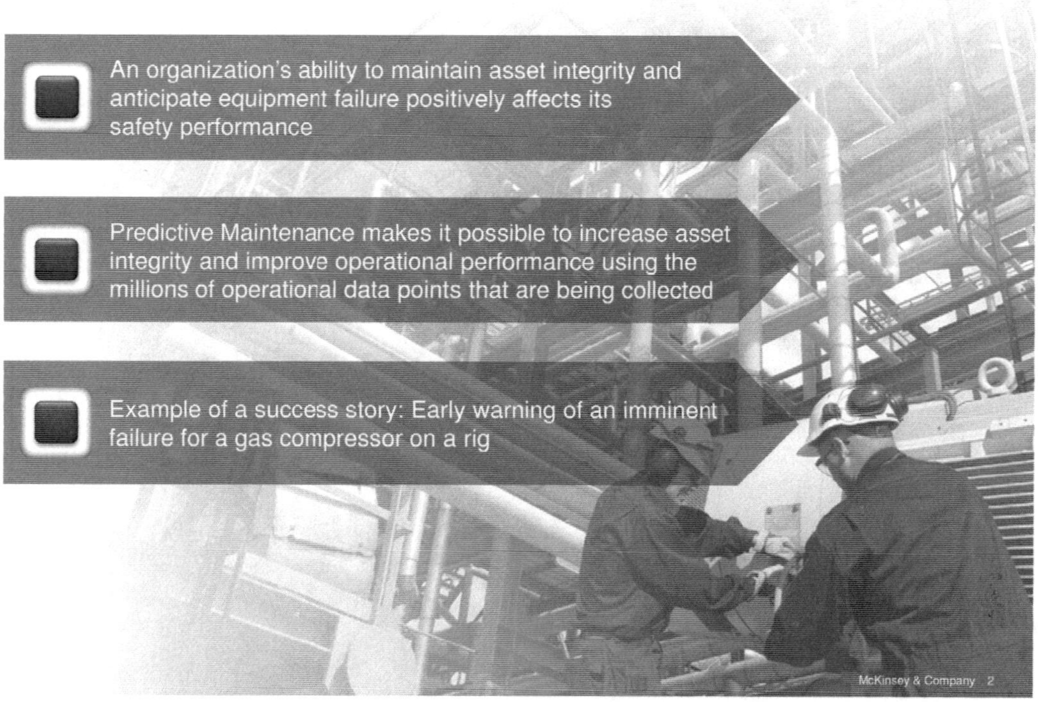

- An organization's ability to maintain asset integrity and anticipate equipment failure positively affects its safety performance
- Predictive Maintenance makes it possible to increase asset integrity and improve operational performance using the millions of operational data points that are being collected
- Example of a success story: Early warning of an imminent failure for a gas compressor on a rig

McKinsey & Company 2

There is a strong correlation between equipment reliability and safety performance

1993 - 2003 (base 100)

EXAMPLE FROM ALUMINUM SMELTING

—— Emergency work on cranes
- - - Safety performance

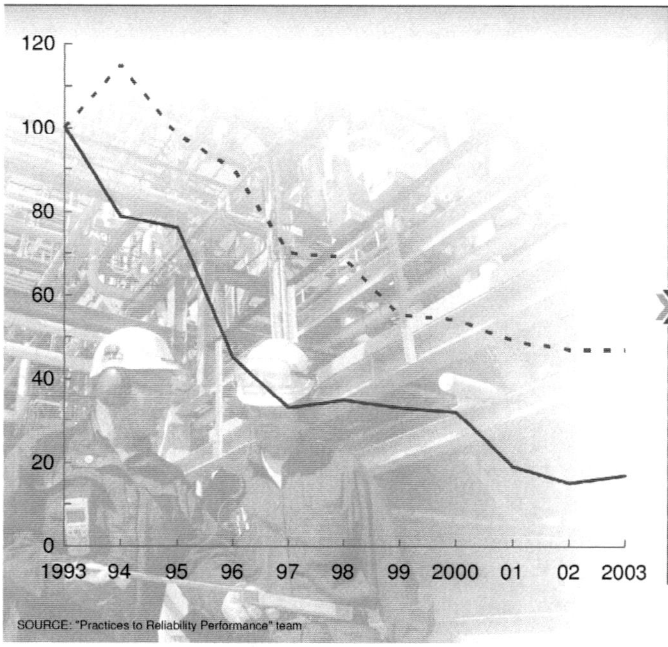

- **Safety** and reliability performance require **control over the plant environment**
- **Unscheduled work** – often in rush – **is by nature much more risky** than planned interventions
- **Reliability and safety performance share multiple common drivers**
  - Importance of **behavioral factors**
  - **Anticipation rather than reaction as a management practice**
  - **Ownership** and care for equipment

McKinsey & Company 4

Industry collects Petabytes of data which can be used for Predictive Maintenance to improve asset integrity and reliability

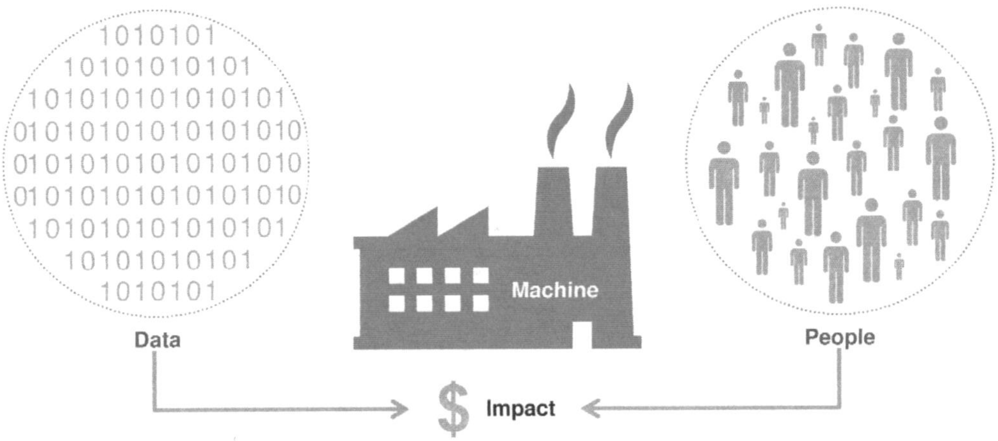

SOURCE: McKinsey

McKinsey & Company 5

Getting from data to impact requires a combination of many different skills

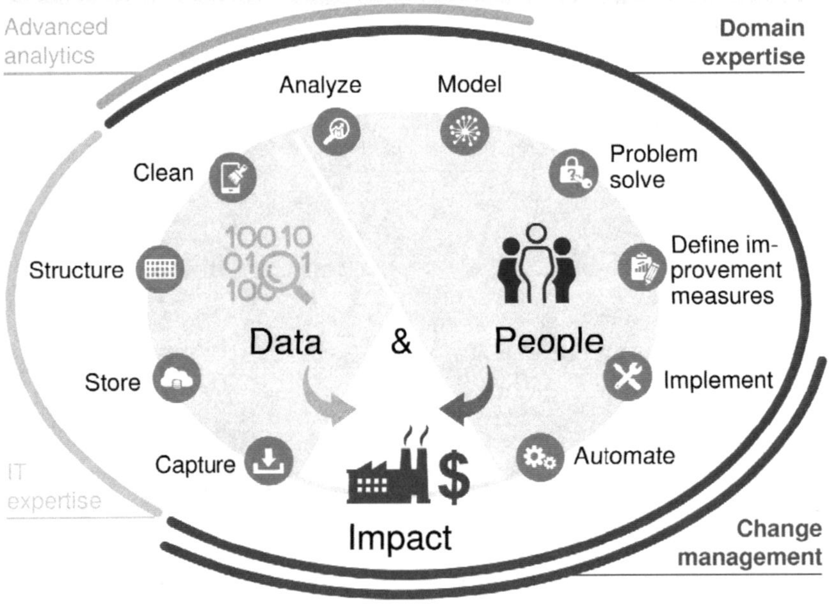

SOURCE: McKinsey

McKinsey & Company 6

Advanced Analytics generates previously untapped insights in Predictive Maintenance from troves of historical data

Predictive Maintenance
is the use of data (structured and unstructured, machine and non-machine based) to generate insights with advanced analytical methods that would not be possible with conventional techniques

Predict

Failure forecasting: Model future asset behavior to anticipate needs *(e.g., HR, spare parts inventory, capital replacement)*
Incipient failure detection: Continuous data acquisition & analysis to detect or predict failure *(e.g., early warnings to allow planned intervention)*

Avoid

Avoid failures
Identification of undesired patterns in data as primary causes of failure (e.g., preventive actions in specific locations of roads with poor conditions to prevent truck failures)

Learn

Leverage specific asset knowledge
Access reliability data from a global equipment database comprised of hundreds of thousands of components across the globe *(e.g. Failure rates, failure mode distributions, causes)* to inform maintenance strategies

SOURCE: McKinsey

McKinsey & Company 7

The main benefits of implementing predictive maintenance include availability increase and cost reduction

Predictive maintenance aims to ...

... detect early signs of problem **for timely correction at minimal costs**

... prioritize and optimize maintenance resources using real-time data

... increase unit availability due to **shorter downtimes**

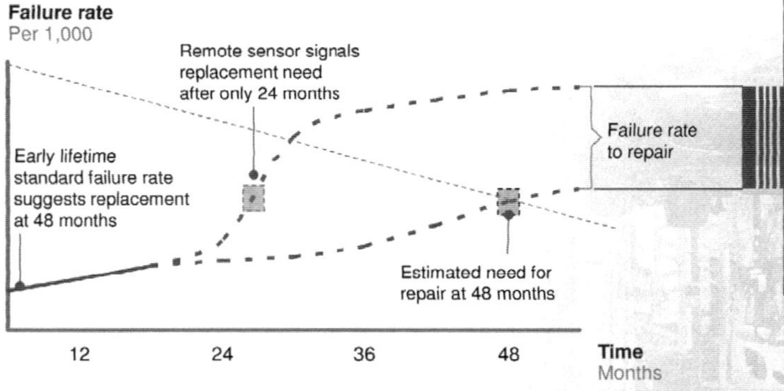

Typical impact of PdM

Machinery breakdowns	-50%
Spare part inventory	-20%
Total machine downtime	-50%
Overtime expenses	-20%
Machine life	+20 - 40%
Overall productivity	**+20 - 30%**

SOURCE: ARC Advisory Group; Department of Defense; expert interviews; GE; Schaeffler; VibroSystM; WiHART Systems

McKinsey & Company 8

Predictive maintenance is proven to add value in many industries

What was Predictive Maintenance used for

✓ Focus of the engagement
Detailed in following pages

Industry	Reduce operating costs	Reduce maintenance costs	Increase machine availability	Impact
1 Mining			✓	▪ 25% downtime reduction
2 Mining *(Copper mine)*	✓	✓		▪ Savings of ~12% in operational expenditures
3 Oil and gas *(Rig operator)*		✓		▪ Reduced maintenance spend by 18 to 27%
4 Oil and gas			✓	▪ ~70% reduction in downtime on critical component
5 Railroads			✓	▪ Reduced failure cost by 20% to 30%
6 Heavy-Motors *(Cruise operator)*	✓	✓		▪ Reduced operating cost by 7%
7 Semiconductor	✓			▪ Increased failure prediction to 87%

We have more than 20 use cases. The remaining cases cannot be shared due to confidentiality.

SOURCE: McKinsey Asset Productivity Service Line

McKinsey & Company 9

PREDICTIVE MAINTENANCE IN OIL & GAS

Summary of the case

Applying advanced analytics in off-shore oil and gas gives early failure warning can reduce production losses due to gas compressor by ~70%

Situation

Sector and site
- Offshore oil and gas operator
- Rig maintenance

Situation
- Gas compressors are **critical equipment**: they account for many failures and a large portion of production losses
- Loss data, maintenance work order data and sensor data is collected

Question
- To what extent can we reduce maintenance spend and improve reliability by predicting when a failure will happen?

Typical failure mode: inlet strainer

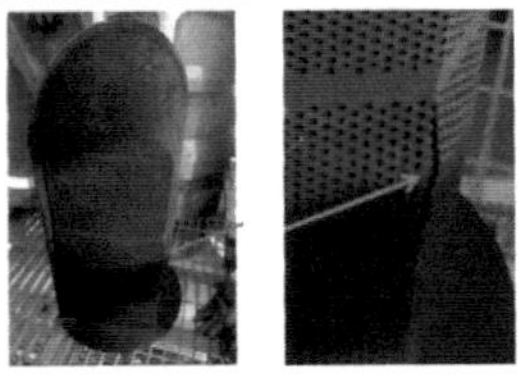

Key Activities

Data collection from several sources
- 120 gigabytes of data
  - ~1,200 sensors
  - 10 years of work orders and deferment data

Identify failures and construct target variable
- Identify if downtime is caused by compressor

Compr | Other

Create a model to predict time to failure
- For any point in time we predict the **time to the next failure**

Inputs 1,200 sensors → Prediction Model → Time to next failure

- **Inputs** read from 1,200 sensors
- **Regression Model** predicts the time to failure with ~80% accuracy

Impact

- Downtime due to gas compressor is expected to reduce by ~**70%** by planning intervention when a failure is imminent
  - **Prediction of time to failure:** if intervened at predicted TTF < 5 days, you would act between 9 and 6 days before the failure actually will happen

Predicted TTF for different failures, Days

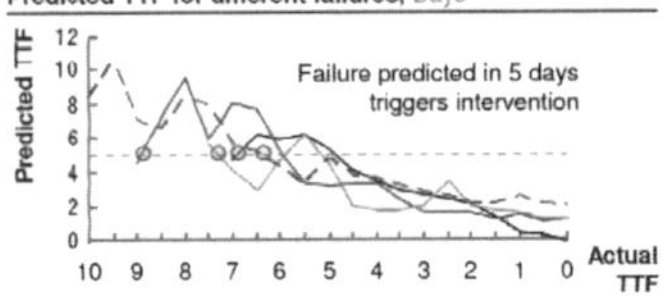

  - **Early warning of an imminent failure** means repairs can be planned. Reduction is on average from 14 to 4 days of downtime

	Breakdown
Old situation	Up → Down 14 days → Up
New situation	Up → Prep. → Down 4 days → Up

1 TTF = Time To Failure, each of the lines in the graph is one failure case

SOURCE: McKinsey

McKinsey & Company 10

Details of the case

Data collection was based on 120 GB of data – including more than 1200 sensors and 10 years of work orders

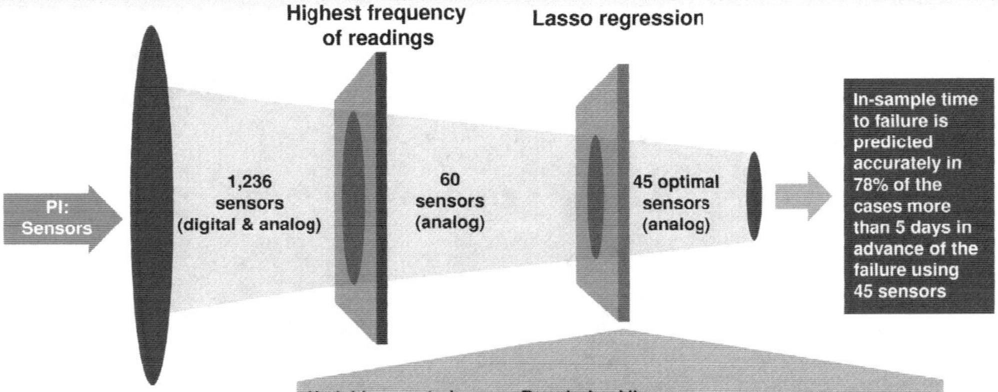

Variables created using sensor data	Regularized linear regression (LASSO)	Time to failure
▪ Moving averages on 6-hour time-windows	▪ Most informative sensors are identified in an automated way ▪ 60 analog sensors with highest frequency of readings reduced to 45 via cross-validation	▪ Binned to 10-min intervals

SOURCE: McKinsey

Details of the case

In order to predict downtimes, data preparation as well as a clear target variable were keys to success

A Identify downtimes caused by the low pressure compressor (LPC)

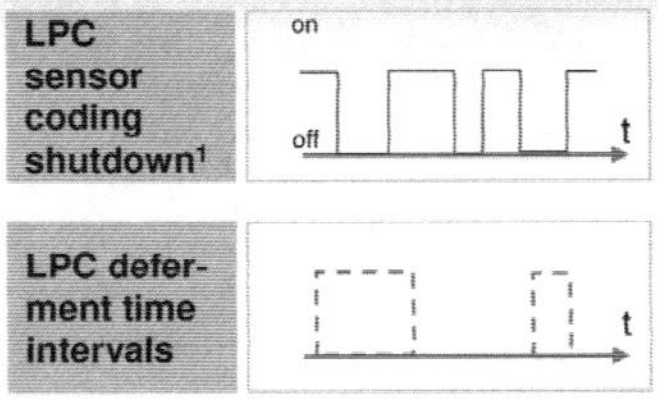

Overlap shutdown and deferments to Identify downtimes caused by LPC or other reasons

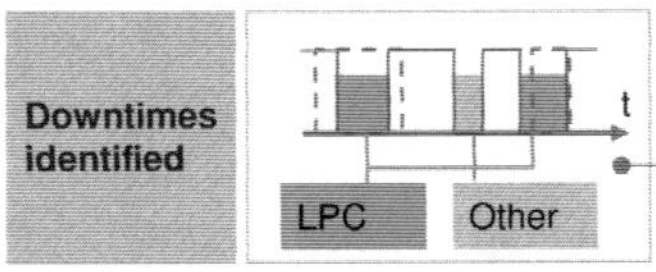

B Constructing the target variable: Time to next LPC failure

When is the next LPC failure?

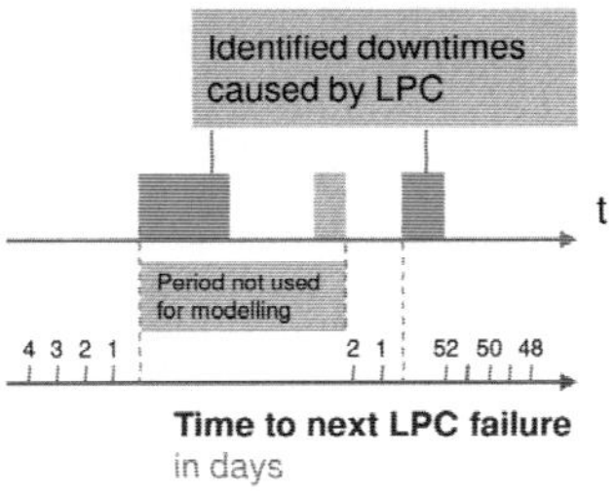

The identification can also be more granular down to a certain failure mode such as "Cooling medium strainer burst"

C Predict time to failure

For any point in time we can use our prediction model to get the **time to the next LPC failure**

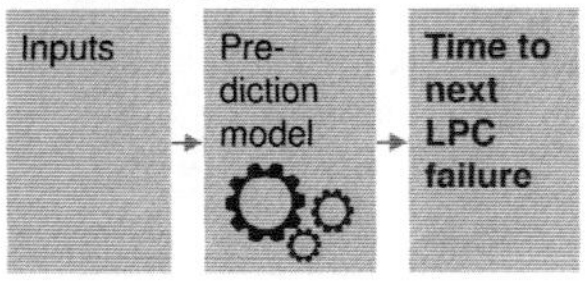

- **Inputs** are sensor readings from PI-system (and further data)
- **Model** is constructed/fitted from historical sensor readings and our **target variable (time to next LPC failure)** via **machine learning** algorithms and **statistical** methods (e.g., regression, decision trees, ...)
- Validation of the model is done by dividing the historic data set into "training" and "test" set (cross-validation)

SOURCE: McKinsey

Thank you

Carl March

Operations Practice Expert

- Predictive Maintenance

Carl.March@mckinsey.com

McKinsey & Company 13

FROM ANALYSIS TO VALUE CREATION

Top down value assessment for predictive maintenance

PRELIMINARY

Digital asset management

- ↑ **Availability 2-5%**
  - ↓ **Planned downtime 10-20%**
    - Decrease in work order frequency
      - Work order cadence to equipment needs instead of OEM recommendation
    - Decrease in work order duration
      - Word order instructions pushed to mobile devices
      - Improved preparation to increase employee wrench time
      - Fully digital and mobile enable maintenance process and cycle
  - ↓ **Unplanned 30-50%**
    - Actively performing routine work orders based on predictive analytics
    - Monitoring conditions of equipment to accurately predict failures
- ↓ **Maintenance cost, 5-25%**
  - ↓ **Operating maintenance costs**
    - Cheaper, routine work orders (predictive > reactive)
    - Improve part and tool inventory strategy with improved database
  - ↓ **Sustaining capex costs**
    - Longer equipment life through planning component replacement extension
    - Adjust to a smaller fleet size

McKinsey & Company 14

I The predictive model combines sensors in a way that cannot be replicated manually

Most important sensors for prediction

10 most significant out of a total of 45 sensors

Rank	Sensor	Unit of measure	Relative contribution Abs. value of coefficient
1	SEAL VENT	BARG	4.2
2	SEAL GAS VENT DP	BAR	2.3
3	FUEL GAS PRESSURE	BARG	1.9
4	COMPRESSOR FLOW DP	MBAR	1.8
5	INLET FILTER DP	MBAR	1.6
6	SEAL GAS SUP PRESS	BARG	1.3
7	ENG BRG2/3 DRAIN	DEG C	1.3
8	FUEL SYS AIR SUPPLY	BARG	1.2
9	FUEL GAS FLOW DP	MBAR	1.1
10	PRI NDE SEAL VENT	BARG	1.1

- The statistical model has identified a total of **45 sensors** to be significant when predicting compressor failures
- The contribution of the sensors is evenly distributed – there is **no one sensor that can replace the predictive model**, which combines a total of 45 sensors
- The absolute value of a coefficient can only be extracted in a linear regression case, and that other metrics should be used for other models e.g. variable importance in random forests

SOURCE: Team analysis, Sensor data

McKinsey & Company 15

II We are able to predict breakdowns at least 5 days before failure

Failure 1 — Failure 2 — Failure 3 - - Failure 4 ○ Point of intervention

Predicted time until failure. 12-hour average of prediction

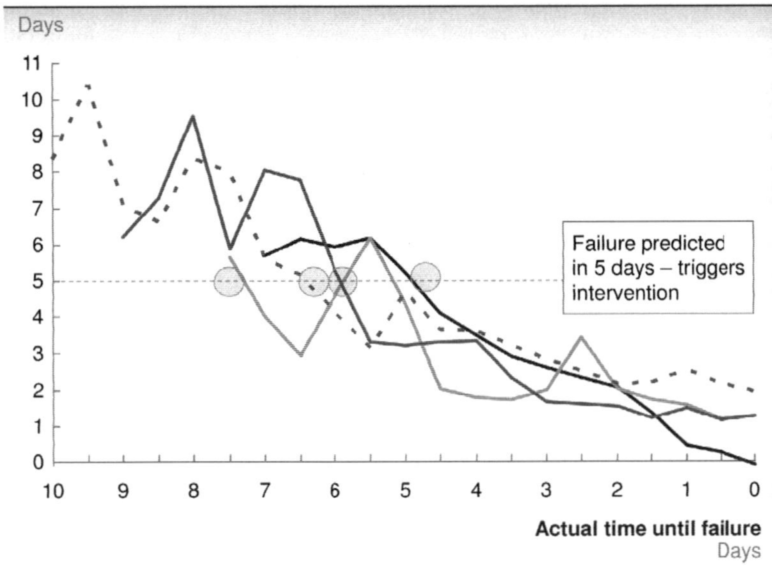

We are able to react prior to failures

- Early indications of failure can be seen 10 days in advance
- In all observed cases an intervention could be triggered more than 5 days prior to failure
- With normal reaction time the maintenance organization will have enough time to inspect and intervene

SOURCE: Team analysis, client data

McKinsey & Company 16

481480 Data visualization Connecting Big Data to Integrate Every Level of the Organization- Messaging, Action and Opportunities!

Abstract:

In an effort to improve data transparency, identify true costs, and most importantly, capitalize on opportunities to improve the safety culture throughout the enterprise, United developed a data visualization program, to communicate information clearly through graphical means including Geo Mapping. By centralizing big data from various departments and organizations, we can mine big data to a level that provide universal messaging through visualization. The ultimate goal was to reduce the frequency and severity of employee injuries and implement the daily use of data to address trends and decrease the number events impacting the health and safety of our employees as well as protecting our aircraft from damage.

United's Corporate Safety team in a collaborative agreement is working with the University of New Haven in changing the way data is processed, analyzed and communicated. Our approach is allowing 24/7 universal access to data and streamlining analyst processing time to a fraction of the time while being more comprehensive in what the data means!

United was able to apply the same logic to employee injuries and aircraft events globally across the network, integrating safety and operational data to allow end users to analyze events down to a specific location (such as a specific gate at an airport); the type of equipment involved; the type of injury; frequency; and impact of such events. We are now applying the same logic to how facilities and equipment are maintained and operated in way to be more predicative.

United Airlines has seen success as frontline leaders continue to rely on the data visualization dashboards to actively manage the health and safety of our employees. These techniques are being utilized by operational division to eliminate waste and improve efficiency throughout the airline.

Leadership in Catastrophic Incident Prevention: Using Organizational Data to Assess and Respond to High Reliability Loss Prevention Precursors

Michael Snyder
DEKRA Insight
113 Campus Drive
Princeton, NJ 08540
Mike.snyder@dekra.com

Prepared for Presentation at
American Institute of Chemical Engineers
2017 Spring Meeting and 13th Global Congress on Process Safety
San Antonio, TX
March 26 – 29, 2017

Leadership in Catastrophic Incident Prevention: Using Organizational Data to Assess and Respond to High Reliability Loss Prevention Precursors

Michael Snyder
DEKRA Insight
113 Campus Drive
Princeton, NJ 08540
Mike.snyder@dekra.com

Keywords: leadership, catastrophic incident prevention, highly reliable organization, early warning signals

Abstract

Catastrophic Occupational & Process Safety Incidents usually occur after a series of early warning signals, often subtle, are overlooked. Some organizations have been very successful in avoiding catastrophic events, even in environments with high risk factors and complex operations. Known as Highly Reliable Organizations (HROs), these operations have developed cultures that embrace the concepts of Anticipation, Inquiry, Execution, and Resilience – leading to significant improvement in the detection of and response to early warning signals.

Contemporary research has identified methods, which allow for identification of incident precursors to Serious & Catastrophic Potential Events, along with diagnostic tools to assess the maturity of an organization's "High Reliability". These tools provide powerful insights, and allow leaders to develop a comprehensive approach to catastrophic incident prevention through the effective use of Organizational Data.

1 Introduction

Four years before BP's Deepwater Horizon became global news, ExxonMobil walked away from a $187 million investment in a similar deep sea well. The well had yet to produce a single drop of oil when engineers detected geologic instability in the seabed. After consulting with geologists and drillers on the project, leaders decided that the best course of action was to cap the well and move on. Industry analysts at the time accused the organization of being too risk averse; after all, no incident had actually occurred. [1, 2]

That narrative changed in 2010, when operators of the Deepwater Horizon lost control of a wellhead that had been experiencing similar disturbances. This time there was no speculation of whether the risk was "real" or not. The resulting explosion and fire killed 11 workers and created an environmental and public relations nightmare. For leaders across industry, the tragedy of the Deepwater Horizon was compounded by the fact that many of the lessons it had to offer had already been "learned" before.

After any serious process safety incident has occurred, a diligent investigator will typically find a long history of less consequential incidents that preceded it. The goal of High Reliability Safety is to understand why these "early warning signals" are overlooked and what can be done in today's manufacturing environment to improve the identification of and response to these signals.

2 Emergence of the High Reliability Organization (HRO)

Throughout the industrial age, human-made complex systems have advanced to provide the world with materials, clothing, fuel, food, and resources. However, these systems are susceptible to catastrophic events that result in multiple fatalities and major property damage. Since the 1980s, academics have worked to define what they call the High-Reliability Organization (HRO). The high-reliability organization typically defined as an enterprise that executes consistent and sustainable operations by creating a culture that anticipates and minimizes variation and implements high-quality decision making and controls.

The HRO paradigm seeks to categorize the characteristics of organizations that stay safe despite operating in high- risk conditions (Figure 1). Much of the HRO organizational research has studied U.S. military organizations, wildfire incident management teams, and nuclear plant operations. In theory, a high-reliability organization would routinely make the same kinds of safety-based decisions that ExxonMobil did back in 2006.

3 Four Disciplines of High Reliability Organizations

Through study of both theoretical models of high performing safety organizations and incident investigations of actual process safety events we have recognized that there are **four aspects of organizational performance** that are key to safe operations (Figure 2). Our studies have identified that successful operations have developed cultures that embrace the concepts of Anticipation, Inquiry, Execution, and Resilience – leading to significant improvement in the detection of and response to early warning signals, resulting in fewer process safety incidents.

3.1 Anticipation

Fostering systems and behaviors that are sensitive to "weak signals" that may be indicative of increased risk of catastrophic events.

Anticipation is about recognizing the early warning signs that help you head off incidents. An organization strong in Anticipation will have mechanisms to capture information from a variety of sources that may be meaningful early indicators of change to exposure. Examples could include

Figure 1: Framework for a High-Reliability Organization (HRO) adapted from Sutcliffe & Weick [3].

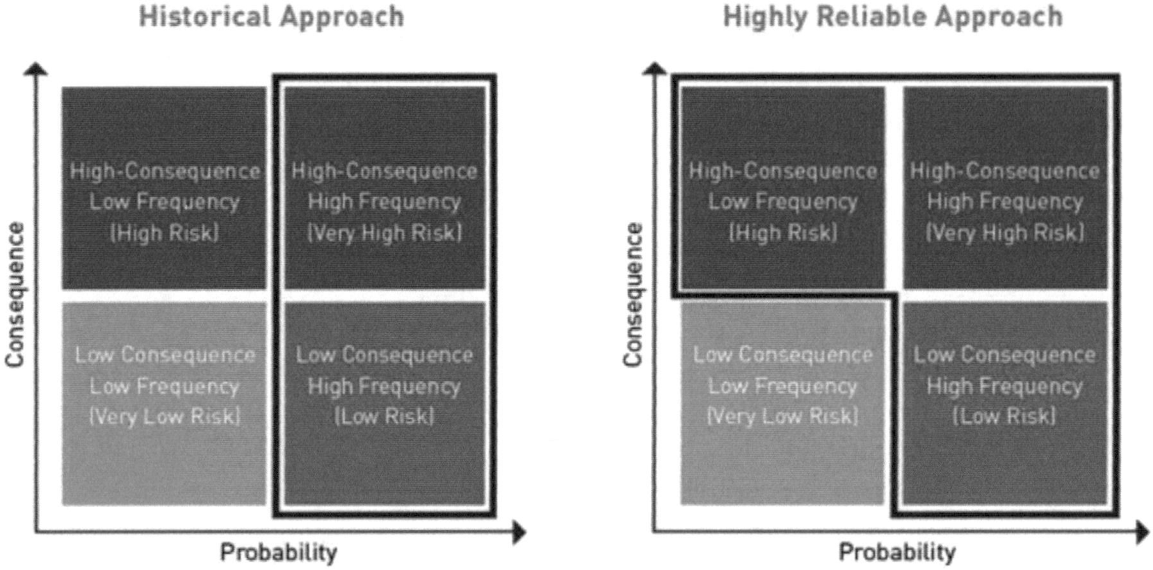

HISTORICAL FOCUS	FUTURE FOCUS
Only high frequency events	**Also high-consequence, low-frequency events**
Lagging metrics	**Leading and "In Process" metrics**
Report only incidents	**Report near misses & other significant events**
Investigate only Recordable events	**Investigate all events with high learning value**
Causal factors: Technical & operational	**Causal factors: Organizational & management system**
Assume past performance predicts future	**Assume the worst case is indeed possible**
Learning environment: Primarily internal	**Learning environment: Includes external view**

process deviations, unusual maintenance requests, and even frontline workers detecting differences in sounds. Strong Anticipation is created by a culture that is always seeking indicators of problems. This means:

- People are encouraged to report and the value of reporting is reinforced
- Data is acted upon and people know that. They won't bother if they perceive that information disappears into a black hole and nothing ever happens
- Leaders understand and accept that there will be lots of information that is not important but it's worth sorting through to be sure we don't miss the information that is critical
- We can identify key leadership behaviors that help create this culture. When those behaviors are practiced we build an organization with strong Anticipation

Organizations with strong cultures have leaders who visibly value the search for early warnings and reinforce the analysis of these indicators, even when this does not result in identification of

Figure 2: The AIER Model

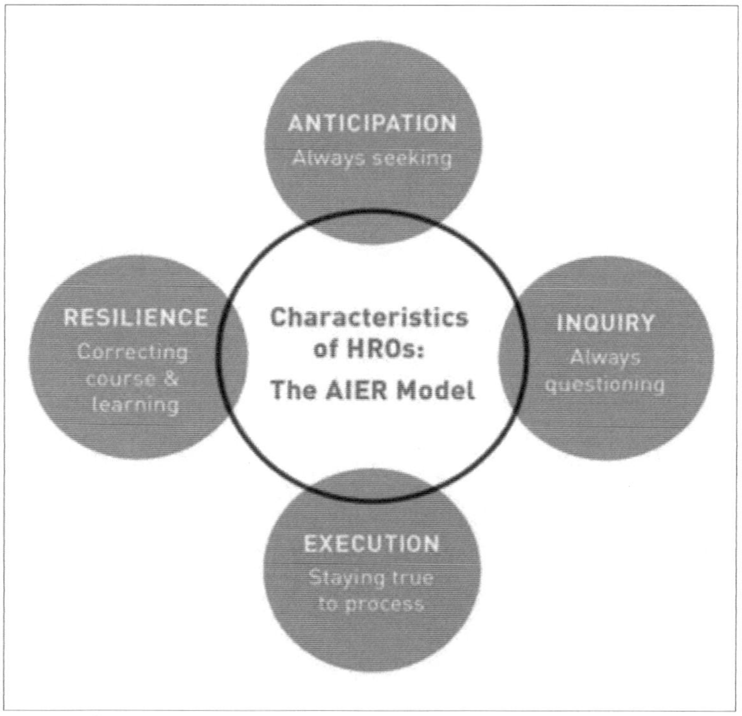

serious risk after it is fully explored. These leaders understand that supporting the detection and investigation of many false positives is worth it, if it results in just one catastrophic event avoided.

3.2 Inquiry

Making effective use of information to analyze, understand, and plan mitigation of risks, while making a conscious effort to overcome bias.

The Inquiry discipline is about preserving the integrity of decision-making and action, specifically by protecting teams from the natural biases innate in all humans and groups. Traditional Process Safety Management (PSM) includes a number of elements (such as process hazard analysis, pre-startup safety review, management of change) designed to evaluate and plan for control of hazards and risks. However, there are common (but often undetected) cultural characteristics that can undermine the effectiveness of these efforts and leave the organization vulnerable.

Cognitive bias refers to the tendency we all have to rely on intuitive, rather than analytical, thinking in order to process information efficiently. Our knowledge and experience allow us to reach conclusions and make decisions quickly and efficiently in many circumstances. However, this can also trap us in poor decisions.

Cognitive bias can also get us into trouble in formal hazard analysis – we can too readily accept what we think we know, e.g., that a particular condition or situation or deviation is not problematic. The tendency to go along with a group's predominant opinion can make this worse.

Biases can include:

- **Confirmation bias.** The tendency to search for or interpret information in a way that confirms one's preconceptions.
- **Normalcy bias.** The refusal to plan for, or react to, a disaster that has never happened before.
- **Availability bias.** The tendency to predict based on how easily an example can be brought to mind.
- **Status quo bias.** The tendency for people to like things to stay relatively the same.
- **Groupthink.** The tendency to do (or believe) things because many other people do (or believe) the same.
- **Risk seeking/risk aversion.** The tendency to make risk-averse choices if the expected outcome is positive, but make risk-seeking choices to avoid negative outcome.

The only way to guard against the insidious effects of cognitive bias is through effective culture and leadership interventions. There are specific leadership behaviors (for example, encouraging the voicing of dissenting opinions) that promote a culture that produces more reliable decision-making. There are also specific skills involved in asking the right question in the right way to get the right data. Leaders must promote and measure the use of these leadership behaviors and skills.

3.3 Execution

Monitoring, reinforcing, and verifying program execution, while staying true to the process.

The Execution discipline is about assuring consistent and reliable use of safety programs and processes. Even the best-designed hazard identification and control efforts are subject to failure or underperformance if the programs and practices are not followed, as intended.

While many organizations use periodic audits to provide a check on implementation, the key to assuring consistent and ongoing activity is developing leaders who frequently monitor, reinforce, and verify effective program execution. Key aspects of these leadership activities are summarized below:

- **Monitoring** involves regularly acquiring information on what subordinates are doing, how they are progressing toward achievement of goals, and what issues or problems they may be encountering. This is not micromanaging; rather it is assuring that the leader has sufficient information to meaningfully recognize good performance, provide support when subordinates need it, and then providing corrective feedback on those (hopefully rare) occasions when subordinates fail to fulfill their responsibilities. A leader's monitoring behavior may take many forms. Depending on the situation and the leader, these may include walking around and observing, informal conversations at the front line, periodic meetings with subordinates to review progress, use of written progress reports, review of appropriate metrics, etc.

- **Reinforcement** involves providing feedback that recognizes good performance. This communicates the importance and priority of the catastrophic event prevention activities and maintains focus on consistent execution. Effective reinforcement is based on effective monitoring, which provides the leader with specific data on which to base reinforcement, avoiding the vague and ineffective "good job" type of feedback to subordinates.
- **Verification** is similar to monitoring, but where monitoring focuses on the performance of subordinates, verification focuses on activities and programs. Assuring that audit findings are resolved in a timely manner is an example of verification, and leaders who do this effectively are more likely to have organizations in which consistent execution is valued as part of the culture.

3.4 Resilience

Developing and exercising the ability to react in ways that prevent upset conditions from becoming catastrophic events — and then learning from the experience.

The Resilience discipline is about developing the agility to recognize and quickly respond to exposures in real time. Upset conditions occur from time to time in any system. A resilient organization is able to react quickly in ways that prevent upset conditions from becoming catastrophic events, and then learning from the experience. This has a major influence on ultimate results. Even where automated control systems are designed to handle upset conditions, it is important that workers understand when and how to intervene, and are not only able but also willing to make appropriate interventions early. An organization that is strong in Resilience is more likely to prevent a small process disruption from becoming a major incident. There are three main questions to consider when looking for aspects for effective Resilience:

- **Knowledge.** ***Do people at various levels have a broad enough understanding of the operation so that they can make good judgments in case of emergency?*** Some organizations approach this through use of extensive sets of rules and procedures. That approach intends to assure consistency and avoid having to rely on technical knowledge at the operating level. However, two problems arise with that approach. First, the range of possibilities that plans for results in a proliferation of procedures and rules that become impractical for anyone to know. Second, this approach assumes that all possibilities (with all variations in every scenario) are identified in advance — something that is unlikely to be true. The alternative is to develop an organization in which people are knowledgeable and taught to make good judgments based on their knowledge and the information at hand.
- **Willingness.** ***Are people more or less likely to take action on their own initiative?*** The answer directly relates to culture. People are less likely to act if they are not confident that the organization will support them. Perceptions of the culture's support for resilience form over a long period and are based on many small actions taken and not taken by

leaders. An organization desiring strong catastrophic event prevention will be sensitive to this and intentionally create the culture that supports resilience.

- **Continuous Learning.** ***Do leaders establish both the expectation that learning is an ongoing process that never stops — as well as promoting the systems and processes to support such continuous learning?*** Resilient organizations make learning routine rather than waiting for an incident to trigger learning mode. They create formal mechanisms (e.g., cross-functional teams, advanced training in incident investigation) to continually capture and respond to risk data. They put the focus on generating new knowledge, not finding fault. Great learning organizations are also characteristically self-aware. They make sure that even the newest employee knows the story of where the organization came from and where it is going. Continuously learning organizations are vigilant about preserving organizational memory and assuring that hard lessons are never forgotten — or repeated.

4 Building a High Reliability Organization

So we come back to our initial question: How can leaders develop a high-reliability organization?

Fundamentally, high reliability is a blended cultural and technical process -- driven by leaders.

Leaders must establish the expectation that attention to exposure is an ongoing learning process that never stops. At the same time, they must also assure that organizational and technical systems, as well as enabling systems, support such continuous learning. Several practices can help an organization move toward high-reliability:

- **Cultural Assessment.** The first step is identifying how the organization is performing now. Rather than look at incident rates (which are poor indicators of actual exposure). The AIER scales identified above are readily measured (Table 1), and can be compared against industry norms. They may be used alongside validated culture indicators shown to predict general safety outcomes.
- **Technical Assessment.** Cultural assessments can be immensely powerful and have been demonstrated to predict safety outcomes. To provide a more holistic picture of organizational reliability, it is advisable to conduct a technical assessment that evaluates the effectiveness of process safety systems, including the identification of risk, robustness of protective barriers, and competency of personnel responsible for system operations.
- **Develop a process safety improvement roadmap**. Based on the findings from the Cultural and Technical assessments, an organization's capability can be assessed against one of several internationally recognized Process Safety Management models. Improvement plans can be developed and paced to meet organizational risk resolution timelines and resource capacity.

- **Adopt a learning orientation.** High reliability organizations are characteristically self-aware. They make sure that even the newest employee knows the story of where the organization came from and where it is going. Continuously learning organizations are vigilant about preserving organizational memory and assuring that hard lessons are never forgotten — or repeated.

Table 1: Sample criteria to evaluate AIER scales

ANTICIPATION	INQUIRY	EXECUTION	RESILIENCE
Always seeking	**Always questioning: Avoiding bias**	**Maintaining control while staying true to process**	**Preparation, response, and learning**
Priority is given to leading measures and weak signals Systems and culture support problem-seeking Input from the front line is given high value High sensitivity to change Reinforcement of reporting	Rigorous analysis systems are in place and constantly used The organization actively works to overcome bias The risk acceptance process is clearly defined and followed Recommendations are applied widely	Leaders set clear expectations Roles and responsibilities are well defined and understood Routine monitoring and checking occurs Follow-up assures findings are addressed in a timely manner	Teams have rehearsed responding to a variety of upset conditions Leadership supports and reinforces employee intervention to minimize impact Post upset activity focuses on learning and putting lessons into action widely

5 Identifying Data to Serve as Early Warning Indicators

While it is intuitive that organizations demonstrating higher performance on the four characteristics (Anticipation, Inquiry, Resilience and Execution) related to excellent catastrophic incident prevention have fewer process safety incidents, research efforts have begun to validate this hypothesis [4]. While this analysis is straightforward for a company with multiple sites using common operational definitions for Process Safety incidents, challenges remain.

In general, comparison among different companies is constrained by the lack of common reporting definitions for Process Safety Incidents, particularly those with much larger potential than the outcome actually experienced. In addition, the use of Tier 1 and Tier 2 metrics from API 754, Process Safety Performance Indicators for the Refining and Petrochemical Industries, generally provides a very limited count of events. While intended for internal company analysis, increased standardization of Tier 3 and Tier 4 metrics may help advance the comparative evaluation of the AIER scales for larger datasets.

Specific performance weakness on an individual AIER scale component have been used to improve focus on the types of leading and in process metrics used to aid in identification of catastrophic incident precursors. Several examples are listed in Table 2 below.

Table 2: Example incident precursor metric focus provided by AIER scales

Incident Attribute	Definition	Example Metrics
ANTICIPATION	**Always seeking**	• **Percentage Process Safety Events (including near misses) that were identified by PHA or Hazard Evaluation.** • **Number of Hazard Reports Filed (Focus on Process Systems – for example, "Valve Inspection is Out of Date")**
INQUIRY	**Always questioning: Avoiding bias**	• **Near Miss / Close Call Reporting (Ratio in respect to actual incidents)** • **Number of Stop Work or Job Pauses Executed** • **Number of Safety Critical Objects that Fail to Pass Preventative Inspection/Test in "As Found" Condition.**
EXECUTION	**Maintaining control while staying true to process**	• **Results of Procedure & Work Process Audits** • **Percentage Process Safety Events (including near misses) that were predicted by PHA or Hazard Evaluation, but had one or more barrier failures (categorize barrier failures)** • **Corrective & Preventive Actions Completed on Site (including audit that actions are effectively implemented)**
RESILIENCE	**Preparation, response, and learning**	• **Local incident learnings applied throughout site or corporation.** • **Emergency Drills on Facility Major Accident Hazards (including follow-up corrective actions)** • **Near Misses with Significant Potential Mitigated by Proactive Employee Action.**

6 Conclusion: Moving Towards HRO Performance

In the high-hazard manufacturing environment, safety expectations have grown well beyond the early charter of reducing costs and injuries. The promise of high-reliability organizations is that they define a path to consistent safe performance, even at the frontiers of industry and technology. As with all innovation, the key to adoption is being able to replicate the conditions and factors that create the phenomenon reliably and predictably. The practical and measureable nature of the AIER scales opens a new way to leaders wishing to create high-reliability organizations of their own.

7 References

[1] S. LeVine. "Exxon: Juggernaut or Dinosaur?, *BusinessWeek*, February 5, 2009, https://www.bloomberg.com/news/articles/2009-02-05/exxon-juggernaut-or-dinosaur

[2] J. Mouawad, "New Culture of Caution at Exxon After Valdez", *New York Times*, July 12, 2010

[3] K. M. Sutcliffe and K. E. Weick "Managing the Unexpected: Assuring High Performance in an age of Complexity", San Francisco: Jossey-Bass, 2001

[4] *The AIER Assessment: Technical Report.* DEKRA Insight/BST, Oxnard, CA, 2015.

Addressing the Dissonance between Corporate and Individual Process Safety Drivers

Angus Keddie
Process Safety Matters Ltd
23 Somerville Rise
Bracknell
RG12 7LW
UK
angus.keddie@processsafetymatters.com

ID# 476110

Prepared for Presentation at
American Institute of Chemical Engineers
2017 Spring Meeting and 13th Global Congress on Process Safety
San Antonio, TX
March 26 – 29, 2017

Addressing the Dissonance between Corporate and Individual Process Safety Drivers

Angus Keddie
Process Safety Matters Ltd
23 Somerville Rise
Bracknell
RG12 7LW
UK
angus.keddie@processsafetymatters.com

ID# 476110

Keywords: Psychology, Human Factors, Learning, Behaviour

Abstract

Conventional wisdom holds that the Process Safety Goals of the Organization and the Individual are roughly in tune. This paper will argue, however, that the drivers of the individual's Brain and the Organization are sufficiently incoherent so as to merit attention. It will explain why this is and suggest some innovative ways to drive alignment.

1 Corporate Process Safety Drivers

What drives Process Safety in Corporations? My 30 years as a Chemical Engineer working in diverse roles in the international Oil and Gas sector leads me to believe that corporate drivers for Process Safety are two-fold: Money and Reputation. Corporations are and should be driven to maximise a return to their shareholders. If corporations didn't do this, they may become uncompetitive and possibly perish. All the benefits that are attributed to them: useful products, employment, shareholder profit would be lost. Additionally, in some cultures they also have an obligation (financial and/or psychological) to their employees and other stakeholders.

Thus I believe, their approach to Process Safety (PS) should be underpinned by these tenets:

- to minimise PS incidents, which cost, on average, 10000+ times the cost of measures which would have prevented them
- to maximise their PS Return on Investment
- to maximise their uptime, which is negatively correlated with Lagging Indicators such as breaching the PS envelope (e.g. popping PSVs)

Consequently, they strive to align the PS related behaviour of their employees with these goals. The following table gives examples of Current/Recent Methods Employed by Corporations to promote PS within workforce:

Company	Method	Workforce Impact
A	Chronic Unease – workforce encouraged to be vigilant at all times	This is unsustainable and, on its own, can lead to increased levels of stress and higher risk of violation
B	Operator Care – workforce is empowered to tour the facility and be alert to anything out of the ordinary	The worker is acknowledged thus making it more likely that he/she will persevere with any behaviour change such that habits form
C	Safety Day - A day (or at least a significant part thereof) is devoted to considering and discussing an key element of Process Safety	If it is led sincerely and competently from the top, individuals will be impressed by the 'talk walking'.
BP (Texas City 2005)	Focus on Lagging Indicators (days without lost time incident) exacerbated by celebrating milestones	Workers able to maintain unconscious competence mode (avoiding trips, slips and falls) while generating psychological sunken costs ahead of a milestone (not reporting lost time incidents)

2 Individual Process Safety Drivers

What drives Process Safety in People? On operating sites, the tenets are different from those of corporations. I would argue that they are complex, but would certainly include protecting self and colleagues.

Why is it that Individuals and a group of Individuals (a corporation) have different drivers? I believe that individuals generally respond to the drivers of their own brains. Human brains are among the most complex organisms in existence. They are very energy draining (25% of body's energy usage vs 2% weight). Consequently, they have evolved to be very energy efficient and will drive to minimise activities which are energy sapping:

- Conflict
- Uncertainty
- Conscious thought

They have also developed senses to protect us against atavistic threats such as attack by wild animals:

I think it is useful here to look at the Learning Paradigm: how we master a new skill. Firstly, we are not aware that we don't have (or need) the skill - for example a child doesn't know (or need to know) how to drive a car. This first phase is known as Unconscious Incompetence. In the Process Industry, a school leaver becoming a plant operator would be in this mode. The Risk for the corporation is Ignorance leading to Mistakes. The remedy is classroom and 'on the job' training.

Secondly, when we attempt to engage with a new skill, we are aware of our incompetence. We are Consciously Incompetence. Here the Risk for corporation is Nervousness leading to Lapses. The remedy is Buddying or Mentoring.

Thirdly, we start to become competent at the new skill. However, we are still aware of our execution actions. Here we are in the realm of Conscious Competence, which is ideal for corporations. Unfortunately, the brain will drive relentlessly to the final mode - Unconscious Competence, as this is the low energy consumption (ideal) state for the brain. Here again there are risks for the corporation Complacency leading to Violations. How are we to counter these?

Let's examine the key individual Process Safety related behavioural drivers. These are:

- Minimising conflict – risk of groupthink
- Minimising uncertainty – fear of change
- Minimising conscious thought – risk of human factor errors: lapses, mistakes, violations (normalisation of deviation)
- Use of senses (sight, hearing, smell, touch, taste) has evolved to protect us from atavistic threats. Less useful when having to manage PS incidents (when nothing happens for a long time)

3 Alignment Tools

How can we address these issues to drive alignment and improve process safety? The following table illustrates potential alignment tools:

Individual Driver	Impact	Alignment Tool
Minimising Conflict	Groupthink – unwillingness to report unsafe situations	Empower employees and lead by example. Encourage external perspective.
Minimising Uncertainty	Resistance to necessary change (especially in culture which rewards success)	Incentivise the hassle to develop new habit. Reward endeavour over success.
Minimising conscious thought	Increased risk of human factor errors: lapses, mistakes, violations (normalisation of deviation)	Consider use of 'safe surprises' e.g. carry out 'on the spot' site audits.
Use of senses (sight, hearing, smell, touch, taste) has evolved to protect us from atavistic threats	Less useful when having to manage PS incidents (when nothing happens for a long time)	Empower employees to tour the facility and be alert to anything out of the ordinary (applying their 6th sense).

Addressing Conflict Minimisation

Groupthink is alive and well in our industry. Look no further than the BP Deepwater Horizon Disaster. The uncontrolled subsea blowout in the Gulf of Mexico in 2011 led to fire and explosion resulting in 11 deaths. It was the biggest US environmental disaster for 20 years. The cost to BP - $50bn. It was partly caused by collective psychological issues including groupthink.

One BP operator reported that when one of the men carried a handful of rubber material to a superior concerned that the rubber seal down in the well had been damaged, he was told, "No, that can't be. We always get that kind of material coming up."

BP had already sunk millions of dollars into the oil rig's production. Changing direction at that point – in other words stopping production because the rubber seal might be broken - was out of the question. BP had become wilfully blind: the evidence of barrier transgression didn't fit with the collective view that the rig was safe and was therefore ignored.

So how can we best counter Groupthink:

1. Anticipate Groupthink in your Risk Plan. While it might sound like planning to fail, ignoring the potential for Groupthink is a failure to plan for a very real risk. And like any risk plan, there must be processes for monitoring and mitigating emerging Groupthink.
2. Size counts. Limit the typical team size to less than 10 and ensure that there are well-defined boundaries for inclusion. Porous team boundaries and widespread casual involvement on teams breeds dysfunction, including pressure towards consensus for the wrong reasons.
3. Invite external perspectives at various stages of the process. Of course, you've got to have the procedures in place to both protect external viewpoints and to find ways to incorporate them into the group's thinking and plans.
4. Lengthen the discussion phase. Use structured discussion to focus on vetting the issues.
5. Develop a second solution. Challenge your team to assume that management will reject their first solution. Develop an alternative and very different second solution and be prepared to defend it.
6. Invite the Devil's Advocate to the party. While a designated Devil's Advocate is a contrived role and everyone knows it, at least someone will be throwing rocks at the groups beautiful picture. Rules on respecting and vetting the DA's perspective are critical to benefitting from this approach.

Addressing Uncertainty Minimisation

How could we start addressing Resistance to Change? Change is uncomfortable and therefore instinctively avoided. We do start off well however. I remember encouraging the effort both my sons made when they started to walk as babies.

Unfortunately, at some point, we seem to segue to only praising skill. One of the consequences of this change is that, as we grow, we are more likely to develop a 'fixed mindset' (one that believes that ability is inate) rather than a 'growth mindset' (one that believes ability can be improved by practice). As Carol Dwerck argues in her book 'Mindset', people with a fixed mindset are more likely to lie and normalise deviation to counter the discomfort they feel at the prospect of making mistakes.

Over half a century ago. Mick Jagger uttered the famous mantra 'I can't get no satisfaction' in the eponymous song. It seems to me the lyric encapsulates a key drive of human behaviour, where ever the end goal is on the Maslow Hierarchy of Needs. If the purpose of life is simply more life, then perhaps the goal of life should be individual satisfaction. For most of the key tenets of our lives, we know and target the scenarios which lead to consumption gratification.

And mostly that works fine. We buy and consume food and drink, we engage with other people for social and sexual benefit, we sleep when we are tired. All of which is likely to make us feel good if they are achieved. It seems our evolution has created rewards, often in the form of the hormones dopamine and oxytocin, which are released to encourage us to abandon the status quo and engage with the barriers between us and these tenets. Often the more significant the tenet and the higher the barrier, the greater the associated reward: for example, you may risk competition and rejection in pursuit of a mate.

However, in the arena of industrial hazards, this relationship becomes very tenuous. In terms of risk reduction, it benefits the group if near misses are reported and acted upon. Statistically, there are many times more near misses than the accidents they may have become. And as the antecedents of both are often the same - for example, the cause of a dropped load leading to an injury or a 'lucky escape' - the learning will reduce the likelihood of the same or similar cause reoccurring. So, it would seem to make sense to encourage the reporting of near misses.

However, this is not always the case. Currently, many operating companies set milestones for lost time incidents. As the milestone is approached, who wants to be the one to report a minor injury which would scupper the celebrations? Also, many companies have incentive schemes which are aligned to productivity, which is superficially negatively correlated with downtime. Furthermore, in companies with a strong safety culture, where near miss reporting is encouraged and the corresponding reduction in incidents means that nothing untoward happens for increasing lengths of time - which is a good thing, but not a satisfying thing.

So what to do? How can we make the promulgation of less incidents be satisfying to the front line worker? One idea I have is take an idea that Hollywood has hijacked and re-imagine it. I don't think I'm alone is watching action movies where heroes prevent bad things happening. Tom, Arnie et al battling to prevent the explosion, toxic cloud, wild beast from doing their worst. We are experiencing the vicarious echo of the actual experience. Risk and Reward. Why not apply the same principle to rewarding front line workers when they report a near miss which could have led to an injury or fatality.

An example I heard recently from an operating company was a propane leak in the vicinity of the air intake for a compressor. 'There was no explosion but for luck and the quick reaction and reporting of the shift supervisor'. In this case the supervisor was thanked. How about enhancing this thanks with the reward of a simulation of the incident his (or her) quick and robust intervention prevented. I could be shown at an end of year event and a copy presented to the worker. You have just created a lasting psychological reward which will reap benefits long after any monetary offering has been forgotten.

In an interview with Entrepreneur.com in 2012, James Dyson, the billionaire engineer and inventor, said 'we have to embrace failure and almost get a kick out of it.' Before successfully developing his eponymous bag-less vacuum cleaner, he had a large number of failures: 5216 to be precise. If he had become discouraged at, say prototype 3682, for example, he wouldn't have ultimately succeeded. How can we harness this spirit in our industry? By incentivising the hassle to develop new habit and rewarding endeavour over success.

Addressing Conscious Thought Minimisation

What about the Normalisation of Deviation? In his 1997 paper, published in the Journal Safety Science, Jens Rasmussen argues that in the presence of strong cultural and psychological forces (Corporate, Safety, Individual) human behaviour are likely to migrate towards the boundary of acceptable performance.

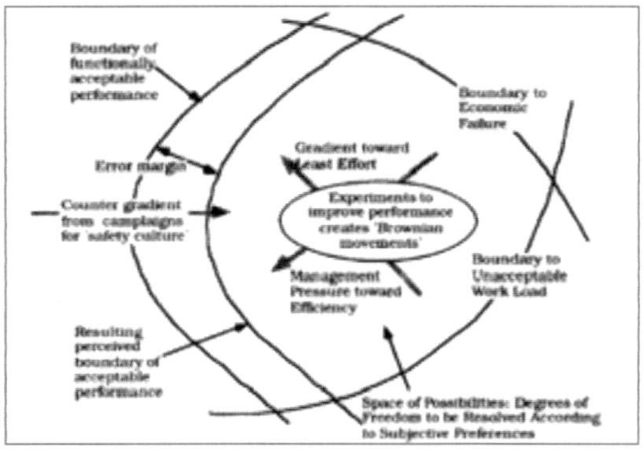

He argued that Experiments to improve performance creates 'Brownian movements' and that individual and corporate drivers will push the operations envelope towards the perceived boundaries of acceptable performance. If we take individual employees, they want to make their working lives as comfortable as possible, generating a gradient of least effort.

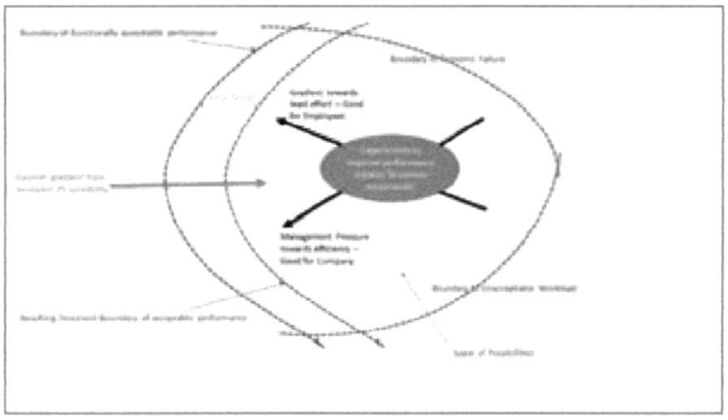

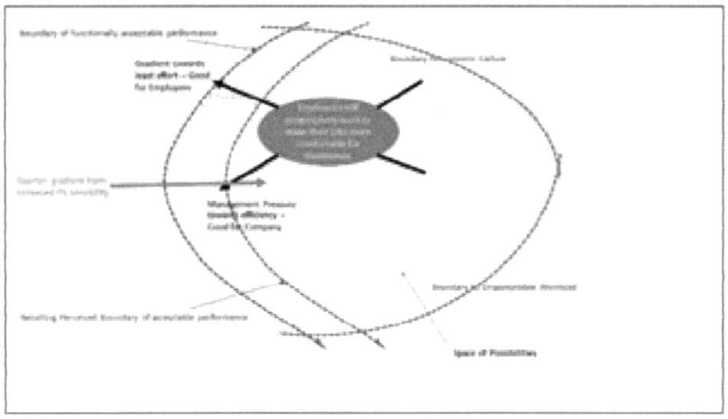

In doing so they may transgress the perceived boundary for acceptable performance.

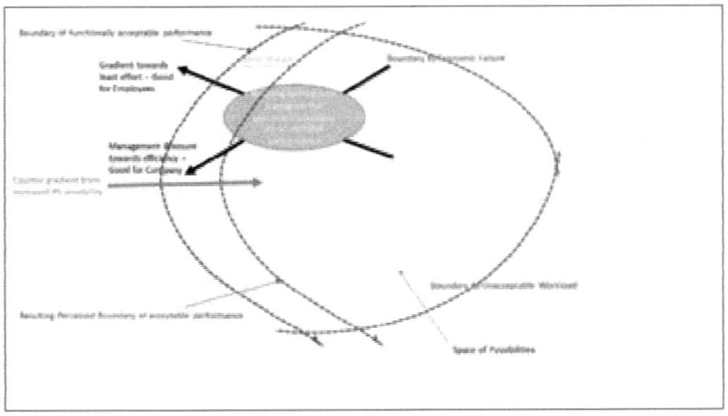

If PS sensibility isn't strong, this may lead to transgressions being normalised.

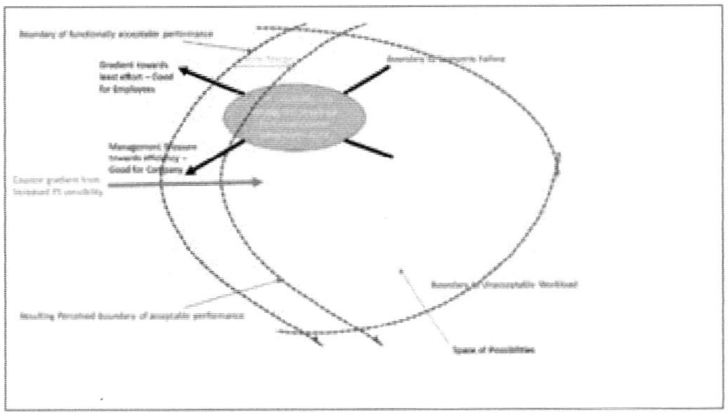

The Perceived Boundary of acceptable performance may recede towards and even through the actual boundary

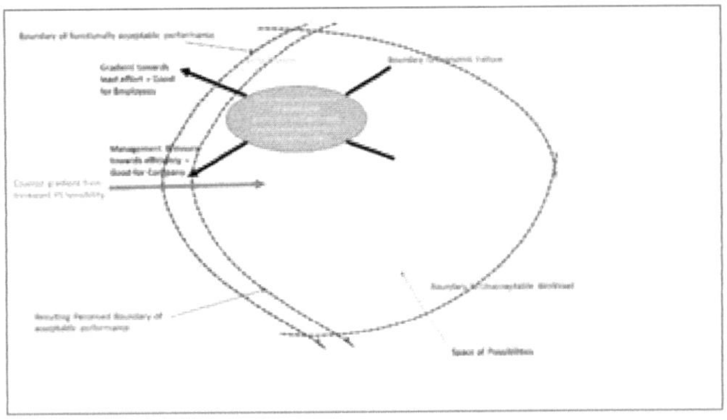

Reducing and eliminating the error margin and leading inexorably to breeching the actual boundary:

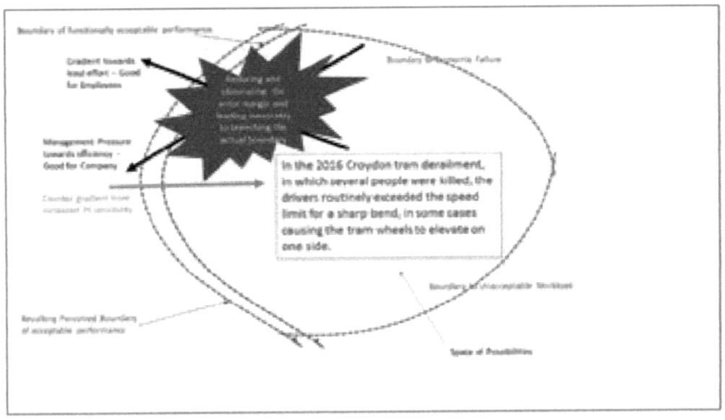

In the 2016 Croydon tram derailment, in which several people were killed, the drivers routinely exceeded the speed limit for a sharp bend, in some cases causing the tram wheels to elevate on one side. The accident took place in the dark and during heavy rain, on a sharp left curve. The curve comes almost immediately after the line emerges from a tunnel. It has been described as a "sharp bend" and has a 20 km/h (12 mph) speed restriction. The tram entered the curve at a speed of 70 kilometres per hour (43 mph) and overturned onto its right side, falling outside the curve on which it was travelling and trapping several people inside. There were seven fatalities with 58 other people injured.

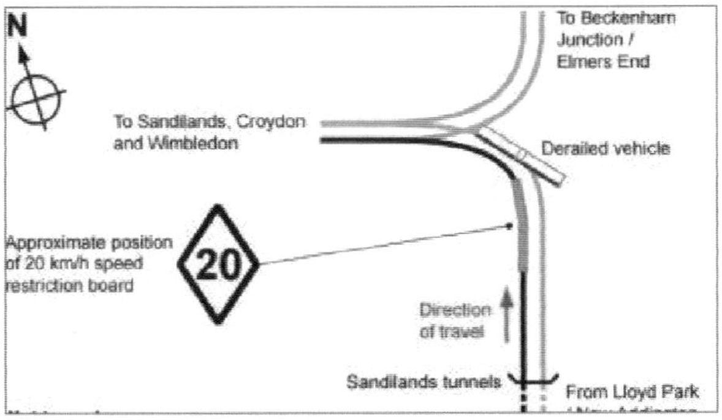

Images from the Aftermath:

However, if we make use of Safe Surprises and Unplanned Alerts, we can help to reset the Perceived Boundary. The Error Margin between Perceived and Actual Boundary of Acceptable Performance is thus re-established.

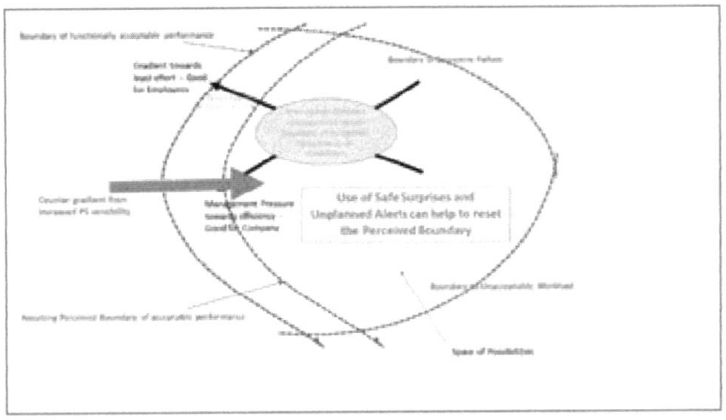

Addressing 'Senseless' Danger

Finally, how can we augment the front line operator's senses to improve PS outcomes? By Providing 6th Sense Tools such as 'Google Glass'. We could upload Key Process Parameter and Condition Based Monitoring Data. The glasses would be able to sense dynamic plant conditions, such as temperature or pressure spikes. The software could then trigger triaged warnings:

- Red – immediate plant shutdown
- Orange – investigate and risk assess
- Yellow – Periodic monitoring

4 Conclusions

So to summarise, this paper has demonstrated that the drivers for Process Safety of individuals and organisations are troublingly distinct, with worrying implications for both employee safety and company profits. The paper has shown that it is possible to better align the drivers and provided a number of tools to effect this, including:

- Groupthink - Develop a second solution (and compare it to the rock damaged initial one)
- Fear of Change - Find ways of rewarding endeavour over success e.g. generate simulations of prevented incidents
- Drive towards unconscious competence - Apply a mix of planned and surprise alerts
- 'Senselessness' of PS Hazards - Purchase technology to enhance front line workers 6th sense
- Use CBT to generate customised message for your workforce

Hopefully, this will lead to a Successful Outcome

5 References

Carol Dwerck. Mindset: How You Can Fulfill Your Potential. ISBN1780332009. 2012.

Nadia Goodman. James Dyson on Using Failure to Drive Success. Entrpreneur.com. November the 5th 2012.

Jens Rasmussen. Risk Modelling in a Society: A Modelling Problem. Safety Science Vol 27No 2/3. Pp 183-213. 1997.

Recipe for a Complete Process Hazard Analysis – Especially Addressing the Key Demands from US CSB

Revonda Tew, Principal Engineer & Instructor
PROCESS IMPROVEMENT INSTITUTE, INC. (PII)
1321 Waterside Lane, Knoxville, TN 37922
e-mail: rtew@piii.com

William G. Bridges, President
PROCESS IMPROVEMENT INSTITUTE, INC. (PII)
1321 Waterside Lane, Knoxville, TN 37922
Phone: (865) 675-3458
Fax: (865) 622-6800
e-mail: wbridges@piii.com

Prepared for Presentation at
13th Global Congress on Process Safety
San Antonio, TX
March 27-29, 2017

UNPUBLISHED

Recipe for a Complete Process Hazard Analysis – Especially Addressing the Key Demands from US CSB

Revonda Tew, Principal Engineer & Instructor
PROCESS IMPROVEMENT INSTITUTE, INC. (PII)

William G. Bridges, President
PROCESS IMPROVEMENT INSTITUTE, INC. (PII)

Keywords: PHAs, Damage Mechanisms, Risk control, Non-routine modes of operations, SOPs, process safety management, PSM

Abstract

Process Hazard Analysis (PHA) is not a new topic. Surprising, more that 80% of PHAs performed today do not comply with the current interpretations by US OSHA, much less the industry best practices. Most PHAs address less than 10% of the hazards during startup, shutdown, and online maintenance[1] and less than about 30% address damage mechanisms such as corrosion, erosion, external impacts, external stresses, vibration, etc. How to address these hazards has been part of the CCPS Guidelines for Hazard Evaluation Procedures since 1991 and the US CSB and US OSHA have noted how these weaknesses have led to many accidents. The citations and comments from US regulators and the CSB are detailed to provide some of the business case for performing more thorough PHAs on these key issues. This paper and presentation also illustrates step-by-step how to address all hazards of the process during ALL modes of operation, during a PHA. The presentation and paper will use examples of process hazards missed and found to illustrate the importance of this step-by-step approach. The results are based on thousands of unit-sized PHAs performed or managed by PII staff.

Background

Process Hazards is a broad term. The relationship between many components define process hazards. For a PHA team, Process Safety Information (information on the chemicals, technology and equipment) and operating procedures are necessary to identify specific components of and analyze a hazard scenario: process deviations (parametric and procedural), their causes, the process safety consequence and various safeguards for detecting deviations and causes, preventing causes, detecting and mitigating consequences and intervening safeguards. A thorough review of incident investigations offers an opportunity to further analyze process hazards.

Over the last twenty years the CSB has brought needed attention, through its investigations of industry incidences, to two significant gaps in many current PHAs; where the teams did not identify and analyze:

- Process hazards unique to non-routine modes of operations resulting in catastrophic consequences and
- Process hazards related to Damage Mechanisms.

Though the intent of the US OSHA PSM regulation has always been for PHA teams to analyze the hazards for all modes of operation, OSHA did not give specific requirements for addressing damage mechanisms and only recently have they increased their enforcement attention on PHA of non-routine modes of operation.

Industry has responded by modifying the *Guidelines for Hazard Evaluation Procedures*, 3rd Edition, 2008,[1] CCPS/AIChE to improve the coverage of these two main issues. API in turn issues guidance on addressing damage mechanisms. Industry best practices now exist for addressing both of these weaknesses. Now, with proper approaches, a team with process safety competencies and process knowledge can thoroughly address the issues during a PHA.

Another less significant gap noted by regulators and by the US CSB has been an incomplete consideration, within the PHA, of previous incidents.

This paper lays out the case from the US CSB and US OSHA on these PHA gaps:

1. Poor or no coverage of hazards during non-routine modes of operations
2. Poor or no coverage of damage mechanisms
3. Poor coverage of prior incidents

This paper also provides proven-in-use best practices for how to close these gaps, while also comparing these best practices to other approaches that have not performed as well.

Although regulatory pressure can be a driver for improvement, data from prior papers on the subject has shown there is near a 1000 to 1 economic return on investment for simply doing a reasonable PHA of non-routine modes of operation (primarily due to the catastrophes avoided).[22]

This paper will **not** cover the other critical aspects of a complete PHA/HAZOP, which can certainly still be problems in PHAs and which PII still finds issues with around the world. The other critical aspects of a fully compliant (excellent) PHA include:

- Excellent team leadership and PHA scribe that are fully competent in all aspects of a Best Practice PHA, including how to efficiently lead and document the PHA
- The most experienced and knowledgeable team members available from operators, operations engineers, and others as needed
- Up-to-date process safety information and access to underlying details, as needed
- Up-to-date, clear, and accurate operating procedures
- Thorough understanding and coverage of facility siting issues and human factors

Increasing Regulatory Pressure Related to PHA/HAZOP of Procedures to Analyzing Process Hazards During All Modes of Operation

Industry initiatives:

Industry has taken some initiatives to help standardize PHA analysis of non-routine modes of operation. One initiative was to improve the focus on PHA of non-routine procedures in *Guidelines for Hazard Evaluation Procedures,* 3rd Ed, 2008, AIChE/CCPS[1]. A new Chapter 9, Section 1, was added that necessitates hazard evaluations of all hazards of the process during all modes of operation. This textbook also explained why, when, and how to perform such analysis of step-by-step procedures.

Many companies have taken the initiatives to do the same, including about 20% of the largest chemical, petrochemical, and refining companies (based on PII data from hundreds of clients) But, the vast majority of companies who should be analyzing step-by-step deviations are not; and the major accidents continue to occur partly because of this. As a result, the US regulators are beginning to bring more pressure for the regulated companies to do a PHA of all modes of operation.

US OSHA Regulation and Enforcement:

The US OSHA PSM regulation requires PHA of all hazards during all modes of operation as well, and several key citations since 1990 have focused on PHA of non-normal modes.

- **Before there was a PSM regulation from US OSHA**, the Agency published CPL 2-2.45 (Systems Safety Evaluation of Operations with Catastrophic Potential).[2] In this guidance document, OSHA stated that a human error analysis should address:
    - o *Consequences of failure to perform a task*
    - o *Consequences of incorrect performance of a task*

- *Procedures and controls to minimize errors.*[3]

This approach is still the fundamental analysis method for PHA of non-normal modes of operation.

- **Phillips 66 "PHA" Citation —A citation with 566 instances was issued to Phillips 66 in Pasadena, TX, following their 1989 disaster** that killed 23 workers.[4] **The citation was related to a violation of the General Duty Clause (Section 5(a)(1) of OSH Act of 1970).** US OSHA cited Phillips against the General Duty Clause, since the PSM standard (29 CFR 1910.119) had not yet been issued. OSHA cited Phillips for not protecting its workers from hazards of fire/explosion by, among others, not performing a PHA that should have included an evaluation of the effect of design modifications on operator performance, and the identification of the source of observed human error and the identification of human factors that could result in incident event sequences. The citation stated, "**This review should result in a systematic listing of the (1) types of errors likely to be encountered during normal or emergency operation, (2) factors contributing to such errors, and (3) proposed system modifications to reduce the likelihood of such errors.**"

  The settlement agreement[5] between US OSHA and Phillips included the following requirements for process hazard analyses (PHAs) of the rebuilt and surviving units:
  - "Phillips will analyze each process…and will include human factors analysis … [and] will be …led by an independent consultant."
    - William Bridges (of JBF Associates at the time, now with PII) led these PHAs. Before these PHAs began, OSHA, Phillips, and Mr. Bridges decided that the best approach for finding all human error scenarios was to perform a HAZOP of deviations of the steps for the procedures governing activities for startup, shutdown, and particularly online maintenance.
  - "Phillips will provide OSHA an independent consultant's evaluation of the adequacy of its settling leg maintenance procedures performed while the polyethylene reactors are in operation…"
    - As part of the settlement to meet this requirement, it was decided by JBFA, Phillips and OSHA to perform a Human Reliability Analysis (HRA) of the Setting Leg online maintenance procedure, to ensure that the statistical risk of the accident recurring is less than the background risk of driving to work.

  The PHA and HRA resulting from the Phillips settlement agreement are presented as a Case Study later in this paper for sake of clarity.

- **Paragraph (e) of the US OSHA regulation on PSM, 29 CFR 1910.119,[6] and similar requirements in US EPA's rule for risk management programs (RMP), 40 CFR 68.24,[7]** specifically require that PHAs consider and address hazards of the process, i.e., all hazards regardless of the mode of operation (routine or non-routine).
  - 29 CFR 1910.119(e)(1) states that the PHA, "shall identify, evaluate, and

control **the hazards** involved in the process"

- 29 CFR 1910.119(e)(3)(i) states that the process hazard analysis shall address **"the hazards of the process"**
- 29 CFR 1910.119(e)(3)(vi) states that the process hazard analysis shall address human factors
- Appendix C to the OSHA PSM standard states that both routine and non-routine activities need to be addressed by the PHA of the covered process.

*There is no qualifier that limits the OSHA PHA requirement to only routine modes of operation. PSM requires that **all hazards** related to the process be addressed, regardless of the mode of operation or activity (routine or non-routine).*

- **OSHA Inspection No. 103490306 (Nov 2, 1992).**[8] In the first major PSM inspection in 1992 using 29 CFR 1910.119, OSHA assessed a serious violation when the PHAs did not address "human factors such as board operator error, line breaking mistakes, and improper lockout and isolation of process equipment," all of which are errors originating from failure to either perform tasks or perform them incorrectly.

- **US OSHA published an internal document on Program Quality Verification of Process Hazard Analysis in 1993 (by Henry Woodcock, of OSHA).**[3] This document states that a PHA should include analysis of the "procedures for the *operation* and *support* functions" and goes on to define a "procedure analysis" as evaluating the risk of "skipping steps and performing steps wrong." The authors concur and PII has found the same true in PHAs that we have performed using various methods; a 2 Guideword HAZOP approach is normally optimal for PHA of procedures.

- **OSHA Inspection No. 123807828 (Nov 18, 1993)**[9] – Ashland Oil, Catlettsburg, KY. Several operators were preparing to ignite a 2-B-3 crude heater after a two week turnaround. The lead operator had two very inexperienced workers helping him light the heater. A large concentration of fuel gas was allowed to enter the heater before the pilot light was ignited. The resulting explosion killed one employee. The operators bypassed safety shutdown features and did not do a visual check of the emergency shutdown system to ensure that the features were closed. In addition, they did not check the firebox to ensure that it was gas-free before lighting the heater.

  The Kentucky OSHA citation read: *The PHA did not address all hazards of the #2 Crude unit..; The PHA did not address the hazards associated with the startup of the crude unit after a turnaround, ...emergency shutdown..., emergency operations and normal shutdown of the unit. The process hazard analysis that was completed by the PHA team for the #2 Crude unit only evaluated the hazards associated with normal mode of operation of the #2 Crude unit.*

Settlement: All procedures were re-written and all PHAs were redone to include a PHA of deviations from procedural steps for all non-continuous modes of operation.

- **Recent US OSHA PSM National Emphasis Programs** for Chemical Processes[10] and also for Refineries[11] underscore the need for companies to identify potential accident scenarios during non-routine modes, and to reduce the frequency and consequences of such errors as part of an overall process safety management (PSM) program.

OSHA recognizes that CCPS/AIChE has added as Chapter 9.1 in the 3rd edition of *Guidelines for Hazard Evaluation (2008*[1]) to further emphasize the need for a PHA to include hazard evaluations of all modes of operation and that this chapter has added best-practice detail on the approach for doing the hazard evaluation of startup, shutdown, and online maintenance modes of operation. Despite the specific OSHA standard that requires PHAs of covered processes to address all hazards, many PHAs still do not address hazards during all modes of operation. Further, many of the regulated community have stated "Well, OSHA did not tell us to perform a PHA of procedures for non-routine modes of operation." On the other-hand, OSHA did not state to do only a hazard evaluation of normal mode of operation and stop there.

To highlight the importance that PHAs address hazards during all modes of operation and activities (routine and non-routine), OSHA is considering issuing a Hazard Alert that would incorporate the concepts in Chapter 9.1 of *Guidelines for Hazard Evaluation Procedures, 2008, CCPS/AIChE.*[1] Also, as stated above, OSHA has an enforcement initiative, CHEM NEP, that utilizes a list of dynamic questions that OSHA compliance officers use to evaluate compliance at facilities covered by the program. **It is possible that future dynamic list questions could address PHAs of all modes of operation, and is further possible that this CHEM NEP update is drafted and waiting for release.**

Pressure from the US Chemical Safety and Hazard Investigation Board (US CSB)

The CSB has commented on the need for PHAs to address all hazards of the process during all modes of operation. Their clearest statement was in the report **2008-08-I-WV-R1[12] for the Bayer CropScience accident in Institute, WV, 2008.** In that report, CSB asks Bayer to:

- Revise the corporate PHA policies and procedures to require:
  a. Validation of all PHA assumptions to ensure that risk analysis of each PHA scenario specifically examines the risk(s) of intentional bypassing or other nullifications of safeguards
  b. **Addressing all phases of operation and special topics including those cited in chapter 9 of "Guidelines for Hazard Evaluation Procedures" (CCPS, 2008),**
  c. Training all PHA facilitators on the revised policies and procedures prior to assigning the facilitator to a PHA team, and

d. Ensure all PHAs are updated to conform to the revised procedures.

Another strong emphasis on PHA of procedures was in their report: ***Tesoro Anacortes Refinery April 2, 2010 Incident CSB Report, "Catastrophic Rupture of Heat Exchanger," Report 2010-08-I-WA, January, 2014.***[13] This accident resulted in 7 fatalities. In that report, CSB states:

- The startup of the NHT heat exchangers was hazardous non-routine work. Leaks routinely developed that presented hazards to workers conducting the startup activities. Process Hazard Analyses (PHAs) at the refinery repeatedly failed to ensure that these hazards were controlled and that the number of workers exposed to these hazards was minimized.
- None of the Anacortes refinery PHAs effectively evaluated and controlled hazards associated with the non-routine work necessary to periodically clean the NHT heat exchangers. The Washington PSM regulations address the need for non-routine operations to be evaluated and require that at least one member of the PHA team has expertise in non-routine tasks.[14] The CCPS describes the importance of PHA evaluations, as well as the hazardous potential and frequent problems of PHAs that lack sufficient analysis of nonroutine work as follows:[1, 15]
- It is not uncommon for initial PHAs of continuous processes to focus only on normal operations, failing to address non-routine, critical operating modes such as startup, shutdown, preparation for maintenance, emergency operations, emergency shutdown, and other activities whose characteristics may differ considerably from normal operations. Experience indicates that many accidents do not occur during "normal" operation but, rather, during such non-routine modes of operation. Consequently, it is important that a PHA evaluate the hazards of a process during non-routine as well as normal (routine) operating modes. The 1996 Shell Oil PHA for the NHT unit did not evaluate or identify any issues related to non-routine hazardous work associated with the frequent NHT heat exchanger cleaning operations. The 2006 Tesoro NHT unit PHA revalidation identified startup as a non-routine operation but noted that existing procedures were adequately addressing non-routine work.

But there was no PHA on the procedures to determine if there were sufficient protection against errors that could (and did) occur during non-routine operations. Simply stating there are procedures equates to No PHA of non-routine operations.

Another mention was in their report: ***DuPont La Porte, Texas Chemical Facility Toxic Chemical Release Interim Recommendations. Investigation: 2015-01-I-TX. Incident Date: November 15, 2014 Issue Date*: September 30, 2015.**[16] This accident resulted in 4 fatalities. In that report, CSB states:

- DuPont's process hazard analyses (PHAs) and relief system design scenarios do not effectively identify hazards from non-routine operations, such as opening valves to connect the liquid methyl mercaptan piping to the vapor waste gas vent header – the piping connection that provided the pathway for the methyl mercaptan release in this incident. Along the methyl mercaptan feed line there were three locations where it was connected by valves to the waste gas vent header piping. At the time of the incident, one of these valves was fully open and a second valve was slightly open.

US EPA's RMP Regulation

In the Risk Management Program rule (40 CFR 68.24)[7] EPA also recognizes the importance of procedural analysis, by defining the purpose of a PHA to **"examine, in a systematic, step-by-step way, the equipment, systems, and *procedures* (emphasis added) for handling regulated substances."**

A well-done PHA should identify all failure scenarios that could lead to significant exposure of workers, the public, or the environment.......For toxics under PSM, however, you may plan to address a loss of containment by venting toxic vapors to the outside air. In each circumstance, a PHA should define how the loss of containment could occur. However, for EPA, the PHA team should reassess venting as an appropriate mitigation measure. (From EPA RMP Guidance, Chapter 7, pgs 7-6 & 7-7; General Risk Management Program Guidance.[26])

Pressure from the Local Regulator – Contra Costa County (CCC) California

CCC implements the state's Accident Release Prevention (ARP) regulations locally and have supplemented the Cal-ARP with an Industrial Safety Ordinance (ISO)[17] for elements or features of process safety that they deemed were not handled appropriately (or were missing altogether) from the ARP and/or from Cal-OSHA regulations on PSM. One element they added was the requirement that PHA of Procedures, which requires a guideword analysis, step-by-step, through each critical non-routine task. **These requirements closely follow Chapter 9.1 of the Guidelines for Hazard Evaluation Procedures, 3rd Edition.** CCC Hazardous Materials Department enforces these requirements.

CCCHM contracted PII to perform a Safety Evaluation of the Chevron Richmond Refinery following the August 2012 accident in the Crude Unit. The audit was refinery-wide and included the review of safety culture and also reviewed refinery implementation of process safety versus industry best practices. One specific finding listed in the Draft Report[18] was:

- **Recommendation 5 from the Draft Report:** Continue and expand the current Procedural PHA program implementation so that all modes of operation are evaluated. Include a PHA of startup, shutdown, and online maintenance procedures, as described in Chapter 9 of the *Guidelines for Hazard Evaluation Procedures*, 3rd edition, CCPS/AIChE, 2008.[1] This will entail using a 2-guideword HAZOP approach on each step for critical procedures and a What-if (no guideword) approach for less-critical procedures. No procedures should be excluded.[1, 22, 23, 24]

Increasing Regulatory Pressure Related to Addressing Damage Mechanism during PHA/HAZOP

Industry initiatives:

From the second edition of the *Guidelines for Hazard Evaluation Procedures* (1991) until 2007, there has been a good explanation and examples of how to cover damage mechanisms during PHA/HAZOP. However, most PHAs did not follow the examples in this textbook and instead provided no or minimal coverage of damage mechanisms such as corrosion, erosion, wrong materials of construction, external impacts, etc.

At the request of US CSB following the large explosion and losses at *Formosa Plastic*[19], the industry has taken some initiatives to improve the coverage of damage mechanism. In particular, additional text was included in the *Guidelines for Hazard Evaluation Procedures,* 3rd Ed, 2008, AIChE/CCPS[3] that emphasized the earlier examples of how to address damage mechanism during PHA team meetings and in the PHA report. In addition, ***API 571***[20] was issued that discussed various damage mechanisms and how these should be addressed by companies.

Many companies have taken the initiatives to do the same, including many of the largest chemical, petrochemical, and refining companies. But, the vast majority of companies do not provide thorough coverage of damage mechanisms during their PHA/HAZOP.

Pressure from the US Chemical Safety and Hazard Investigation Board (US CSB)

The CSB has made numerous references to the need for PHA/HAZOPs to address all damage mechanisms. Based on interviews of senior staff at CSB by PII staff, they recognize this as the second largest deficiency in current PHAs and have discussed this in presentations at the Global Congress on Process Safety (GCPS) over the past few years. In addition, they have provided in-depth analysis of a few accidents that stemmed from failure to address damage mechanisms, including the 2012 release and fire at Chevron's Richmond, Ca, Refinery and the catastrophic rupture of the heat exchanger at Tesoro Anacortes Refinery, 2010.

One of the first and strongest emphasis by US CSB on considering damage mechanisms during PHAs was in the ***Tesoro Anacortes Refinery April 2, 2010 Incident CSB Report, "Catastrophic Rupture of Heat Exchanger,"*** *Report 2010-08-I-WA, January, 2014.*[13] This accident resulted in 7 fatalities (note that the Anacortes Refinery was owned and operated by the Shell Oil Company prior to 1998). In that report, CSB states:

- The 1996 Shell Oil Naphtha Hydrotreating (NHT) unit PHA simply cited ineffective, nonspecific, judgment-based, qualitative safeguards to prevent equipment failure from high temperature hydrogen attack (HTHA). However, the effectiveness of these safeguards was neither evaluated nor documented; instead the PHA merely listed general safeguards. Had the adequacy of the safeguards been verified, improved safeguards intended to protect against HTHA failure could have been recommended.

- The 2001 and 2006 Tesoro PHA revalidations did not address or modify the analysis performed in the 1996 Shell Oil PHA. The Tesoro 2010 NHT unit PHA failed to identify HTHA as a hazard for the shell of the B and E heat exchangers. For the 15 years before the April 2010 incident, assumptions used by PHA teams at the Anacortes refinery contributed to ineffective safeguards, ineffective hazard identification, and ineffective control of hazards to prevent equipment failures from HTHA damage, such as the E heat exchanger in the NHT unit.
- Shell Oil completed a PHA in 1995 related to process modifications that could increase the hydrogen partial pressure in the NHT heat exchangers. However, when managing this change no consideration, evaluation, or recommendations were made to address the potential for HTHA damage to the NHT heat exchangers.
- Shell Oil and Tesoro periodically performed damage mechanism hazard reviews (DMHRs), called corrosion reviews. However, these reviews did not identify HTHA as a credible failure mechanism for the B and E heat exchangers. These reviews were weakened by primarily relying on design operating data for these heat exchangers rather than data from actual process operating conditions.

Note that the DMHRs were not done as part of a PHA or with a similar PHA team structure.

Another strong emphasis by US CSB on considering damage mechanisms during PHA was in the **U.S. Chemical Safety and Hazard Investigation Board (CSB), *Interim Investigation Report: Chevron Richmond Refinery Fire (August 6, 2012).***[21] In that report, CSB states that failure to do damage mechanism hazard reviews and follow-through on recommendations related to know damage mechanisms led in part to the accident.

Pressure from the Local Regulator – Contra Costa County (CCC) California

CCC implements the state's Accident Release Prevention (ARP) regulations locally and have supplemented the Cal-ARP with an Industrial Safety Ordinance (ISO)[17] for elements or features of process safety that they deemed were not handled appropriately (or were missing altogether) from the ARP and/or from Cal-OSHA regulations on PSM. One element they added was the requirement that DM reviews be completed for each unit and that DM be considered in each PHA. CCC Hazardous Materials Department enforces these requirements.

CCCHM contracted PII to perform a Safety Evaluation of the Chevron Richmond Refinery following the August 2012 accident in the Crude Unit. The audit was refinery-wide and included the review of safety culture and also review refinery implementation of process safety versus industry best practices. Recommendation 3 and the background observations from the Draft Report are listed below:

- …since specific/individual damage mechanisms from API 571 "Damage Mechanisms Affecting Fixed Equipment in the Refining Industry"[20] are not considered at each process section (node) within the process, it is likely that unique hazardous scenarios have been missed (best practice, as stated in *Guidelines for Hazard Evaluation Procedures*, 2008[1], is to consider damage mechanisms within each equipment node). It is also likely that some necessary safeguards such as remote isolations, leak detection, etc., for pipe sections, pumps, vessels, and columns have been missed.

- [Refinery] management reports that the Design Engineering Group piloted a specification- break review in 2013 (specification "spec" breaks are where the materials of construction or pipe thickness change). Also, each PHA is intended to identify any potential spec break concerns within the unit (as is typical for PHAs that follow industry best practice). It is understood that a policy and procedures to codify spec-break reviews will be implemented. This approach to damage mechanisms related to materials specification is considered a best practice. Most competent PHA teams in the industry ensure a discussion of each spec break during a PHA.
- **Recommendation 3 from the draft report**: To avoid missing causes and necessary safeguards against loss of containment, be sure hazardous scenarios starting from pertinent damage mechanisms are reviewed during each unit's PHA (pertinent damage mechanisms include corrosion, erosion, seal failure, pump failure, external impact, external fire, material defect, improper maintenance, drains/vents left open, etc.). In particular, review external impact as a damage mechanism for each node. A review of such loss of containment scenarios in each node is recommended, as described in the *Guidelines for Hazard Evaluation Procedures*, 3rd Edition, 2008, CCPS/AIChE.[1] Adding a loss of containment deviation would double-check against missing standard damage mechanisms (such as corrosion and erosion) during unit-wide reviews. It would also improve the current PHAs by providing a review at each node for external impacts and control of drains/vents. Review and incorporate mechanical integrity data during this analysis.

Best Practice for PHA/HAZOP of Procedures to Analyze Process Hazards During All Modes of Operation

Overview of Methodology for Hazard Evaluation of Non-Routine Modes of Operation

The hazard evaluation of non-routine modes of operation involves reviewing procedures using a HAZOP, simplified HAZOP, or What-if analysis to uncover potential accident scenarios associated with non-routine operations, for continuous or batch operations. Human error is more likely and more critical during non-routine operations. By analyzing procedural steps where human error is more likely, and where human error or component failure could lead to a consequence of interest, risk can be reduced. The hazard evaluation team's objective is to evaluate the risk associated with skipping steps and performing steps wrong.

FMEA cannot be applied to procedure-based deviations, unless you create a "human" component, in which case you have simply merged HAZOP deviations for "steps" into FMEA. Pre-Hazard Analysis (P_rHA) and other hazard evaluation methods are not applicable for accomplishing a detailed hazard evaluation of non-routine modes of operations.

Checklist of human factors issues (see an earlier paper[22, 23, 24] and also in the *Guidelines for Hazard Evaluation Procedures*[1]) can be very useful after the detailed hazard evaluation of deviations of steps. Such analysis can indicate where generic weaknesses exist that can make errors during any mode of operation more likely, or that can make errors during maintenance more likely. Such human factors checklists are normally used at the end of the analysis, they can be done piecemeal during an analysis (on breaks from the meetings) by individuals on the team, and then the results of each individual review can be discussed as a team at the end.

Purpose of Hazard Evaluation of Procedures-Based Modes of Operation

Although incorporating human factors considerations into hazard evaluation studies of continuous operation is straightforward by asking why the human might make a mistake that leads to a parametric deviation, this approach only addresses a small fraction of the potential human errors that can affect process safety. Many analysts have tried to find accident scenarios in non-routine modes of operations by adding generic guide words such as "deviations during startup" and "deviations during maintenance/sampling" to the hazard evaluation of equipment nodes/sections. Unfortunately, this only catches a fraction of the accident scenarios that can occur in non-routine modes since a hazard evaluation team is focused on "continuous" mode of operation during HAZOP or What-if of equipment sections/nodes.

From an informal survey of more than 100 companies, most do not currently perform process hazard evaluations of procedures, although many do perform some type of job safety analysis (JSA). The JSA is an excellent starting point for an evaluation of procedures because a JSA identifies the tasks that workers perform and the equipment required to protect workers from typical industrial hazards (slips, falls, cuts, burns, fumes, etc.). Unfortunately, a typical JSA will not usually identify process safety issues or related human factors concerns. For example, from a JSA perspective, it may be perfectly safe for an operator to open a steam valve before opening a feed valve; however, from a process safety perspective, the operator may need to open the feed valve before the steam valve to avoid the potential for overheating the reactor and initiating an exothermic decomposition. The primary purpose of a JSA and other traditional methods for reviewing procedures has been to ensure that the procedures are accurate and complete (which is required of employers in 29 CFR 1910.119(f)(3)).[6]

By contrast, the purpose of a hazard evaluation is ***not*** to ensure the procedures are accurate and acceptable, but instead, to ***evaluate the accident scenarios if the procedures are not followed***. Even the best procedure may not be followed for any number of reasons, and these failures to follow the prescribed instructions can and do result in incidents. In fact, in the chemical industry and most other process industries the chance of an operator or other worker making a mistake in following a procedure is greater than 1/100, and in some cases much greater. When taking into account common human factor deficiencies that accompany non-routine operations, such as fatigue, lack of practice, the rush to restart and return to full production, etc., the probability of errors can climb to 1/10 chances per task (a task being about 1 to 10 detailed steps).[28, 29]

The purpose of a hazard evaluation of non-routine modes of operation (governed by written procedures) is to make sure an organization has enough safeguards for the inevitable instance when a step is either performed wrong or skipped (inadvertently or due to shortcutting or other reasons)

Industry has found that a HAZOP or what-if analysis, structured to address procedures, can be used effectively for finding the great majority of accident scenarios that can occur during non-routine modes of operation.[1, 22, 23, 24] Experience shows that reviews of non-routine procedures have revealed many more hazards than merely trying to address these modes of operation during the P&ID driven hazard evaluations.

To reinforce the need for and to explain the method for analysis of deviations of steps in a procedure, Section 9.1 was included in the 3rd Edition of *Guidelines for Hazard Evaluation Procedures,* 2008[1]; this was one of the major changes to the hazard evaluation procedures.

HAZOP Method for Analyzing Deviations of Procedural Steps

The Hazard and Operability (HAZOP) method has two major variations; one for the continuous mode of operations (where the team brainstorms what would happen if there were deviations of parameters) and procedure-based (where the team brainstorms what would happen when the steps of a procedure are not followed correctly). The procedure-based variation of HAZOP is the oldest form of HAZOP (from ICI in 1960s).[1, 25] It was an expansion of a Hazard Evaluation method based strictly on asking:

- **What happens if the step is skipped?**
- **What happens if the step is performed wrong?**

In turn, the "pre-HAZOP" method for brainstorming accident scenarios from not following procedures (including because the procedure is itself wrong) is based on the understanding that human errors occur by someone not doing a step (errors of omission) or by doing a step incorrectly (errors of commission). So, simply asking what would happen if the operator omitted a step or performed a step wrong is one way to structure a hazard evaluation of a step-by-step procedure. The usefulness of this simple approach to hazard evaluation of steps will be discussed later.

Seven (7) Guide Word Method

In an effort to be more thorough, the inventors of HAZOP (at ICI) broke these two types of errors into subparts and agreed on using the following 7 Guide Words:

Omission:	**Skip (or Step Missing)** **Part Of**
Commission:	**More**

Less
Out of Sequence
As Well As
Other Than
Reverse

In the early 1990s, the guide word Skip was augmented by adding the option of discussing "are there any steps **missing** from the procedure."[24]

Table 1: Definitions of 7-8 Guide Words for HAZOP of Procedure-Based (Non-Continuous Mode) Operation

Guide Word	Meaning When Applied to a Step
Missing (optional guide word)	A step or precaution is missing from the written procedure prior to this step (similar to "Out of Sequence", except the missing step is not written)
Skip (No, Not, Don't)	The specified intent of this step is not performed
Part-of	A portion of the full intent is not performed. Usually only applies to a task that involves two or more nearly simultaneous actions ("Open valves A, B, and C".)
More	Too much of the specified intent is done (does not apply to simple on/off; open/close functions); or it is performed too fast
Less	Too little of the intent is done, or it is performed too slowly
Out of sequence	This step is performed too early in the sequence
As well as	Something happens, or the user does another action, in addition to the specified step being done correctly (could be a short cut)
Other than (or Reverse)	The wrong device is operated, selected, read, etc., or operated in a way other than intended. Or the wrong material is selected or added. "Other than" errors always imply a "Skip" as well.

To apply HAZOP to procedural steps for startup, shutdown, online maintenance, and other modes of operation, the facilitator (or team) first divides the procedure into individual actions. This is already done if there is only one action per step. Then, the set of guide words or questions is systematically applied to each action of the procedure resulting in procedural deviations or what-if questions. The guide words (or procedural deviation phrases) shown in Table 1 above were derived from HAZOP guide words commonly used for analysis of batch processes. The definition of each guide word is carefully chosen to allow universal and thorough application to both routine batch and non-routine continuous and batch procedures. The actual review team structure and meeting progression are nearly identical to that of a process equipment HAZOP or what-if analysis, except that active participation of one or more operators is even more important and usually requires two operators for a thorough review; a senior operator and a junior operator. For each deviation from the intention of the process step (denoted by these guide words applied to the process step or action), the team needs to dig beyond the obvious cause, "operator error," to identify root causes associated with human error such as "inadequate emphasis on this step during training," "responsible for performing two tasks simultaneously," "inadequate labeling of valves," or "instrument display confusing or not readable." The guide word *missing* elicits causes such as "no written procedural step or formal training to obtain a hot work permit before this step," or "no written procedural step or formal training to open the discharge valve before starting the pump."

Two (2) Guide Word Method for Analyzing Deviations of Procedural Steps

A more streamlined guide word approach has also proven very useful for (1) procedures related to less hazardous operations and tasks and/or (2) when the leader has extensive experience in the use of the guide words mentioned previously and can therefore compensate for the weaknesses of a more streamlined approach. The two guide words for this approach (as defined in Table 2 below) encompass the basic human error categories: errors of omission and commission. These guide words are used in an identical way to the guide words introduced earlier. Essentially "omit" includes the errors of omission related to the guide words "skip," "part of," and "missing" mentioned earlier. The guide word "incorrect" incorporates the errors of commission related to the guide words "more," "less," "out of sequence," "as well as," and "other than" mentioned earlier. Note that these two guide words (Table 2) fill the basic requirements for a human error analysis as outlined in OSHA's CPL 2-2.45.[2]

Table 2: 2 Guide Word (Guide Phrases) for Modified-HAZOP of Procedure-Based Operation

Guide Phrase	Meaning When Applied to a Step
Step not performed	The step is not done or part of the step is not done. Some possible reasons include the employee forgot to do the step, did not understand the importance of the step, or the procedures did not include this vital step
Step performed wrong	The employee's intent was to perform the step (not omit the step), however, the step is not performed as intended. Some possible reasons include the employee does too much or too little of stated task, the employee manipulates the wrong process component, or the employee reverses the order of the steps.

Table 3: Example of 2 Guide Word HAZOP of a Critical Step in a Procedure

Drawing or Procedure: SOP-03-002; ***Cooling Water Failure***	**Unit:** HF Alkylation	**Method:** 2 Guide Word Analysis	**Documentation Type:** Cause-by-Cause

Node: 23	**Description:** STEP 2: Block in olefin feed to each of the 2 reactors by blocking in feed at flow control valves				
Item	**Deviation**	**Causes**	**Consequences**	**Safeguards**	**Recommendation**
23.1	Step not performed	Operator failing to block in one of the reactors, such as due to miscommunication between control room operator and field operator; or control valve sticking open or leaking through	High pressure due to possible runaway reaction (because cooling is already lost), because of continued feeding of olefin (link to 11.7 - High Rxn Rate; HF Alky Reactor #1/#2) High pressure due to high level in the reactor, because of continued feeding olefin (link to 11.1 - High Level; HF Alky Reactor #1/#2)	High temperature alarm on reactor High pressure alarm on reactor Field operator may notice sound of fluid flow across valve Flow indication (in olefin charge line to reactor that is inadvertently NOT shutdown) Level indicator, high level alarm, and independent high-high level switch/alarm	
		Operator failing to make sure bypass valve is also closed, since this precaution is not listed in the written procedure; or the bypass valve leaks through	High pressure due to possible runaway reaction (because cooling is already lost), because of continued feeding of olefin (link to 11.7 - High Rxn Rate; HF Alky Reactor #1/#2) High pressure due to high level in the reactor, because of continued feeding olefin (link to 11.1 - High Level; HF Alky Reactor #1/#2)	High temperature alarm on reactor High pressure alarm on reactor Operator skill-training requires checking bypasses are closed, when blocking control valves Field operator may notice sound of fluid flow across valve Flow indication in olefin charge line (but likely not sensitive enough for small flows) Level indicator, high level alarm, and independent high-high level switch/alarm	
		Operator failing to close low control valve manually from the DCS because the phrase "block in" is used instead of the word "close"	Valve possibly opens full at restart, allowing too much flow to reactor at restart, resulting in poor quality at startup and/or possibly resulting in runaway reaction and high pressure	Control room skill training requires always manually commanding automatic valves closed before telling field operator to block in control valve	37. Implement best-practice rules for procedure writing, which includes using common terms.
23.2	Step performed wrong	Operator closing the olefin charge flow control valves before shutting down the charge pump, primarily because the steps are written out of the proper sequence	Deadheading of charge pump, leading to possible pump seal damage/failure and/or other leak, resulting in a fire hazard affecting a small area (link to 5.12 - Loss of Containment; Olefin Charge Line/Pump)	Step 3 of procedure that says to shutdown charge pump The step to shut down the charge pump (Step 3) is typically accomplished before Step 2 (in practice)	41. Move Step 3 ahead of Step 2.
		Field operator closing both upstream and downstream block valves	Possible trapping of liquid between block valve and control valve, leading to possible valve damage (due to thermal expansion)	Field operator skill training stresses that only one block valve should be closed	

What-if Method for Analyzing Deviations of Procedural Steps

The What-if method for analyzing procedure-based modes of operations is free brainstorming without the aid (or constraints) of guide words. This method is described in detail in the *Guideline for Hazard Evaluation Procedures (CCPS).*[1] The hazard evaluation team using this method would read the procedure and then answer the question: "What mistakes will lead to our consequences of interest?" The team would list these mistakes and then brainstorm the full consequences, causes, and existing safeguards – the same analysis approach described for the guide word approaches mentioned earlier in this section. What-if brainstorming **is not** applied to each step of the procedure, but rather covers the entire task (procedure) at one time.

Choosing the Right Method for Analysis of Non-Routine Modes of Operation

Obviously the What-if approach takes far less time than the 2-Guide Word method, and the 2 Guide Word method takes much less time than the 7-8 Guide Word method of HAZOP of procedures. Experience has shown that hazard evaluation facilitators, newly trained in the three techniques above, tend to overwork an analysis of non-routine procedures, so a tiered approach is best. In this tiered approach, the first step in choosing the right method of analysis in the hazard evaluation of procedures is to screen the procedures and select only those procedures with extreme hazards. These procedures should be subjected to a detailed HAZOP analysis (7-8 guide word set) presented above. The 2-Guide Word set is efficiently used for less complex tasks or where the consequences are lower. The What-if method is applicable to low hazard, low complexity, or very well understood tasks/hazards.

Experience of the leader or the team plays a major part in selecting the procedures to be analyzed, and then in deciding when to use each guide word set.

Figure 2 shows the typical usage of the three methods described above for a typical set of operations procedures within a complex chemical plant or refinery or other process/ operation. Most of the procedures are simple enough, or have low enough hazards to warrant using the What-if method. Currently, the 7-8 Guide Word approach is used infrequently, since most tasks do not require that level of scrutiny to find the accident scenarios during non-routine modes of operations.

Figure 1. Relative Usage of Techniques for Analysis of Procedure-Based Modes of Operation [1, 22, 23, 24]

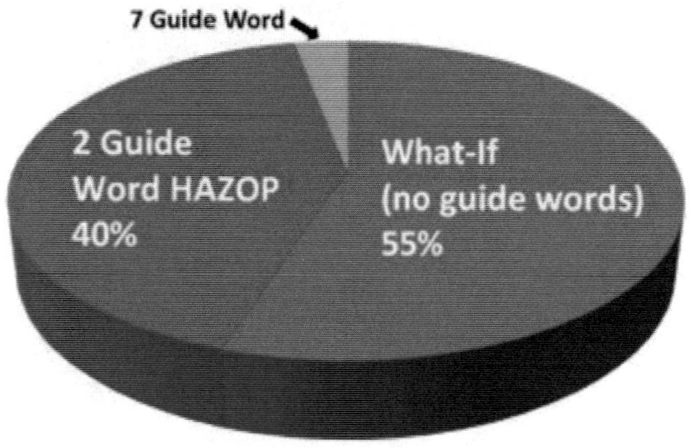

The experience of the leader or the team plays a major part in selecting the method to use for each task/procedures to be analyzed. However the first decision will always be "are these procedures ready to be risk reviewed?" If the procedures are up-to-date, complete, clear, and used by operators, then the best approach for completing a complete hazard evaluation of all modes of operation, including routine modes of operation, is shown in Figures 2A and 2B below:

Figure 2A: PHA of a ALL Modes of Operation for a CONTINUOUS Process

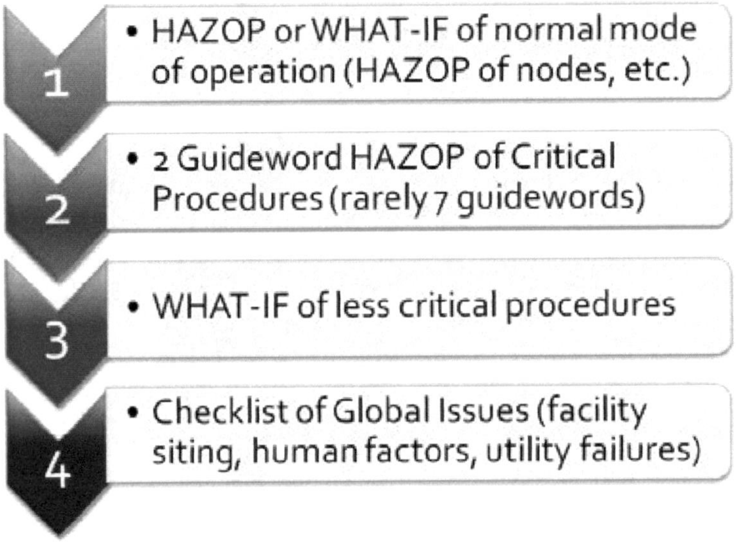

Figure 2B: PHA of a ALL Modes of Operation for a BATCH Process

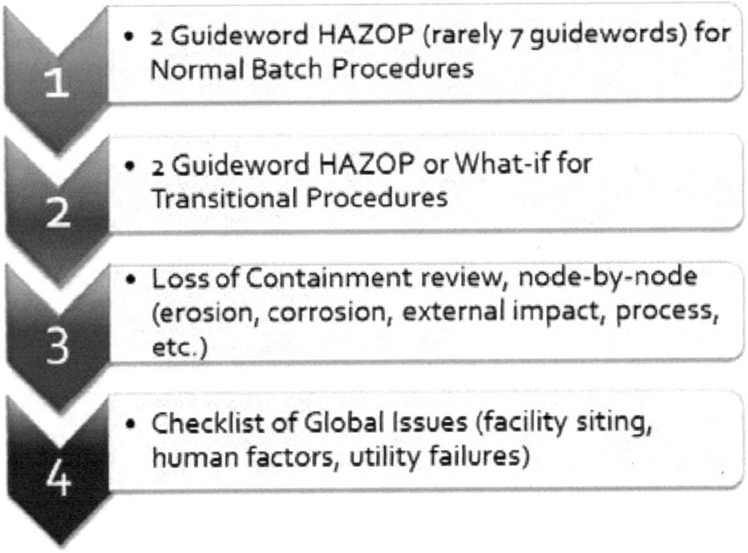

If your procedures are not at least 90% accurate, then the best approach is to develop accurate and up-to-date procedures as quickly as possible and afterwards do a PHA of the newly issued procedures.

Any procedure (even a computer program) can be analyzed using these techniques. Reviews of routine procedures are important, but reviews of non-routine procedures are even more important. As mentioned earlier, the nature of non-routine procedures means that operators have much less experience performing them, and many organizations do not regularly update these procedures [though this should change as companies comply with 29 CFR 1910.119(f)][6]. Also, during non-routine operations, many of the standard equipment safeguards or interlocks are off or bypassed.

Using the approaches above, a company doing a complete hazard evaluation of an existing unit will invest about 65% of their time to evaluate normal (e.g., continuous mode) operation and 35% of their time for evaluating the risks of non-routine modes of operation.[1, 22, 23, 24]

Many companies do **not** perform a thorough analysis of the risk for startup, shutdown, and on-line maintenance modes of operation; the reason normally given is that the analysis of these modes of operation takes "too long." Yet, actually, the hazard evaluation of the normal mode is taking too long and so the organization feels it has no time left for the analysis of procedures for startup and shutdown modes of operation. But, if these hazard evaluations for the normal mode of operation are **optimized** (such as using rules presented elsewhere[23]), the organization will have time for thoroughly analyzing the non-routine modes (typically discontinuous modes) of operation and the organization will still have a net savings overall! This point is critical since 70-80% of catastrophic accidents occur during non-routine modes of operation.[22] Figure 4 illustrates (for a continuous process unit) the typical split of meeting time for analysis of routine mode of operation versus non-routine modes of operation.[1, 22, 23, 24]

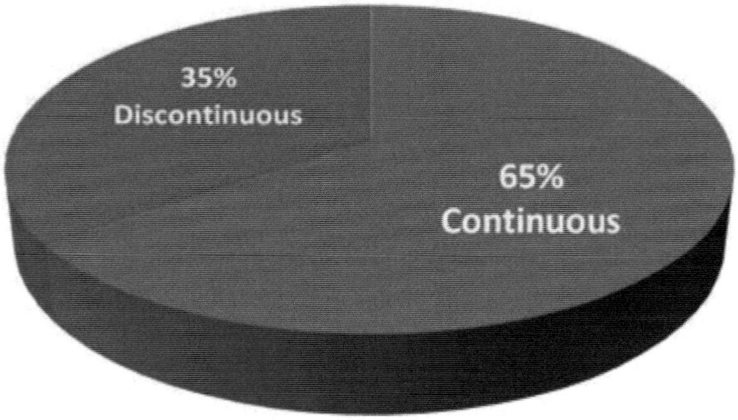

Figure 3. Relative Amount of Meeting Time Spent for Analysis of Routine and Non-routine Modes of Operation for a Continuous Process[1, 22, 23, 24]

General Guidelines for Analyzing Non-routine Modes of Operation or Batch (Step-by-step) Processes

- Define the assumptions about the system's initial status. "What is assumed to be the starting conditions when the user of the procedure begins with Step 1?"
- Define the complete design intention for each step. "Is the step actually 3 or 5 actions instead of one action? If so, what are the individual actions to accomplish this task?"
- Don't analyze safeguard steps that start with ensure, check, verify, inspect, etc., or where the consequence of skip is "loss of one level of safeguard/protection against ….." There is no reason to analyze these steps since they will show up as safeguards of deviations of other steps. This approach is similar to not analyzing a PSV during a HAZOP of continuous mode (parametric deviation analysis); instead the PSV is shown as a safeguard against loss of containment.
- Together with an operator before the meeting, identify the sections of the procedures that warrant use of:
    - 7-8 Guide Words (extremely large consequences can happen if deviations occur)
    - 2 Guide Words (the system is complex, mistakes are costly, or several consequences could occur)
    - On others, use What-If (no guide words or guide phrases; for use on simpler or lower hazard systems)
- Decompose each written step into a sequence of actions (verbs)
- Apply guide words directly to the intentions of each action

The Following Preparation Steps May Also Be Needed:

- Walk through procedure in the plant with one or more operators to see the work situation and verify the accuracy of the written procedure. This is optional and should have also been performed as part of validation of the procedure after it was originally drafted
- Determine if the procedure follows the best practices for "presentation" of the content; the best practices will limit the probability of human error
- Discuss generic issues related to operating procedures, such as:
    - staffing (normal and temporary)
    - human-machine interface
    - worker training, certification, etc.
    - management of change
    - policy enforcement
- Review other related procedures such as lock out/tag out and hot work.

IF the procedures are NOT >90% accurate, then redo procedures first!

Best Practice for Addressing Damage Mechanisms During PHAs

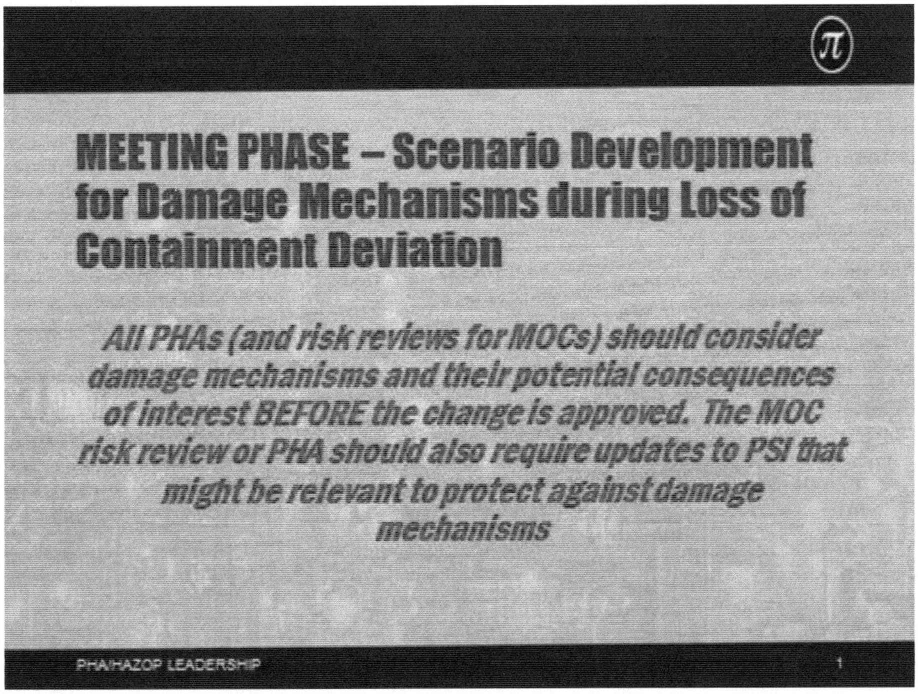

Figure 4. Meeting Phase – Review of Damage Mechanisms

Recently, US CSB and more and more regulators are making analysis of Damage Mechanisms (DMs) a PHA requirement. This requires that potential process damage mechanisms and their potential consequences of interest be identified. Damage mechanisms can be broken down into three main categories. These categories and examples of each are shown in the following table.

Table 4: Types of Damage Mechanisms

Mechanical	• Mechanical loading failure • Mechanical fatigue • Buckling • Cracking • Embrittlement • Ductile fracture • Brittle fracture
Chemical	• Corrosion o Uniform o Localized o Pitting
Physical	• Thermal failures o Creep o Thermal fatigue o Transformation

The team may use aids such as piping specs, established literature and standards, any applicable MOC documents, etc. to help identify DMs. The PHA leader has several options of how to address damage mechanisms:

- Discuss during HAZOP under "Loss of Containment" deviation
    - **This is PII's preferred approach and is one we recommend to all clients, so this is discussed in more detail below.**
- Note damage mechanisms revealed during discussions of other process deviations
    - Pressure
    - Temperature
    - Concentration

  The consequence of these deviations is ultimately loss of containment. Therefore causes of pressure, temperature and concentration deviations can be damage mechanisms and ultimately lead to loss of containment.
- Include Damage Mechanisms as a separate node (for HAZOP technique), *such as proposed by Risk Management Professionals.*[26]
- Perform a "mini" Mechanical Integrity checklist reviews under each process node
- Divide P&ID (or PFD) into separate services for purposes of DM hazard review (DMHR) discussion considering:
    - Process
    - Chemical
    - Spec changes
    - Flow through
    - Operating parameters

  This approach is used by some refineries to accomplish the requirements of a DMHR as recommended in API RP 571.[20] *From PII's experience, this approach has merit and finds some issues that the preferred PII does not always find, BUT overall the PII approach is superior in finding the key scenarios related to damage mechanisms and related safeguards because the review is specific to each Node of process equipment.*

Damage mechanisms can (and sometimes ***should***) be reviewed prior to the PHA. For example, all MOCs should consider damage mechanisms and their potential consequences of interest before the change is approved. The MOC should also require updates to process safety information which might be relevant to protect against DMs.

Coverage of Damage Mechanism for Each Major Section (HAZOP section or What-If node) – as recommended by PII

As mentioned above this is PII's preferred method for thoroughly covering all DM within a PHA/HAZOP of an entire unit. The benefits are that the team can more easily catch changes in DM from section to section and they can more easily identify when unique safeguards, such as Remote Isolation Valves, segregated containments (dike), and unique materials of construction are needed. To facilitate this, a list of generic causes of loss of containment and a generic list of safeguards against loss of containment are covered in the Loss of Containment (LOC) deviation of each section. Below are tables that summarize the typical causes of loss of containment and typical safeguards against loss of containment.

Table 5 Typical Causes of Loss of Containment (and Tube Leak/Rupture) to Discuss in Each Major Process Section (includes Damage Mechanisms)

- Corrosion – internal, including steady corrosion and also stress cracking
- Corrosion - external (including corrosion under insulation
- Erosion
- Fatigue failure (such as due to vibration)
- Improper material of construction (including wrong grade of welding rod/wire)
- Improper maintenance/repair
- Drain valve open
- Vent valve option
- Instrument line breaks
- External impact or external load (crane, load dropped, fork truck, backhoe, failure of structural supports, etc.)
- External fire (flame impingement and heating)
- Thermal expansion (while process is blocked-in)
- Lightning

➢ Any deviation that links into loss of containment as a cause, such as high process temperature or high process pressure

Table 6 Examples of IPLs Protecting Against or Mitigating Damage Mechanisms

RELIEF VALVES

- Relief valve designed for the limiting scenario which vents to a safe location or destruction system (can prevent loss of containment due to high pressure events cause by process upsets or external fire)
- Rupture disks (same criteria and functional protection as relief valve except cannot re-seat)
- Conservation vent (same criteria and functional protection as relief valve)
- Vacuum relief valve (same criteria and functional protection as relief valve, except protects against negative pressure)
- Thermal relief valve (online protects against thermal expansion in a closed system)

ISOLATION

- Remote isolation valves that either operate automatically or can be remotely closed manually and the loop qualifies as a Human IPL
- Locally operated manual valves that can be reached quickly and safely (typically not a valid IPL case)

- Remote shutdown of pumps, blowers, etc. to limit amount of material released

CONTAINMENT & MITIGATION

- Double-walled piping for lines that run under rivers and underground and that carry toxics
- Process contained in enclosed building
- Dikes are provided for secondary containment of spills from major vessels; drain valves on diked areas are kept closed/can be closed quickly
- Process areas are drained to sump
- Rainwater runoff containment
- Fire insulation that protects metal structures for 4 hours

SUPPRESSION AND EXTINGUISHING SYSTEMS

- Deluge system under or over tanks or storage vessels
- Manual pull station fire alarms
- Fire extinguishers in the Power Distribution Center
- Fire extinguishers in the control room rated for electrical fire

Table 7 Examples of Non-IPL Safeguards Protecting Against or Mitigating Damage Mechanisms

PIPING

- Applicable design codes are reflected in equipment design
- Welding done by qualified welders using qualified procedures
- Regular inspection (e.g., ultrasonic, X-ray) for wall thickness
- The wall thickness inspection frequency is increased in critical areas (e.g., areas where erosion or corrosion are likely)
- Systems are leak checked (pressurized with air and soap bubble applied at flanged joints) before starting up after systems are opened for maintenance
- A winterization list is used to check freeze-protection features before the onset of cold weather
- Piping and components that are fabricated on site and installed in the process are pressure tested before they are installed as necessary

PUMPS

- Written specifications in standard maintenance procedures (SMPs) for pumps to help ensure that pumps are maintained, repaired, and replaced in a way that meets original specifications

VALVES

- Fugitive emissions monitoring program that includes monitoring valves for packing leaks and

flange leaks
- Fail-safe settings for control valves upon loss of instrument air, nitrogen, or electric power

CORROSION DETECTION AND PREVENTION
- Periodic inspection of piping, equipment, and structures
- Application of exterior paint and protective coatings when needed on piping, equipment, and structures
- Positive material identification (PMI) of 100% of components and welds in the field/unit

EQUIPMENT MAINTENANCE
- Periodic walkthroughs by personnel inspecting for leaks or other abnormalities, and checking local instrumentation
- Preventive maintenance systems, including:
    - Lubrication schedule
    - Operators inspect pump seals for visible leaks
    - Records of equipment failures
- Periodic inspection of heat exchanger tube bundles

INSTRUMENTATION
- All controllers are indicating controllers
- Testing and calibration of critical process instrumentation

PERSONNEL PROTECTION
- Personal protective equipment (PPE), including:
    - Self-contained breathing apparatus
    - Respirators
    - Protective clothing
- Periodic inspection and testing of PPE and personal safety systems, including:
    - Self-contained breathing apparatus
    - Safety showers
- Availability of:
    - First aid kit
    - Specially trained Emergency Medical Technicians

EMERGENCY RESPONSE
- Remote emergency communication systems, including:
    - Handheld, two-way radios
    - Buddy system for critical or hazardous operations
    - Emergency communications procedures
    - Telephones
    - Emergency reporting stations strategically located in the plant
- Cooperative fire brigade
- Plant wide alarm system
- Facility and unit contingency plan

- Emergency procedures for special incidents
- Trained HAZMAT response teams with emergency equipment
- Spill containment and control guidelines; all operators receive hazardous waste operations and emergency response (HAZWOPER) training

The Leader and Scribe must consciously discuss the typical damage mechanisms listed in Table 5 and discuss at each LOC deviation (so, once in each section).

Documentation of Results of Damage Mechanism Discussions in PHA Meetings

For each node (each line, each vessel, each column, etc.), the PHA team should discuss and document each damage mechanism listed in Table 5 (as a cause of loss of containment), consequence of the failure if the damage mechanism occurs, and the safeguards in place to prevent the damage mechanism, detect the mechanism before failure, prevent the release, detect and response to the release, and mitigate or contain the release.

The documentation style varies between PHAs. Figure 5 show two different styles that have been acceptable in the past.

- One style (Example A) uses a reference to a summary table of typical causes (this summary table is not shown but is similar to Table 5 of this paper) instead of listing each individual damage mechanism. In Example A, the same approach is used for safeguards. This approach saves redundant text and some time, but it requires the leader and scribe to be diligent to cover everything in the Typical tables that were referenced in the Cause and Safeguard columns. A modified approach is to relegate some generic causes and safeguards to Typical tables for reference and then to carefully list the specific cause and/or safeguards of interest for LOC for each specific node.
- Example B does not use or reference a generic or Typical list of causes (damage mechanism) or Typical safeguards, but instead develops a specific listing for the LOC deviation of each node. This style has proven easier to justify to regulators and other outside reviewers but takes more documentation effort.

Regardless of method, the team leader and scribe must ensure the team rigorously discusses all of the damage mechanisms in Table 5 and adequately documents the results of the damage mechanism review in the PHA/HAZOP analysis tables.

Figure 5. Excerpts from Select PHA Reports Showing Various Acceptable Styles for Documenting Damage Mechanism Review Results Within the LOC Deviation

Example A – Referencing Generic Tables for Typical Causes and Typical Safeguards

Dev#	*Deviation*	*Causes*	*Consequences*	*Existing Safeguards*	*Recommendations*
9.10	Loss of containment	Accelerate corrosion External fire High pressure (linked from 9.7) Typical causes of loss of containment (see Table A.1)	Release to atmosphere leading to potential injury of workers and/or community	Pressure Relief Valve's 231A, B, C on Reactor and no valves in line to reactor Most lines / connections are welded construction; only a few flanges Drills conducted each year on evacuation, rescue and isolation Emergency Response personnel are trained at SABIC FTC Generic safeguards protecting against or mitigating process material releases (see Table A.2)	Safety 7. Consider changing the ITPM schedule for managing most PSVs whose inspections are too infrequent based on industry standards and best practice. For instance, the current inspection frequency for PSV-8220 on the Ammonia Receiver (I-2005G) is 9 years, whereas, consensus codes typically recommend testing/inspection every 1-4 years for PSVs in highly toxic services.

Example B – Specific Listing of Causes and Safeguards

1.9	Loss of containment	Corrosion/erosion External fire and/or flame impingement Gasket, packing, or seal failure Improper maintenance Material defect Operator failing to close or inadvertently opening a valve to the atmosphere (e.g., a valve at a hose connection) Railcar inadvertently derailed Valve leaking to the atmosphere High pressure (linked from 1.5) Acid corrosion caused by high concentration of water (linked from 1.8) High ambient temperature External impact (such as from a mini-engine or another railcar)	Catastrophic release of chlorine from a ruptured railcar Steady release of chlorine from a ruptured connection Steady release of chlorine from a leaky connection High pressure caused by thermal expansion of liquid chlorine if railcar is also over-full	Chlorine repair kit Derailer and warning flag to prevent impact by a mini-engine or another railcar Limited vehicular access to area Maintenance/operator response as required, including isolation if needed Operator periodically monitoring the railcar valves while unloading Personal protective equipment in the area Plugs installed in all chlorine valves to the atmosphere when the valves are not in use Relief valve on each railcar for mitigating releases caused by overpressure Supplier maintenance of railcars (per strictly enforced US DOT requirements) Video monitoring of the unloading area Concrete crossties on rail spur Dike preventing any combustibles spilled nearby from reaching the unloading rack area Concrete railroad ties in chlorine unloading area to prevent fires near railcar	10. Consider installing a chlorine detection system in the unloading and vaporizing area to help detect chlorine releases (especially at likely release points) 11. Verify that periodic maintenance and inspections are being performed in accordance with Chlorine Institute recommendations 12. Review the drainage system for the unloading are, and identify the areas that may be affected by a large chlorine release 13. Consider prohibiting the use of heavy equipment (e.g., cranes) in the unloading and vaporizing area unless special precautions to prevent equipment damage are enacted 33. Consider providing a high pressure alarm for each vaporizer 40. Consider providing a water deluge system in the unloading area to help mitigate chlorine releases from the railcar

Best Practices for Addressing Previous Incident

Information on process hazards is also found in incident investigation reports and are often discounted. PHAs predict what can go wrong. Incidents are actual sequences of human errors and equipment failures. Don't miss this opportunity to thoroughly review previous incidents. There is no industry or regulatory guidance regarding how to incorporate previous incidents into HEs. The intent is to review near misses and accidents to ensure that potentially catastrophic events are identified and analyzed.

Typically, this is done by listing and then discussing relevant incidents in the industry and in similar plants to:

- Help identify potential causes or consequences of interest
- Evaluate related safeguards in the process, and
- Make recommendations for additional layers of protection if needed to prevent similar incidents related to the system.

Know Your History

- What caused past incidents? *This includes near misses and accidents and also includes incidents in similar systems. If the company has a good system for getting near misses reported and any incident investigated to root causes, then the PHA team will have a deeper understanding of what can go wrong and why. Root cause analysis of incidents is just as important as PHA. The team should volunteer incidents that may not have been reported formally and these should be kept confidential so that more "unreported" incidents are reported within the PHA team setting.*
- What was done in response? *Were any changes made to the system? Did these improve the system (reduce the error or failure rates)? Has the incident reoccurred?*
- If the incident occurred in a system other than the one under review by the PHA team, should similar actions be applied to the system under evaluation?
- Did the previous countermeasures work? *Did these prevent similar incidents?*

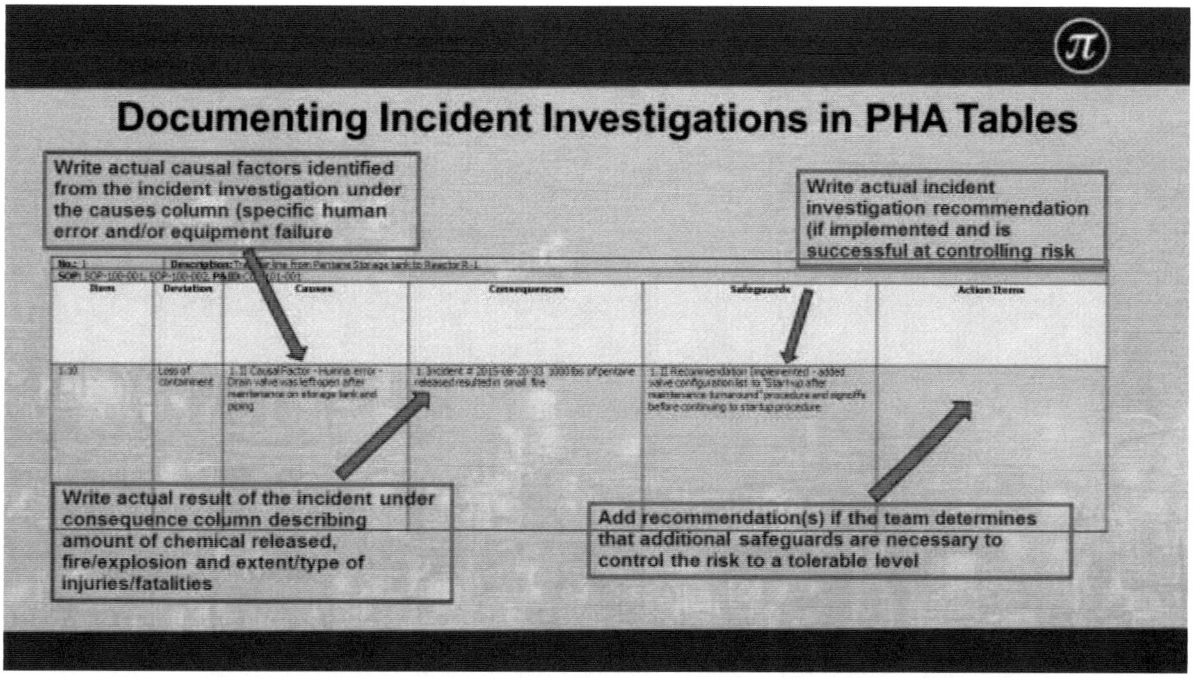

Figure 6. Documenting Incident Investigations in PHA Tables

Conclusions

Qualitative analysis of non-routine operating procedures is an extremely powerful tool for uncovering deficiencies that can lead to human errors and for uncovering accident scenarios during all modes of operation. This approach of step-by-step HAZOP and/or What-If analysis is not new to industry, and regulators have required similar approaches for decades. And regulators continue to note lack of analysis of the risk of non-routine operations and lack of risk review of changes to procedures.

> ***From the Wall Street Journal[27] referencing the presidential commission investigating the Deepwater Horizon accident of April 2010:*** *BP had rules in place governing procedural changes, but its workers didn't consistently follow them, according to BP's September [2010] internal report on the disaster and the report released earlier this month [January 2011] by the presidential commission on the accident. "Such decisions appear to have been made by the BP Macondo team in ad hoc fashion without any formal risk analysis or internal expert review," the commission's report said. "This appears to have been a key causal factor of the blowout."*

> ***From CSB Report on August 2008 Bayer CropScience Explosion:[12]*** *"The accident occurred during the startup of the methomyl unit, following a lengthy period of maintenance ... CSB investigators also found the company failed to perform a thorough Process Hazard Analysis, or PHA, as required by regulation...In particular, for*

> *operational tasks that depend heavily on task performance and operator decisions, the team should analyze the procedures step-by-step to identify potential incident scenarios and their consequences, and to determine if the protections in place are sufficient."*

Since about 80% of major accidents occur during non-routine modes of operation,[1] the PHA of the deviations from the steps of the procedures that govern these modes of operation is certainly critical. But, of the 20% of the major accidents that occur during normal operation, most of these arise from **damage mechanisms** such as corrosion, material defects, gasket failures, external impacts and the like (based on PII data). PHAs can effectively cover DM while discussing LOC.

So, the PHAs must cover both of these concepts (PHA of all modes of operation and addressing DM) thoroughly to fulfill the role of the PHA in controlling risk. **More regulatory pressure is sure to follow on these two issues**

But of course, the PHA must be strong in all aspects, including consideration of lessons learned from previous incidents.

Acronyms Used

AIChE– American Institute of Chemical Engineers
API – American Petroleum Institute
CCPS – Center for Chemical Process Safety (a division of AIChE)
CSB – US Chemical Safety and Hazard Investigation Board
DM – Damage mechanism
DMHR – Damage mechanism Hazard Review
HAZOP – Hazard and Operability Analysis
HAZWOPER – Hazardous and Waste Operater (US OSHA 29 CFR 1910.120)
HTHA – High Temperature Hydrogen Attack
IPL - Independent Protection Layer
LOC – Loss of Containment
LOPA – Layer of Protection Analysis
MOC – Management of Change
NHT – Naphtha Hydrotreater
OSHA – Occupational Safety and Health Administration, US Department of Labor
PHA – Process Hazard Analysis
PII – Process Improvement Institute, Inc.
P&ID – Piping & Instrumentation Diagram
PSI – Process Safety Information
PSM – Process Safety Management
SOP – Standard Operating Procedure

References

1. *Guidelines for Hazard Evaluation Procedures*, 3rd Edition, 2008, CCPS/AIChE.

2. U.S. Department of Labor: Systems Safety Evaluation of Operations with Catastrophic Potential. Occupational Safety and Health Administration Instruction CPL 2-2.45, Directorate of Compliance Programs, September 6, 1988.
3. Woodcock, Henry C., "Program Quality Verification of Process Hazard Analyses (for instructional purposes only)," US OSHA, 1993.
4. OSHA Inspection Number 106612443 - Phillips 66 Company, Houston Chemical Complex, Citations 1-1 through 1-566, Issued 4/19/1990.
5. *Stipulation and Settlement Agreement,* Phillips 66 Company ("Phillips") and Lynn Martin, Secretary of Labor, United States Department of Labor, 8/22/1991.
6. "Process Safety Management of Highly Hazardous Chemicals; Explosives and Blasting Agents," US OSHA Final Rule, 29 CFR 1910.119, February 24, 1992.
7. "Risk Management Programs for Chemical Accidental Release Prevention," US EPA Final Rule, 40 CFR 68, 1994-2000.
8. OSHA Inspection Number 103490306, Issued November 2, 1992.
9. OSHA Inspection Number 123807828; Issued November 18, 1993.
10. U.S. Department of Labor: PSM Covered Chemical Facilities National Emphasis Program. Occupational Safety and Health Administration CPL 03-00-014, Directorate of Enforcement Programs, November 29, 2011.
11. U.S. Department of Labor: Petroleum Refinery Process Safety Management National Emphasis Program. Occupational Safety and Health Administration CPL 03-00-010, Directorate of Enforcement Programs, August 18, 2009.
12. "Investigation Report: Pesticide Chemical Runaway Reaction Pressure Vessel Explosion, at Bayer CropScience, LP, Institute, WV, on August 28, 2008", US Chemical Safety Board, Report No. 2008-08-I-WV, January 2011.
13. U.S. Chemical Safety and Hazard Investigation Board (CSB). *Investigation Report: Catastrophic Rupture of Heat Exchanger (Seven Fatalities), Tesoro Anacortes Refinery, Anacortes, WA, Report No. 2010-08-I-WA, Draft for Public Comment*, January 29, 2014.
14. Washington Administrative Code (WAC) 296-67-291 Appendix C--Compliance guidelines and recommendations for process safety management (nonmandatory) http://www.lni.wa.gov/WISHA/Rules/hazardouschemicals/default.htm#WAC296-67-021 (accessed December 3, 2013).
15. Center for Chemical Process Safety (CCPS). *Revalidating Process Hazard Analyses*. 2001; pp 31-32.
16. U.S. Chemical Safety and Hazard Investigation Board (CSB), *DuPont La Porte, Texas Chemical Facility Toxic Chemical Release Interim Recommendations Investigation: 2015-01-I-TX* Incident Date: November 15, 2014 Issue Date: September 30, 2015
17. Contra Costa County Hazardous Materials Programs, *Contra Costa County Industrial Safety Ordinance (ISO)*, by CCHMP. June 15, 2011.

18. Draft Initial Report "*Safety Evaluation of the Chevron Richmond Refinery.*" November, 2015, by Process Improvement Institute, Inc., for Contra Costa County Hazard Materials Department.
19. U.S. Chemical Safety and Hazard Investigation Board (CSB). "CASE STUDY *Fire at Formosa Plastics Corporation: Evaluating Process Hazards,*" July 20, 2006.
20. API RP 571. "Damage Mechanisms Affecting Fixed Equipment in the Refining Industry." 2nd ed., April 2011.
21. U.S. Chemical Safety and Hazard Investigation Board (CSB), *Interim Investigation Report: Chevron Richmond Refinery Fire,* U.S. CSB, August 6, 2012.
22. "Necessity of Performing Hazard Evaluations (PHAs) of Non-normal Modes of Operation (Startup, Shutdown, & Online Maintenance)", W. Bridges and Mike Marshall (US OSHA), *18th Annual International Symposium, Mary Kay-O-Connor Process Safety Center, College Station, TX, October 2015 and 12th Global Congress on Process Safety, April, 2016.*
23. "How to Efficiently Perform the Hazard Evaluation (PHA) Required for Non-Routine Modes of Operation (Startup, Shutdown, Online Maintenance)," W. Bridges and T. Clark, *7th Global Congress on Process Safety*, Chicago, AIChE, March 2011.
24. Bridges, W.G., et. al., "Addressing Human Error During Process Hazard Analyses," *Chemical Engineering Progress*, May 1994
25. Rasmussen, B. "Chemical Process Hazard Identification," *Reliability Engineering and System Safety*, Vol. 24, Elsevier Science Publishers Ltd., Great Britain, 1989.
26. "Effectively Addressing New PSM/RMP Damage Mechanism Review Requirements with an Integrated PHA (iPHA)," Nour, Maher, Schutz, all from Risk Management Professionals, 11th GCPS, *April 2015.*
27. Wall Street Journal, January 29, 2011.
28. Gertman, D.; Blackman, H.; Marble, J.; Byers, J. and Smith, C., "The SPAR-H Human Reliability Analysis Method," NUREG/CR-6883, U.S. Nuclear Regulatory Commission, Office of Nuclear Regulatory Research, Washington, DC, August 2005.
29. Swain, A. D., Guttmann, H. E., "Handbook of Human Reliability Analysis with Emphasis on Nuclear Power Plant Applications, Final Report," NUREG/CR-1278, 1983, US Nuclear Regulatory Commission.

When the Fail Open Valve Fails Closed: Lessons from Investigating the "Impossible"

Brenton L. Cox
Exponent Failure Analysis Associates
4580 Weaver Parkway, Suite 100, Warrenville, Illinois 60555
bcox@exponent.com

Sean J. Dee, Russell A. Ogle
Exponent Failure Analysis Associates
4580 Weaver Parkway, Suite 100, Warrenville, Illinois 60555
sdee@exponent.com, rogle@exponent.com

Prepared for Presentation at
American Institute of Chemical Engineers
2017 Spring Meeting and 13th Global Congress on Process Safety
San Antonio, TX
March 26 – 29, 2017

When the Fail Open Valve Fails Closed: Lessons from Investigating the "Impossible"

Brenton L. Cox
Exponent Failure Analysis Associates
4580 Weaver Parkway, Suite 100, Warrenville, Illinois 60555
bcox@exponent.com

Sean J. Dee, Russell A. Ogle
Exponent Failure Analysis Associates

Keywords: Incident investigation, hazard assessment, management of change

Abstract

Process hazard analysis is recognized by many safety professionals and regulatory groups as the key provision of a process safety management system. This analysis evaluates what can go wrong, and what safeguards are needed to prevent incidents. One challenge in process hazard analysis is the reliance on humans to determine reasonable, expected, or anticipated failure modes. The implementation of safeguards may also foster a lost sense of vulnerability to potential process upsets. These factors in combination can lead to a disregard for failure modes considered "unlikely" or even "impossible". Unfortunately, this problem is often revealed when the "impossible" leads to an incident or accident. In this paper, a case study is used to motivate a discussion of incidents involving hazards that were disregarded in process hazard analysis due to their low probability, or supposed impossibility. Tools, techniques, and guidance will be discussed to improve process hazard analysis when evaluating low probability failure modes.

1 Introduction

Process hazard analysis (PHA) is one of the key aspects of a robust process safety management system. PHA is one of the fourteen elements of the Occupational Safety and Health Administration Process Safety Management Standard (OSHA PSM). [1] PHA is also interwoven throughout several elements of a Risk Based Process Safety (RBPS) approach to process safety management. [2] While regulatory compliance is critical for facilities covered under the OSHA PSM standard, PHA is an important best practice for all facilities that aim to identify and manage the risk associated with hazards at their facilities. Therefore, in the context of this manuscript, PHA will be discussed synonymously in the context of the Hazard Identification and Risk Analysis (HIRA) element of RBPS.

The HIRA element focuses on three questions:

- Hazard - What can go wrong?
- Consequence – How bad could it be?
- Likelihood – How often might it happen?

One could argue that the Hazard component of HIRA is almost a prerequisite for the actual analysis. Without an understanding of what can go wrong, it is very challenging to determine a consequence or likelihood for the analysis. Incident investigation teams often revisit pre-incident hazard analyses as a basis for understanding why an accident occurred. Some may argue that facilities feel most vulnerable when faced with responding to a failure that was not identified in their hazard analysis. However, even more concerning are accidents that occur after a facility identified the hazard scenario, but ultimately concluded it was extremely unlikely, or even impossible in their hazard analysis.

This manuscript will discuss the issues facilities face when unexpected failure modes result in accidents. A specific case study involving an unanticipated valve failure is used to guide a more general discussion. This example is particularly illustrative because of the relatively simple function – and failure – of the valve. Lessons learned and opportunities for improvement in PHA will be presented.

2 Process Hazard Analysis

As previously discussed, PHA is intertwined within several elements of a RBPS approach to process safety management. While PHA is part of the foundation of the HIRA element of RBPS, it is also a key component that underlies several other elements including, but in no way limited to, Process Safety Culture, Asset Integrity and Reliability, Management of Change, and Incident Investigation. Thus, facilities that are adept and proficient in PHA can benefit not only in the HIRA element of RBPS, but also in several additional elements. This pushes a facility closer to the goal of implementing a reliable system to manage the process' risk to people, equipment, and the environment.

Hazard identification and analysis has been a focus of process safety management for over 30 years. The first project undertaken by the Center for Chemical Process Safety (CCPS) focused on the development of a comprehensive text to improve process safety performance in the industry. [3] This project resulted in the publication of *Guidelines for Hazard Evaluation Procedures* in 1985, with updates issued in 1992 and 2008. Hazard evaluation is the cornerstone of a process safety management program and is performed throughout the life cycle of a process and facility. Therefore, improvements and advancements in process hazard analysis are an important component of mitigating and preventing process incidents.

Process incidents can be simplified into three components:

1. Existence of a process hazard

2. Determination of a potential initiating cause

3. Occurrence of an incident outcome

Commonly, process hazard analysis is performed to understand and evaluate each of these three elements. First, an analysis aims to identify and understand the potential hazards associated with a process. Next, scenario based or non-scenario based methods are used to identify potential failure modes that could occur. Then, both unmitigated and safeguard mitigated outcomes are considered to determine the potential loss event or impact of a particular failure mode.

Shortcomings in analysis of any of the three elements of a process incident can introduce a vulnerability in the process safety management system. Incidents can occur when facilities either fail to identify or misunderstand a particular hazard in their process. For example, an insufficient understanding of the chemical equilibrium established between elemental sulfur and hydrogen sulfide may result in a failure to identify personnel exposure hazards during material transfer activities. [4] Similarly, hazard evaluations can mistakenly overlook a mechanical failure mode that could initiate a process safety incident. Additionally, the outcome from an initiating event can be much worse than anticipated if its propagation is not characterized, or if controlling safeguards are not properly implemented or maintained.

This paper will focus on a case study related to a valve failure that resulted in an incident at a chemical facility. The facility had identified and understood the hazards involved in the incident, but their hazard analysis and safety system design did not include the valve failure mode that initiated the incident. As a result, several key learnings were developed to improve the facility's hazard analysis, operator response, and safety system design.

3 Case Study

The following case study examines an incident that occurred when a fail open valve failed closed. Heating and cooling were both provided to the facility's polymerization reactor using a jacketed system as shown in Figure 1. When a batch was started, heating was applied to the reactor using steam to initiate the reaction. After the reactor reached the desired temperature, an automatic control chilled water valve would modulate cold makeup water into the jacket circulation loop to maintain the reactor at the appropriate temperature. An operator in the field could also open a manual emergency cooling bypass valve around the automatic control chilled water supply valve in the event that the reactor needed additional cooling.

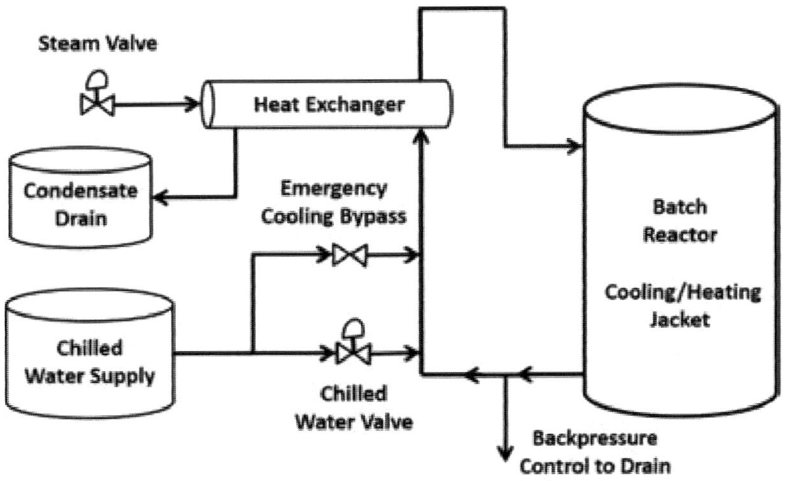

Figure 1. Heating/Cooling Jacket for Batch Polymerization Reactor

On the day of the incident, the temperature within the polymerization reactor began to increase, and triggered the high temperature alarm. The operator at the control station acknowledged the alarm, and noted that the HMI screen indicated the chilled water valve was already 100% open and the temperature of the water supply to the jacket inlet was slowly decreasing. The temperature in the reactor continued to increase, returning an additional error message related to insufficient cooling water flow. The operator confirmed that the chilled water valve was still 100% open on the HMI and that the jacket supply temperature was continuing to decrease indicating that the jacket was receiving cold makeup water. The operator then began to troubleshoot the cooling system. Over the next six to seven minutes, the operator tried to cool the reactor contents by injecting cold water into the reactor. The reactor temperature continued to increase, eventually reaching the high/high alarm set point for the emergency shutdown system (ESS). The ESS then intervened, aborted the batch and dumped the reactor contents to a secondary containment pit. While no one was injured, there was a recordable environmental release, which was reported to the local authority having jurisdiction.

After securing the area and verifying it was safe to enter, operators examined the jacketed system on the reactor. Surprisingly, they found the water supply valve to the reactor water almost completely closed, even though the control station showed the valve as 100% open. An engagement pin from the valve's feedback positioner mechanism was found on the floor next to the chilled water valve. The valve had been designated by the site as a "fail open" valve. However, the valve had failed in a mostly closed position, limiting the flow of cooling water to the jacket circulation loop.

3.1 Chilled water control valve design and operation

The reactor jacket chilled water control valve is shown in Figure 2 and consists of three parts: the valve body, the actuator, and the positioner. As shown in Figure 3, flow through the valve body is regulated by the position of the valve stem, which moves an internal plug up and down in the valve body to open and close the valve. The valve stem is connected to the actuator stem as shown in Figure 4. The entire stem assembly is pushed up to open the valve with an internal actuator spring that pushes against a diaphragm. The stem assembly is pushed down to close the valve by air pressure, which is applied to the top of the diaphragm and overcomes the actuator spring force. The chilled water valve is an air-to-close valve, meaning that the valve requires air pressure to close. Upon loss of air or pneumatic signal, the actuator spring is the sole force acting on the valve stem; in this condition, the valve is forced fully open corresponding to full cooling flow to the reactor jacket.

The positioner shown in Figure 5 controls the valve stem position. An instrument input pressure signal is supplied to the positioner, which expands the bellows. As the bellows pushes up on the "D" shaped beam, the beam rotates about the input axis (vertical) and the flapper assembly moves closer to the nozzle. As the flapper approaches the nozzle, the nozzle pressure increases, which increases the output pressure supplied to the actuator by the relay. As the pressure in the actuator increases and pushes against the diaphragm, the stem assembly and attached travel pin move downward, lowering the rotary shaft arm. The rotary shaft arm then rotates the cam, which pivots the "D" shaped beam about the feedback axis (horizontal) moving the flapper assembly away from the nozzle. This movement decreases the nozzle pressure until it equilibrates for the given bellows position and supply pressure.

Figure 2. Jacket cooling water control valve from reactor.

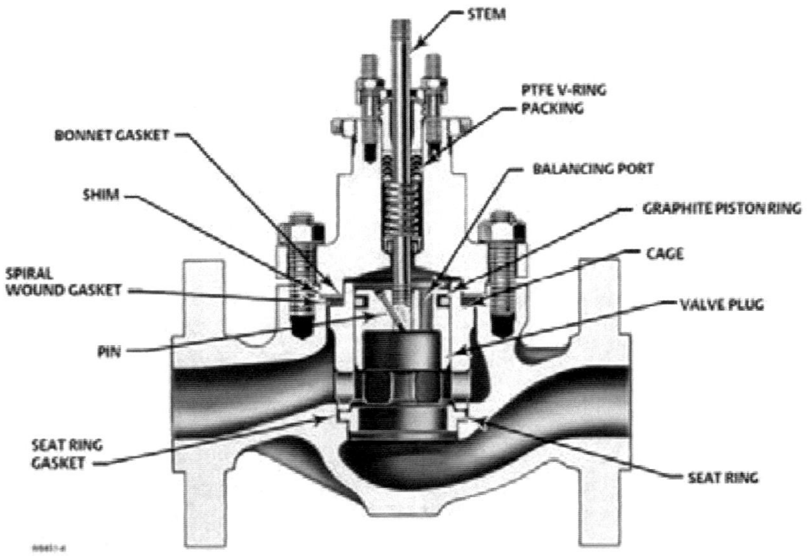

Figure 3. **A cutaway diagram for the valve body of the jacket chilled water control valve.**

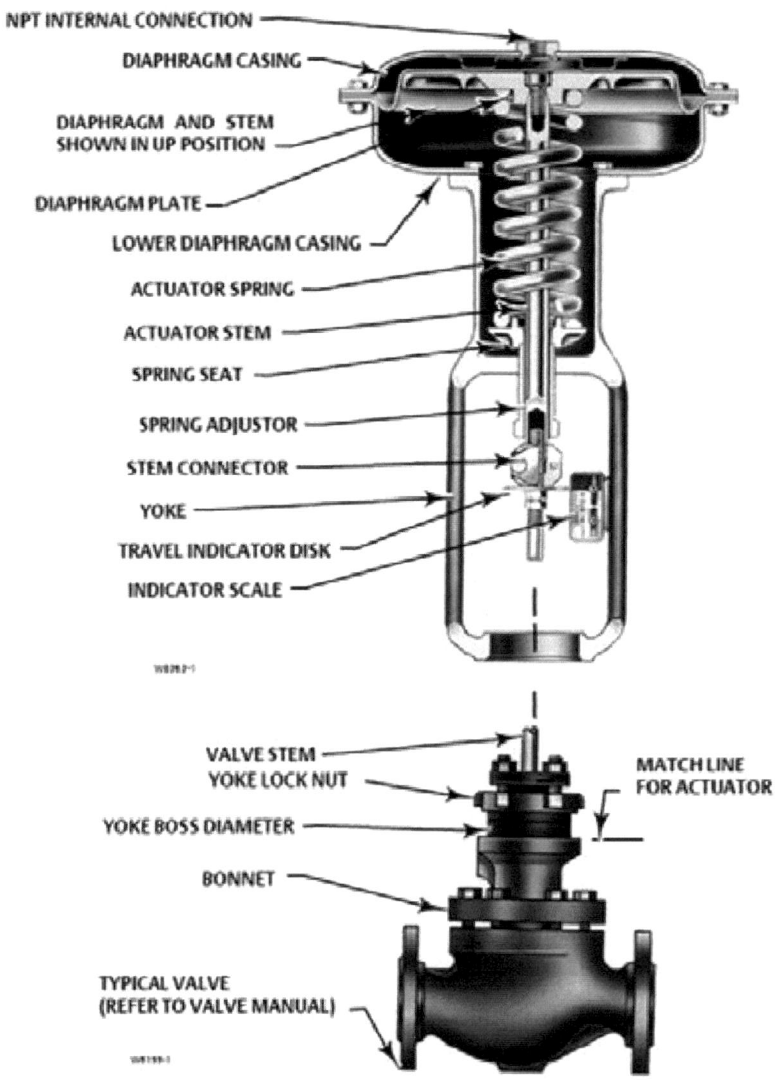

Figure 4. Valve actuator diagram. The actuator moves the valve stem up and down by counter balancing upward and downward forces on the stem.

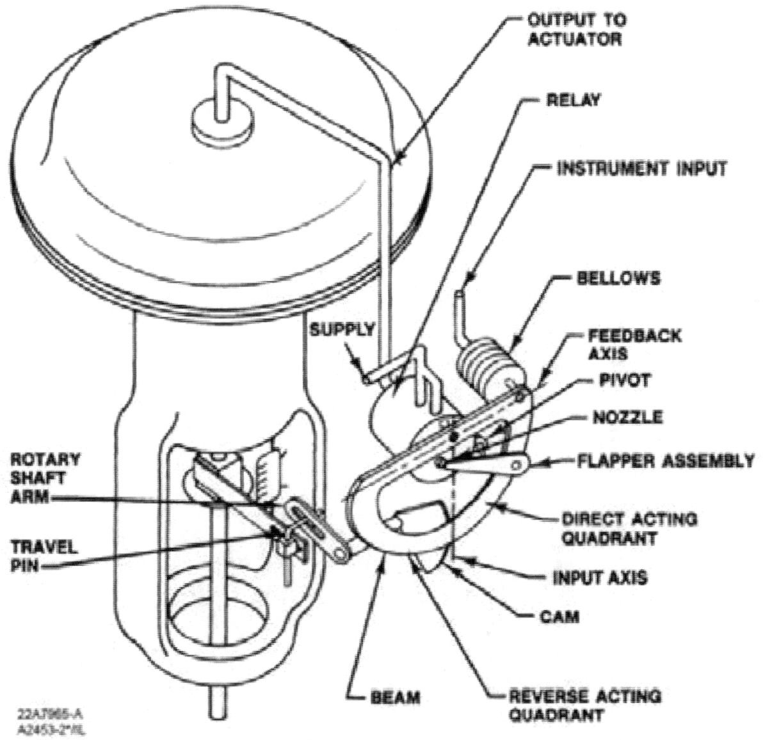

Figure 5. Valve Positioner diagram showing the components of positioner mechanism and the mechanical linkage that attaches to the valve stem.

3.2 Failure mechanism of the chilled water valve.

The incident batch was dumped to the containment pit because the contents of the reactor overheated due to insufficient cooling water introduction into the jacket circulation loop. In the post incident, investigation the travel pin was found disengaged from the rotary shaft arm on the reactor chilled water control valve. With the travel pin disengaged, the rotary shaft arm faulted to a full upward position corresponding to a fully open valve. In fact, the mechanic troubleshooting the issue after the reactor evacuation noted that the input signal pressure gauge on the positioner indicated 3.0 psig (corresponding to the DCS command signal for a 100% open valve). However, with the travel pin disengaged, the valve was found mostly closed. When the travel pin was reinstalled, the valve operation was verified to be within the calibration specified for the valve.

Testing was conducted to examine the effect of disengagement of the travel pin on the open/close state of the chilled water control valve, and is shown in Figure 6. The chilled water valve is calibrated to begin closing from a fully open position at 4.2 psig until it reaches a fully closed position at 9.6 psig. After verifying this calibration, the valve was commanded to a fully closed position (9.6 psig) and the pin was disengaged from the rotary shaft arm. The rotary shaft arm sprung up to the fully open position, but the valve stem did not move. The instrument input pressure was fully released (0.0 psig), and the valve fully opened. Then, a 3.0 psig instrument input pressure was applied and surprisingly, the valve closed. The valve remained closed until the instrument input pressure decreased to approximately 1.0 - 1.5 psig.

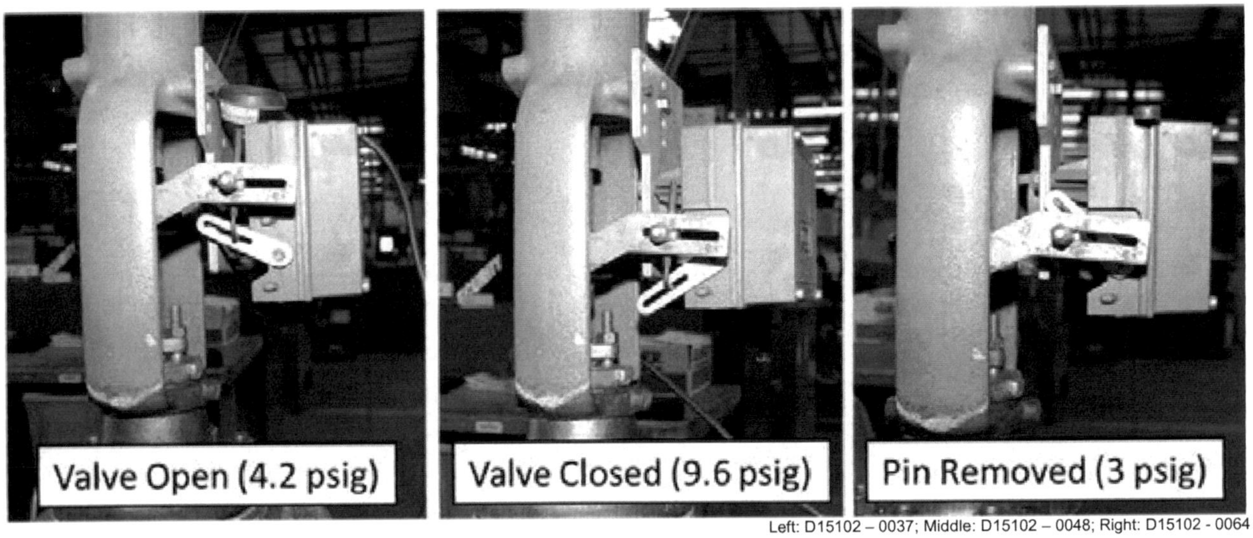

Figure 6. Chilled Water Control Valve positions during testing.

Failure of the chilled water control valve with a disengaged travel pin in a mostly closed state at 3.0 psig was consistent with witness statements, process data trends, and independent testing. When the travel pin disengaged from the rotary shaft arm, the arm moved to the fully open position. This movement rotates the cam, such that the flapper assembly is at the maximum distance from the nozzle about the feedback axis. However, the input pressure associated with a fully open command (3 psig) still pushes the "D" shaped beam about the input axis pivoting the flapper assembly toward the nozzle. The resulting failure state supplies almost enough input pressure to close the valve.

Thus, while a 3 psig instrument input signal was insufficient to close the valve with the travel pin engaged, disengagement resulted in movement of the flapper assembly closer to the nozzle. This increased the nozzle pressure and the pressure output to the actuator causing the valve to fail in a mostly closed position. This mostly closed state prevented the cooling water flow to the jacket supply, and inhibited the reactor cooling function. Without sufficient cooling, the reactor overheated and the SIS intervened to evacuate the contents to the emergency containment area.

Upon visual inspection of the control valve, it was determined that three factors were responsible for disengagement of the travel pin:

- Misalignment of the connector arm;
- Wear of the sliding contact surfaces between the travel pin and the rotary shaft arm; and
- Hydraulic forces caused by the transient, turbulent flow through the valve during normal operation

In a discussion with the valve manufacturer, the representative indicated that the disengagement of the travel pin was an extremely rare event. The site had several exemplar valves for additional reactors with jacketed heating/cooling systems. Inspection of the other chilled water control valves in service at the facility revealed that the mechanical components were not identical in appearance and varied in both wear and engagement distances as shown in Figure 7, Figure 8 and Figure 9. In fact, each mechanism had a unique combination of dimensions, orientations, and tolerances. The preventative maintenance for the control valve at the time of the incident did not address any measures to prevent pin disengagement. The manual for the valve did not offer any specific guidance for preventative maintenance procedures that would prevent travel pin disengagement.

Figure 7. Wear pattern observed on bottom side of rotary shaft arm.

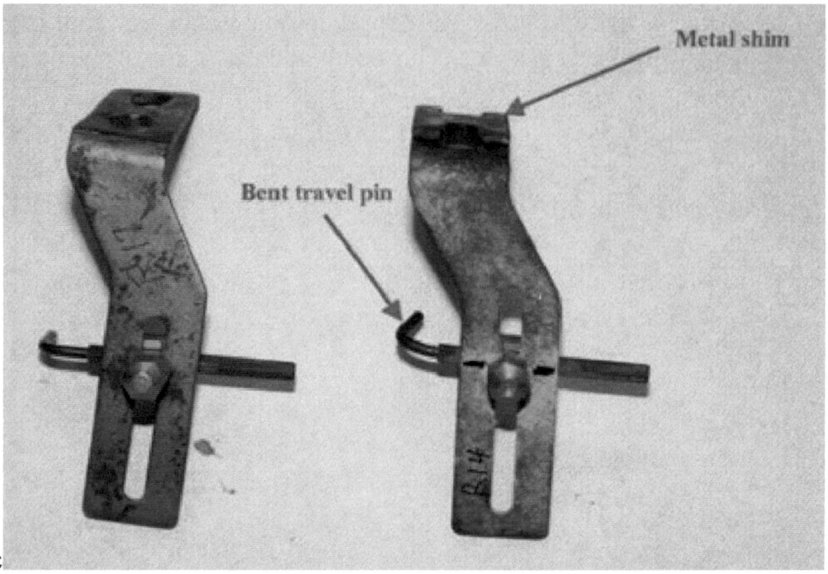

C

Figure 8. Travel pin and bracket assemblies on two Cooling Jacket Control Valves at the facility. The assembly on the right was found to have a bent pin, and a metal shim welded to the bracket.

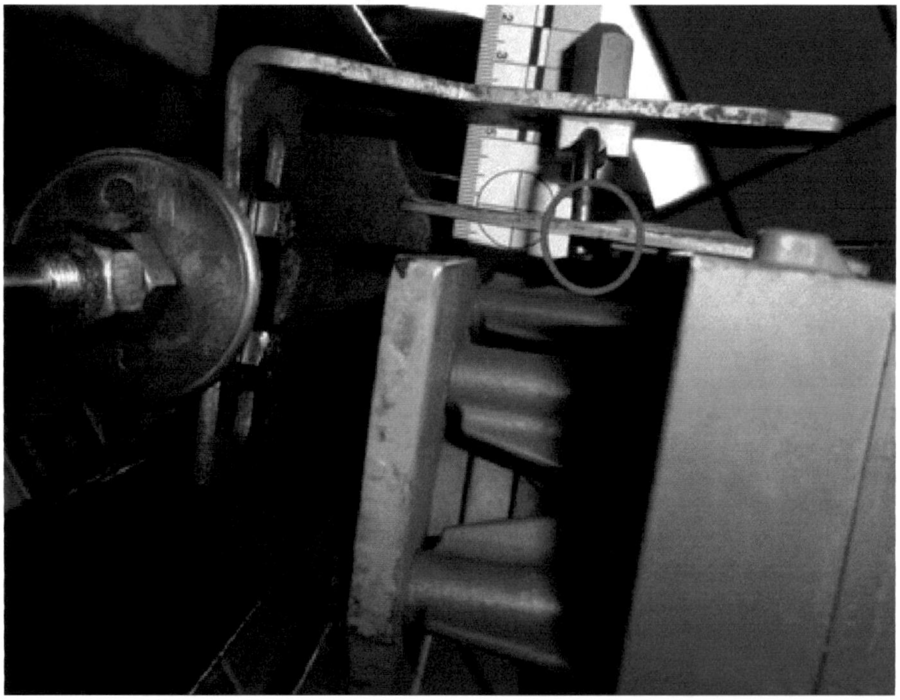

Figure 9. Pin engagement distance

3.3 Operator intervention based on conflicting information and insufficient time.

The impact due to a failure of the chilled water valve can be mitigated with operator intervention. However, in this incident, conflicting information on the control screen resulted in the wrong course of action in response to the high temperature alarms. The operator screens indicated that the chilled water control valve was fully open, yet the reactor contents continued to increase in temperature. Plant personnel indicated that this situation occurs because the operator screen only displays the command from the system to the control valve, not an actual feedback position from the valve. The operator response was further complicated by the observed decrease in temperature on the jacket supply. The valve did not fail completely closed, and in its impaired state, just enough cooling water was being supplied to decrease the jacket supply water temperature. Operators did not recognize that the decrease in jacket supply temperature was much lower than the decrease in temperature observed with a 100% open valve.

Based on this conflicting information, the operator chose to attempt to manually cool the reactor, rather than order the field operator to open a manual emergency bypass cooling valve. The attempts to cool the reactor were ineffective, and the ESS activated before the operator realized that the manual emergency cooling bypass valve needed to be opened. Given the short response time (6-7 minutes) for the alarm, it was important that the instruction for responding to the high temperature alarms be clear, straight forward, and concise. The timeframe was not sufficient to troubleshoot and diagnose the cooling system performance issues. Instead, opening the emergency cooling bypass valve was viewed as the preferred first response to the high temperature alarm.

4 Lessons Learned

After the incident, the facility identified several key lessons learned regarding their hazard analysis, mechanical integrity/maintenance program, and operational discipline.

4.1 Hazard Analysis

The use of terminology such as "fail open" or "fail close" is common in process engineering. Unfortunately, this terminology does not include the initiating cause for the failure state. As a result, process hazard analysis teams risk misinterpreting a piece of equipment as "fail open" or "fail close" for a wide variety of initiating causes. For example, many components are designated for a specific failure state upon loss of utilities. However, equipment malfunctions, human errors, or external events can also influence a component's operation and cause a failure state that is different than the one observed upon loss of utilities. Therefore, it is important that hazard analysis teams continue to consider additional potential initiating causes and failure states for equipment.

In the presented case study, the facility thought that under all potential failure conditions, the chilled water valve would fail open. As a result, the control valve was also used as a first response emergency cooling valve. When the valve failed, the system not only lost the ability to control temperature in the reactor, but also lost the first response of the emergency cooling system. After the incident, the facility installed a redundant emergency cooling valve. The redundant valve was independent of the control valve and ensured that the failure of the control valve would not impede the first response emergency cooling system.

4.2 Mechanical Integrity and Maintenance

The facility recognized that a valve mechanical linkage failure could be mitigated through preventative maintenance and inspection. After the incident, the facility added inspection of the mechanical linkage on the cooling water control valve to the quarterly preventative maintenance for the process. This inspection included evaluation of the linkage components and the engagement distance of the travel pin into the rotary shaft arm. The goals of the program were to recognize linkage component that were deteriorating and replace them prior to failure, and to make adjustments if parts of the linkage began to move out of alignment.

It is often challenging for a facility to continue to service aging equipment. The original manufacturer of the equipment may go out of business, or elect to stop offering replacement parts. After the incident, the facility reevaluated its procurement process for replacement parts in control valves. It reached out to the valve manufacturer to confirm that the replacement parts it had obtained were approved for use in the valve's mechanical linkage. It also added a dimension verification step to the procure process to confirm that the ordered parts matched the equipment specifications.

4.3 Operational Discipline

As highlighted in the case study, operator intervention is one of the controls that a facility may rely upon to safely navigate process upsets. Using procedures as an administrative control can be particularly challenging because they must be effectively written, implemented, and performed. [5] For this reason, procedures often represent a potential area for improvement in many safety management systems.

After the incident, the facility questioned why the manual emergency bypass line was not opened. The operator in the control room stated that when he responded to the alarm, there were two observations he made: 1. the emergency cooling valve was 100% open; 2. the jacket supply temperature was decreasing. Based on that information, the operator moved to troubleshooting the issue, and did not realize that the reactor temperature was still increasing. The operator also commented that there were several temperature alarms and they tripped infrequently. Given the short response time (~ 6-7 min); it was also possible that a field operator would not have had sufficient time to locate the necessary tools and open the bypass cooling valve prior to the incident.

The facility developed a job aid for the control room operator to improve operator response to the alarms. The job aid was a one-page chart that detailed critical alarms that required opening the bypass cooling valve. The chart also include the alarm set points, and the appropriate response for each alarm. The facility also tied a wrench to the emergency cooling bypass line, to ensure that if the valve needed to be opened, the field operator would have all the necessary tools available at the valve.

5 Conclusions

Process hazard analysis is a fundamental component of several aspects of process safety management systems. When a facility struggles to perform process hazard analysis, there is a potential for risks to go unrecognized, or under mitigated.

A case study was presented that examined an incident related to a "fail open" cooling water valve failing in a partially closed state. The valve failure disabled not only the temperature control function on an exothermic reactor, but also the primary emergency response system. Administrative controls were ineffective as implemented at preventing the incident following failure of the valve. The reactor overheated due to insufficient cooling water flow, and the contents were evacuated to a secondary containment pit resulting in a reportable release. Following the incident, the facility identified several key learnings related to improving their hazard analysis. The facility also implemented corrective actions to mitigate the potential failure of other cooling water control valve, and improve the reliability of the emergency cooling system and operator performance.

6 References

[1] United States Department of Labor, Occupational Safety and Health Administration. Process Safety Management of Highly Hazardous Chemicals, 29 CFR 1910.119.

[2] Center for Chemical Process Safety, Guidelines for Risk Based Process Safety, John Wiley & Sons, Inc., New Jersey, 2007.

[3] Center for Chemical Process Safety, Guidelines for Hazard Evaluation Procedures, John Wiley & Sons, Inc., New Jersey, 2008.

[4] M.J. Viz, R.A. Ogle, S.J. Dee and B.L. Cox. Hydrogen Sulfide Exposure from Molten Sulfur – A Forgotten Hazard? Proceedings of 9th Global Congress on Process Safety, San Antonio, Texas, April 28-May 2, 2013.

[5] S.J. Dee, R.A. Ogle, S.D. Lakhiani. Procedural Safeguard Reliability. Proceedings of 11th Global Congress on Process Safety, Austin, Texas, April 27-April 29, 2015.

PROCESS HAZARD ANALYSIS (PHA) – ARE STUDIES IN ITSELF GOOD ENOUGH TO REDUCE RISK?

Palaniappan Chidambaram
DuPont Sustainable Solutions
Singapore
Palaniappan.Chidambaram@dupont.com

Prepared for Presentation at
American Institute of Chemical Engineers
2014 Spring Meeting
2017 Spring Meeting and 13th Global Congress on Process Safety
San Antonio, TX
March 26 – 29, 2017

Keywords: Process Hazard Analysis, HAZOP, Risk Recognition, Risk Governance, Operational Risks, Risk Management

Abstract

Despite the advances in processing technologies, construction techniques, inspection methods, and materials, the number of process safety incidents are still increasing, and often the same types of incidents are being repeated. While regulators are enforcing stringent requirements in regards to safety management systems, process safety incident statistics reflects ineffective implementation. Upon closer look at the implementation of safety management systems, it becomes evident that they were not based on robust risk management system. When asked how risks are managed, a common response in the industry is that we have completed the hazard analysis and risk assessment studies. This paper will help understand why studies alone are not sufficient to manage risks effectively. Typical observations based on our client experience will be highlighted in the areas of risk governance, risk thinking culture, risk communication and risk awareness which goes beyond risk recognition/identification and assessment.

Introduction

Operational risks can be defined as the potential loss due to possible or foreseeable worst case consequences with a realistic probability of occurrence during facility operation, where such risks span across the entire value/supply chain. Typical operational risks include employee and public health and safety, community economic impact, environmental damage, asset damage, and business interruption. These risks are managed effectively in an organization when events leading to such risks are identified, well understood, and barriers required to prevent/mitigate them are implemented and maintained. It is essential to recognize risks associated with the process and operations through the lifecycle of the facility and implement systems to manage them effectively.

Organizations that understand the importance of implementing Process Safety Management systems do not always realize the benefits despite significant effort and leadership commitment. This can be primarily attributed to the following factors:

- Ineffective risk governance and lack of risk ownership at different levels;
- Inadequate risk recognition and risk assessment;
- Lack of holistic risk based thinking and risk based implementation;
- Poor risk awareness and risk communication; and
- Lack of robust risk assurance program;

Figure 1 depicts a model to represent operational risk management in a comprehensive manner and is designed to provide organizations with a better understanding and highlight the importance of risk governance and recognition in managing operational risks. In this paper, we will highlight our observations and findings based on our client engagement experience in the areas of risk governance, risk culture, risk recognition and assessment, risk awareness and risk communication. This will help demonstrate that PHA studies alone are not sufficient to manage or reduce risk.

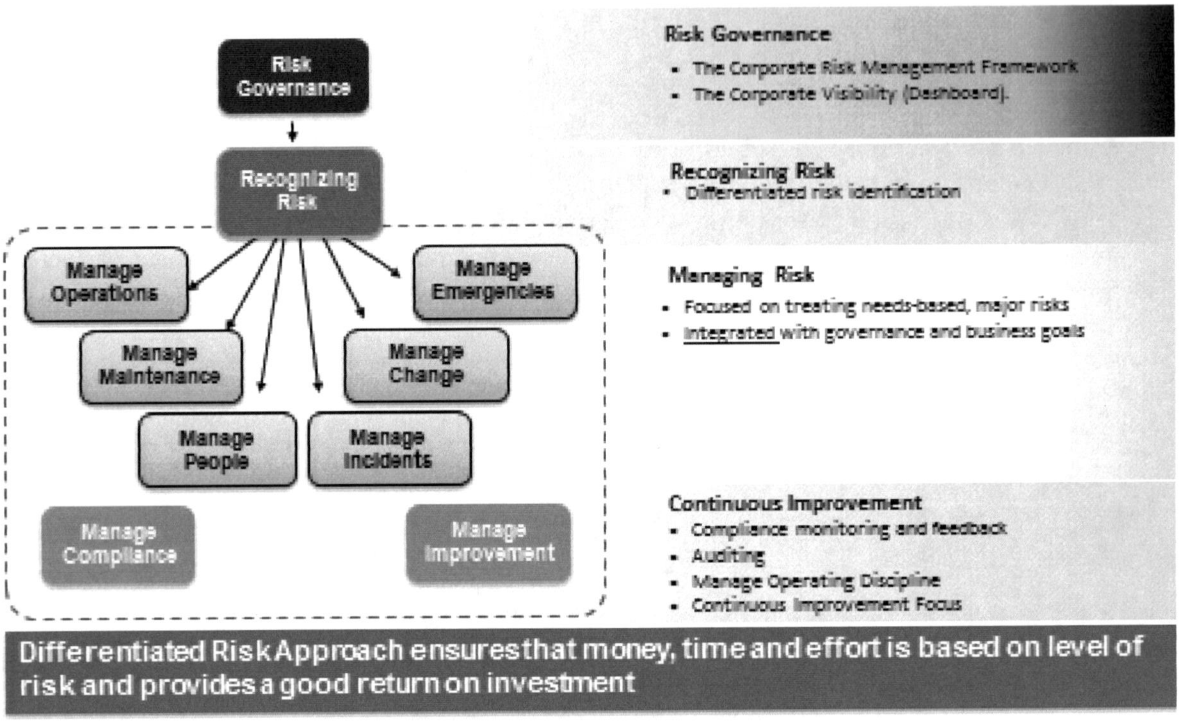

Figure 1 – DuPont Sustainable Solutions' approach to managing Operational Risks
© 2017 DuPont

Risk Governance

To consistently manage risks and make informed decisions, it is essential to have a well codified risk management process and governance structure to execute it. One of the key elements of risk governance is defining the acceptable risk criteria, along with risk ownership and accountability at different levels of the organization. In most cases, risk criteria are represented in the form of a qualitative/semi-quantitative risk matrix. A qualitative or semi-quantitative risk matrix is used during What If/Checklist reviews and Hazard and Operability (HAZOP) reviews. Other common areas where risk based decisions are made include task risk assessments, prioritization of inspection and maintenance activities, as well as recommendations from auditing and incident investigations.

In the absence of a robust risk governance structure and risk thinking culture, risk recognition and assessment efforts are not leveraged effectively (e.g. definition of emergency scenarios, identification of PSM critical elements, training on SOPs, etc.) to drive effective implementation of an operational risk management program. Some of the observations indicating poor risk governance include:

- Inconsistency in the risk matrix utilized for ranking risks at enterprise level and during risk assessment for facilities and sites.
- Identified risks and treatment actions are generic to business and not specific for a given risk event;
- Specific risks identified in hazard evaluation and risk assessment studies are not included in the risk register; and
- Risk escalation and approval mechanism is not in place to address the emerging risks due to changes (for example: impairment/bypass of safety critical devices) or incidents;

This often leads to inconsistent risk based decisions in the organization despite being risk based, leading to underestimation of risk (worst case) and a false sense of security. In addition, efforts and resources are likely to be invested where it is not required.

Risk Culture

It is essential to embed a risk thinking culture into an organization at all levels to achieve effective, systematic and consistent management of operational risks. More often, organizations are task focused with a compliance mindset instead of outcome focused with a commitment towards managing risks. This orientation is often reflected in their behavior and focus on what needs to be done and how, instead of why and how well it needs to be done. Significant effort is invested in activities, whereas outcome of such activities does not result in risk reduction and a shift in risk thinking mindset and behaviors.

Third party consultants are often engaged by organizations to perform their hazard evaluation and risk assessment studies. Organization with task focused mindset tends to complete studies primarily to fulfill insurance, regulatory and corporate requirements. Due to lack of inadequate and competent resources, results and outcome from such studies often resides with a function or department in the organization. Assessment reports are often not reviewed for completeness, understood well, communicated and utilized effectively. Without a strong culture, the outcome of such studies is not reviewed, understood, communicated and utilized in managing risks effectively. Such an organization tends to equate completion of such studies and compliance to local regulations and standards with effective risk management.

Organizations that are unable to respond to the following questions confidently and consistently at all levels of the organization often have difficulty managing risk:

- What are the top risks for the site, sector, plant and section and barriers associated with them?
- Are identified barriers effective and will they function on demand?
- What actions should be taken if barriers do not function?
- What is my role in recognizing and managing risks and consequences of not managing them?

Often responses for each question vary depending on the individuals' risk perception. Based on our client experience, the following paradigms need to be understood and internalized by each employee and contractor:

- Recognition of risks by itself does not result in risk reduction and effective risk management;
- Risks are not static, but dynamic and need continuous and consistent management;
- Not all risks are the same, thus require differentiated risk management;
- Past performance in managing risks does not define future performance; and
- Risk perception varies among individuals (contractors and employees) resulting in varying risk tolerance requiring the need to sensitize everyone to view a given risk consistently.

Cultural traits listed below often lead to risks being overlooked and ineffectively and inconsistently managed in an organization:

- Lack of openness to ideas, opinions and perspectives due to directive & hierarchal culture and lack of "what could go wrong?" and high risk tolerance mindset;
- Inability to foster conviction and commitment to make risk based decisions and help individuals understand the benefit of doing so ("what is in it for them?");
- Silo mentality resulting in risk ownership being limited to their respective area of work or activities;
- Communication rituals & channels being one way monologues; Not recognizing the importance of two-way risk communication internally in raising risk awareness and instilling risk ownership at all levels of organization in the process;

- Race against time (everything is priority, everything is urgent) driving a behavior of ad-hoc quick fixes without understanding of unintended consequences; and
- Lack of ability (will & skill) to say "Positive NO" and challenge at-risk decisions at peer and leadership levels.

Risk Recognition

Process Hazard Analysis (PHA) is a framework for identifying the process hazards, hazardous events and their consequences, evaluating the adequacy of the existing safeguards and determining the need for additional safeguards to reduce the risks to an acceptable/tolerable level of an organization. PHA is often carried out during the design stage by most of the organization and is often limited to HAZOP and SIL assessment studies, in addition to studies required by regulations. Key observations and gaps found based on client experience in diverse sectors are:

1. Initial PHA done by project team without involving operations, resulting in an incomplete assessment;
2. Lack of risk assessment or use of different risk matrices that is not aligned with the organization risk tolerance level without understanding implications;
3. Process Technology or Process Safety Information not handed over by the project team to the operations team or the information is not kept up to date;
4. HAZOP considered as the beginning and end of PHA and lack of augmenting with appropriate quantitative studies, especially for high consequence and low frequency events;
5. PHA outputs not considered in designing the remaining PSM elements;
6. PHA not done periodically or when there are changes in process and/or equipment, resulting in increase of potential unmitigated risks;
7. Lack of Rigor in implementing PHA recommendations;
8. PHA conducted without consideration to internal and external incidents; and
9. Not considering PHA in the early stages of project design.

Similarly, task based risk assessments which are carried out for non-routine activities and as part of permit to work systems are found to be inadequate due to following reasons:

- Viewed as paper exercise and often copied from a previous assessment without detailed review for its validity, correctness and relevance for a given task;
- Assessments are carried from a compliance mindset without giving regards to WHY it is important;
- Not reviewing the tasks with an open mindset and linking to potential consequences beyond the work areas in case of deviation. This limits the severity to the impact on the work crew or individuals directly involved in the task;
- Risk assessments are not communicated effectively (often, one way) to work crew and is assumed that they are familiar with and have understood them;

- Lack of clarity on risk ownership and roles as to what needs to be done, who will do it and how it will be done; and
- Poor risk assurance to ensure that all risks have been identified and barriers are functional and effective during the execution.

Risk Identification

Process Hazard Analysis is often limited to HAZOP and SIL assessment studies in many organizations and jurisdictions. Quantitative Risk assessment and Safety case is developed to satisfy regulatory requirements to obtain the license to operate. The most common method applied in the industry for hazard evaluation is HAZOP.

Based on a review of HAZOP studies in different industry sectors and regions, the quality of HAZOP needs to be improved in terms of application of methodology, level of details and accuracy and completeness based on the below or at organization risk tolerance level. Observations are listed below:

- HAZOP team discussions are mostly limited to few standard parameters.
- "All the causes for each deviation" as well as "all the consequences for each documented cause" are not identified and documented;
- Ultimate consequences, such as impact on people (including occupied plant buildings), environment and business, are not consistently identified resulting in underestimation of consequence severity and risk and in inadequate identification of protection layers;
- Lack of application of organizational risk matrix consistently resulting in identification of protection layers based on team experience;
- Lack of availability and provision of required process safety information to HAZOP team members to enable them to review and provide inputs;
- Not covering non-routine operations such as start-up, shutdown, catalyst change over isolation of equipment for handover to maintenance, etc.;
- HAZOP reviews do not comprehensively capture both the "hazards" and "operability" issues pertaining to process. In some studies, focus is more on "operability" issues, while some are more focused on "hazards".
- Validation of effectiveness, adequacy and independence of layer of protection for which credit is given during qualitative risk assessment. For example,
    - Availability of sufficient time for the operator to intervene upon alarms (e.g. LAH, LAL, TAH, etc.) and specific actions to be undertaken including effectiveness of the action in preventing the consequence; and
    - Review of Safety Valves Sizing Cases (e.g. verification that all overpressure scenarios identified during the HAZOP have been adequately covered in the PSV Datasheet).

Risk Communication & Risk Awareness

Most often, risk awareness among employees is primarily driven by their experience and individual perceptions. Terms such as "Hazard", "Risk", and "Consequence" are often misunderstood and used incorrectly by individuals at different levels of the organization.

Risk communication internally is predominantly absent and ineffective in most organizations. This is reflected in the high degree of variance in risk awareness. Communication is often limited to written email communication or posters and primarily focus on What needs to be done and to some extent on the How. It does not cover Why and How Well it needs to be done, and what are the benefits and consequences of not doing it well.

Absence of two-way communication and ineffective employee engagement results in lack of risk ownership and accountability. Often communication is assumed to have happened in the following areas:

- Policies, principles, standards, procedures, work instructions across the organization;
- Condition of the equipment, unavailability, impairment or bypass of safety critical equipment;
- Changes or modifications to the facilities, personnel, procedures and impact;
- Hazards, consequences, risks and barriers to manage them as part of permit to work system;
- One way tool box sessions and sharing of incident summary and lessons learned;
- Shift change communication;
- Sharing of good practices and lessons learned limited to their respective department, function, unit and sites;

A robust mechanism to assess the quality and effectiveness of risk communication and an attempt to measure is non-existent in most cases. Thus, it is difficult to ensure given risks are understood and will get the same level of treatment and attention across all functions and levels within the organization and that resources are allocated appropriate to the risk level.

CONCLUSION

Technical studies and process safety management systems alone will not help in achieving effective risk management. The organization must embed risk thinking culture at all levels and ensure risks are communicated effectively and seamlessly. This will help manage risks and enable individuals to take ownership for the risks and hold each other accountable for the risks that are prevalent in the organization.

481540 Dan Crowl on Fire - Basics of Flammability *Dan Crowl*

Abstract:

Flammable materials are common in industrial facilities - either as a feedstock, solvent, product or fuel. It is important to understand the flammability characteristics of these materials in order to handle them safely. This talk will provide an introduction to flammability. This will include the nature of fires and explosions, the properties of flammable materials, and how to use these properties to handle flammable materials safely

481541 How Do Flame Arresters Work?

Abstract:

Processes involving the storage, transport, separation, & destruction of flammable vapors are ubiquitous in the chemical & petrochemical industries. These operations present inherent explosion risks that are often addressed through the application & use of Flame Arresters. However, many practicing engineers & recent graduate engineers are unfamiliar with Flame Arresters, which can lead to dangerous misapplications. This introductory presentation details the basic fundamental principles of flame arresters, the industry standards governing these devices, & common flame arrester applications.
Key technical concepts related to flame arresters, such as the MESG, the dynamics of propagating explosions, & arrester element design will be described. The critical regulatory & industry requirements of NFPA 67, NFPA 69, & 33 CFR Part 154 will be cited. A generous cross section of typical flame arrester applications will be provided.

481542 BLEVE Basics, A Primer on Boiling Liquid Expanding Vapor Explosions

Abstract:

Boiling Liquid Expanding Vapor Explosions (BLEVEs) are a special case of bursting pressure vessels in which the pressure vessel is filled with a liquid that is above its normal atmospheric boiling point. When the vessel fails, the liquid boils and adds significant energy to the bursting vessel event. The thermodynamics of these events can be very complicated and interesting and the consequences of these events can be catastrophic. This session will provide new practitioners with a working knowledge of BLEVE phenomena, potential consequences, prediction methodologies, and some sample event sequences that can lead to a BLEVE. The focus will be on developing fundamentals, discussing common misconceptions, and presenting how to deal with the most common scenarios and the session will include a worked example of a BLEVE calculation to illustrate the techniques.

482027 Industry 4.0 - Challenge and Opportunity for the Chemical Industry

Abstract:

Disruptive trends come whether industry is prepared to accept them or not. Today we are observing rapid increases in computing ability, application of data analytics, implementation of machine connectivity, and other platforms that are influencing the future designs of manufacturing. Manufacturing is moving from the optimization of the individual unit, plant, or production line to the optimization of the entire production eco-system. These trends represent the 4th, and latest, manifestation of the industrial revolution, the so-called *Industry 4.0*. Industry 4.0 has been defined as the growth of cyber-physical systems within the constructs of the Internet of Things (IoT), big data and ancillary capabilities. In the chemical industry, Industry 4.0 is manifesting itself in rapid advances in enterprise analytics, robotics and other key elements that could combine to create a strategy for a chemical producer and has spawned many Government-Academic-Industry consortia such as the Digital Manufacturing and Design Innovation Institute (DMDII), the Smart Manufacturing Leadership Coalition (SMLC), among others. This paper will reflect on the challenges that Industry 4.0 sets before the chemical industry, and then explores the Dow Chemical Company's plans to turn these challenges into opportunities.

Speaker biography:

Billy B. Bardin is the Global Operations Technology Director. His responsibilities include driving technology and innovation strategy within Manufacturing and Engineering (M&E) and oversight of all commercial technologies as well as development of technical talent across manufacturing and supporting operations.

Billy began his career in 2000 with Union Carbide/Dow in the Catalyst Skill Center in South Charleston, WV where he led alternative feedstock and catalytic process development programs. In 2001, he joined the Chemicals Sciences group within Core R&D, leading programs in the areas of alternative feedstocks, paraffin functionalization, and olefin partial oxidation. While in Chemical Sciences, Billy relocated to Dow facilities in Midland, Michigan; Horgen, Switzerland; and Terneuzen, The Netherlands as leader for the development of Dow's high throughput heterogeneous catalyst testing capability.

In 2007, Billy was named leader for the North American Production Support Group within the Hydrocarbons and Energy business in Freeport, TX. He led efforts in cracker plant support and developed R&D programs in next generation cracker technology. The following year, he accepted the combined role of R&D Director for Chemicals and The Dow Technology Licensing and Catalyst (DTLC) Business where he drove both internal R&D programs and technology/catalyst development efforts for licensee sales. In 2010, Billy became R&D Director for the Feedstocks, Hydrocarbons, and Chemicals businesses, leading the technology strategy for Dow's hydrocarbons and alternative feedstock efforts and developing options to provide cost advantaged chemical feedstocks to Dow's downstream product portfolio. He was named to his current position in 2014.

Billy holds a Bachelor of Science in Chemical Engineering from North Carolina State University (1995), and a Master of Science (1997) and a Ph.D. (2000) in Chemical Engineering from the University of Virginia. He is a Registered, Professional Engineer (PE) with the WV State Board of Registration for Professional Engineers. He is chair of the Industrial Advisory Board for the School of Chemical Engineering at Purdue University and a member of the advisory board for the Dept. of Chemical Engineering at the University of Virginia.

479971 Big Data Vs. Little's Law: A Systems-Based Framework Changing the Risk-Reward of Large Scale Facilities

Abstract:

The current rush to adopt Big Data ignores the fact that *the data is not the system*. Ignoring system dynamics in the interpretation of thedata often produces the opposite of the intended result. Fortunately, Little's Law ($L = \lambda W$) provides a framework to properly interpret the dynamic nature of data and diagnose the true underlying systemic problems. Little's Law is especially useful in identifying and eliminating *hidden factories* in the system that erase profits, increase systemic risk, and erode morale and culture. Dr. Carrier will explore examples from high-risk environments that have saved hundreds of millions of dollars per year while improving safety. In addition, the amount of data required to make and implement informed decisions is dramatically decreased through the use of Little's Law. The approach will be demonstrated with several examples in Oil and Gas, Refining, and Fine Chemicals worked on by the author. The methodology will also be demonstrated for use in maintenance management and the radical improvement of a mid-sized refinery's MoC (management of change process). Finally, the role of culture in actively the dynamic response of our systems will be discussed.

479057 Big Data Doesn't Necessarily Mean Big Intelligence

Abstract:

Big data is the IT phrase of the moment in almost every industry. The concept of collecting huge amounts of data to learn, manage and drive your business forwards is attractive to all managers, especially as without it you may feel that you're being left behind. But without knowing how to collect data effectively, knowing what data has value, and how to effectively and efficiently use the volume of information, Big Data can be a hindrance, not a help. In Process Safety, we rely on data coming from the field in many instances. Collecting this data can be time consuming and costly, so making the most of the data you have is paramount, as is ensuring the data you collect is seen as a valuable asset by everyone in your organisation. But turning that data into intelligence to guide you on your path to Process Excellence is what it's really all about. Data is as valuable an asset as any other. Look after it. Nutrue it. Value it. And make sure you use it. In this paper we examine the core principles of making PSM data work for you to give you the best usable, actionable intelligence possible.

479724 Fluidization: Bubble, Bubble, Toil and Trouble

Abstract:

A quick dip into the open literature concerning fluidization quickly leaves one with a headache. The open literature is rife with empirical correlations relating pressure drop to fluid flow, material bed height, material particle diameter, and density differences. The same literature is also replete with "rules of thumb" for the onset of fluidization and for the various stages of fluidization. Undoubtedly, companies employing fluidized bed technology have similar documentation describing their own correlations and their own "rules of thumb".

In this presentation, we derive the pertinent relations for describing fluidization using Dimensional Analysis. We also discuss scaling fluidized processes from laboratory-sized equipment to commercial-sized equipment.

481004 Managing Heat Loss in Pilot Plant Operations to Simulate Full Scale Operation

Abstract:

Pilot plant operations present unique challenges regarding heat loss to the environment. The volume of chemicals and sizes of the equipment typically result in a high surface to volume ratio that results in excessive heat losses. Understanding and managing these heat losses (or gains for cold services) is critical to successful scale-up as the losses or gains from the atmosphere can be a significant part of the plant's energy balance.

For example, a 2" distillation column with no heat tracing is likely to lose heat to the atmosphere, resulting in cooling on each tray and giving a false result when interpreting the separations data and trying to fit it to process models. Heat tracing the column on the other hand, if not done correctly, will result in hot spots in the column and similarly poor matches to process models. If the thermophysical properties are adjusted to match the columns in these conditions, the result could be a poor model that is used for scale-up to the full commercial process. The result can easily be a commercial column that doesn't perform as expected.

This presentation will explore similar past experiences with heat transfer to the atmosphere in small scale equipment and some of the more successful methods to understand or mitigate these effects.

480842 Solids Handling Challenges during Process Scale-up

Abstract:

Solids handling operations are sometimes not given enough emphasis during planning and design stages. This is often based on the perception that solids will relatively easily move from bins, silos, hoppers, transfer chutes and feeders. However, this is not the case many times, especially when you are scaling up operations. During the pilot-plant stage, due to the low production volumes, it is possible to rely on manual intervention for assisting solids handling. However, when the pilot plant successfully demonstrates the concept and it is time for scale-up, one cannot rely on manual intervention. The solids handling operations need to perform reliably, on their own. Many studies have shown that when handling solids, the time required to start-up a plant can be significantly more than the planned time, due to challenges involved in solids handling.

This presentation will review different solids handling operations at a typical production plant, the challenges associated with reliably handling the solids and what can be done to avoid solids flow issues.

479208 Laboratory-Scale Integrated ARP Filter Test

Abstract:

The Savannah River Site (SRS) is currently treating radioactive liquid waste with the Actinide Removal Process (ARP) and the Modular Caustic Side Solvent Extraction Unit (MCU). Recently, the low filter flux through the ARP of approximately 5 gallons per minute has limited the rate at which radioactive liquid waste can be treated. Salt Batch 6 had a lower processing rate and required frequent filter cleaning. There is a desire to understand the causes of the low filter flux and to increase ARP/MCU throughput.

One plausible cause of the low filter flux is plugging of the primary (i.e., crossflow) or secondary filters. Much of the testing that was conducted to design and develop the filtration process in ARP focused on one aspect of the process (e.g., filtration, washing, chemical cleaning) rather than the entire process.

This work attempted to simulate the entire ARP process, including multiple batches (5), washing, chemical cleaning, and blending the feed with heels and recycle streams. The objective of the tests was to determine whether one of these processes is causing excessive fouling of the crossflow or secondary filter. The authors conducted the tests with feed solutions containing 6.6 M sodium Salt Batch 6 simulant supernate with no MST.

This presentation will describe the test design and the test results.

481908 Benefits for Having a PE According to a Young Professional

Abstract:

Earning a PE can do a wide variety of things for a person's career. It can help to show that they are knowledgeable about chemical engineering and to open up many future possibilities. These benefits apply to both younger and older engineers, but are especially critical to younger professionals.. In this discussion, I will walk through why as a young professional hoping to ascend the career management ladder, I became and will maintain my PE license.

Reduced Energy Requirements for Direct Chlorination of Ethylene to Ethylene Dichloride

Kyle Self, Chemetry, Moss Landing, CA, USA

Michael Weiss, Chemetry, Moss Landing, CA, USA

Ryan Gilliam, Chemetry, Moss Landing, CA, USA

Marshall Lee, TechnipFMC, Boston, MA, USA

Vincent Welch, TechnipFMC, Boston, MA, USA

Abstract

The eShuttle™ process eliminates the need for chlorine gas generation in the synthesis of ethylene dichloride. The new process uses the same feedstocks (brine and ethylene) to produce ethylene dichloride, caustic soda, and hydrogen. The main benefit of the new process is reduced electrical power consumption. By replacing the conventional chlor-alkali anode reaction with a transition metal oxidation reaction, the electrolysis cell operates at substantially lower voltage and without the safety issues associated with chlorine gas generation, compression and storage. Over the past 36 months, the eShuttle™ process has been successfully implemented and tested at Chemetry's pilot scale facility in Moss Landing, CA. The overall process economics and fundamental theory of technology are also discussed.

Introduction

The existing vinyls value chain:

The chlor-alkali reaction is the most widely used electrochemical production process in the chemicals industry. In this reaction, chlorine and caustic are generated via the electrolysis of an aqueous sodium chloride brine:

$$2NaCl + 2H_2O \rightarrow Cl_2 + 2NaOH + H_2$$

Because the Gibbs Free Energy for the above reaction is positive, driving the reaction forward requires external work in the form of electricity. Thus, the process is best described as two independent half-cell reactions, each shown below with their voltages vs. the standard hydrogen electrode (SHE):

Anode Reaction:	$2Cl^- \rightarrow Cl_{2(g)} + 2e^-$	(-1.36V vs. SHE)
Cathode Reaction:	$2H_2O + 2e^- \rightarrow H_{2(g)} + 2OH^-$	(+0.83V vs. SHE)

Combining the voltages of the two reactions in an electrolysis cell results in a theoretical minimum voltage of 2.2V. However, commercial membrane cells operate at voltages above

3.0V, where the additional voltage is required to compensate for losses such as membrane resistance, solution resistances, and required over-potentials.

A major consumer of chlorine is the polyvinyl chloride (PVC) value chain, which represents about 34% of the downstream market.[1] To produce PVC, the next step is the manufacture of ethylene dichloride (EDC) which is made through two commercial pathways. The first pathway is direct chlorination, where chlorine reacts with ethylene to produce EDC:

$$C_2H_4 + Cl_2 \rightarrow C_2H_4Cl_2$$

Oxychlorination, the second pathway, uses ethylene, oxygen, and hydrochloric acid (HCl), the latter typically formed as a by-product of other processes (see below):

$$C_2H_4 + 2HCl + ½O_2 \rightarrow C_2H_4Cl_2 + H_2O$$

EDC formed from both reactions is then thermally cracked to produce vinyl chloride monomer (VCM):

$$C_2H_4Cl_2 \rightarrow C_2H_3Cl + HCl$$

Finally, the VCM is then polymerized to form PVC for end use. The HCl byproduct from EDC cracking is typically recycled to the oxychlorination unit to produce more EDC. As a result, all the chloride atoms contained PVC were originally sourced as chlorine in the chlor-alkali process. The entire vinyls chain is shown below in Figure 1:

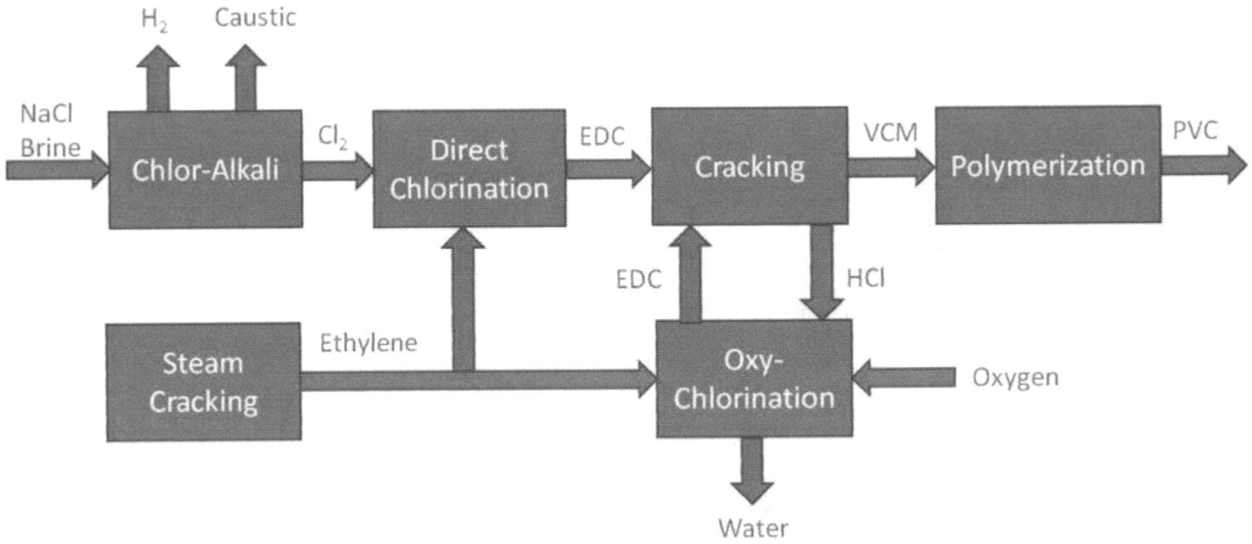

Figure 1: Major processes and products in the PVC value chain. Note that all chloride atoms leaving in PVC were originally sourced via the chlor-alkali process.

The opportunity:

The ability to replace reactions requiring chlorine gas with safer, lower-energy alternatives represents a great opportunity. Chlorine production from the chlor-alkali process

[1] See for example, https://www.ihs.com/products/chlorine-sodium-chemical-economics-handbook.html

is electrically intensive because two electrons must be transported over a 3V potential difference for every molecule of chlorine generated. As a result, chlor-alkali power demands often represent a significant fraction of the electrical grid consumption in developed countries. For example, chlor-alkali energy usage (just over 4 GW) represents approximately 1% of total US grid consumption[2]. In addition to the recognized direct costs associated with energy demand, chlorine production has significant hidden costs associated with safety risks in storage and transport.[3] Finally, because of the high power demand, conventional chlorine production results in a high carbon footprint due to associated power plant CO_2 emissions. As a result, replacement of conventional chlorine production methods remains an important environmental, safety, and economic objective.

The eShuttle™ Process

eShuttle™ Process Chemistry:

Chemetry's eShuttle™ process uses a recirculating, aqueous stream of copper to shuttle chloride ions from NaCl to EDC. This eliminates chlorine generation from the PVC value chain by replacing the combined chlor-alkali and direct chlorination processes as shown below in Figure 2:

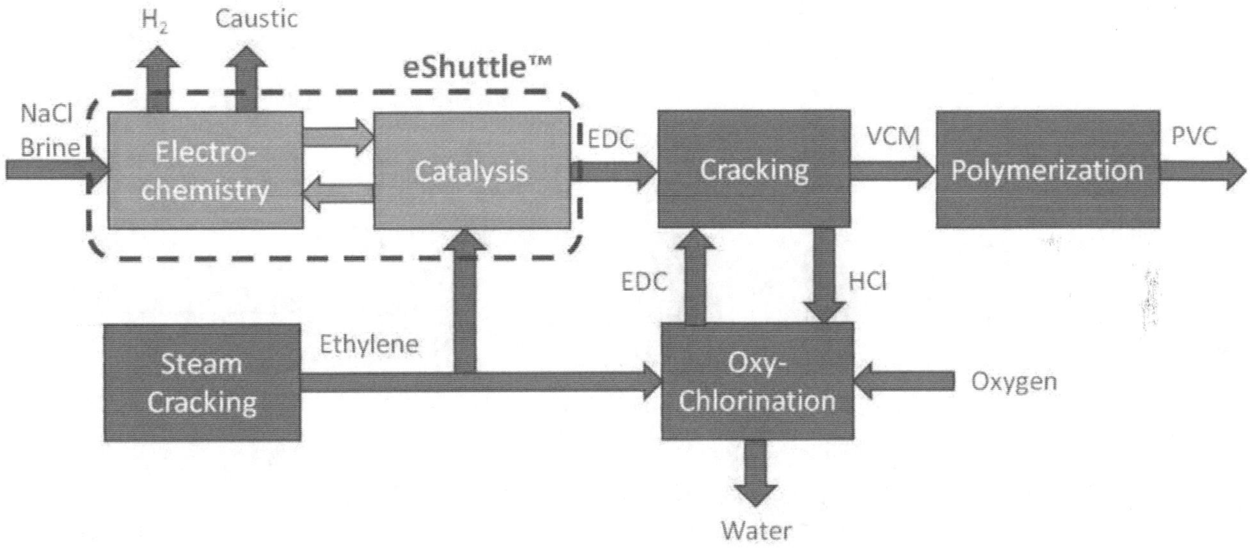

Figure 2: The eShuttle™ process replaces the chlor-alkali and direct chlorination processes with a single process that operates at lower energies and avoids chlorine as an intermediate.

Comparing Figure 1 with Figure 2 illustrates that the eShuttle™ process and the conventional chlor-alkali/direct-chlorination processes have the same feedstock inputs (NaCl

---

[2] Chlor-alkali energy consumption based on internal TechnipFMC calculations. US energy grid consumption of 447GW from https://en.wikipedia.org/wiki/List_of_countries_by_electricity_consumption

[3] See for example the US Center for Disease Control (CDC) report: "Top Five Chemicals Resulting in Injuries from Acute Chemical Incidents — Hazardous Substances Emergency Events Surveillance, Nine States, 1999–2008" available at https://www.cdc.gov/mmwr/preview/mmwrhtml/ss6402a6.htm.

brine and ethylene), product outputs (hydrogen, caustic and EDC) and, consequently, the same overall reaction:

$$2NaCl + 2H_2O + C_2H_4 \rightarrow C_2H_4Cl_2 + 2NaOH + H_2$$

The difference is that the chlorine gas intermediate has been eliminated in the eShuttle™ process.

As shown below in Figure 3, the eShuttle™ technology uses a recirculating stream of aqueous copper chloride. In the electrochemistry portion of the process, (shown on the left below), the cathode reaction remains the same as the chlor-alkali process. However, the anode reaction has now been replaced with the oxidation of cuprous chloride to cupric chloride. A second membrane, for anion exchange, prevents the copper ions from migrating into the brine stream. The design of the middle brine compartment, the development of the anion exchange membrane, and the engineering of the anode itself all represent major advances in reducing the cell voltage further improving the power advantages of the process.

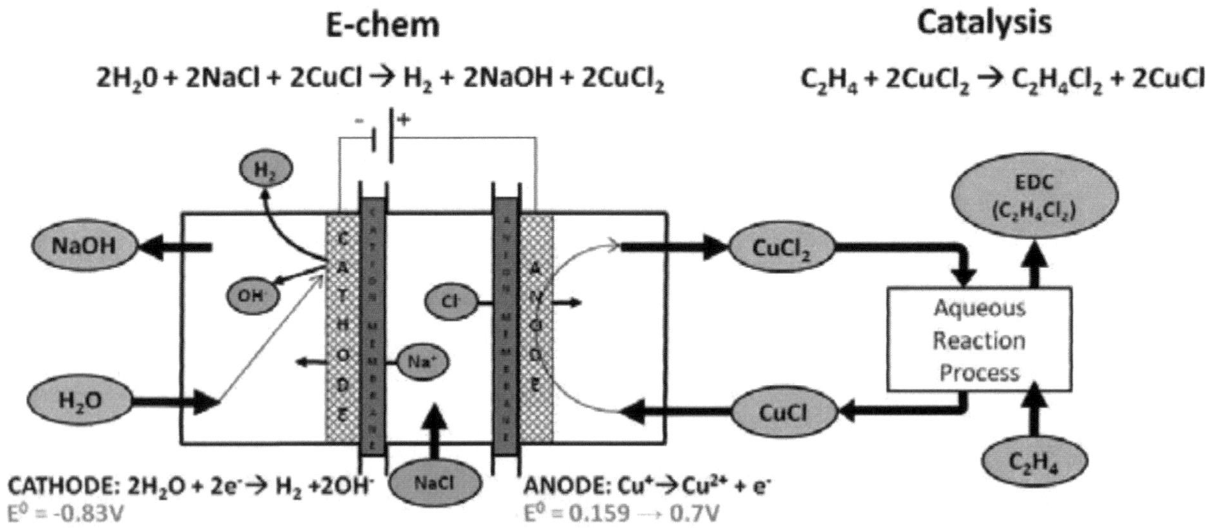

Figure 3: The eShuttle™ process circulates a continuous stream of aqueous copper chloride salts in a closed loop transporting Cl between the electrochemical cell and the process reactor.

The cupric chloride stream is then directed to a reaction system where it is contacted with ethylene to produce EDC under pressure and at elevated temperature. The optimal temperature and pressure is determined by balancing a combination of several variables including selectivity, space-time yield (STY) and capital costs.

A major breakthrough of the eShuttle™ process is the replacement of the chlor-alkali anode reaction, substituting chlorine generation (-1.36V vs. SHE) with a metal redox couple. The current version of eShuttle™ utilizes copper(I) oxidation as the anode reaction:

Anode Reaction: $Cu^{+}_{(aq)} \rightarrow Cu^{2+}_{(aq)} + e^{-}$ (-0.16V vs. SHE)

Thus, the theoretical minimum for this cell reaction should be about 1V. However, the copper speciation in the eShuttle™ process differs from the dilute ions assumed in the standard reaction above. That is the half cell reaction for the eShuttle™ process is more correctly written as the oxidation of cuprous chloride[4]:

Anode Reaction: $CuCl + Cl^- \rightarrow CuCl_2 + e^-$ (~ -0.70V vs. SHE)

Thus, the minimum voltage is actually 1.5V. Factoring in the losses due to membrane resistances and over potential, Chemetry operates its cells at about 2.4V, based on economic optimizations for commercial scale. As shown below in Figure 4, compared to state-of-the-art membrane electrolysis, the net result is a savings of 400kWh/tonne savings (on a caustic basis) – a major step change in energy performance.

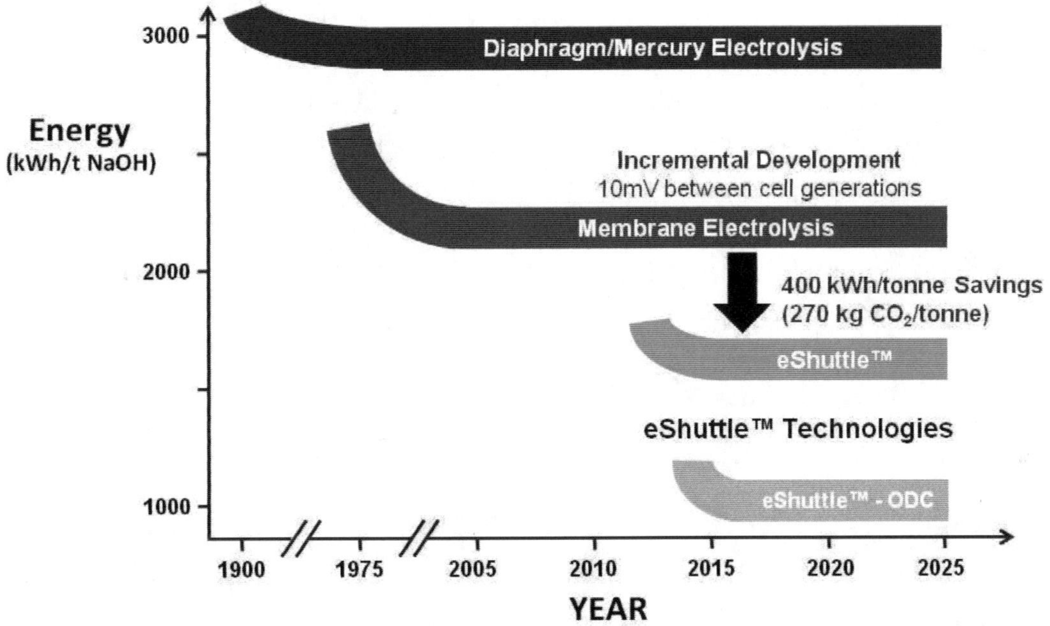

Figure 4: eShuttle™ is a significant advance in energy demand vs. traditional electrolysis routes[5].

eShuttle™ Process Thermodynamics:

The thermodynamics of the eShuttle™ process provide interesting insight into the source of the energy advantage over chlor-alkali/direct chlorination. Because both processes produce the same outputs from the same inputs, the minimum energy requirements are the same as defined by the Gibbs Free energy change for the overall reaction (see above):

$$2NaCl + 2H_2O + C_2H_4 \rightarrow C_2H_4Cl_2 + 2NaOH + H_2 \qquad \Delta G_{298K} = +274.5 \text{ kJ/mol}$$

[4] The reaction shown is for illustrative purposes. In fact, copper speciation is much more complex and the equilibrium speciation is dependent on a number of factors beyond the scope of this paper.

[5] The eShuttle™ ODC technology is the replacement of the traditional chlor-alkali cathode with an oxygen depolarized cathode (ODC). It is beyond the scope of this paper.

Because the beginning and ending states are the same, the energy savings must come from an improvement in efficiency. As will be discussed below, this efficiency improvement is fundamentally different than efficiencies achieved for the conventional chlor-alkali process, which have been small, engineering-based increments over a long period of time.

When $\Delta G > 0$, the Gibbs Free Energy change defines the minimum external work, electrical work in this case, that must be provided to a steady-state, continuous process. When the work is in the form of electricity, the above ΔG is better considered in electron-volts. If the reaction were a single-electron process, the minimum voltage required would be +2.84 V. Because the reaction is a two electron process, the minimum voltage for a cell supplying energy to the reaction is +1.42V (=2.84/2), a number that is quite close to the 1.5V theoretical voltage for the eShuttle™ process. Thus, the eShuttle™ process represents a very close approximation to the minimum required electrical energy for the combined processes. As with chlor-alkali, additional efficiency gains may still be realized through engineering improvements in membranes, cell design, and electrode performance, but these will be improvements on a superior, thermodynamically optimal baseline.

Another way of viewing the same information is shown in Figure 5 below. The y-axis is the Gibbs Free Energy of the system (based on the Gibbs Free Energies of formation for the components). The x-axis represents a nominal reaction coordinate. The top line follows the trajectory of the traditional chlor-alkali and direct chlorination processes, left side and right side respectively, and the lower line follows the trajectory for the eShuttle™ process.

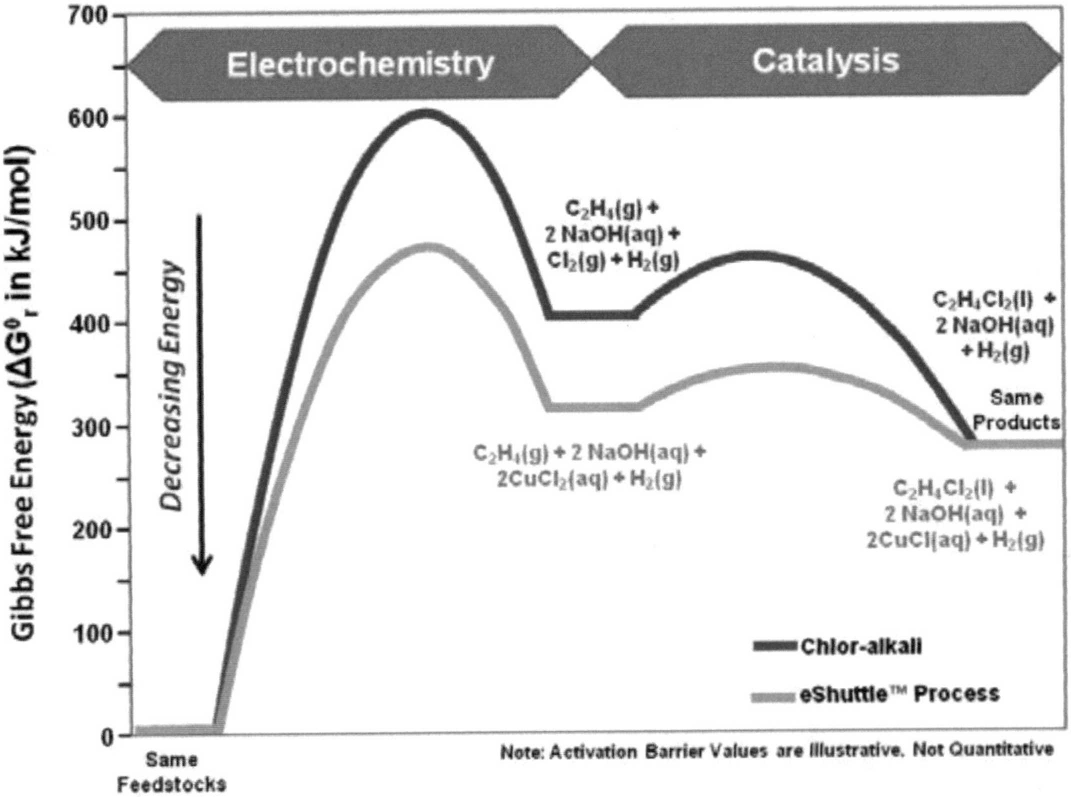

Figure 5: Thermodynamic comparison of chlor-alkali/direct chlorination and eShuttle™.

As noted previously, because both processes have the same inputs and outputs, the net change in going from left to right must be the same. As is clear in the above diagram, the voltage usage for the conventional chlor-alkali process leads to an unnecessarily high energy intermediate (i.e. chlorine). This extra energy can only be recovered as heat via downstream steam generation. Put another way, the steam recovered in direct chlorination is a direct result of the unnecessary voltage required in the chlor-alkali process. Because the intermediate state in the eShuttle™ process is much closer to the theoretical minimum electrical input, it represents the more efficient path.

Scale-up:

Over the past 36 months, the eShuttle™ process has been successfully implemented and tested at Chemetry's pilot scale facility in Moss Landing, CA. The pilot unit (shown below in Figure 6) is a fully automated plant designed with significant flexibility to test various process and equipment configurations. The electrochemical cell utilized is a full commercial height cell which allows for representative testing membranes and electrodes under commercial hydraulics. The packed bed process reactor is about 20 ft and can be operated in both trickle bed and bubble column flow regimes.

Figure 6: Chemetry's Moss Landing Pilot Plant. For scale, note the operators in the lower left corner.

During the operation, we have successfully demonstrated all phases of the integrated process including control systems, thermodynamic and transport data, component lifetimes, materials of construction, product quality, and environmental and safety systems.

Process Economics

The primary advantage for the eShuttle™ process is in the electricity savings. However, a full evaluation of the process economics must also account for differences in steam production and ethylene consumption between the combined chlor-alkali/direct chlorination processes and the eShuttle™ process. The economics should also account for regional price differences in these variables. As one example, conversion of an existing diaphragm-based chlor-alkali facility with direct chlorination to the eShuttle™ process would result in a net OpEx savings between $37/MT and $95/MT depending on regional costs as shown below in Table 1.

Table 1: Costs for electricity, steam and ethylene on a regional basis

Assumptions	Region/Market		
	US	Europe	SE Asia
Energy ($/MWh)	50	75	120
Steam ($/ton)	12	32	19
Ethylene ($/ton)	600	1100	1100
OpEx Savings $/MT EDC	**30**	**77**	**73**
OpEx Savings $/MT NaOH	**37**	**95**	**91**

As expected OpEx savings are greatest in regions where the power prices are highest. Capital equipment (CapEx) expenditures for the eShuttle™ process are comparable for chlor-alkali and direct chlorination in grassroots installations. As a result, the eShuttle™ process has the highest return on investment and is the most economical route to EDC.

Retrofit cases, which are beyond the scope of this paper, are better described on a case-by-case basis.

Conclusion

Replacing the standard chlor-alkali anode reaction for chlorine generation with a copper oxidation step reduces the cell operating voltage to near the theoretical minimum, which in turn leads to a dramatic reduction in overall power costs. As a result, Chemetry's eShuttle™ process provides the highest return on investment for EDC production when compared with conventional chlorine generation/direct chlorination technologies. In addition, the eShuttle™ process provides the added benefits of improved safety and lower carbon footprint. Over the past 36 months, the process has been demonstrated at the pilot plant scale at Chemetry's facility in Moss Landing, CA. Chemetry and TechnipFMC are actively pursuing demonstration plant opportunities as the next step in process commercialization.

479893 Application of LNG Technology Solutions for Optimisation of Liquefaction Processes

Abstract:

The application of proven, specialised equipment, efficient natural gas liquefaction process technology and modular construction is enabling the export of US natural gas by challenging the paradigm that a large baseload facility provides the best economy of scale. The concept is widely known as mid-scale LNG but the story doesn't have to end there. The principal reason for cryogenic liquefaction is storage and transportation and the same criteria that make the model work for LNG can be applied to other gases increasingly in demand; specifically ethylene, ethane, liquid air for energy storage, nitrogen, hydrogen and helium.

Key to cryogenic processing is the brazed aluminum heat exchanger (BAHX). A BAHX is a highly efficient, custom designed, compact heat exchange device that provides a heat transfer area significantly greater than conventional heat exchanger technology and also has an inherent multi-stream capability where 8 or more process streams in a single composite is common. Minimizing power consumption is a key design criterion for liquefaction processes, creating an inherent requirement for highly efficient heat exchange. Process efficiency is gained by minimizing temperature approach between the hot and cold streams. As the temperature approach decreases, the required surface area increases. With that in mind, the heat exchanger of choice must offer high surface area economically.

To ensure the robustness of the BAHX state-of-the-art software is coupled with in-house engineering expertise and experience to perform advanced transient thermal analyses. Exchangers are studied three-dimensionally from available operating data or dynamic simulation process software to determine the internal fluid and metal heat exchanger temperatures and the associated thermal stresses imparted on the exchanger. These analyses can be carried forward to predict potential fatigue damage and life expectancy of the equipment. Dynamic simulation is used to simulate operation of the BAHX and plant upset conditions to ensure process technologies are designed with proper controls to mitigate situations that introduce thermal stresses.

Typically refrigeration cycle configurations are focused on two; nitrogen cycle and mixed refrigerant cycle. Each cycle has its advantages and disadvantages when evaluated for use in a specific application. Usually designers study the trade-offs between power consumption and equipment utilization. This analysis allows them to optimize the liquefier for the total plant cost (capital plus operating costs). Comparison of operating costs should focus on the energy consumption of the refrigeration system, as the balance of plant should be essentially the same for all processes.

Nitrogen cycles are simple to operate and eliminate the need for the use and storage of hydrocarbons as refrigerants. They are ideally suited to remote areas that do not have easy access to hydrocarbons for refrigerants. However, simplification comes at the expense of higher specific energy consumption.

Mixed Refrigerant (MR) processes are based on mixtures of light hydrocarbons and nitrogen and process optimization is achieved by varying the mixture of refrigerant components and operating pressures in a manner that essentially allows the heating curve of the refrigerant to mirror the cooling curve of the gas to be liquefied in a tight temperature approach. The brazed aluminum exchanger accommodates the surface area requirements resulting from the tight approach temperatures and log mean temperature differences (LMTD). Its inherent multi-stream capability also allows the process engineer to add or remove refrigeration along the process thermal gradient, wherever needed, to optimize process efficiency.

Because all the capital equipment and technology required is already proven and deployed globally in LNG service, it represents a no risk solution in their application to other liquefaction processes.

A Process Synthesis Approach to Natural Gas Liquefaction

Baraka Celestin Sempuga, Diane Hildebrandt
Material and Process Synthesis (MaPS) University of South Africa (UNISA), PO Box 392, Unisa 0003, South Africa

Introduction

Natural gas is one of the most important commodities in the global market. Oil is the leading source of energy; however coal and natural gas contribute almost 50% of the energy source (S. A. Al-Sobhi et al., 2009) . Natural gas is becoming more attractive due to its environmental advantages compared to other fossil fuels such as coal and oil. CO_2 is produced from coal and oil at an approximate rate of 1.4 to 1.75 higher than that produced from natural gas. Furthermore NO_X produced from burning natural gas are approximately 20% less than those produced from burning coal and oil (Kidnay and Parrish, 2006). Natural gas is often found in remote locations and sometimes requires long distance transportation to utilization sites. It is either transported as a gas by pipeline or as Liquefied Natural Gas (LNG) using insulated tankers. The mode of transportation used is much dependent on the distance over which natural gas must be transported as well as on the cost of transportation.

LNG is produced by condensation of natural gas. Methane is the principle element of natural gas but may also contain ethane, propane and heavier hydrocarbons. Small quantities of nitrogen, oxygen, carbon dioxide, sulfur compounds and water may also be found in natural gas. The typical natural gas composition is approximately 82% methane. LNG is produced by a sequence of processing steps that transform natural gas from gas phase to liquid phase while getting rid of the undesirable substances. The final composition for LNG is usually determined by the market specifications(S. Al-Sobhi et al., 2009).

The LNG Process

The cooling and liquefaction of natural gas is generally done in a sequence of refrigeration cycles. The number of cycles is a key factor in the success of liquefaction process. To make the cold temperatures required for the LNG, work must be supplied to the refrigeration cycles and heat must be rejected from the cycles. The process is work intensive, and thus energy efficiency is important as feed gas is consumed in order to carry out the liquefaction process (Remeljej and Hoadley, 2006). Most of the existing LNG technologies use closed cycles, where external source of refrigeration is used to cool the natural gas in a number of steps until it liquefies. The external source of refrigeration uses a refrigerant (mixed or pure component) to create the cooling effect by compressing and expanding the refrigerant in a cycle. Among the known existing LNG technologies include APCI Propane pre-cooled mixed refrigerant process, Phillips optimized cascade process, Black & Veatch PRICO process, Statoil/Linge mixed fluid cascade process and Axens Liquefin.

The basic principle behind all the above mentioned LNG processes consists of matching as closely as possible the cooling and the heating curves of natural gas and the refrigerant, so as to reduce thermodynamic inefficiency. Figure 1 shows the cooling curve for natural gas and the heating curves for a refrigeration system using a mixed components refrigerant and a cascade refrigeration system using three different pure component refrigerants. We can see from Figure 1 that the mixed refrigerant system comes closest to the natural gas cooling curve and therefore results in reduced thermodynamic inefficiencies because it minimizes the temperature different between the two fluids compared to pure component refrigerants (Kidnay and Parrish, 2006).

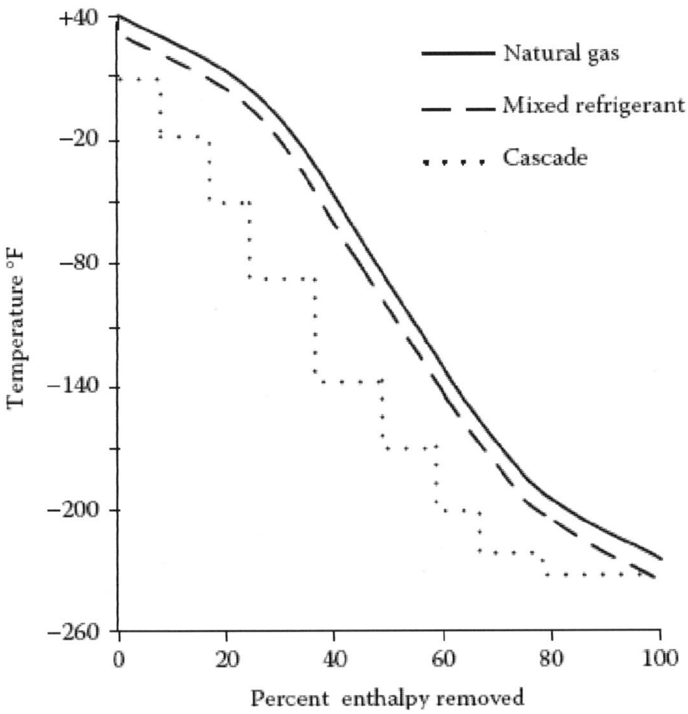

Figure 1: Cooling curve for natural gas and the corresponding heating curves for the mixed refrigerant and cascade cycle with pure component refrigerants (Kidnay and Parrish).

This manuscript presents an alternative way of analyzing LNG processes. The approach used is based on fundamental thermodynamic concepts and a graphical representation which can provide insight and information regarding the LNG process. The approach helps determining the targets for the process and can provide an idea of how these targets can be achieved. This approach is not only applicable to LNG processes but to any process that requires refrigeration.

LNG Process in the gh-diagram

It has been shown that Region 2 (that is 2A and 2B) of the gh-diagram (Figure 2) represents processes that run below ambient temperature and reject heat and require a refrigeration system. LNG is one of these processes as it requires refrigeration systems in order to liquefy natural gas. The ΔH axis of region 2 on the gh-diagram gives the amount of heat to be absorbed, the ΔG axis gives the minimum amount of work needed to absorb the heat and the temperature scale gives the reversible refrigeration temperature. This can be shown in Figure 2; a process located at point m (given by ΔH and ΔG across the process) would require that 14 kJ/mol of work be supplied, and this can be achieved by removing 10 kJ/mol of heat at 123 K; if this is done using a reversible refrigeration system then the net amount of work that would be consumed is 14 kJ/mol. The amount of work will increase as the temperature at which heat is removed decreases. For example at point n 10 kJ/mol of heat is removed at 80 K, therefore the minimum work required is 27 kJ/mol given by the intersection of line o-n and line m-n; we can see that if the same amount of heat needs to be removed at 0 K then an infinitely large amount of work will be required, this is shown by line o-q which can intercept line m-n only at an infinitely large value of ΔG.

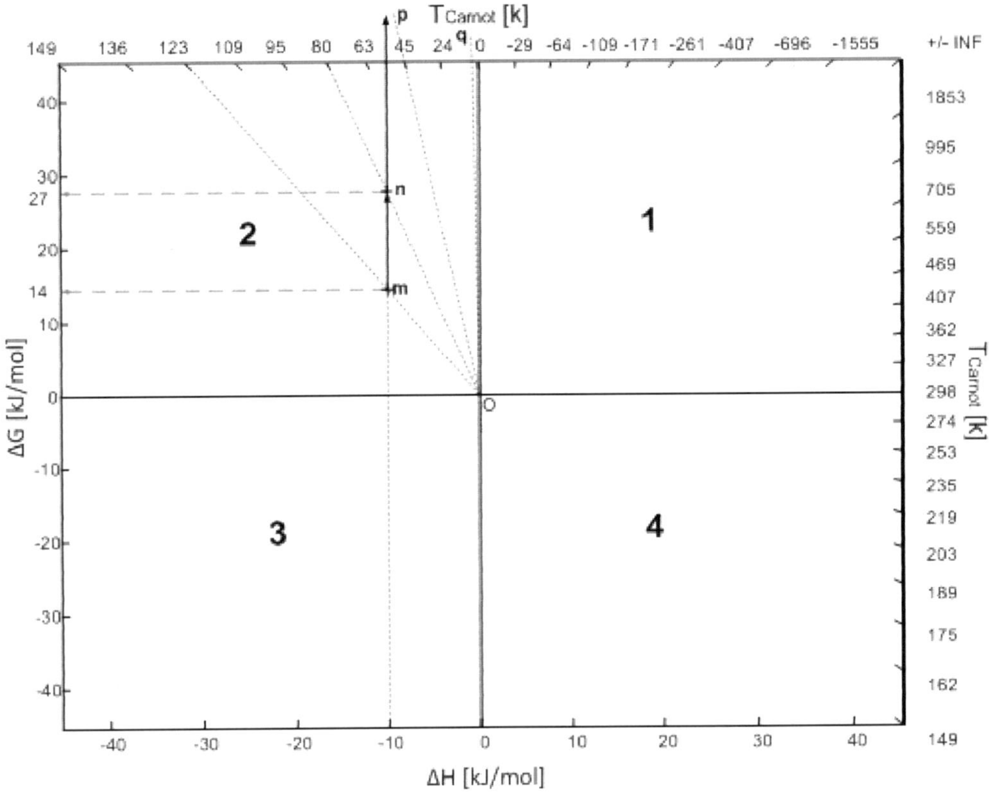

Figure 2: Example of a refrigeration process represented in region 2. Points m and n shows the amount of work needed (ΔG axis) to remove heat (ΔH axis) at the Carnot temperature (Temperature scale)

So far the gh-diagram has been used to represent processes where the input and output streams are at ambient conditions of temperature and pressure, T_O and P_O respectively. In this case the amount of work required in order for the process to proceed is given by the change in Gibbs free energy at T_O and P_O (ΔG (T_O, P_O)) (Denbigh, 1956). However at any other inlet and outlet temperature and pressure (T1, P1) and (T2, P2), the work required to transform A to B is given by the change in exergy (ΔE) across the process (Remeljej and Hoadley, 2006).

$$\Delta E = \Delta H\left(1 - \frac{T_O}{T}\right) \tag{1}$$

If the input and the output of the process are given, then ΔE and ΔH are known and therefore there is a unique value of T at which Equation (1) will hold. We will refer to this temperature as the Carnot temperature (Sempuga et al., 2010) it is the unique temperature at which the heat requirement must be supplied in order to satisfy the work requirement of the process. In a similar way as processes are represented in ΔH-ΔG diagram (gh-diagram), we can also represent processes in the ΔH-ΔE diagram (eh-diagram); here the input and output to the process are not restricted to ambient conditions. We can also identify the same thermodynamic regions in the eh-diagram as in the gh-diagram. We will now see that when we represent the LNG process in the eh-diagram, this can provide insight regarding the process and also provide a framework where alternatives can quickly be evaluated. Since natural gas is mainly composed of methane, in the following discussion we will use the thermodynamic properties of methane to represent natural gas. When we represent natural gas at various temperatures and pressures in the ΔH-ΔE domain this will form a surface which will be located in region 1 and region 2 of the eh-diagram. However, in this discussion we will only look at the part of the surface which is in region 2 as shown in Figure 3.

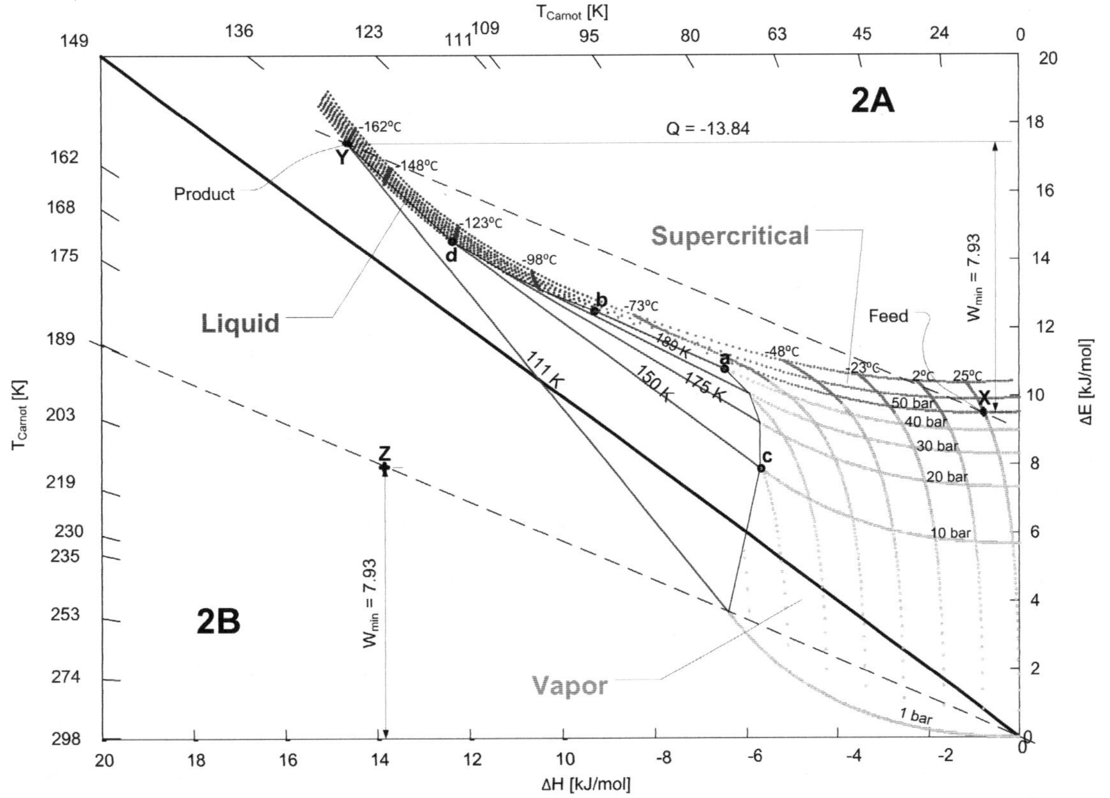

Figure 3: Natural gas at different temperatures and pressures represented on the eh-diagram. The basis is the natural gas enthalpy (H) and entropy (E) are zero at TO and PO. Colors are used to distinguish the different phases. Liquid, vapor, and supercritical

We will explore the usefulness of the eh-diagram (Figure 3) in the LNG process by looking at an example. Let us suppose that the feed to the process is natural gas (NG) at 25°C and 50 bar; this is located at point *X* on the eh-diagram in Figure 3. The product is liquefied natural gas (LNG) at -161°C and 1 bar; this is located at point *Y* on the eh-diagram. The minimum amount of work that the process requires to move from *X* to *Y* is given by the change along the ΔE axis and is W_{min} = 7.93 kJ/mol. This represents the work target for any route chosen to go from *X* to *Y*.

W_{min} can be supplied to the process by removing heat at the Carnot temperature. This can be seen on the eh-diagram in Figure 3; the Carnot temperature is obtained by subtracting *X* from *Y* to get point *Z*, and then the temperature is found at the intersection of line *O* – *Z* with the temperature scale. Note that line O – Z is parallel to line *X* – *Y*. Thus the Carnot temperature can be alternatively obtained by drawing a line parallel to *X* – *Y* from the origin to the temperature scale. This will give a temperature of 189 K (-84°C). This means that if we are able to remove the process heat (-13.83 kJ/mol) at the Carnot temperature (189 K), we would be able to reversibly liquefy NG. However we might need to manipulate the process stream to get it to the final state (111 K and 1 bar) by using heat exchangers and expanders and the process must be fully heat and work integrated internally such that only the process heat (-13.83 kJ/mol) at the Carnot temperature (189 K) is exchanged with an external refrigeration system as shown in Figure 4. On the other hand, in order for the refrigeration system to be reversible it must absorb the heat from the process at the process temperature and heat must be rejected from the refrigeration system at ambient temperature. In that case the refrigeration system will require an amount of work equivalent to W_{min}.

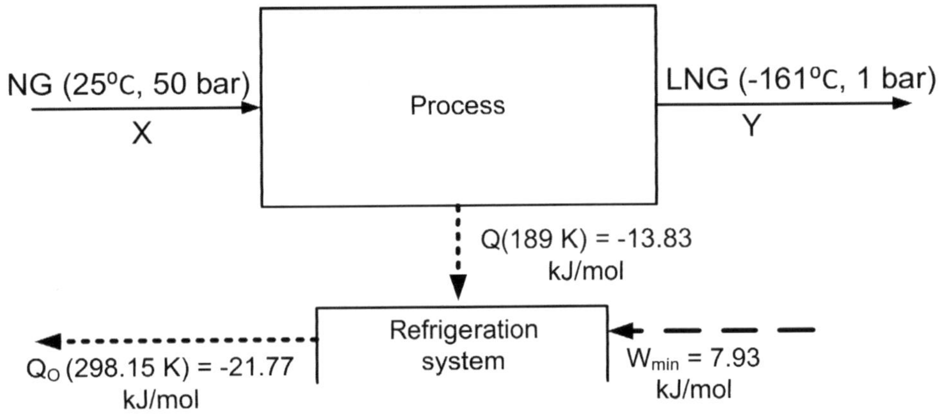

Figure 4: Schematic diagram Supplying work to a LNG process using a refrigeration system: Minimum work is supplied to the LNG process by removing heat at an appropriate temperature

Note that heat can be transferred at single temperatures only in the two-phase region of the natural gas envelop in Figure 3. Thus for the process in Figure 4, heat must be exchanged with the refrigeration system along line $a - b$ (at 189 K) on the eh-diagram (Figure 3) and the process route will be as follows: $X - a - b - Y$. However the process must be manipulated internally such that the amount of heat between $X - a$ and between b – Y are rejected from the process only along line $a - b$; this might require a certain combination of recycles, heat exchangers, expanders and flashes as well as multiple points of heat transfers with the refrigeration system all happening at 189 K. Alternatively the process could be split into multiple processes using different refrigeration systems. For example the process could be split into three processes (Figure 4); the first process would go from X to c, the second from c to d and the third process, from d to Y. Each process with its own refrigeration system as shown in Figure 5.

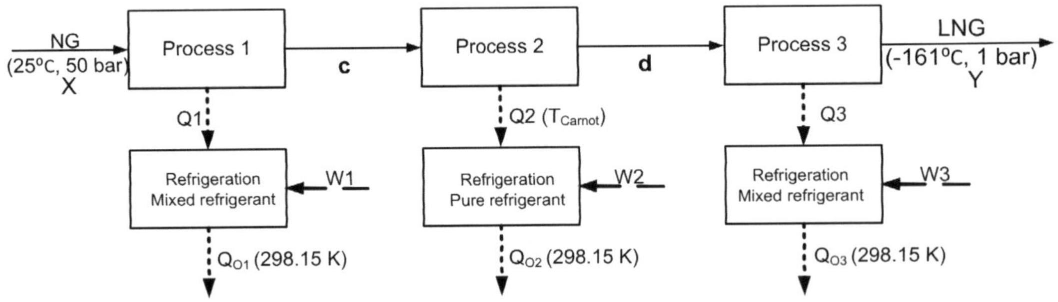

Figure 5: The LNG process split into three different processes, each with its own refrigeration system

Operations in Process 1 might involve expanding and cooling NG to move it from X to c. However cooling in this region must be done over a range of temperatures; this could be a source of irreversibility. In light of the diagram in Figure 1 it would be preferable to use a refrigeration system with a mixed refrigerant in order to reduce irreversibilities. Process 2 would operate in the two-phase region (*c* – *d*) where heat is removed at a single temperature. Thus in order for the process to be reversible, it would be preferable to use a refrigeration system with a pure refrigerant, since both the refrigerant and the process gas would change phase at a single temperature. Process 3 will also require cooling over a range of temperatures to move from *d* to *Y*. Thus in a similar way as in process 1, a mixed refrigerant system would be preferable in order to reduce irreversibilities.

Note that the temperature of any point in the two-phase region on the eh-diagram corresponds with its Carnot temperature; this can be seen in Figure 6: line $e - Y$ is the vapor-liquid line of NG at 1 bar and 111 K, this line is parallel to line $O - p$ indicating that each point of line $e - Y$ has a Carnot temperature of 111 K. Similarly the vapor-liquid line at 10 bar and 150 K, line $c - d$, is parallel to line $O - q$ indicating that each point on line $c - d$ has a Carnot temperature of 150 K. This means that it is only in this region where work can be transferred reversibly to the process by removing heat at a single temperature and in one step. In other words, the two-phase region is the only region in the envelope where cooling can be done reversibly; thus a pure refrigerant system must be used in order to transfer heat at a single temperature. The other regions (liquid, vapor, and supercritical), will require cooling to be done over a range of temperatures, and therefore mixed refrigerants that approximate the cooling curves in these regions must be used in order to reduce irreversibilities.

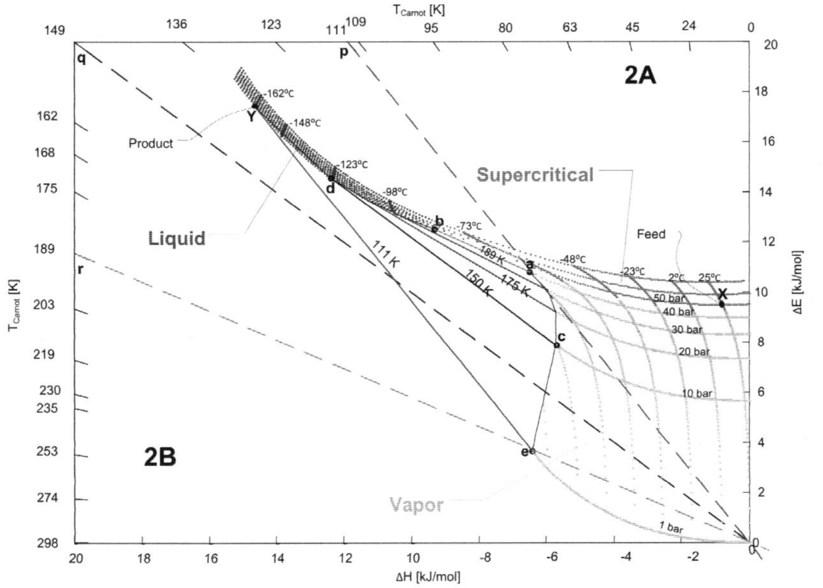

Figure 6: Natural gas on the eh-diagram: The temperatures in the two-phase region correspond to the Carnot temperatures

Furthermore minimum work requirement for refrigeration systems is achieved when the cycle operates in the two phase region at all times. Consequently the refrigerant must be partially evaporated before compression and partially liquefied before expansion. Irreversibilities occur when liquefaction and evaporation are not conducted at a single temperature.

Figure 7 (A) is a schematic representation of existing refrigeration cycles. The refrigerant in vapor phase is compressed to a pressure P_h and reaches a temperature T_{h1}; the gas is then liquefied by cooling it to a temperature T_{h2}. The liquefied refrigerant is then expanded, through a valve, to a pressure P_C where it evaporates partially. The mixed phase is then evaporated completely (usually at constant temperature) reaching a temperature T_C and pressure P_C before it is recompressed. The major source of inefficiency in these processes is in the liquefaction part since it occurs over a range of temperatures. We can show that the resulting temperature (T_{hot}) at which heat (Q_{out}) is removed during liquefaction is given by the log mean temperature between T_{h1} and T_{h2} (Equation (2)).

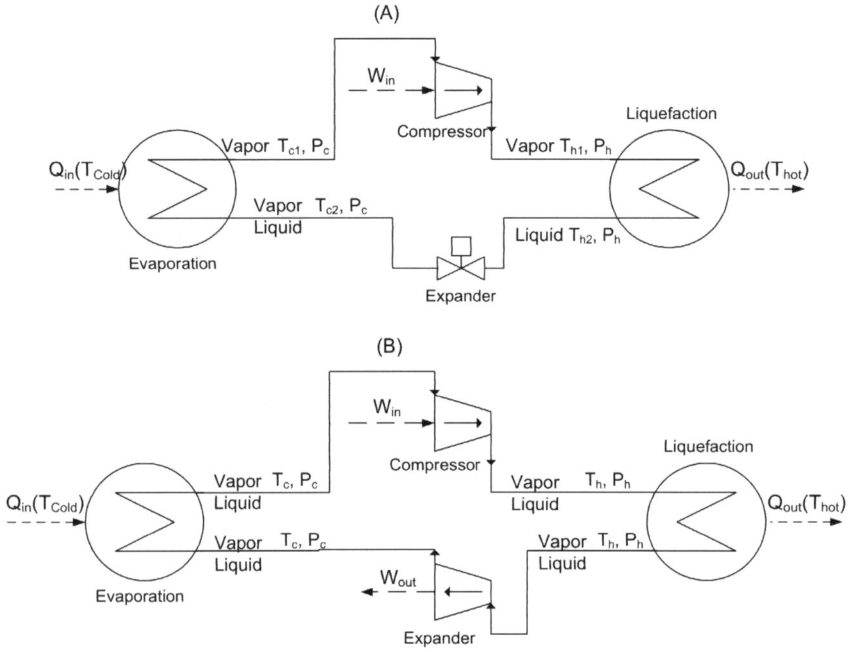

Figure 7: Schematic representation of refrigeration cycles. (A) In conventional refrigeration cycle the refrigerant is compressed in the vapor region consequently liquefaction and evaporation are conducted at multiple temperatures. (B) Compression and expansion are conducted in the two phase region consequently liquefaction and evaporation are conducted a single temperature and thus minimum work requirement is achieved

$$T_{hot} = \frac{\left(T_{h1} - T_{h2}\right)}{\ln \frac{T_{h1}}{T_{h2}}} \tag{2}$$

In most cases irreversibility arises because T_{hot} does not match the Carnot temperature (T_{Carnot}) of the liquefaction unit. T_{Carnot} has been previously defined as the temperature at which heat must be transferred for the process to be reversible and can be obtained by rearranging Equation (1) as follows:

$$T_{Carnot} = \frac{T_O}{1 - \frac{\Delta E}{\Delta H}} \tag{3}$$

Where T_O is the ambient temperature and ΔE and ΔH are the change in exergy and change in enthalpy across the unit respectively.

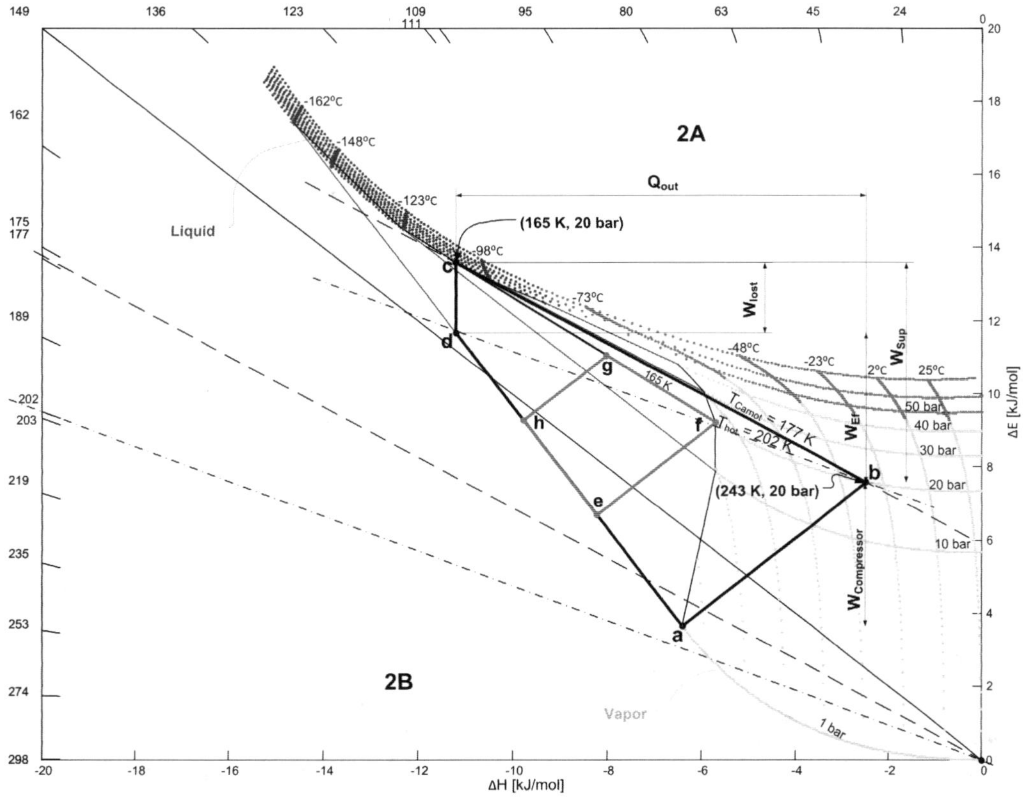

Figure 8: Refrigeration cycle on the eh-diagram. For path a-b-c-d, work is lost when cooling is done at multiple temperatures between b and c T_{Carnot} does not correspond to the actual temperature T_{QH} at which heat is removed, therefore more work (W_{Sup}) than necessary (W_{Ef}) is supplied to the process, and the process becomes irreversible. Every point in path e-f-g-h corresponds with the Carnot temperature; therefore no work is lost and the process is reversible.

As an illustration consider a refrigeration cycle represented on the eh-diagram in Figure 8 where the refrigerant is methane. The refrigeration path, a-b-c-d corresponds to the path on the diagram in Figure 7 (A). The gas is compressed isentropically from 1 bar (a) to 20 bar (b) reaching a temperature $T_{h1} = 243$ K. The gas is then cooled at constant pressure and liquefies at a temperature $T_{h2} = 165$ K (c). The Carnot temperature between b and c is $T_{Carnot} = 177$ K and the log mean temperature is $T_{hot} = 202$ K. We therefore see that the work supplied to move directly from b to c is $W_{Sup} = 6.01$ kJ/mol which correspond to heat ($Q_{out} = -8.75$ kJ/mol) being removed at 177 K. However in the actual process heat (Q_{out}) is being removed at 202 K and this corresponds to the effective work requirement of the process $W_{Ef} = 4.17$ kJ/mol. The work lost during liquefaction is $W_{lost} = 1.84$ kJ/mol. This work lost is the penalty for moving from b to c in one step. The work lost could be minimized if the gas is cooled progressively between b and f until it reaches the dew point, followed by liquefaction at constant temperature of 165 K.

On the other hand the refrigeration path e-f-g-h (Figure 8) is in the two phase region and corresponds to the path on the diagram in Figure 7(B). We can see that liquefaction and evaporation occur at constant temperatures 165 K and 111 K, and these temperatures correspond to their respective Carnot temperatures, therefore there is no work lost in the path. However to ensure that minimum work requirement is achieved one must use an expander that allows recovering the work of expansion which would otherwise be lost if a simple valve is used as in Figure 7(A) and thereby increasing the process irreversibility.

The major problem that could arise in implementing refrigeration in the two phase region is that most existing compressors and expanders operate efficiently only with vapor, and could result in major work losses when used in the two phase region. However screw compressors and expanders have shown to be able to handle two phase fluids at a relatively high efficiency (75%) (Smith et al., 2001). With further studies these equipment have the potential to be adapted to refrigeration systems. More studies (Hundseid et al., 2008) are also being conducted on centrifugal compressors and expanders for their possible efficient use in handling two phase fluids.

Conclusion

We have introduced a graphical approach to analyze LNG processes. The use of the eh-diagram can allow determining the optimum work target for the process and how this can be achieved as well as determine the optimum temperatures at which heat must be removed from the process. Furthermore the approach can be useful in scanning different routes from NG to LNG and help determine the appropriate refrigerant for each step of the process in a more systematic way. The eh-diagram can be used in the analysis and design of refrigeration cycles and can be used to match refrigerants in cascade refrigeration for optimum operation. We have also shown, by using the eh-diagram, that one of the possible ways for reaching minimum work requirement in refrigeration cycle is to operate in the two phase region at every point in the cycle. While most existing compressors and expanders cannot handle two phase fluids, numerous studies are being conducted in this regard and we see that the outcome of these studies could potential lead to much improved refrigeration systems.

References

Al-Sobhi, S.A., Alfadala, H.E., El-Halwagi, M.M., 2009. Simulation and Energy Integration of a Liquefied Natural Gas (LNG) Plant. In: Proceedings of the 1st Annual Gas Processing Symposium. pp. 131–135.

Al-Sobhi, S., Alfadala, H., El-Halwagi, M., 2009. Simulation and energy integration of a Liquefied Natural Gas (LNG) plant. In: Proceedings of the 1st Annual Gas Processing Symposium.

Denbigh, K.G., 1956. The second-law efficiency of chemical processes. Chem. Eng. Sci. 6, 1–9.

Hundseid, Ø., Bakken, L.E., Grüner, T.G., Brenne, L., Bjørge, T., 2008. WET GAS PERFORMANCE OF A SINGLE STAGE CENTRIFUGAL COMPRESSOR. In: Proceedings of ASME Turbo Expo 2008. pp. 1–10.

Kidnay, A.J., Parrish, W.R., 2006. Fundamentals of natural gas processing. CRC Press Tayllor Fr. 418.

Remeljej, C.W., Hoadley, a. F. a, 2006. An exergy analysis of small-scale liquefied natural gas (LNG) liquefaction processes. Energy 31, 1669–1683.

Sempuga, B.C., Hausberger, B., Patel, B., Hildebrandt, D., Glasser, D., 2010. Classification of chemical processes: A graphical approach to process synthesis to improve reactive process work efficiency. Ind. Eng. Chem. Res. 49.

Smith, I., Stosic, N., Kovacevic, A., 2001. Power recovery from low cost two-phase expanders. Trans. ….

479496 A Proposal for Improving the Efficiency, Effectiveness, Quality, and Usefulness of PHA Identification of Controls

Abstract:

Process Hazard Analysis (PHA) is an important discipline for identification and understanding of hazard causes, consequences, consequence severity, controls / safeguards, end consequence likelihood, risk assessment, recommendations for reducing risk, and prioritization of recommendation resolution.

Controls / safeguard discussion and documentation is fraught with potential problems. If controls identified by members of the PHA team are ***not*** adequate and are ***not*** included in the list, team members may reduce their participation as they become increasingly discouraged and disengaged. If controls that are ***not*** adequate ***are*** included in the list, the sheer number of listed, inadequate controls may result in: a false sense of sufficiency of the listed controls; a risk likelihood and risk rank that are less than actual; and reduced recommendation resolution priority, if a recommendation is made at all.

Accordingly, consistency, discipline, and rigor of controls discussion and documentation is recommended for useful, quality PHAs, whatever methodology is employed. Unfortunately, the time required for useful, quality PHAs employing consistency, discipline, and rigor can also result in tedium, reduced participation of the members, and increasing discouragement and disengagement anyway.

This paper outlines a proposal in which leaders of PHA studies list and annotate controls suggested by members of the PHA Team to be reviewed in the study for quick, instructive, and collaborative disposition. The controls listed for each scenario can be sorted and discussed by the annotations which may be modified in the finalizing discussion.

Using such an approach would encourage participation by all members of the PHA Team and reduce occasion for discouragement and disengagement. Application of the approach supports: quicker, improved evaluation and documentation of the effectiveness of controls; improved quickness and quality of evaluating likelihood and risk rank; better evaluation of the need for recommendations; and proper prioritization of recommendation resolution.

Effective Human Factors Tools to reduce Plant Costs and the Number and Severity of Injuries

Dennis Attwood, P.Eng., Ph.D.
Senior Consultant, Midland Engineering Limited
Midland, MI
dennis@hf-apps.com

Prepared for Presentation at
American Institute of Chemical Engineers
2017 Spring Meeting and 13th Global Congress on Process Safety
San Antonio, TX
March 26 – 29, 2017

Effective Human Factors Tools to reduce Plant Costs and the Number and Severity of Injuries

Dennis Attwood, P.Eng., Ph.D.
Senior Consultant, Midland Engineering Limited
Midland, MI
dennis@hf-apps.com

Human Factors, ergonomics, valve design, risk analysis, HAZOPs, PHA, Critical tasks, procedures

Abstract

Human Factors is defined as the "systematic process of designing for human use". This process systematically applies our knowledge of the capabilities, limitations, expectations and needs of people to the equipment they use, the environments they operate in, the tasks they perform and the management systems that guide the safe operation of their plants.

This paper describes easy-to-use and cost-effective Human Factors tools for the solution of major issues in process plants. Considering the breadth of the field of human factors and the limited time available for presentation, the paper limits discussion to the following four applications:

- Assessing the physical risk of operating manual valves to reduce maintenance costs.
- Improving the design of operations procedures.
- A method for applying Human Factors to Process Hazard Analyses.
- Critical task identification and analysis.

Each tool is thoroughly explained and its use is demonstrated with a practical example from the process industries with a summary of Benefits. References for papers explaining each tool in more detail are provided.

Introduction

The Human Factors discipline is continually evolving. Many in the field are conducting basic research to determine how humans respond to musculoskeletal, physiological and cognitive stressors. Others are developing tools to analyze plant situations and recommend solutions. This paper focuses on some of the many methods developed for analyzing Human Factors in Process Plants. The methods selected herein, can be used to solve a problem that has been identified, for example, to rank the musculoskeletal risk of manual valve operation. They can be used to determine if a Human Factors problem exists, for example, whether procedures are designed to make users make mistakes. Or they can be used to analyze systems during process design to identify opportunities for human error and to provide a proactive solution, for example, to identify critical operator tasks and analyze them to identify appropriate safeguards to mitigate risks.

The methods on which this paper will focus include:

1. Assessing the physical risk of operating manual valves to reduce maintenance costs. The tool ranks each valve in terms of musculoskeletal risk. Plant management can use this ranked list to lower the long term cost of ownership by scheduling maintenance on the highest risk valves first and reducing human injury
2. Improving the design of operations procedures. Statistics indicate that the majority of safety-related incidents in process plants are the results of not following operating procedures and/or not having complete and easy to follow procedures. This set of tools assesses procedures for inherent risk and provides guidance on redesign. Having improved procedures that are easy to follow and understand reduces the probability of human error/injury.
3. A method for applying Human Factors to Process Hazard Analyses. The paper reviews the development of the API Standard designed to include Human Factors in Process Hazard Analyses and provides an improved format for plant use. The benefit of Including Human Factors in the PHA will make sure the key Human Factor risks are addressed and mitigated
4. Critical task identification and analysis. Critical (high risk) process tasks have the potential for personal injury, equipment or facility loss, environmental release, community exposure or business interruption. The tool provides a systematic method of identifying high-risk tasks, analyzing the tasks for inherent risks and identifying appropriate safeguards to mitigate risks. The benefit is to reduce EHS incidents

Four Human Factors Application Tools

1. Assessing the Physical Risk of Operating Manual Valves to reduce Maintenance costs

Every process plant contains a large number of process valves with a wide diversity of designs that control the flow of feedstock, products and service liquids and gases. The ergonomic issues associated with the design, operation and maintenance of valves include:

- Physical stress to open and close them
- Potential for injuries and subsequent related costs
- Lack of access to process critical valves
- Difficulty to remove and replace them
- Potential for process upsets when valves cannot be operated in the time required

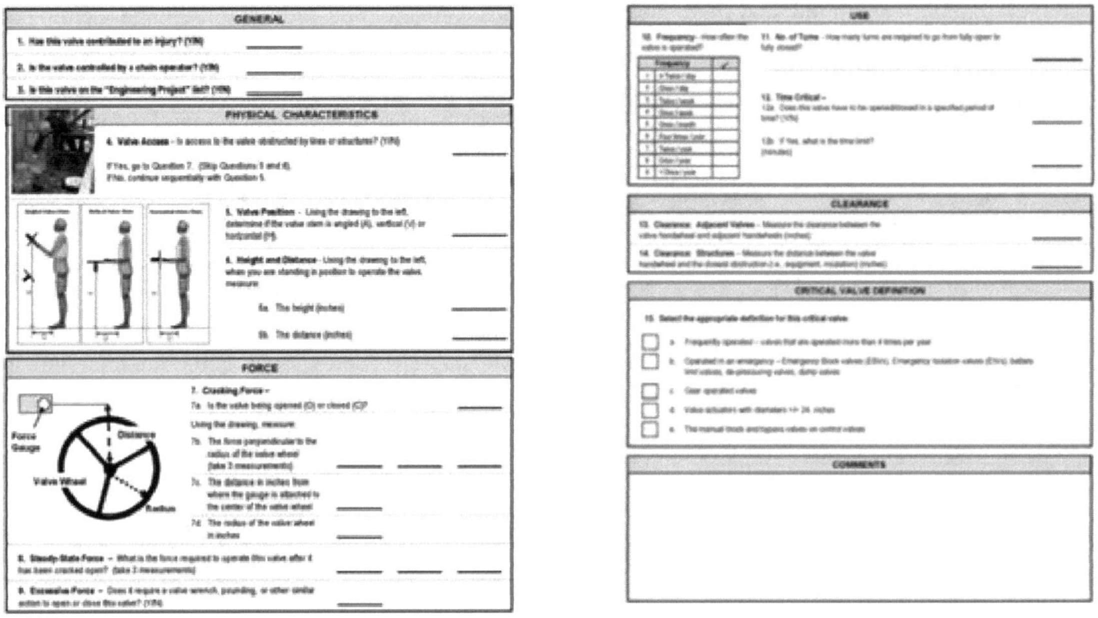

Figure 1: Valve risk assessment forms

As a result of injuries suffered while operating valves, a major U.S. petroleum refinery company initiated a program to identify high risk valves and to modify them to reduce the operating risk of injury.

We developed a systematic approach Attwood and Tys (2013) to identify and assess the operations and maintenance issues associated with process significant valves within the refinery. The approach consisted of eight distinct activities:

1. Develop plant standards for human factors requirements in the design, operation and maintenance of valves.
2. Train a group of refinery personnel, consisting of operators, maintenance techs, plant engineers and safety, health and environment staff, in ergonomic issues associated with valve operations and maintenance.
3. Lead the group in developing the definition of "process significant" valves.
4. Identify the valves that meet the definition in each unit of the refinery.
5. Assist refinery personnel to define the characteristics of valves that determine their risk of personnel injury or their potential for causing process upsets.
6. Develop a quantitative scale for each of the characteristics in (5) above.
7. Field assess each of the "process significant" valves against the characteristics.

8. Calculate the sum of the risk characteristics for each valve and develop a risk prioritization ranking list. This will permit the refinery to repair or replace valves, as business conditions permit, according to their position in the ranking list.

The characteristics that were identified to assess the risk of valve operation are shown in Figure 1. We have used this methodology in two different refineries with excellent results. The data were able to distinguish between high and low musculoskeletal risk valves. Figures 2A and 2B, for example, illustrate and describe both a HIGH and a LOW risk valve and recommends practical improvements for each.

Figure 2A: High risk valve

High Risk Valve

- Risk score = 79 (out of 112 possible)
- Has contributed to injuries
- Third choice location
- Cracking force > 246 lb.
- Steady state force >149 lb.
- Requires 42 turns to open
- It is recommended to increase the height of the actuator to waist height and to increase the size of the actuator to reduce the steady state rim force required

Figure 2B: Low risk valve

Low risk valve

- Risk score = 51 (out of 112 possible)
- Frequently operated
- Cracking force > 181 lb.
- Steady state force > 144 lb.
- Time critical operation (16 turns in 5 minutes)

It is recommended to adjust heights of actuators so pinch points are eliminated and to increase the diameter of the actuator wheels to reduce rim force required

Our valve risk ranking was reviewed by the operations team that provided the data and accepted as a viable risk-ranking technique that can be used to schedule valve maintenance so that the valves with the highest risk are maintained first. Moreover, our recommendations for improving each high risk valve were both practical and economical.

Practical Benefits of this Application

- Reduces the potential for musculoskeletal injuries and days off from work
- Reduces maintenance costs by targeting valves with the highest musculoskeletal risk

2. *Improving the design of operations procedures*

The term "procedure" refers to a written description of the steps that a process worker needs to follow to complete a task. The purpose of a procedure is to guide a process worker, step-by-step, through a task. Not every task requires a procedure (Haas, 1999). If the task has few steps, if it is performed frequently and if the consequence of error is minor, it will not likely require a procedure. However, if the task is complex, if it is performed infrequently and if the consequence of making an error is high, a procedure will likely be required. Failure to follow required procedures or failure to follow them correctly accounts for as much as 70% of the error-related incidents made by process workers (HSE, 2005). Incidents include both injuries and errors that affect process safety.

Evaluation Techniques

The effectiveness of a procedure program can be evaluated in a number of different ways. One of the most effective is a site audit to record any gaps between the existing on-site procedure program and the recommended guidelines. The audit might include:

- Observing multiple operators performing the procedure and comparing their performance against the documented procedure to identify deviations
- Evaluating the written procedure with a checklist. The checklist published in Attwood et al, 2003, contains detailed questions that can be used to evaluate written procedures
- Computing a Readability Score which is an open-source index that includes several well-known indices of readability
- Implementing Simplified English. We are proposing to use a powerful new tool to evaluate and develop your procedures. Simply put, simplified English or simplified technical English (ASD, 2005) is a standard set of English words and vocabulary rules that:
    - Reduces ambiguity
    - Improves comprehension for people whose first language is not English
    - Makes translations cheaper and easier
    - Facilitates automatic checking to ensure that procedures and manuals are written according to the rules and vocabulary
    - Improves understanding among users whose first language is not English

We have evaluated the procedure systems in a number of companies and have concluded that the resulting procedures are easier to understand and use and that the risk of making errors is greatly reduced.

Figure 3 illustrates the some common errors that were identified from one company's procedures based on the evaluation checklist identified above.

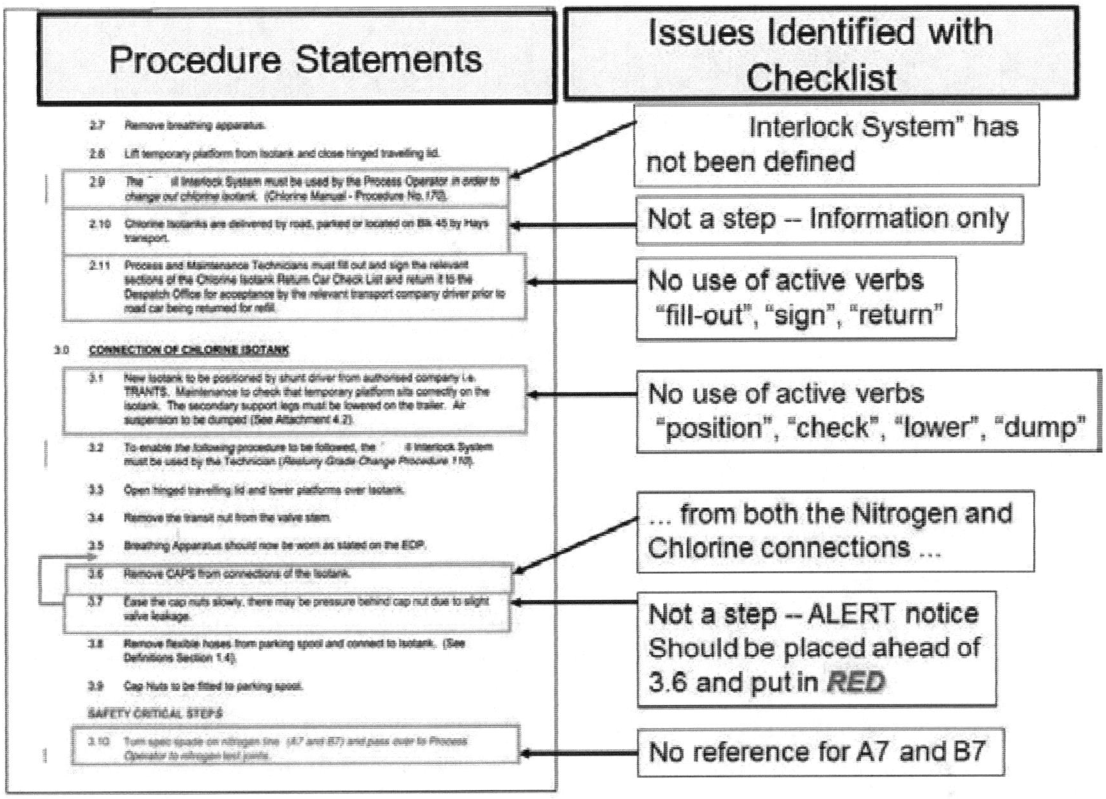

Figure 3: Procedure design issues

<u>Practical Benefits of this application:</u>

Our procedures improvement methodology ensures that clearly defined processes are in place to guide the development and implementation of effective procedures. We need to ensure that process considerations such as those below are considered for any procedures improvement program:

- Jobs that require procedures are identified
- Procedures reflect the way work is done and are technically accurate
- Procedures are written at the reading level of the users
- The detail in the procedure is not so complex that novice employees won't understand it, nor so limited that experienced employees will not read it
- Users can quickly locate the correct procedure for the job
- Users are instructed in the application of each procedure
- Procedures are reviewed and updated when the task, equipment or process changes

- Shared knowledge between operations, maintenance and engineering is routinely incorporated into procedures updates
- Users are advised when a procedure changes and instructed on the changes

3. *A method for applying Human Factors to Process Hazard Analyses*

The principle objective of the HAZOP is to identify process hazards, especially those that affect containment integrity (Attwood and Froehlich, 2003). Some of these hazards, however, are a result of human factors i.e. either with the process equipment itself or with instrumented process control. Many HAZOP protocols are less systematic than they should be for addressing HF concerns such as Human-Machine-Interface (**HMI**). They rely on the use of a checklist without recommending a process for applying it. Since, Human Factors is not the prime reason to perform a HAZOP review, the HAZOP leader will often elect not to review the checklist during the HAZOP proceedings especially during a HAZOP for major capital projects. More likely, the checklist will be reviewed at the end of the HAZOP study resulting in a less critical and systematic approach.

The objective of this tool is to assist the HAZOP team to identify Human Factors (HF) concerns associated with particular pieces of equipment while conducting the HAZOP.

The HAZOP is a sequential process -- a line by line and equipment by equipment review. It requires that each line and/or process system on the Process and Instrumentation Drawing (P&ID) be followed and checked and that the hazards associated with the equipment that is located at each node be identified and discussed. The HF issues need to be highlighted at these nodes as an **integral** part of the HAZOP discussion and not at the end of the HAZOP when the team has completed each P&ID.

The key to this tool is the use of "Screening Questions" for each of 12 pieces of equipment. The screening questions are a simple checklist that identifies the most important HF characteristics of the targeted equipment and /or system so the team can quickly decide whether a potential problem exists and what action to recommend. If a Human Factors concern is identified, it will be recorded on the HAZOP worksheet. It is possible that sufficient information and knowledge is available to resolve the issue during the HAZOP. However, the resolution of the issue will most likely happen outside the HAZOP.

The following equipment and HF issues are included in the screening lists:

1. Valves: sequential; critical; frequent
2. Blind and Blanks, weight, lifting; location; frequency; clearing, draining
3. Pumps and Compressors; maintaining, draining, start-up; shut-down
4. Field displays; critical; start-up; shut-down

5. Firefighting and deluge systems; critical; accessing; escaping; directing
6. Field instrumentation; rodding; clearing; critical
7. Sample points, exposure; PPE
8. Reactors/Driers, critical; switching/regenerating; clearing; entering; catalyst dumping; lifting
9. Vessels (including exchangers); cleaning; exposure; PPE; lifting; entering

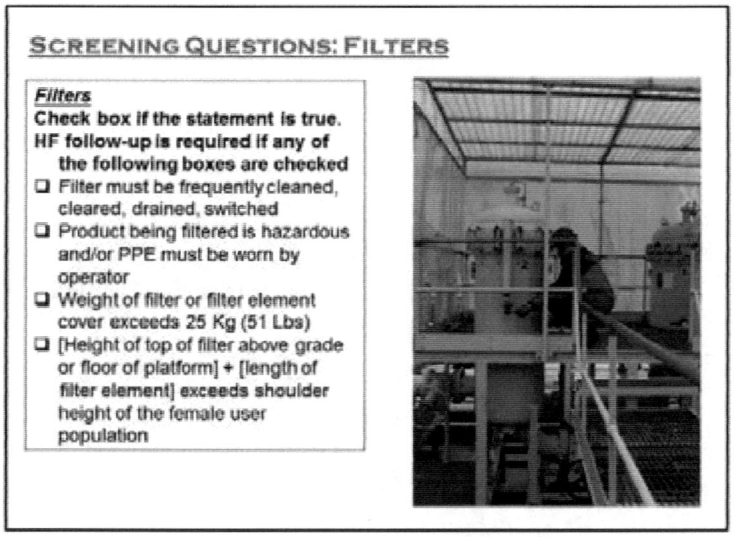

Figure 4: Screening Questions - Filters

10. Furnaces and Fired Heaters; light-off; decoking, flame controlling; switching; PPE; entering
11. Filters; switching; cleaning; clearing; PPE; weight; lifting; entering
12. Loading and Unloading Facilities; height; space; exposure; weight; clearing; disconnecting

Screening questions -- Example Format

An example of a "screening list" for Filters is given in Figure 4. This tool was prepared for the API and was published in 2005 (API, 2005). Our company has revised the original tool to make it easier to use. We have also developed a short introductory awareness program to show HAZOP participants how the tool will work in practice

Practical Benefits of this Application

The major benefit of this application is to ensure that Human Factors Issues are identified and discussed as they are encountered during the HAZOP reducing the probability of an incident.

4. Critical task identification and analysis.

Critical Task Identification Analysis (CTIA) is a systematic method of identifying critical tasks within a process facility, prioritizing their importance, analyzing those tasks that are considered most critical and identifying appropriate interventions to mitigate the risk. Critical in this context means high risk. The tasks that may have the potential for personal injury, equipment or facility loss, environmental release, community exposure, or business interruption are defined as "critical". These tasks represent a hazard and the goal is to identify the hazards and correct them before they turn into incidents.

The phases in the CTIA process are presented in Figure 5. The following paragraphs describe each phase, record examples of the products of each phase and provide insights on the most efficient methods to obtain information.

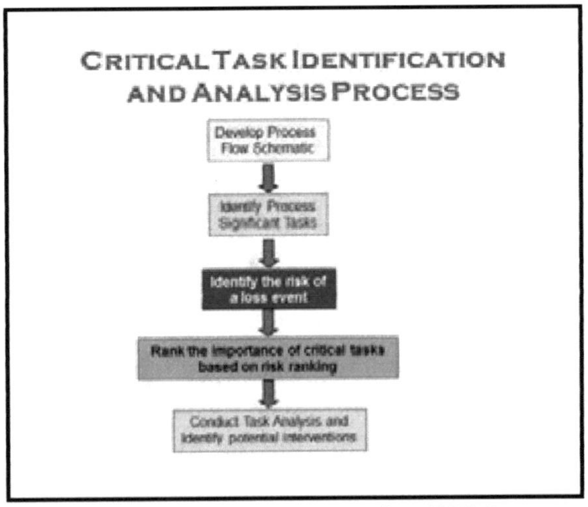

Figure 5: Phases of a CTIA

Develop Process Flow Diagram

The analysis starts by specifying the facility processes and equipment and creating a process flow diagram. A portion of an actual process plant flow diagram is shown in Figure 6.

Identify process significant tasks

The Critical Task Identification process is a 'brainstorming' technique that is designed to encourage the site team to identify the 'critical' tasks that they could be required to perform in each area of the process facility. A critical task could occur during a process upset or other abnormal situation or it could be a task that is difficult to perform or performed frequently. At each block of the PFS, the team is asked to identify the 'critical' tasks that he or she performs by placing them into one of seven different categories including:

1. Safety-critical tasks, e.g., responding to the failure of a pump flange
2. Quality-critical tasks, e.g., responding to an out-of-limits lab analysis
3. Production-critical tasks, e.g., responding to the loss of an exchanger which reduces output
4. Most frequently performed tasks, e.g., filter changes
5. Difficult to perform or complex tasks
6. Time critical tasks, e.g. responding to tasks that, in a short period of time, will cause a larger upset. An example is a compressor shutdown.
7. Environmentally-critical tasks, e.g., responding to an out-of-limit environmental deviation. For example, a high SO_2 reading on the stack of a fired heater.

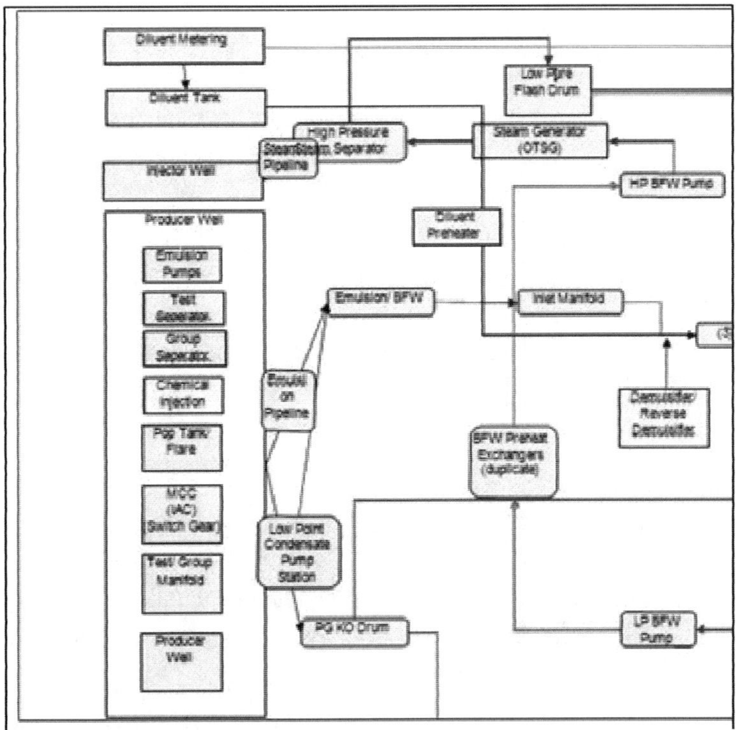

Figure 6: Partial Process Flow Diagram

In many cases the same 'critical' tasks can be categorized in different ways. The purpose of the exercise is to identify the major tasks (it is not important which list they appear in).

Identify the Risk of a Loss Event

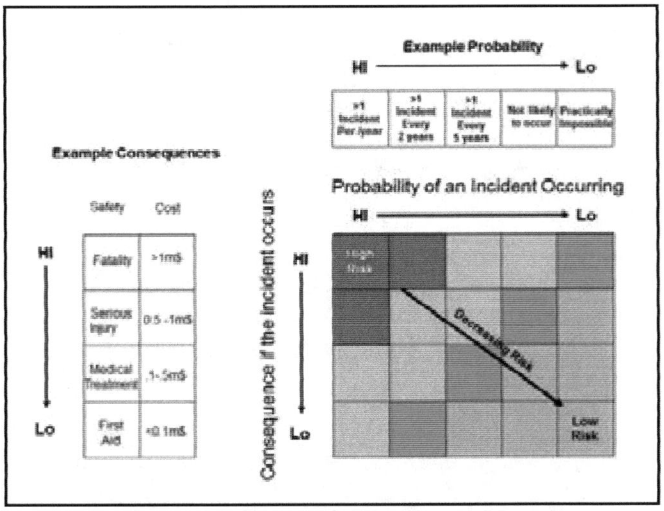

Figure 7: Simplified Risk Matrix

No company can afford to spend money and commit resources on issues that have little benefit to the operation. Thus, it is essential to identify the tasks which, when improved, will improve safety, reduce cost or improve the environment. The approach used at this stage is to estimate the risk associated with each task identified in the above exercise. Many companies have developed their own risk matrix. The next step in this exercise is to prioritize each 'critical' task in terms of its risk to the facility.

Figure 7 illustrates a generic risk matrix that is

designed to help workers prioritize tasks. It is presented for illustration purposes only and is not intended for use in formal risk assessments. Experience has shown that a site team can easily categorize tasks into the simplified consequence/probability risk matrix shown in Figure 7. Alternatively, the team can use the risk matrix developed for their company.

Identify the highest risk tasks

The tasks that most require analysis are those with the highest risk in Figure 7 (shown in RED). The analysis could also be extended to Medium risk tasks shown by the ORANGE sections and the lower risk tasks shown in the GREEN section of Figure 7

Conduct Task Analysis and Identify Potential Interventions

Finally, a task analysis is conducted on the highest risk tasks. Task analyses can take many forms and are explained in several texts and papers. The one that we use is detailed in CCPS (2010).

Table 1 provides, as an example, the activity analysis of a Feed Water Knockout Drum Sampling Task. Column 1 lists the sequential activities performed by the worker, column 2 lists the Human Factors Concerns for each human activity, column 3 lists the potential errors that can occur for each activity, and column 4 lists the potential solutions that can be used as task design recommendations.

Table 1: Activity Analysis for Feed Water Knock-out Drum Sampling

Activity	Ergonomic / Human Factors Concern	Error Possibility and Consequences	Potential Solution / Mitigation
Check sample drain tank for level by looking at overflow at sample position	Inability to check level	Tank overflows	Add level indication on sample drain tank with level controller in loop for tank pump
Approach sample station	Lighting Access	Trips/falls while carrying plastic bottles Poor lighting	Provide adequate lighting Bring sample station to ground Ensure proper head clearance
Ensure ventilation is working	Toxicity Air quality Ventilation area	Injury due to exposure (both inside and by ventilation)	Ensure ventilation is installed Provide safe ventilation area outside

Remove cap from sample point on common header	No human factors concern	Poor quality samples	Design issue
Displace and purge line	Force Fatigue Proximity	Arm strain Body stress	Ensure sample location and valve are located within visual proximity Ensure valve spring tension is low enough for human use Ensure handle is designed for constant pressure
Line up bottle with spout and take sample	Lighting Access Location of sample valves	Poor lighting Take wrong sample	Ensure labeling on valve easily corresponds to bottle (redundant coding with labels/colors) Mount lights outside of
Put bottle into carrier	Bending Lifting	Body stress Dropping samples	Provide space on sampling platform to place carrier
Select next bottle and repeat sampling process	Bending Lifting	Body stress Dropping samples	Provide space on sampling platform to place carrier
Close valve, recap	No new Human Factors	No new hazardous	No recommendations
Take samples to lab	No new Human Factors	No new hazardous	No recommendations

Practical Benefits of this Application

Identifies and analyzes tasks performed by humans that can impact on the overall integrity of the unit, the operating efficiency and reliability of the processes and can influence the safety of humans interacting with the process or facility

Task analysis techniques can be applied to existing tasks or tasks that do not yet exist. The task analysis team is made up of all stakeholders in the plant including engineering, safety and environmental personnel.

Summary and Conclusions

The effectiveness of a Human Factors program in a process plant is enhanced by the use of applications that can identify and solve human-related issues quickly and increase plant profits or reduce expenses or serve to reduce the number and severity of injuries.

The four applications described in this paper have proven, through previous use, to reduce maintenance costs, reduce the time required to launch projects, and make operations more efficient with lower incident rates. Readers who are interested in learning more about Human Factors analysis techniques are invited to read CCPS (2010) and Attwood et al. (2005).

References

1. ASD Simplified English Specification ASD-STE100, Aerospace and Defence Industries Association of Europe, 2005.

2. ATTWOOD, D.A., DEEB, J.M., and DANZ-REECE, M.D. "Ergonomic Solutions for the Process Industries". Gulf Professional Publishing (Elsevier), Burlington, MA, December 2005. ISBN 0-7506-7704-X

3. ATTWOOD, D.A. and TYS, A.T. (2013) "A method for ranking the musculoskeletal risk of operating valves in process industries." Proceedings of the Society of Petroleum Engineers, SPE-165690-MS, Pittsburgh, PA, August 2013

4. ATTWOOD, D.A., and FROEHLICH, B., "A Tool to Identify Human Factors Concerns during the HAZOP Process." ASSE Proceedings, Denver, June 2003

5. ATTWOOD, D.A. "Critical Task Analysis" Report prepared for the Surmont II project, Conoco Phillips Canada, Project No. 06-267, November 2007.

6. [1]CCPS "A Practical Approach to Hazard Identification for Operations and Maintenance Workers." Center for Chemical Process Safety Publication, John Wiley and Sons Inc., (Publishers), New York, 2010. ISBN 978-0-470-63524-7.

7. [2]CCPS "Human Factors Methods for Improving Performance in the Process Industries." Center for Chemical Process Safety (CCPS) publication. John Wiley and Sons Inc. (publishers), New York, 2007. ISBN 13 978-0-470-11754-5.

8. Haas, P. M. (1999). "Human Performance Engineering: A Practical Approach to Application of Human Factors." 1999 National Petrochemical Refiners Association (NRPA) National Safety Conference, Paper no. NSC-99-108.

9. HSE (2005) "HSE Human Factors Briefing Note Number 4: Procedures." United Kingdom, Health and Safety Executive.

[1] D. Attwood, contracted co-author
[2] D. Attwood, contracted co-author

479344 Selection of Cases for Atmospheric Dispersion Modeling

Abstract:

In many industrial facilities, toxic and/or flammable substances are discharged directly to the atmosphere. To ensure compliance with health, safety and environmental regulations, it is important to determine the potential consequences to the public and to personnel working in the facilities. Atmospheric dispersion modeling is a useful tool to provide plant owners the information to make decision in such cases. However, there are numerous potential releases from multiple scenarios, and hence it can be impractical to perform modelling for all cases. Typical industrial standards do not provide sufficient guidance for users about how to effectively choose the representative cases to model. This paper offers the guidance on selecting the relevant cases to perform atmospheric dispersion modeling based on the configuration of relief devices, the nature of discharged fluid, and points of interest in facility.

479804 Energy Savings & Distillation

Abstract:

Distillation is a significant energy consumer in process industries.

Tremendous energy savings have been achieved in recent times thanks to technologies such as heat integration and thermanlly coupled dividing wall distillation columns.

This paper examines the various energy saving schemes for distillation and absorption towers, combinations of other separation technologies with distillation, and the current and future trends in energy efficient distillation.

479812 Acid Gas Absorption Fundamentals

Abstract:

Absorption is a standard unit operation that has been studied by chemical engineering students for decades. One of its primary uses is for the removal of acid gases, like CO_2 and H_2S, from natural gas streams. Historically, aqueous amines and hot carbonate solutions have been used to perform these absorptions. Many different amines, solution strengths, activators and passivators have been tested and employed industrially. Approximately half of the industrial towers are trayed; half, packed. Foaming adversely affects the hydraulic performance in these systems.

Acid gas absorption is receiving appreciable new attention as concerns rise regarding CO_2 in the atmosphere. In addition, the United States has recently experienced significant growth opportunities in shale gas production which require improvements in gas treating technologies. This presentation reviews acid gas absorption and regeneration fundamentals. Several PFD's will be shown to illustrate important concepts. New absorbent solutions will be discussed, including ionic liquids. Flue gas applications will be discussed, including the difficulties associated with column pressure drops and very large column diameters. Rate-based modeling approaches will also be reviewed.

479816 Process Investigations Using Radioisotope Techniques

Abstract:

In all chemical plants and refineries there generally is not enough instrumentation in the right places in our processes. During the design phase, decisions were made between knowing more about the process, which means buying more instruments, and saving capital funds. When a problem appears with the operation of a vessel, there are always gaps in the knowledge of what is happening due to a lack of instruments in the places where knowledge is not routinely required. Process engineers and operations personnel need to know what is happening so they can take corrective action. When knowledge is missing, they often take actions that make matters worse.

For distillation and separation processes Chemical Engineers have historically relied upon plant process measurements such as flow rates, temperatures, and pressures and model or simulation results for data with which to solve troubleshooting or re-design problems. An array of on-line diagnostic services are available that can provide additional data offering real-time information on how pieces of process equipment are actually operating. The most common application is Gamma Scanning. This test is primarily applied, but not necessarily limited to, distillation or separation columns. Gamma scans provide a density profile of the internal process of operating distillation columns and other process vessels. The density profile can be used to diagnose the hydraulic operating conditions of mass transfer devices such as damage to internals, flooding, degree of entrainment or weeping, liquid levels on trays and distributors, liquid distribution through packed beds, etc. The presenter will show case studies where scanning revealed vital process information that helped solve an operating problem, or helped make a revamp/re-design successful. Additional derivatives from gamma scanning of detecting flow distribution patterns will be demonstrated. The ThruVision Scan technique is a specialized horizontal gamma scan used to generate a topographic profile of the internal cross-sectional density of process equipment. This profile is useful for the detailed study of liquid flow distribution through packed columns as well as having applications beyond distillation columns. A hybrid application involving tracers and scanning on fixed bed catalytic reactors will also be discussed.

Another technique involving radioactive materials is the Tracer Test, in which a radioactive tracer is introduced into the process for investigating the process flows. External radiation detectors are placed at strategically chosen locations outside vessels and piping to monitor the tracer flow through the process. Either liquid, gas or a solid tracer can be utilized depending on process conditions and the fluid to be studied. This tracer technique can identify liquid or gas maldistribution in fixed or fluidized bed reactors. The residence time through a tank or reactor can be measured and compared with the calculated residence time. Observed differences can relate directly to the degree of mixing or channeling inside the reactor. These tracer techniques have also been used in identifying leakage of heat exchangers and valves.

Paper 34a
(Proceedings 479206)

Vacuum Distillation:

Wash Zone Configuration

Andrew W. Sloley
Advisian (WorleyParsons Group)
Energy Center II
575 N Dairy Ashford St
Houston, TX 77079

Prepared for Presentation at the
American Institute of Chemical Engineers
2017 Spring Meeting
San Antonio, TX March 26-30, 2017

UNPUBLISHED

1. Summary

Refinery vacuum units include a deentrainment, or wash, zone for removal of heavy material from desired products. Wash zone reliability problems often shut vacuum units down, causing significant losses. Factors that strongly influence wash zone reliability include vapor and liquid distribution, upstream flow regimes, and wash zone efficiency and pressure drop. For fuels vacuum units, the wash zone should have the minimum amount of packing consistent with acceptable deentrainment. The minimum wash bed has lower capacity and higher reliability.

One major factor that must be evaluated on a case-by-case basis is phase separation in the transfer line. Even small amounts of phase separation will feed superheated vapor to the wash zone. Superheated vapor dramatically raises the amount of wash oil required for wetting the wash bed. Properly installed and operated, some vacuum towers have demonstrated trouble-free performance for 20+ years. Others, with improper understanding of the operation, suffer from one to two year run lengths, dramatically reducing refinery profits.

2. Introduction

This discussion has the intent of covering process considerations for wash zone operations. However, equipment constraints and process requirements interact. Process conditions must remain within equipment operating constraints for reliable and effective operation. Equipment constraints will be mentioned as required to explain the process limits and their importance. Previous work has focused on equipment constraints and includes more detail on them [1].

3. Operating Objectives

Refinery main fractionators include a wash section immediately above a two-phase feed. These towers have very high inlet velocities and high kinetic energy in the feed. The high inlet velocity and energy dissipation in the feed zone of the tower easily creates small liquid droplets. The droplets entrain into the rising vapor. Without a wash zone, the liquid droplets end up in the products above the feed.

The wash zone is a deentrainment device to remove the liquid droplets. The vacuum distillation tower works in a severe operating environment. Figure 1 illustrates a typical refinery vacuum tower and its wash zone. The wash zone is a packed bed above the tower feed.

The wash section has the following major objectives:

1. Deentrain liquid from the rising vapors
2. Maximize yields by maintaining low pressure drop and low wash liquid rate
3. Maximize reliability
4. Fractionate volatile contaminants (metal compounds)

These objectives compete with each other. Wash zones that meet some objectives tend to have problems with the other objectives.

4. Terminology

Figure 2 shows the relevant components of the wash bed, and the external and internal streams. The definitions of the streams for this discussion are:

Table 1 Wash Zone Streams

Stream	Definition
Wash (or wash oil)	Liquid placed on top of the wash bed to irrigate the bed to prevent coking.
Overflash	Distillate liquid leaving the bottom of the wash bed. This is material that was vaporized in the flash zone.
Entrainment	Liquid droplets moved up the tower by high vapor velocity. This is material that was not vaporized in the flash zone.
Slop wax [1]	The combination of entrainment and overflash that is collected on the tray below the wash bed. Slop wax = overflash + entrainment
Condensate	Liquid condensed on the bottom of the collector tray because the overflash is colder than the feed vapor. This stream is shown for completeness. In most units it has a very low flow rate.

If the entrainment rate is high, most of the liquid gathered by the slop wax collector is entrainment, not a recoverable distillate.

1 The term slop wax comes from lubricants vacuum units. The bottoms lubricant cut had a maximum allowed viscosity or viscosity index. The bottoms of the tower went to processing for bright stock (cylinder oil) recovery with a minimum viscosity specification. The slop wax was a drag stream between the two products that removed material with too high a viscosity for the lubricant draw but too low a viscosity for the bright stock.

5. Fractionating Versus Deentraining Wash Zones

Most refinery vacuum units are for fuels production. The vacuum tower products go to downstream processing for producing fuels. Many fewer vacuum units produce lubricant base-stocks. Lubricant production is only a small fraction of total vacuum unit capacity (<10%). This discussion focuses on fuels vacuum units. Nevertheless, many of the concepts apply to lubricants vacuum units.

Fuels vacuum units may have two purification objectives for the wash zone:

1. Remove mechanical entrainment of small liquid droplets from the vapor, or
2. Fractionate volatile materials to improve the lower product (HVGO) quality.

Mechanical entrainment puts vacuum residue in the distillate product. This can cause large problems with asphaltenes, metals, and carbon residue qualities of the HVGO.

The most common volatile contaminants of interest are vanadium and nickel porphyrins. High porphyrin levels are most common in heavy crudes (API gravity <= ~21.9°). Lighter crudes tend to have much lower porphyrin concentrations. Crudes well known for volatile metal compounds include Bachaquero, Leona, Isthmus, and Maya, as well as many others.

When the wash bed objective includes fractionation, fractionation requirements set the packed bed efficiency and liquid requirements. This efficiency and liquid rate tend to be much higher than the requirements for a wash bed intended to meet a deentrainment specification.

This discussion focuses on wash beds for deentraining services.

6. Conventional Configuration

Figure 3 shows a conventional configuration for a wash zone. Generic elements include:

1. Wash oil distributor to put liquid on the top of the wash zone packing.
2. Wash zone packing to act as a deentrainment device.
3. Collector under the wash zone to collect overflash and distribute vapor entering the wash zone.

Other elements in Figure 3 may vary between different conventional vacuum units but have little effect on the overall analysis of wash zone operation. In this case:

1. Overflash is recycled to the tower boot with an external pump. Other options include recycling the overflash upstream and allowing internal overflow of the overflash. Overflash disposition has a negligible effect on the current analysis.
2. The vacuum tower has no stripping trays. Stripping trays use steam to increase overall product yield. The presence of stripping trays has only a minor effect on the wash zone analysis here.

7. Conventional Operations

7.1. Ideal Wash Zone

Figure 4 compares an ideal flash zone to an achievable flash zone. Actual flash zone performance never achieves perfect deentrainment. Close-to-perfect deentrainment requires low vapor velocity to allow droplet settling and very few small liquid droplets (< 100 microns). Capital costs keep flash zone sizes small. Some entrainment always occurs.

A deentrainment or wash section above the feed removes the entrainment. To prevent coking, the deentrainment equipment requires net liquid flow at all points. The distillate liquid entering the wash zone is wash oil. The liquid leaving the wash zone is material that had to be flashed "over" the requirement to meet the yields observed. Hence the name overflash for distillate liquid leaving the wash zone.

With deentrainment as the primary objective, the ideal wash zone would have:

1. Minimum fractionation efficiency (zero if possible).
2. Minimum pressure drop (zero if possible).

No physical device can attain perfect performance. The comparison basis used here has been defined as:

1. One theoretical stage flash zone.
2. Minimum pressure drop (2 mmHg) across the slop wax collector tray.
3. Minimum pressure drop (1 mmHg) across the wash zone.

Various curves are presented to show how yield and unit capacity change with wash zone operation. Yield is quantified by cutpoint, the atmospheric equivalent temperature that would have the same distillate yield as the operating unit as on the ASTM D2892 [2] test. High cutpoints equal more distillate yield from the same feed. Capacity factors are relative unit feed rate. High capacity factors equal a higher feed rate to the vacuum unit.

7.2. Operating Limits

Minimum wash oil rates vary as a complex function of:

- Tower diameter
- Fractionation across the wash zone
- Mass flow rate
- Pressure drop across the wash zone
- Choice of mechanical equipment used

The objective of the wash oil is to create a targeted overflash rate. The liquid rate leaving the bottom of the bed strongly correlates with coking rates and reliability. Due to energy balance and fractionation effects, the wash oil rate will always be more than the overflash rate. The wash bed dries out from the top to the bottom [3]. The wash bed has the highest liquid rate at the top of the bed and the lowest liquid rate at the bottom of the bed. The dry-out ratio is the ratio of the liquid volume at the top of the bed to the liquid volume at the bottom of the bed:

$$dry-out\ ratio = \frac{liquid\ volumetric\ rate\ at\ the\ top\ of\ the\ wash\ bed}{liquid\ volumetric\ rate\ at\ the\ bottom\ of\ the\ wash\ bed}$$

The dry-out ratio is often quoted using standard liquid volume rates. However, here the actual flowing liquid volumes are used.

If the overflash rate is too high, yield drops as excess overflash increases the bottoms rate. If the overflash rate is too low, the wash bed can coke and the unit is forced to shut down. Both problems have a high economic cost.

This creates the fundamental operating challenge for vacuum tower wash zones:

1. The economic cost of using too high a wash rate is high.
2. The economic cost of using too low a wash rate is high.
3. Operating conditions are severe.

Overflash can be defined as a fraction of feed to the unit or a liquid load based on wash bed cross-sectional area. Older references often take the first approach, using a defined percent of unit feed. The most quoted range of overflash rates are based on Watkins [4, p. 110] and use 1 to 2 % overflash for conventional crudes and up to 4% overflash for high-metal content crudes. It is defined as:

$$overflash\ percent = 100\frac{overflash\ rate, std\ volume}{feed\ rate, std\ volume}$$

Many authors have used 2% as the starting point for unit optimization studies. This is consistent with the design curve included in Watkins.

The second approach uses overflash rates defined by actual flowing volumes and tower diameter. Various authors have suggested minimum overflash loads varying from 0.125 to 0.180 gpm/ft^2 of cross section area at the bottom of the wash bed (5.09 to 7.33 lpm/m^2) based on volume at conditions. The examples used here are based of 0.150 gpm/ft^2 (6.11 lpm/m^2) overflash load. The examples use a maximum vapor rate equivalent to a C-factor of 0.45 [1].

Figure 5 plots overflash percent versus overflash rate. The plot shows that a liquid load of 0.15 gpm/ft^2 (6.11 lpm/m^2) corresponds to 4.3% overflash for the example case used here. Thus, operation at the minimum recommended liquid load has 2.15 times the conventional recommended overflash percent. Clearly, the recommendations are highly inconsistent.

Operating pressure is the likely source of the difference. The worked example in Watkins uses 2% overflash and 42 mmHg flash zone pressure. Many modern vacuum towers run down to 20 mmHg (or lower) flash zone pressure. At low pressures, vacuum tower diameter increases. The larger vacuum tower needs a higher volume of overflash to get the minimum required rate at the bottom of the bed.

Extensive industrial experience has shown that targeting a minimum load (gpm/ft^2) is a fundamentally more sound approach to unit design and operation. Overflash percents should not be used.

Figure 6 and Figure 7 show operating changes versus overflash rate. As overflash rates increase, capacity decreases and yield drops.

7.3. Fractionation

The more fractionation in the wash bed,

1. The larger the composition gradient across the bed.
2. The larger the temperature gradient across the bed.
3. The larger the dry-out ratio from thermal effects.
4. The higher the revaporization rate and the larger the tower diameter for the same capacity or the lower the capacity for the same diameter.
5. The higher the wash oil rate for the same overflash rate.
6. The lower the HVGO draw temperature for the same yield (minor effect).

In general, an ideal wash bed would deentrain but have zero fractionation efficiency. Figure 8 illustrates changes in dry-out ratio, tower capacity, and relative wash oil rate with changes in the number of theoretical stages in the wash zone. Higher fractionation efficiency increases dry-out ratio, reduces unit capacity, and increases the needed wash oil rate. Figure 9 shows cutpoint (yield) versus wash zone stages. The more wash zone stages, the lower the cutpoint.

The largest change is moving from zero theoretical stages to one theoretical stage in the wash zone. While more stages do increase the wash oil rate for the same overflash, the increase is minor.

7.4. Pressure Drop

Pressure drop with no liquid contacting the rising vapor increases the amount of vapor superheat at the top of the bed. Higher vapor superheat increases the wash oil rate required. Some of the wash oil must vaporize to get the vapor down to its dew point.

Higher pressure drop has two boundary conditions:

1. Higher pressure drop with a fixed flash zone pressure causing the pressure at the top of the wash bed to be lower.
2. Higher pressure drop with a rising flash zone pressure and a fixed pressure at the top of the wash bed.

The first case, higher pressure drop with a fixed flash zone pressure, approximates a design case with a target yield defined by flash zone pressure and temperature. Figure 10 and Figure 11 show the dry-out ratios, relative capacity, and cutpoint for varying wash zone pressure drops for the first case.

The second case, higher pressure drop with a rising flash zone pressure, approximates an operating case limited by vacuum system capability. Figure 12 and Figure 13 show the dry-out ratios, relative capacity, and cutpoint for varying wash zone pressure drops for the second case.

The effect of wash zone pressure drop on dry-out ratio is small. The design case with a lower pressure at the top of the wash zone has a large impact on unit capacity (or size). The hydraulic limit in the wash zone is nearly always at the top of the bed. Therefore, reducing the pressure at the top of the bed creates a large capacity drop.

Figure 12 shows increasing capacity with pressure drop across the wash bed. The calculations include changes in transfer line operation with changing flash zone pressure. This capacity increase is due to changes in the transfer line. Section 7.8 includes more detail on transfer line effects.

7.5. Chimney Tray

Figure 1 and the other figures all show vacuum towers with a collector tray underneath the wash zone packing. Some vacuum towers do not have a collector tray. Without the collector tray, yields tend to be lower and deentrainment performance worse.

The collector tray:

1. Prevents the colder slop wax from quenching the flash zone vapors. This increases overall yield.
2. Reduces slop wax re-entrainment. This improves product quality.
3. Improves vapor distribution to the packing. This reduces local areas with high vapor rates and helps reduce localized coking rates. The result is more reliable operation.
4. Collects slop wax for possible metering and sampling. This gives the plant potential tools for improved optimization and increased yields.

The collector tray should be installed. However, this comes with more than an equipment cost. The collector tray creates pressure drop. In general, the higher the pressure drop, the better the vapor distribution above the tray. The collector tray pressure drop can be calculated using an orifice equation. These calculations are usually accurate for the HVGO and LVGO distributors above the wash section. However, for the slop wax collector, measured pressure drops are routinely twice the calculated value.

The higher measured value is caused by vapor flow maldistribution. Computational fluid dynamic analysis of tower entry and collector systems has shown complex flow patterns. These patterns often have severe local variations in velocity and direction [5]. If the chimney tray pressure drop is low and the feed entry poor, parts of the chimney tray may even have vapor flowing down instead of up through the chimneys. A better feed entry device and a higher collector tray pressure drop can dramatically improve vapor distribution to the bed above.

With a good feed entry device, the target pressure drop for the collector tray should be 1 mmHg. In service, the expected pressure drop is closer to 2 mmHg. The author has seen units with as much as 10 mmHg pressure drop across the collector tray.

Pressure drop across the collector tray has nearly identical consequences as extra pressure drop across the wash zone packing.

7.6. Loading (C-factor)

Vapor-to-liquid mass ratio affects the dry-out rate. Section 7.2 discusses the reduced dry-out ratios with higher liquid rates. This is because the higher liquid rate has a higher L/V ratio in the wash section. Less of the wash oil vaporizes to absorb the energy from the cooling vapor.

Lower vapor rates have the same results. Low vapor rates give low C-factors in the packing. Figure 14 and Figure 15 show the dry-out rate, capacity factor, and cutpoint for a constant 0.150 gpm/ft^2 (6.11 lpm/m^2) overflash rate at different C-factors. Lower C-factor reduces the dry-out ratio but also reduces the yield.

Figure 16 shows how the column area, wash oil rate, and overflash rate change as the C-factor changes for a constant feed rate. Higher C-factors are equivalent to higher mass flow rates. The change in mass flow rate is the underlying cause of the change in dry-out ratio. While the higher C-factor produces a higher dry-out ratio, it also requires a smaller area. The change in area is more important than the change in dry-out ratio.

This is also the reason why the higher operating pressure mentioned in Section 7.2 gives a lower overflash percent. The change in wash oil rate is the product of two factors:

1. Higher operating pressure increases the density and increases the max rate per unit area, increasing the dry-out ratio.
2. Higher operating pressure decreases the column cross section area, decreasing the irrigation area.

$$Wash\,Oil\,Volumetric\,Rate = Overflash\,Load \times Dry-out\,Ratio \times Cross\,Section\,Area$$

Dry-out ratio and cross section area change in opposite directions when pressure changes. The change in cross section area is more important than the change in dry-out ratio. Hence, as pressure increases, the total volume flow of wash oil drops.

7.7. Entrainment

Entrainment can be the source of liquid on the wash bed. In some units the overflash rate is approximately zero (< 0.02 gpm/ft^2, 0.81 lpm/m^2). In spite of this, multiple-year runs without coking the bed were achieved.

With no overflash, the slop wax is 100% entrainment. In these units, the slop wax rate was measured and was found to be very high (over 10% of the total residue rate from the bottoms). The crude was clean, light, and thermally stable.

Under these conditions, entrainment irrigated the bottom of the wash bed. The small amount of overflash used irrigated the top part of the bed. Entrainment reached far enough into the wash bed that it overlapped the region where overflash was present. The entire bed was irrigated and did not coke up. Figure 17 illustrates this operation. Intentionally operating in this mode is not recommended. The probability of coking the bed is high. Only small changes in either wash rate or entrainment rate can lead to coking or severely off-specification HVGO product.

7.8. Transfer Line Phase Separation

In the early 1990s the author and others reviewed several severe coking incidents where everything indicated that the wash bed should have been sufficiently irrigated. However, plant data was showing very little overflash. In these cases, the transfer line allowed for phase separation between the vapor and the liquid.

With phase separation, the depressurizing vapor becomes superheated. The superheat is not available to vaporize more liquid in the transfer line. The limited surface area between the vapor and liquid limits the amount of heat transfer. Figure 18 illustrates this operation.

The basic technique of accounting for phase separation in the transfer line has been referred to as the non-equilibrium model for transfer lines [6]. The author uses a derivative of the basic approach and prefers to call this the phase-slip model because phase-slip avoids confusion about the topic. The term equilibrium model is already in wide use in the discussion of equilibrium-stage distillation models versus rate-based distillation models.

As well as reducing yield, the superheated vapor entering the vacuum tower increases the dry-out ratio. The superheated vapor vaporizes some of the overflash as the vapor-liquid mixture reaches equilibrium. Figure 19 and Figure 20 shows the effect of phase slip.

Phase slip index (φ) has been defined as:

$$\varphi = \textit{phase slip index} = 100\,\frac{\textit{pressure drop total} - \textit{pressure drop before phase separation}}{\textit{pressure drop total}}$$

Transfer line phase separation is one of the most difficult factors to evaluate when designing a unit. With data, the results of phase separation are clear. Getting the data, however, can be very challenging. The base configuration used for all the comparisons had a transfer line with a $\varphi=29$. Values of 10 to 30 are typical for many operating units.

Based on data, unit operations can cover the entire range. Some units have achieved φ of 0. Others have demonstrated yields and required wash oil rates consistent with φ of over 90. Phase separation is the single largest factor in uncertainty in understanding wash zone operation on many units.

Transfer line phase pressure drop and phase separation are highly dependent on transfer line geometry, overall pressure profile, and operating conditions. Nevertheless, factors that make transfer line separation more likely include:

1. Low heater outlet pressure.
2. Low vacuum tower pressure.
3. Large diameter transfer lines.
4. Reduced rate operation.

7.9. Residence Time

Time and temperature combine to create coke. Keeping the oil hot for a longer period of time increases the risk of coking the wash bed. Residence time is not directly included in ideal calculations, but should never be overlooked. Higher irrigation rates (higher wash oil rates) reduce the residence time in the wash bed.

7.10. Interactions

The author has covered major process issues in the wash zone that affect performance. Each has been covered independently. In actual units, the factors combine. Usually, higher efficiency beds:

1. Have higher pressure drop (use more, or more efficient, packing).
2. Have more liquid hold-up because of the higher pressure drop resulting in longer residence time.
3. Use smaller packing sizes and aggressive surface treatments (to increase efficiency) so they are more likely to trap coke in the packing.
4. Have more stages, so they are more subject to local dry out problems due to limits in vapor and liquid distribution.

Combined with transfer line phase separation and problems with feed characterization [7] and molecular weight prediction [8], dry-out ratios can be high. Field data have shown some units have dry-out ratios as high as 15:1.

8. Hot-Wash Configuration versus Cold-Wash Configuration

Hot wash units use the direct liquid drawn from the collector above the wash section as the wash oil. This is normally an HVGO stream. The hot wash is a bubble point liquid on the collector tray and remains a bubble point liquid when it is returned to the top of the wash packing.

Cold wash units use a liquid that comes from downstream in the HVGO cooling circuit as the wash oil. The cold was stream is a subcooled liquid when returned to the top of the wash packing.

Figure 21 and Figure 22 illustrate the hot wash and cold wash configurations.

As long as the liquid rate leaving the bottom of the wash bed is the same, the amount of liquid on the top of the bed is not particularly relevant to meeting process objectives. However, a number of practical concerns favor hot wash.

The higher liquid rate to the top of the bed in hot wash allows for larger openings in the liquid distributor. This makes the distributor less likely to plug or foul. Additionally, subcooled liquids sprayed into a rising vapor cause some of the vapor to condense. The condensing vapor can pull the liquid cones in and collapse part of the cone. Spray cone collapse can lead to portions of the top of the bed having less liquid than expected. Liquid rates in the wash packing are so low that even small deviations in liquid distribution can cause accelerated coking in the wash bed. Hot wash should be used instead of cold wash.

9. Wash Oil Distributors

Two types of liquid distributors are used in wash oil services:

1. Spray header distributors.
2. Trough distributors.

Mechanical and service details of these have been previously covered [1]. In addition to mechanical considerations, process performance of the wash oil distributors affects wash zone performance.

Data currently available do not clearly support one type over the other. Both can be used. Both have benefits and costs. Overall, the spray distributor is usually cheaper to purchase, and easier and cheaper to install. Unless a specific reason is present, the spray distributor is the recommended choice.

Table 2 summarizes the distributor characteristics. Major factors include:

1. Entrainment of distributor liquid before it reaches the packing.
2. Distribution quality and fractionation performance of the distributor.
3. Heat-transfer performance of the distributor.
4. Size of distribution holes.

Table 2 Distributor Characteristics

	Spray Distributor	Trough Distributor
Entrainment of wash liquid	Small droplets created at spray nozzles. Liquid splashing when entering packing.	Vapor channeling due to high vapor velocity when it passes between troughs can entrain liquid from the packing.
Distribution Quality	Low quality distribution reduces fractionation effectiveness (desired) but can lead to low liquid in some areas that might accelerate coking.	Smooth distribution of liquid increases fractionation effectiveness. This is undesired unless fractionation for metals control or viscosity control is required.
Heat-Transfer Performance	Sprays are excellent heat transfer devices. This is undesired.	No heat transfer above the packing. But the good liquid distribution increases heat transfer in the packing.
Size of Holes	Holes in spray nozzles slightly larger.	Very small holes for most services. This is a potential problem even with deflector plates used to increase hole sizes.
Other	Lower cost and relatively simple installation.	Higher cost. Installation can be labor intensive in large-diameter columns. More likely to be damaged by pressure surges.

10. Spray Chambers

In a column without internals, there is nothing for coke to attach to. Spray chambers use one to three rows of spray nozzles. The wash liquid contacts the rising vapor. The small rising droplets combine with larger droplets coming from the spray nozzle. Liquid drops agglomerate by direct impact and by Brownian motion. Delayed coker main fractionators routinely use spray chambers. The delayed coking service is an extremely severe coking service.

Figure 23 illustrates a vacuum tower with a spray chamber. When trays were still used in vacuum towers, having a spray chamber could significantly reduce flash zone pressure. Additionally, trays have a poor record of rapid coking and relatively high entrainment. Trayed towers often used wire-mesh pads [9]–[15]. The mesh pads then created their own problems with coking, high-pressure drop, and collapse. With annual turnarounds common in the 1950s, plants could tolerate the mesh pads coking. With current multiple-year runs, mesh pads have no place around the wash section of vacuum towers.

In spite of the problems with trays and mesh pads, they had better deentrainment performance than spray chambers. To get adequate deentrainment, spray chambers need to operate with an excellent inlet feed distributor and low C-factors. An upper limit on C-factor of 0.20 should be used with spray chambers. A more prudent limit would be a C-factor of 0.18 or below. A spray chamber is really a separator drum. For comparison, most separators without internals are designed for equivalent C-factors of 0.15 to 0.175.

11. Recycle Wash Systems

Some units have recycle wash systems to increase the amount of wash oil while reducing the overflash rate. These systems take some of the slop wax and add it back to the wash oil. Figure 24 shows a one-stage recycle wash system. Figure 25 shows a two-stage recycle wash system. The vacuum tower has two separate wash beds. The upper wash bed uses HVGO as the wash oil, the same as a conventional system. The lower bed uses 100% slop wax as the wash oil. The wash oil from the upper bed falls down into the lower bed. This creates a purge for the system.

The refiner must keep in mind that the recycle contains entrainment. Potential problems with this include:

1. Increased residence time at a high temperature for the recycled entrainment – possibly accelerating coking.
2. Fouling and plugging of the packing or distributors with solids (scale, clay, rust, iron sulfide) contained in the entrainment.
3. Impurities in the HVGO from entrainment of the wash oil return.

Recycle systems have few benefits. The situations where they work well include units running relatively clean and thermally stable crudes. With very few exceptions, the recycle wash systems analyzed by the author could have been replaced with better or lower cost options.

12. Case Study 1

A refiner converted a trayed vacuum tower into a packed vacuum tower. Figure 26 illustrates the history of modifications made. The trayed tower had three trays in the wash zone. The original configuration is shown in detail '0' in Figure 26. The tray spacing was:

Table 3 Vacuum Tower Tray Spacing

Tray	Elevation	Spacing	Spacing
3	31'-0"		
		5'-6"	1.676 m
2	25'-6"		
		5'-6"	1.676 m
1	20'-0"		
		7'-6"	2.286 m
Feed	11'-6"		

The large tray spacing between the second and first tray was intended to reduce the amount of entrainment. Deentrainment improvements with extra tray spacing dramatically decrease once tray spacing exceeds 4 ft (1.219 m). The extra spacing probably had minimal effect on tower performance.

12.1. Revamp 1: Replace Trays With Packing

The refiner was attracted by the benefits of lower pressure drop and higher capacity of structured packing. Trays 1-3 were removed and replaced with a combined packing-grid bed. Total bed depth was approximately 11' (3.35 m). The bottom 5' (1.52 m) was grid and the upper 6' (1.83 m) was structured packing. A spray header was installed for liquid distribution. Based on expected packing efficiency the bed was estimated to have four ideal stages of separation. The estimate of four theoretical stages assumed that the spray header would perform as a high-quality liquid distributor.

The installation was mechanically convenient. The lower elevation was the existing Tray 1 support. No slop wax collector was installed. Figure 26 shows these modifications in detail '1'.

The wash oil rate was set based upon a 2:1 dry-out ratio. This was a common basis used for older vacuum towers with trays. No data were available from the unit as overflash flowed internally into the bottoms.

Performance did not meet the expected goals. Entrainment became severe at a C-factor of 0.38 compared to the predicted capacity limit of 0.5. Additionally, after approximately 16 months of operation, bed pressure drop started to rise. Entrainment rapidly increased as the pressure drop increased. The unit was shut down after 24 months.

12.2. Revamp 2: Add a Slop Wax Collector and Slop Wax Recycle

The possible causes of the rapid coking and lower than expected capacity were a matter of contentious debate. No data were available on actual overflash rates. The decision was made to install a slop wax collector, increase the wash oil rate, and recycle the slop wax to the heater feed. Figure 26 detail '2' shows these modifications.

The reasoning included:

1. A higher wash oil rate would reduce the coking rate.

2. Recycling the slop wax would allow for recovery of distillate in the slop wax. Therefore, the higher overflash rate would not reduce yields.
3. The collector and pump would allow for sampling the slop wax. This would give the plant better data for optimization.
4. The chimney tray would reduce re-entrainment of liquid by separating the slop wax from the vapor in the flash zone.

That the chimney tray might improve wash zone performance by improving vapor distribution was not considered. Later analysis would show that the improved vapor distribution was important.

The modifications did improve the operation. Capacity rose to a C-factor of 0.42 at the same level of entrainment seen after the 1st set of modifications. While improved, this was still not the promised value of 0.5.

Samples revealed nearly zero overflash in the slop wax. Both an asphaltene balance and a sodium balance showed that the slop wax collected was over 95% entrainment. This was true with double the wash oil rate used in the previous run.

If the slop wax is nearly 100% residue, then it contains no distillate to recover by recycling it to the furnace feed. Operating events supported this analysis. The slop wax pump had an operating upset and the pump had to be removed from service. Unit yields did not change when the pump was out of service.

Based on matching a measured overflash in the slop wax, the wash bed was estimated to have a 15:1 dry-out ratio. This seemed very high. Referring to Figure 8, at three theoretical stages the dry-out ratio is less than 7:1 and the incremental change per stage is already very low. It was difficult to understand how 15:1 could occur. The most popular suggestion was that small errors in laboratory technique were causing the dry-out ratio to be over-estimated.

At the start, gas oil quality was very good. The gas oil had a 2 color and no significant asphaltene content. Quality stayed at this level until the pressure drop started to increase.

No significant improvement in run length was seen. After approximately 16 months, pressure drop started to rise again.

12.3. Revamp 3: Reduce Wash Bed Height

At this point the system was clearly not understood and the path forward was not clear. Several other units were suffering from similar wash zone problems at this time. Different strategies were being used at different plants to understand the problem.

Since product quality had been so good at the start of the run, the idea was that less packing was needed. The perceived benefit was that less packing would reduce residence time in the bed and reduce the coking rate.

Figure 26 detail '3' shows the modification made. Wash bed height was reduced from 11' (3.35 m) to 5' (1.8 m). The bed included roughly 40" (1 m) of grid and 20" (0.5 m) of structured packing.

This change resulted in a lower initial pressure drop (less packing), and HVGO quality still met downstream processing requirements. The quality was better than that seen after the first revamp, but not as good as after the second revamp. This was as expected.

Run length did increase. The unit ran for approximately 20 months before pressure drop started to rise. The unit reached 36 months before needing to be shut down for packing replacement. The extra run-length came from a combination of three factors:

1. A lower starting pressure drop (less packing depth).
2. A longer duration before pressure drop started to rise (20 months versus 16 months).

3. A slower rate of increase (due to lower coking rate from lower residence time).

Sampling of the slop wax showed a 7:1 dry-out ratio. This seemed high for the one theoretical stage expected from the reduced packing height.

Modeling with calculations forced to match the dry-out ratio showed that the maximum operation had a 0.45 C-factor, not a C-factor of 0.42. The higher dry-out ratio creates a higher calculated C-factor at the top of the bed.

Experience from other units supported the 0.45 C-factor limit for acceptable entrainment. The original 0.5 C-factor estimate for this crude was over-optimistic.

During this run, data from other units was showing that phase separation in the transfer line was becoming important. Due to the cost of modifying the transfer line, no modifications were proposed.

12.4. Revamp 4: Relocate Wash Oil Distributor

The packing still had insufficient wash oil. Calculations with an assumed phase slip of 50 (φ) matched a 7:1 dry-out ratio. Modifications were planned to increase the wash oil rate to the 7:1 dry out ratio. Spray header modifications were implemented in the 4th revamp.

During this work, it was noticed that the inside of the tower showed evidence of significant liquid spray on the vessel wall. The previous revamp (#3) had reduced the packing height, but not relocated the spray header. A quick piping modification allowed the spray header to be dropped to the correct elevation above the packing.

Plant data after these modifications showed the expected dry-out ratio to match the prediction. Gas oil quality was excellent and the full desired run length of 5 years was achieved.

12.5. Revamp 5: Update Feed Entry

To reach the maximum capacity, a new feed entry device was proposed for the fifth shutdown. The idea was that if less entrainment was carried into the wash bed, vapor rates could be pushed to a higher C-factor. Some entrainment would occur, but product quality would remain acceptable.

After installation of the new feed entry device, rates could be pushed as high as a C-factor of 0.47. However, minor operating changes would cause the tower to flood rapidly. The decision was made to limit feed rates to the 0.45 C-factor.

The new feed entry device improved the HVGO quality slightly and also reduced slop wax rates a small amount.

12.6. Overview of Case 1

Modifications unfolded over a number of years. At the start, the potential for wash zone coking was not well understood. The unplanned factors that contributed to wash zone coking were phase separation in the transfer line and high efficiency in the packed bed. The phase separation required higher wash oil rates than originally planned. The wash bed was reduced in height to reduce the number of theoretical stages. After modifications were made, run length met the original targets. Further developments that improved operations included adding an improved feed entry device.

13. Case Study 2

A refiner upgraded a vacuum tower wash zone from trays and wire-mesh mist eliminators to a packed bed using a 3rd generation random packing. The packing was 6' (1.83 m) of rings. The planned operation was to run with 0.15 gpm/ft^2 (6.11 lpm/m^2) of overflash. The slop wax was collected and recycled to the furnace feed. The overflash pumps always had reliability problems and rarely remained in service for a complete 4-year run of the vacuum tower.

The packing coked during every run. Plant data showed high pressure drops in the packed bed starting approximately 6-8 months after the run started. The bed was always coked when examined at the end of a run. The bed was replaced in-kind in subsequent runs. In spite of the coking, the wash oil rate was not increased.

Even though the bed coked, the rings had superior performance (lower pressure drop and higher capacity) than the wash trays and mist eliminator pads that were removed.

The current work was intended to upgrade the vacuum tower from the 3rd generation random packing to a mix of structured packing and grid. The wash bed was upgraded during modifications that added a diesel product to the vacuum tower. The new wash bed used ~4 ft (1.22 m) of a mix of grid and structured packing. A new collector tray, slop wax draw nozzle, and slop wax pumps were installed. The slop wax pumps were converted to use variable frequency drives to control flow rate instead of using a control valve. This improved pump reliability by reducing average suction specific speed of the pump [16]. Wash oil rates were re-set based on an approximate estimate of a φ of 50 and a targeted overflash rate of 0.15 gpm/ft^2 (6.11 lpm/m^2).

After modifications, laboratory testing of the overflash showed rates slightly above 0.15 gpm/ft^2 (6.11 lpm/m^2). Over a period of weeks the rate was optimized. Assuming a φ of 30 matched the plant data.

Operation has been consistent and no evidence of coking has been seen from the operating data.

14. Conclusions

Wash zones remain areas where refiners can make, or lose, large profits. Reliable operation allows for high capacity, yield, and quality. With sufficient wash oil, packed beds are very reliable in wash zone service. Factors influencing the right wash oil rate remain unknown to many operators. This review covered a partial list of the factors including:

- Efficiency
- Pressure drop
- Operating rate
- Transfer line phase separation
- Proper liquid and vapor distribution

Properly installed and operated some vacuum towers have demonstrated trouble-free performance for 20+ years. Others, with improper understanding of the operation, suffer from one to two year run lengths, dramatically reducing refinery profits.

15. References

[1] A. W. Sloley, "Refinery Vacuum Towers: State of the Art," presented at the AIChE National Meeting, San Francisco, CA, 2013.

[2] ASTM, "D2892-05, Standard Test Method for Distillation of Crude Petroleum (15-Theoretical Plate Column)," West Conshohocken, PA, 2005.

[3] A. W. Sloley, "Atmospheric Distillation Process: Fundamental Concepts," presented at the AIChE Spring Meeting, New Orleans, LA, 2014.

[4] R. N. Watkins, *Petroleum Refinery Distillation*, 2nd ed. Houston, TX: Gulf Publishing Company, 1969.

[5] K. Kolmetz, W. Kiong, K. Ng, Y. T. Senthil, A. W. Sloley, and T. M. Zygula, "Optimize process operation with new vacuum distillation methods," *Hyd Proc*, vol. 84, no. 5, pp. 77–85, May 2005.

[6] S. W. Golden, V. Shah, and J. P. Kovatch, "Improved Flowsheet Topology for Vacuum Distillation Simulations," presented at the Canadian Chemical Engineering Conference, Calgary, ALB, Canada, 1994.

[7] D. C. Villalanti, J. C. Raia, and J. B. Maynard, "High-Temperature Simulated Distillation Applications in Petroleum Characterization," in *Encyclopedia of Analytical Chemistry*, Chichester, UK: John Wiley & Sons, 2000, pp. 6726–6741.

[8] K. H. Altgelt and M. M. Boduszynski, *Composition and Analysis of Heavy Petroleum Fractions.* New Orleans, LA: Marcel Dekker, Inc., 1994.

[9] A. H. Orr, "Use of Wire-Mesh Sections in Asphalt Entrainment Control in Vacuum Pipe Stills, Part 1... Case History - Cosden Petroleum Corp.," *OGJ*, pp. 84–85, Nov. 1953.

[10] D. H. Stormon, "Use of Wire-Mesh Sections in Asphalt Entrainment Control in Vacuum Pipe Stills, Part 1... Case History - General Petroleum Corp., Torrance, Calif.," *OGJ*, pp. 87, 121–122, Nov. 1953.

[11] H. S. Fourer, "Use of Wire-Mesh Sections in Asphalt Entrainment Control in Vacuum Pipe Stills, Part 1... Case History - Sincalir Refining Co., Marcus Hook, Pa.," *OGJ*, pp. 86–87, Nov. 1953.

[12] J. S. Corlew, "Use of Wire-Mesh Sections in Asphalt Entrainment Control in Vacuum Pipe Stills, Part 1... Case History - Sinclair Refining Co., Sinclair, Wyo.," *OGJ*, pp. 82–84, Nov. 1953.

[13] J. W. Russell, "Use of Wire-Mesh Sections in Asphalt Entrainment Control in Vacuum Pipe Stills, Part 2... Case History - Ohio Oil Co., Robinson, Ill.," *OGJ*, pp. 95–96, Nov. 1953.

[14] C. O. Anderson, "Use of Wire-Mesh Sections in Asphalt Entrainment Control in Vacuum Pipe Stills, Part 2... Case History -Rock Island Refining Co. Indianapolis," *OGJ*, p. 99, Nov. 1953.

[15] J. H. Fisher and L. R. Bunger, "Use of Wire-Mesh Sections in Asphalt Entrainment Control in Vacuum Pipe Stills, Part 2... Case History - Sinclair Refining Co., E. Chicago," *OGJ*, pp. 96–99, Nov. 1953.

[16] A. W. Sloley, "Consider VFDs for Centrifugal Pumps," *Chem Proc*, vol. 76, no. 4, pp. 44–45, Apr. 2014.

16. Tables and Figures

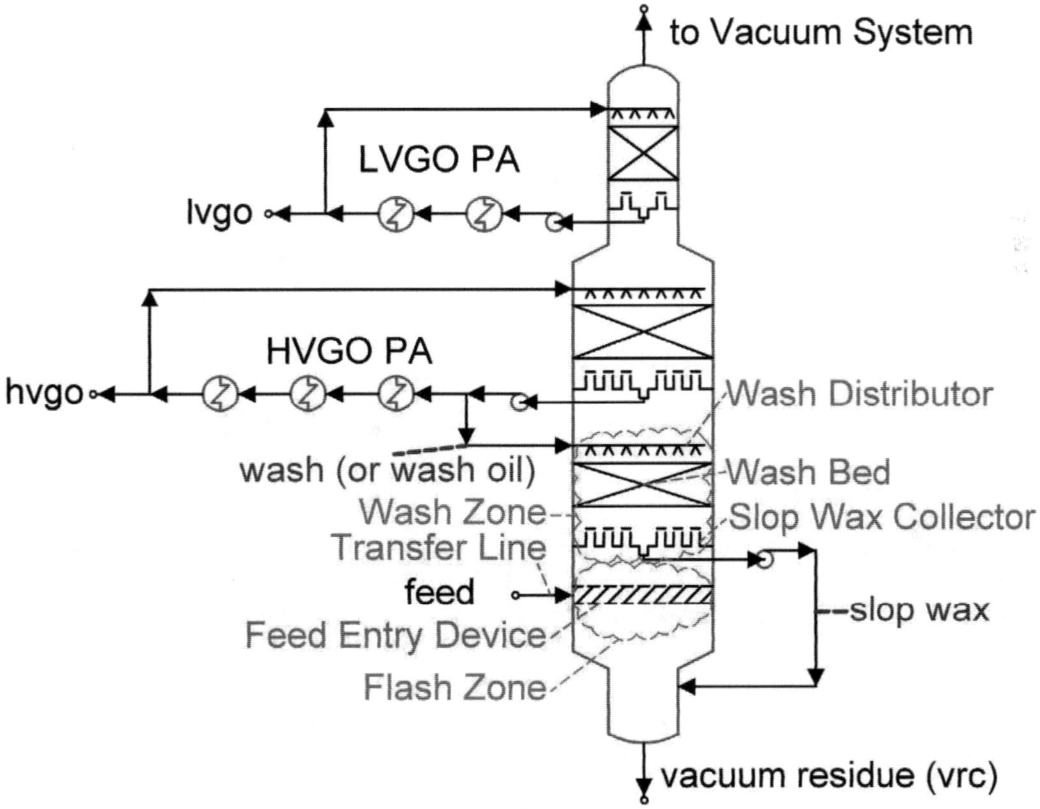

Figure 1 Conventional Vacuum Tower

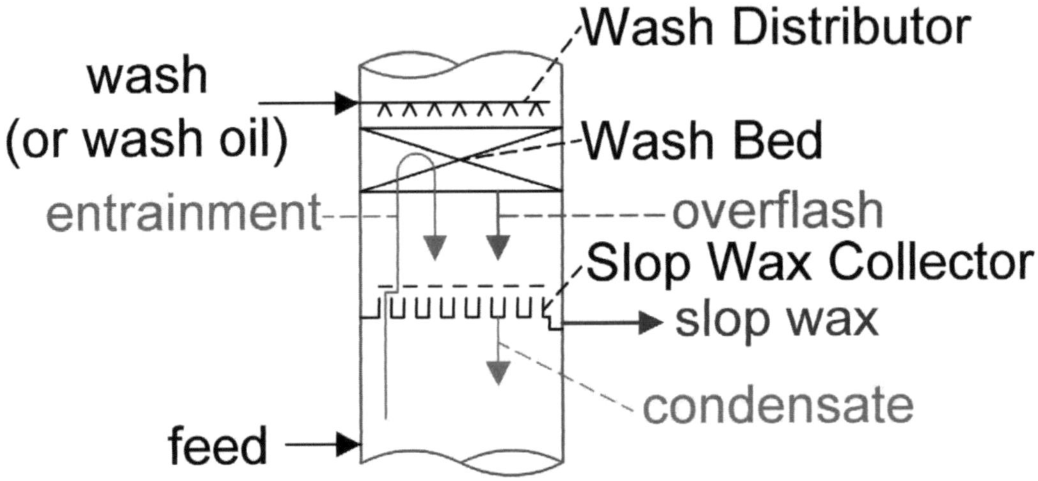

Figure 2 Wash Zone Configuration and Terminology

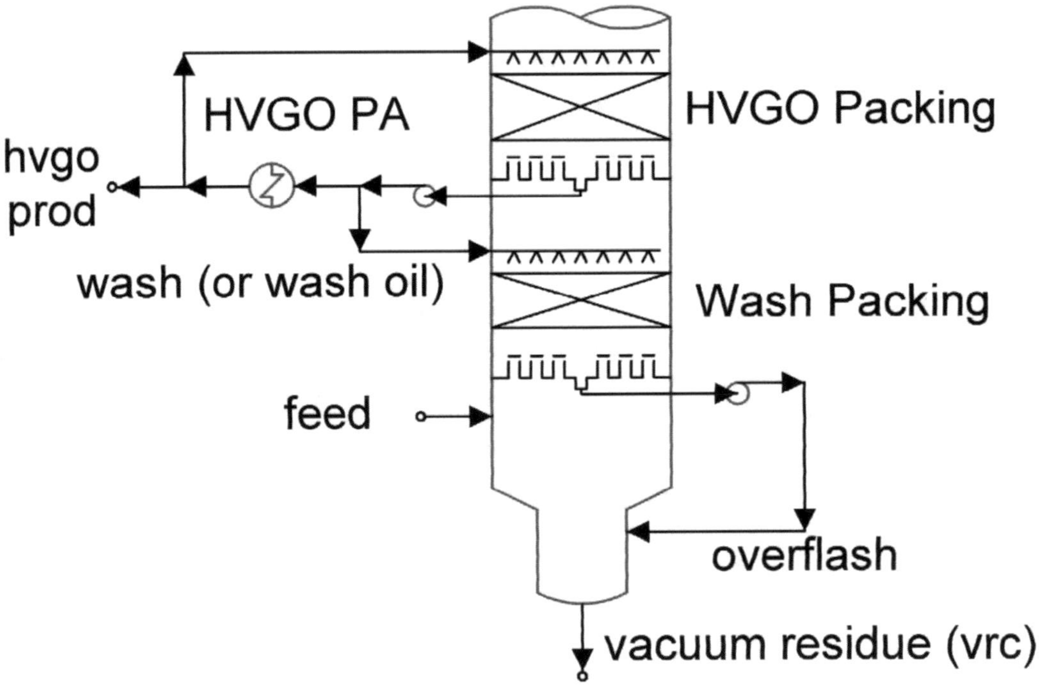

Figure 3 Conventional Wash Configuration

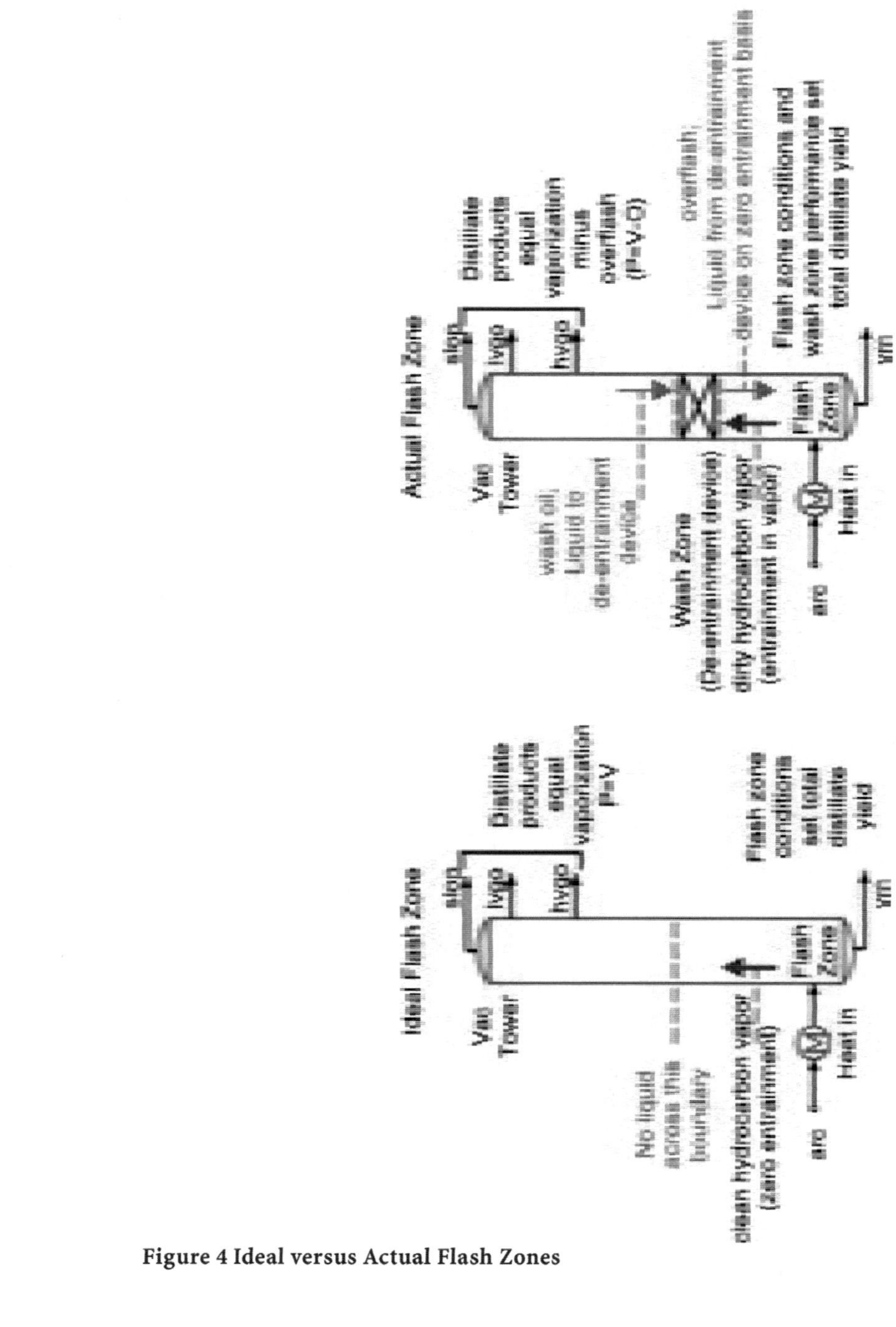

Figure 4 Ideal versus Actual Flash Zones

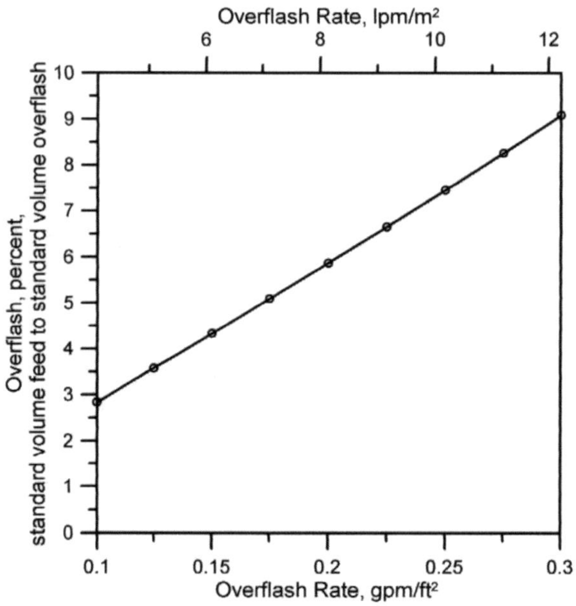

Figure 5 Overflash Percent versus Overflash Rate

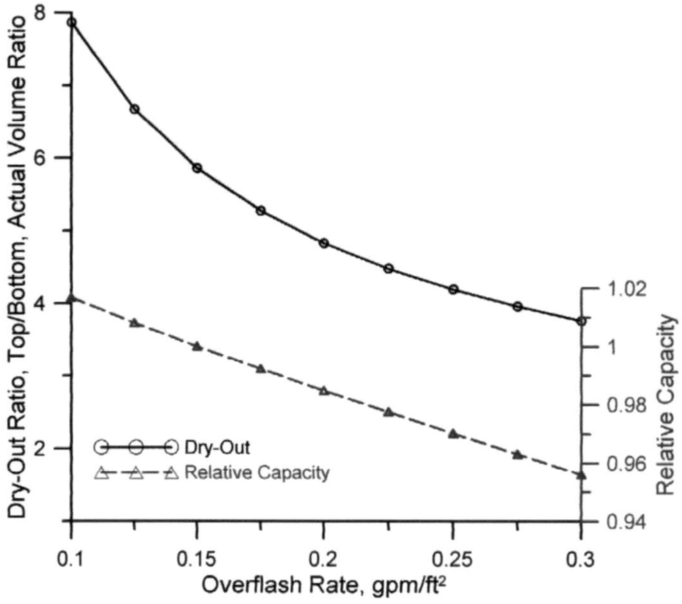

Figure 6 Dry-Out and Capacity versus Overflash Rate

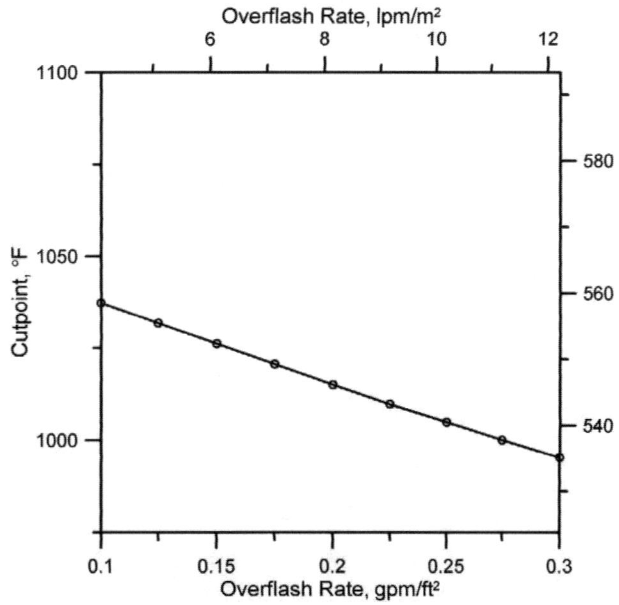

Figure 7 Cutpoint versus Overflash Rate

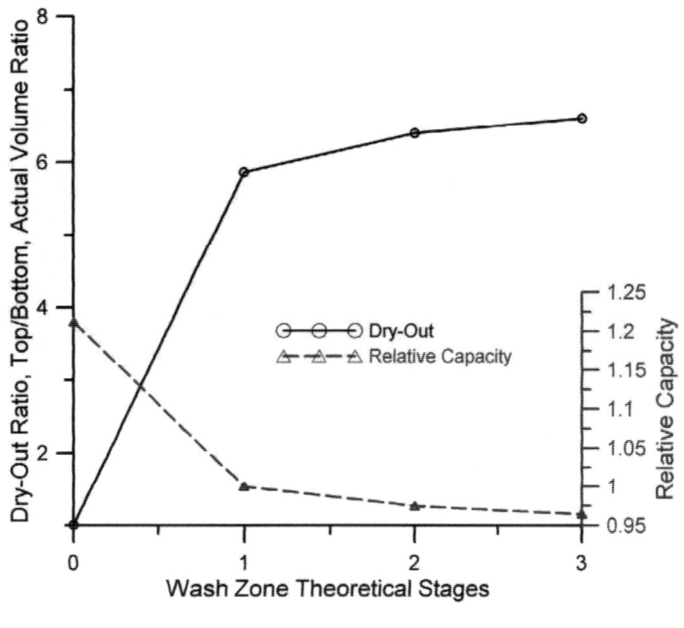

Figure 8 Dry-out and Capacity versus Wash Zone Stages

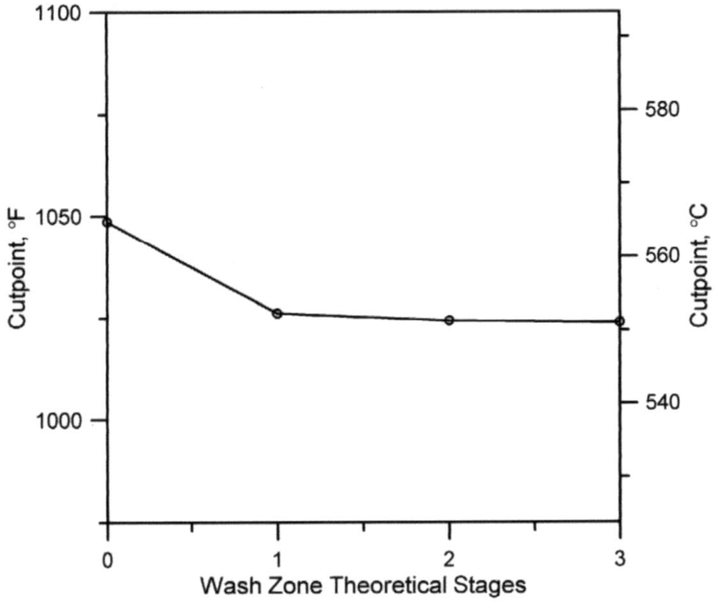

Figure 9 Cutpoint versus Wash Zone Stages

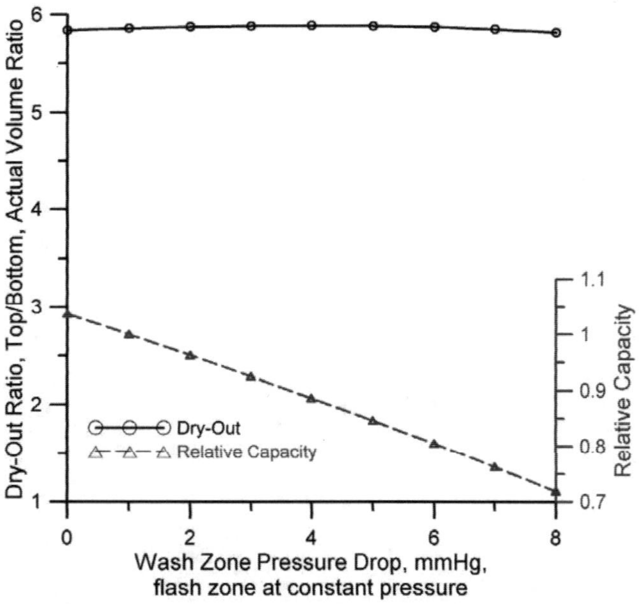

Figure 10 Dry-Out and Capacity versus Wash Zone Pressure Drop

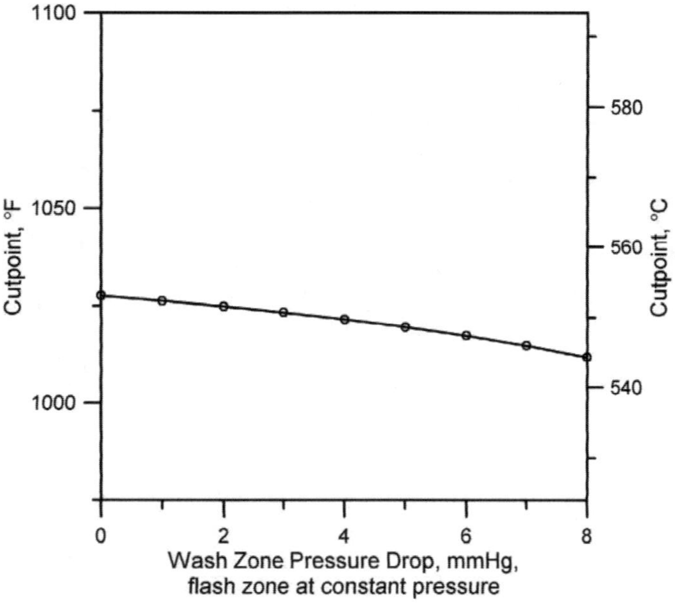

Figure 11 Cutpoint versus Wash Zone Pressure Drop

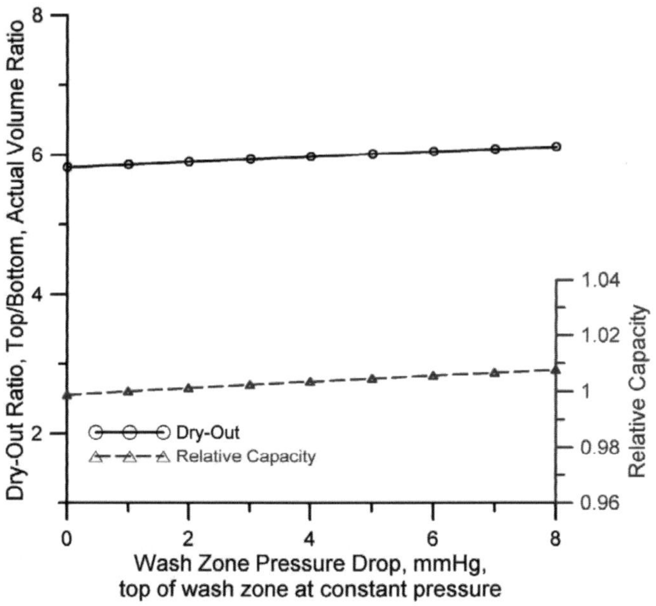

Figure 12 Dry-Out and Capacity Versus Wash Zone Pressure Drop

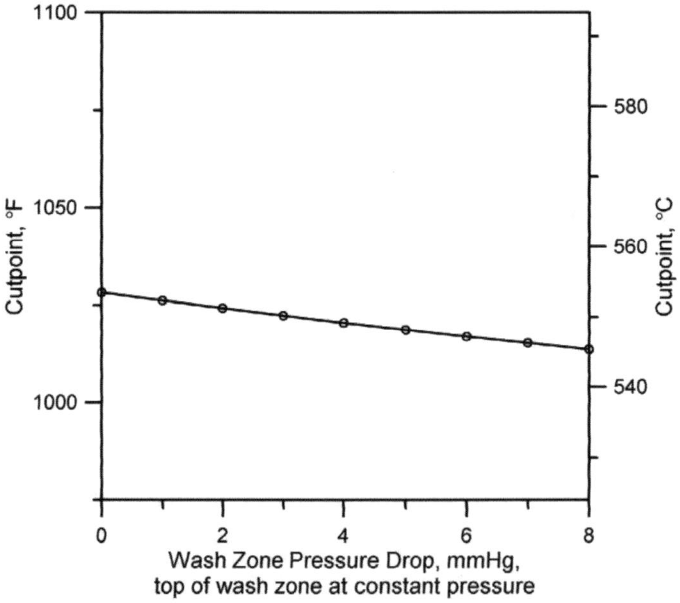

Figure 13 Cutpoint Versus Wash Zone Pressure Drop

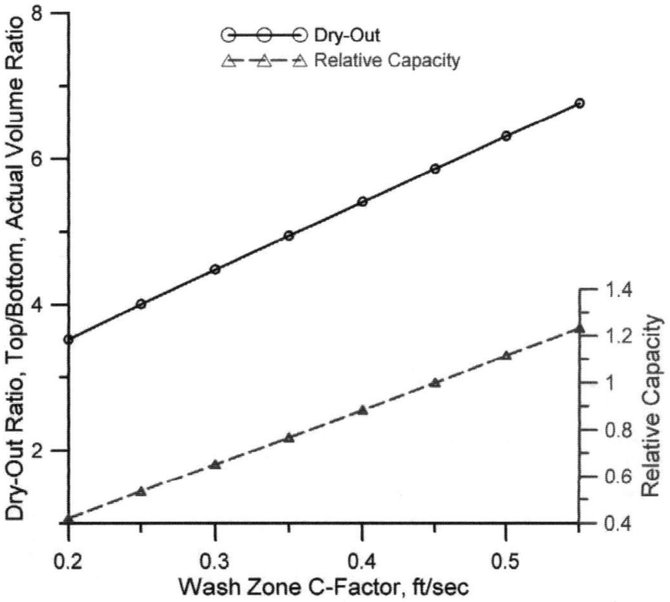

Figure 14 Dry-out Ratio and Capacity Versus C-factor (Constant Feed Rate)

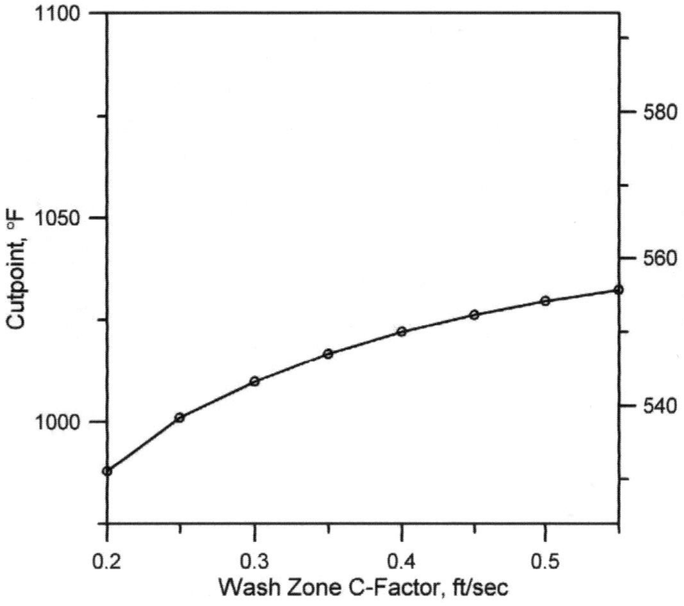

Figure 15 Cutpoint Versus C-factor (Constant Feed Rate)

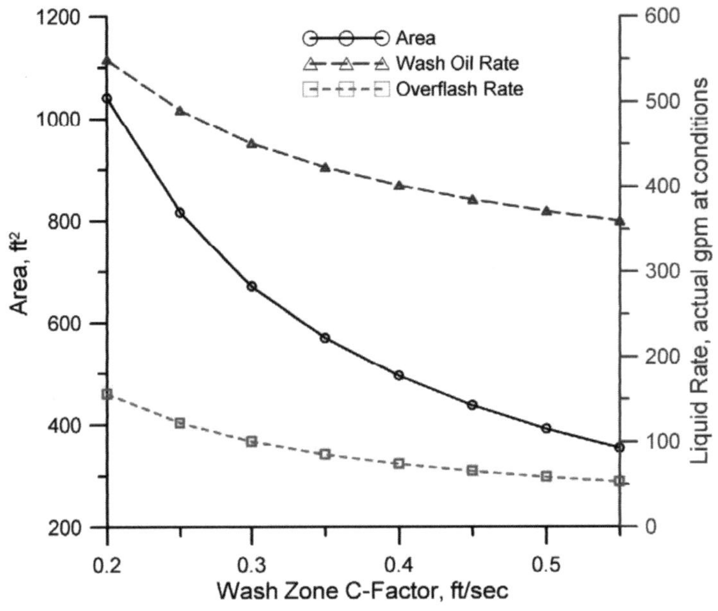

Figure 16 Area and Liquid Rates Versus C-factor

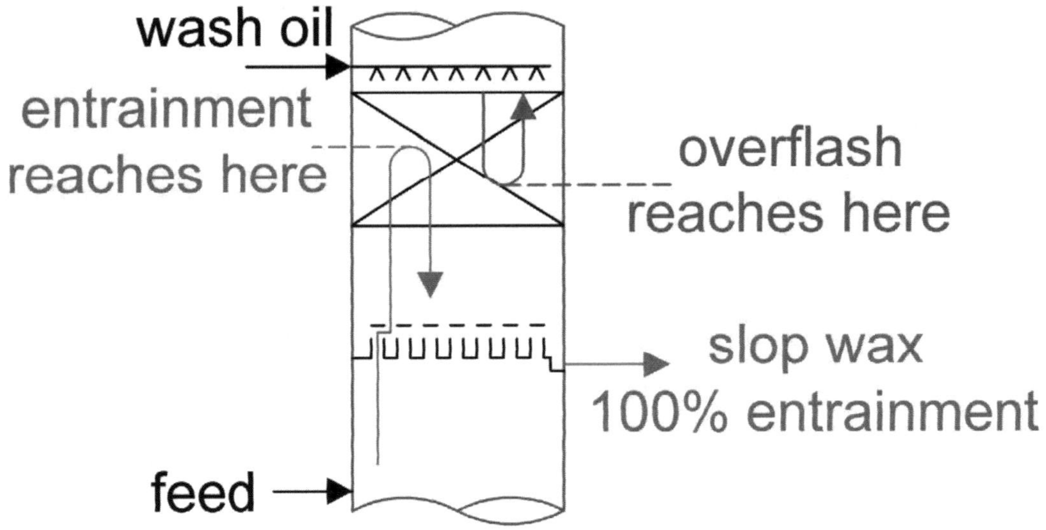

Figure 17 Zero Overflash Operation: Entrainment and Overflash Overlap

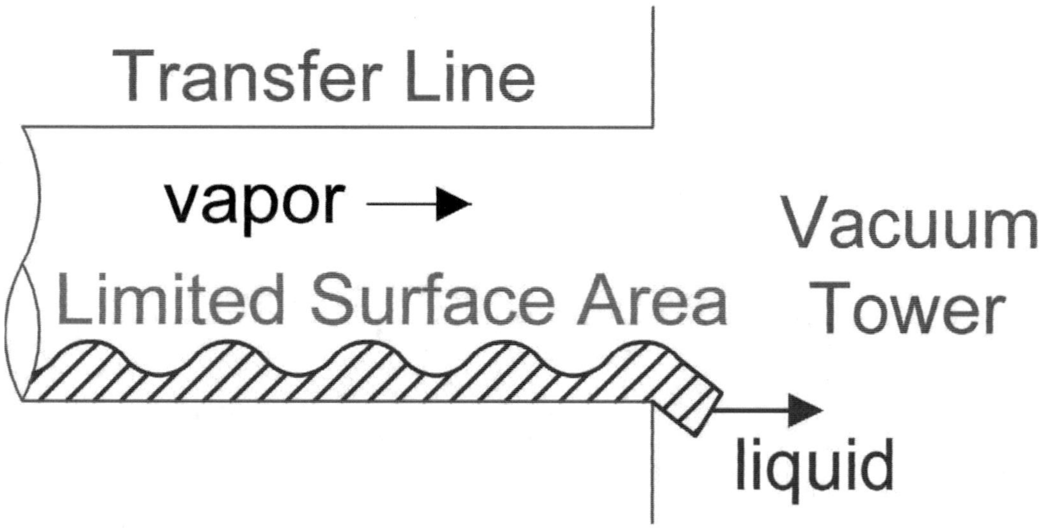

Figure 18 Transfer Line Phase Separation

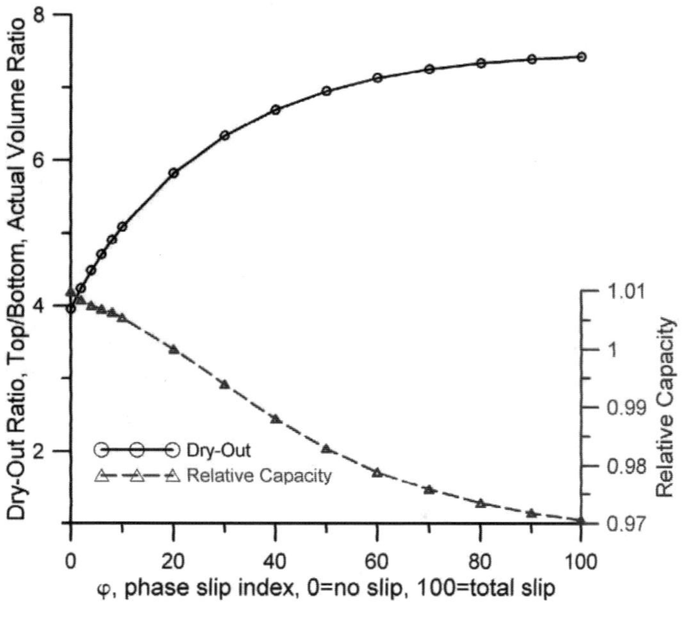

Figure 19 Dry-out Ratio and Capacity Versus Transfer Line Phase Slip

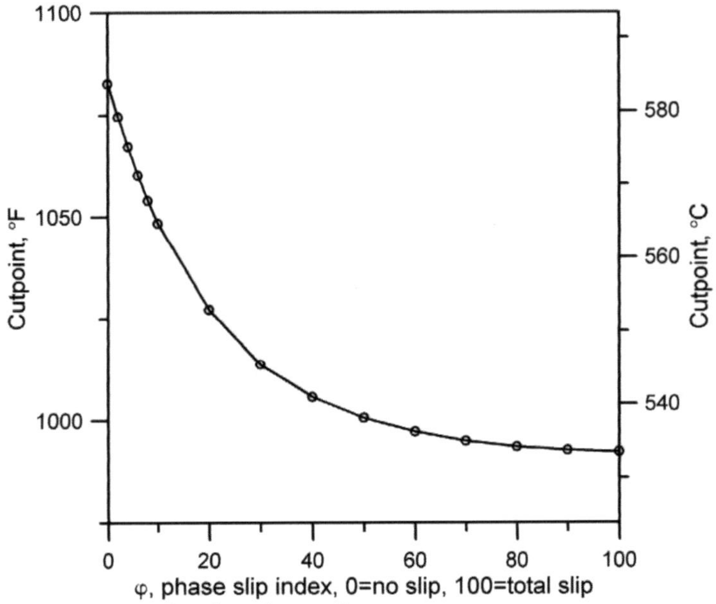

Figure 20 Cutpoint Versus Transfer Line Phase Slip

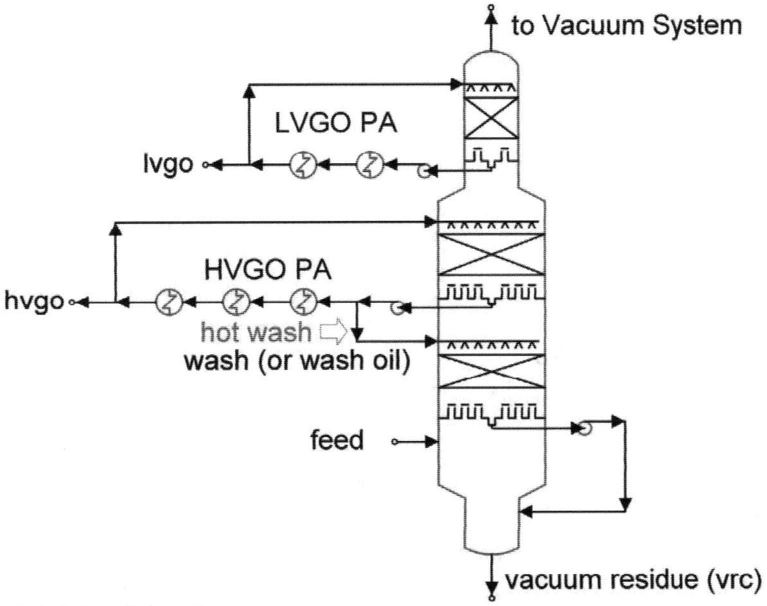

Figure 21 Hot Wash Configuration

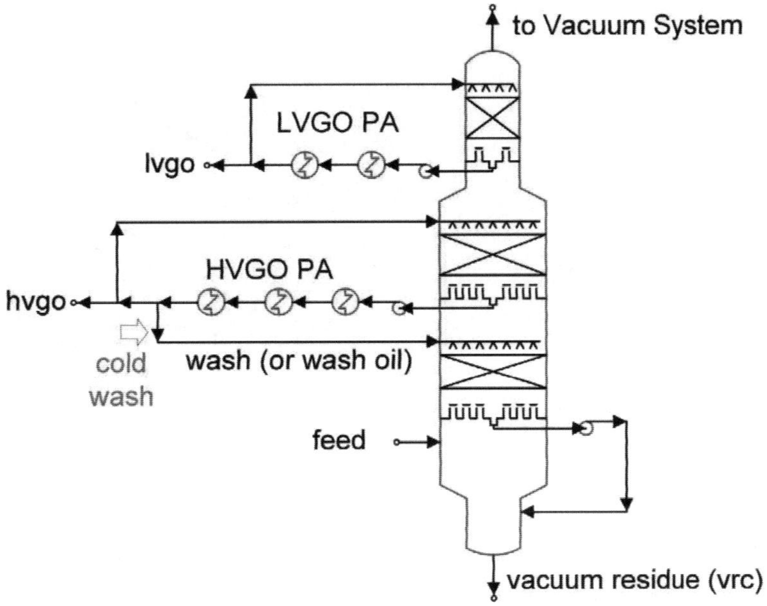

Figure 22 Cold Wash Configuration

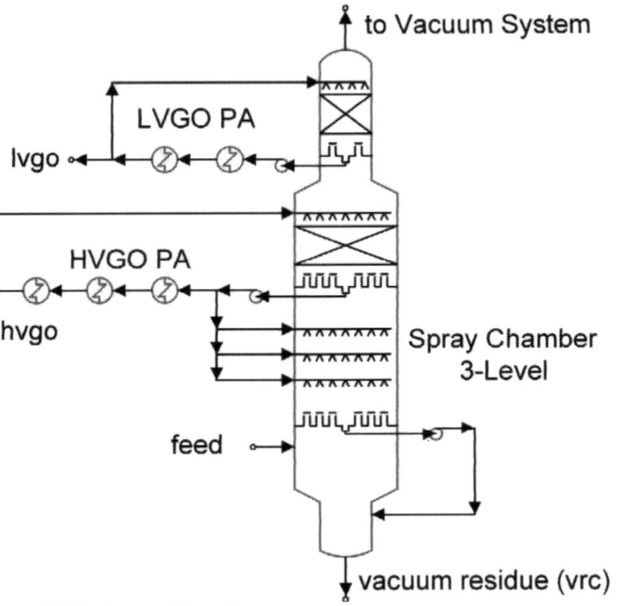

Figure 23 Vacuum Tower With Spray Chamber

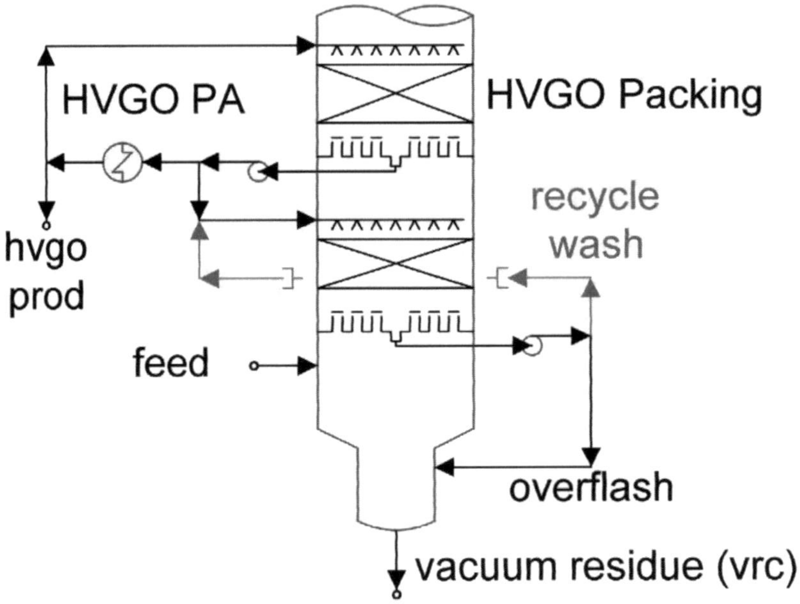

Figure 24 Recycle Wash System: One Stage

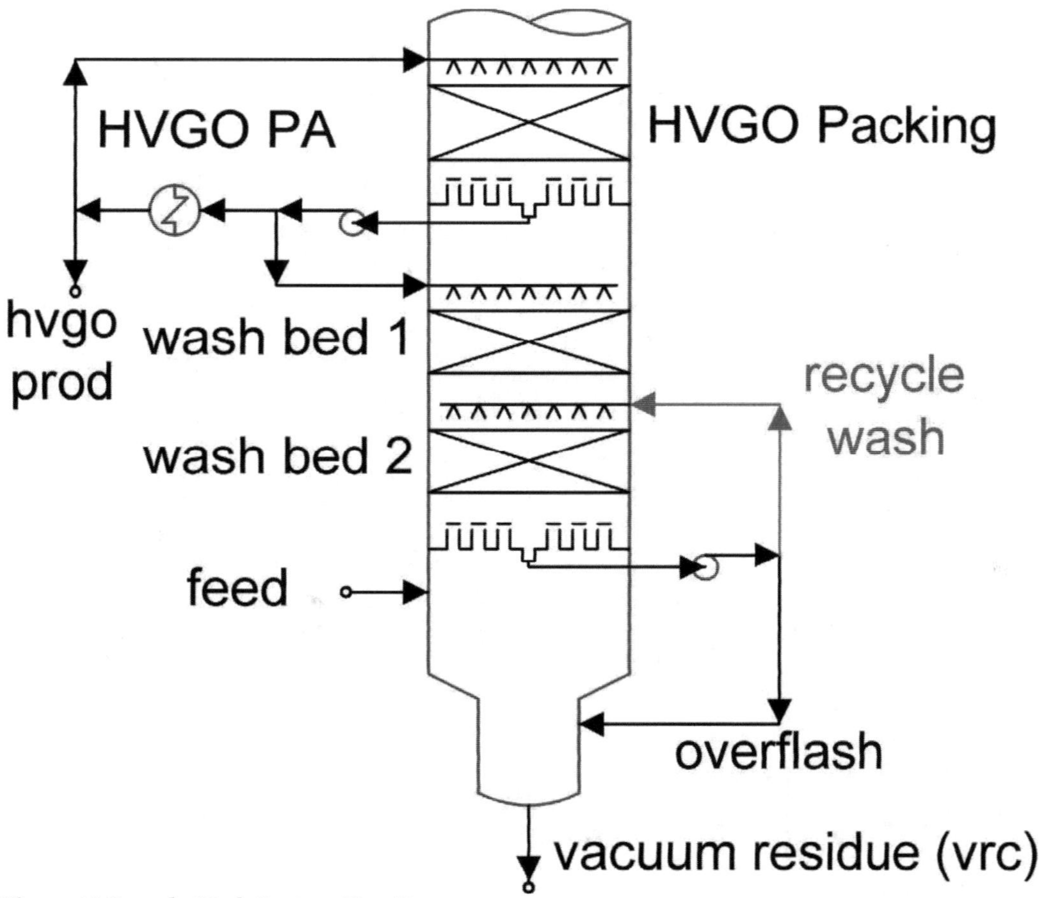

Figure 25 Recycle Wash System: Two Stage

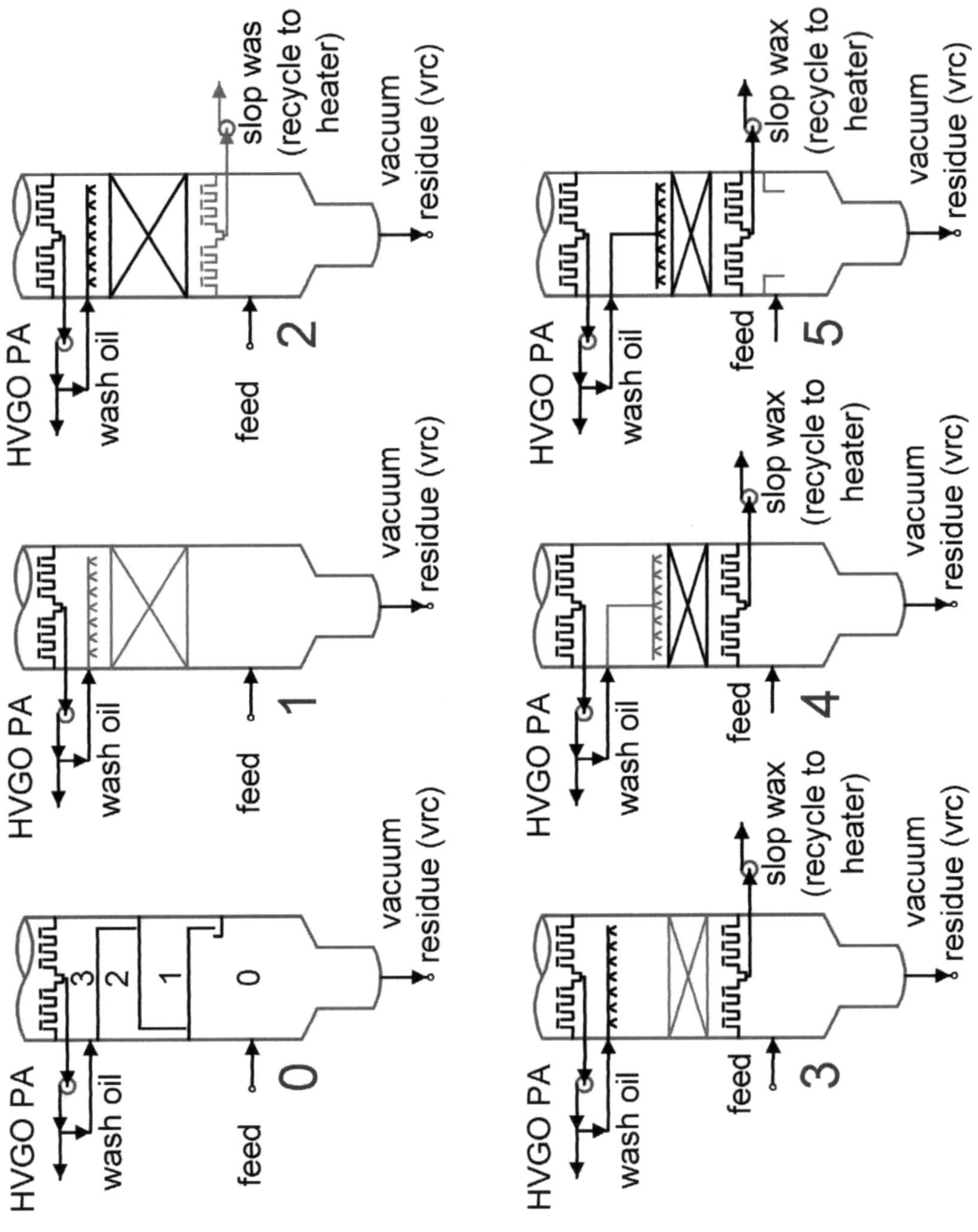

Figure 26 Case History 1: Upgrading a Trayed Vacuum Tower

Optimum Pumparound Balance between Energy & Fractionation Efficiency

Soun Ho Lee – GTC Technology US LLC
2017 AIChE Spring Meeting / Kister Distillation Symposium
San Antonio TX March 27, 2017

GTC Technology – Engineered to Innovate

Refinery Multi-product Fractionator Pumparound Method

GTC Technology – Engineered to Innovate

- Refinery multi-product fractionator
  - Multiple product draws
  - Crude distillation unit, FCC(RFCC), hydrocracker, coker
- Pumparound circuits
  - The most common inter-condenser design
  - Minimizing traffic variation
  - Maximizing heat recovery
  - Overhead condenser size reduction
  - Additional trays/packing for heat transfer: higher column T-T elevation

Pumparound Balance
Fractionation vs. Energy Efficiency

GTC Technology – Engineered to Innovate

- Improving fractionation efficiency
- Downgrading energy efficiency

- Improving energy efficiency
- Downgrading fractionation efficiency

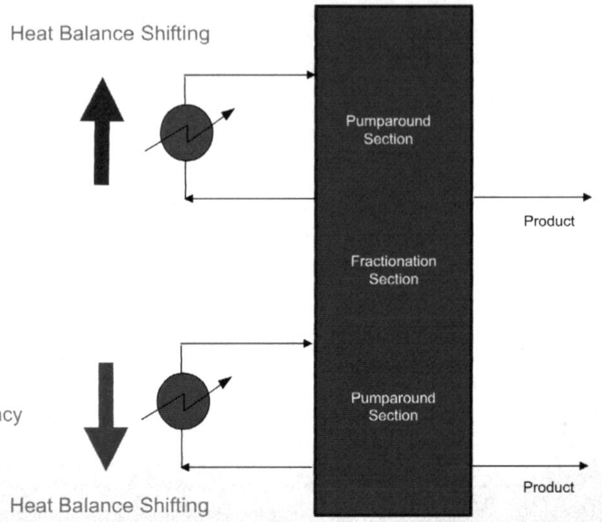

Refinery Multi-product Fractionator
Pumparound / Fractionation Optimization

GTC Technology – Engineered to Innovate

- Non-optimum pumparound balance can be formed if actual process conditions are largely changed.
- Non-optimized pumparound balance can influence performance adversely
- Optimum pumparound balance necessary for target performance

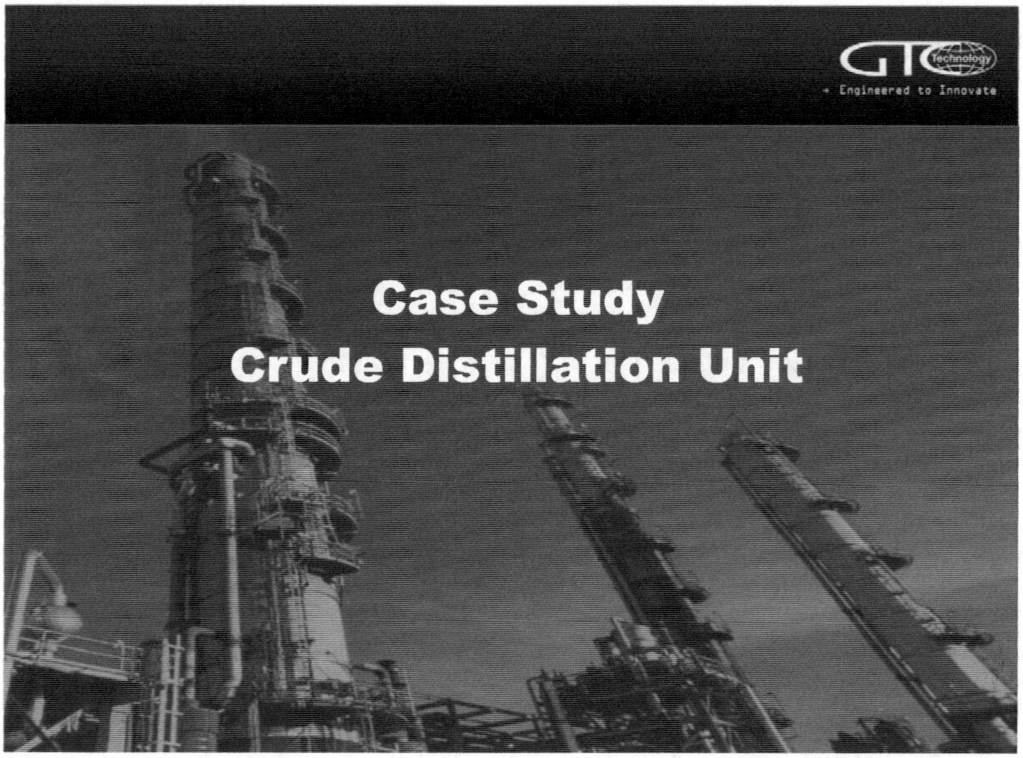
Engineered to Innovate
Case Study
Crude Distillation Unit

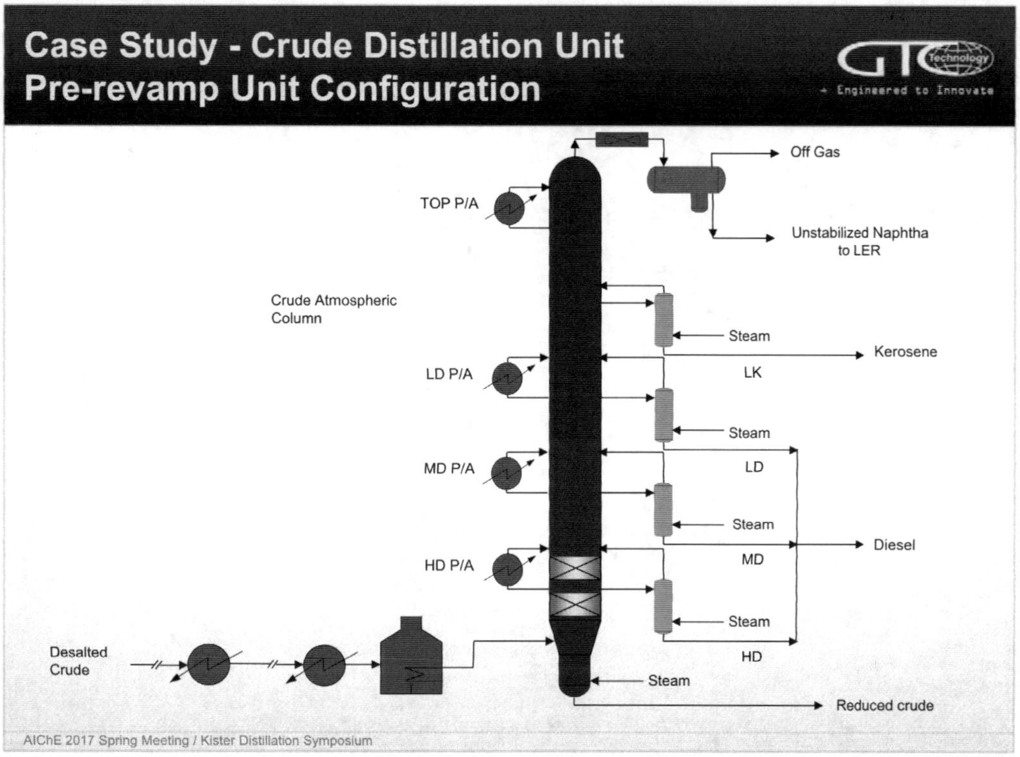
Case Study - Crude Distillation Unit
Pre-revamp Unit Configuration
Engineered to Innovate
Off Gas
TOP P/A
Unstabilized Naphtha to LER
Crude Atmospheric Column
Steam
Kerosene
LD P/A
LK
Steam
LD
MD P/A
Steam
Diesel
HD P/A
MD
Steam
HD
Desalted Crude
Steam
Reduced crude
AIChE 2017 Spring Meeting / Kister Distillation Symposium

Case Study - Crude Distillation Unit
Background / Pre-Revamp Test Run

GTC Technology – Engineered to Innovate

- Limitation
  - Kerosene yield limitation while crude slate containing high kerosene boiling range materials was processed
- Test run basis
  - Crude slate: containing high kerosene boiling range materials
  - Charge rate: maximum without fractionation loss
  - ASTM D-7169 for HD and reduced crude stream
- Observation
  - High pressure drop through top P/A and HD P/A section

Case Study - Crude Distillation Unit
Feed Composition Comparison

GTC Technology – Engineered to Innovate

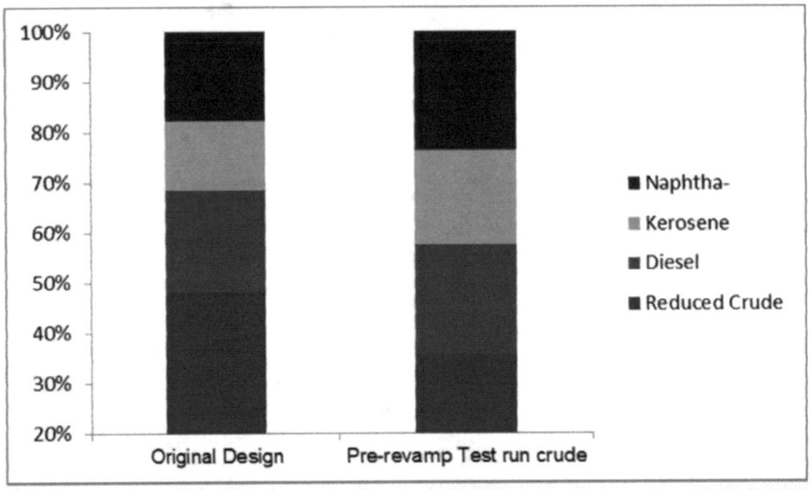

Case Study - Crude Distillation Unit
Pre-Revamp Pumparound Capacity Profile

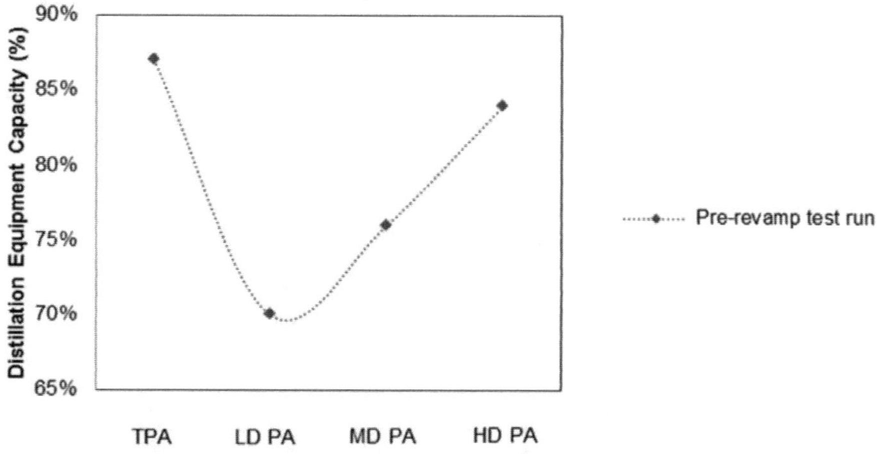

Case Study - Crude Distillation Unit
Revamp Summary

- Re-arranging pumparound circuits for actual feed
  - Mismatch between pumparound locations & feed composition
  - LD P/A was converted to Heavy Kerosene P/A
- Pre-heat train performance was reviewed per new heat balance
  - Heat exchanger modification
- Crude Atmospheric Column
  - Implementing GT-OPTIM™ High Performance Trays

Case Study - Crude Distillation Unit
Post-revamp Unit Configuration

GTC Technology — Engineered to Innovate

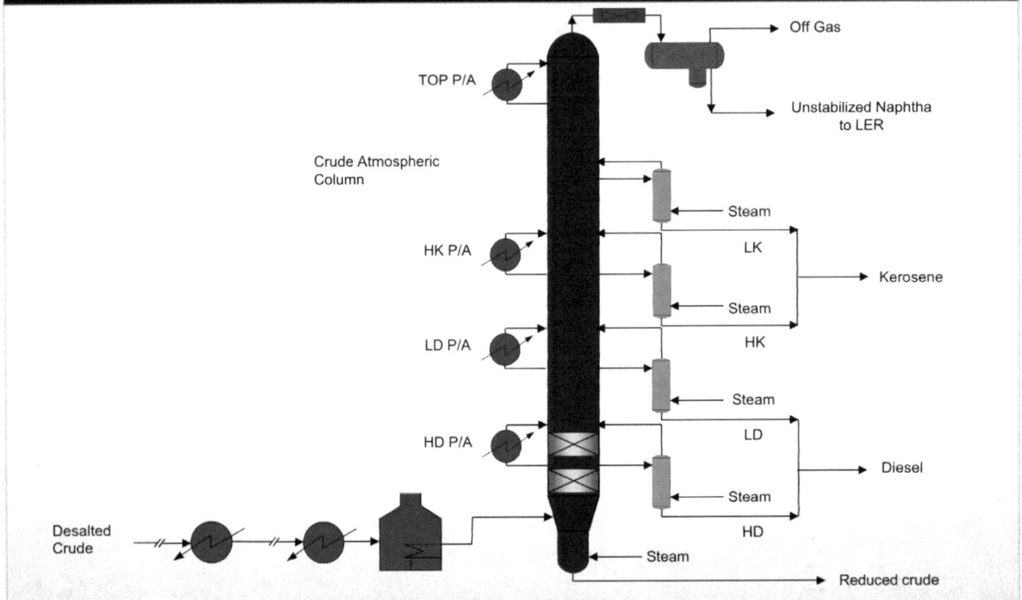

AIChE 2017 Spring Meeting / Kister Distillation Symposium

Case Study - Crude Distillation Unit
Pre- & Post-Revamp Product Yield Pattern

GTC Technology — Engineered to Innovate

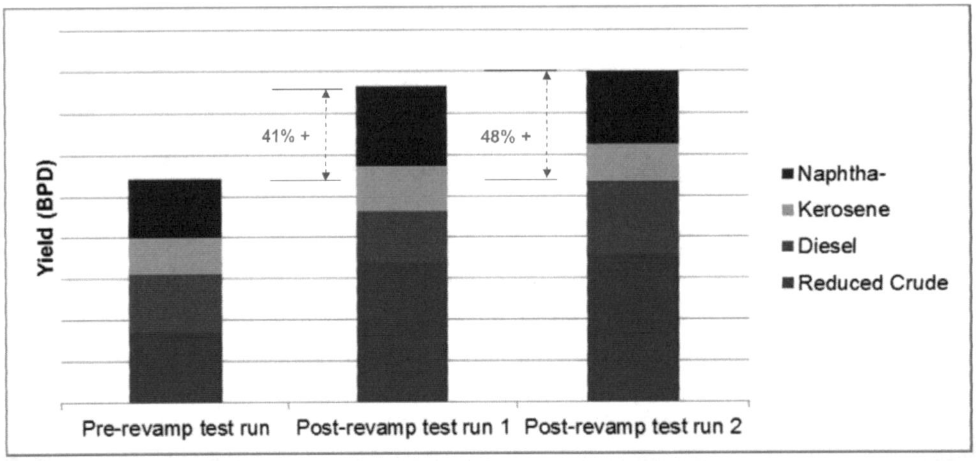

AIChE 2017 Spring Meeting / Kister Distillation Symposium

Page 12

Case Study - Crude Distillation Unit
Pre- & Post-Revamp Distillate Yield Pattern

GTC Technology – Engineered to Innovate

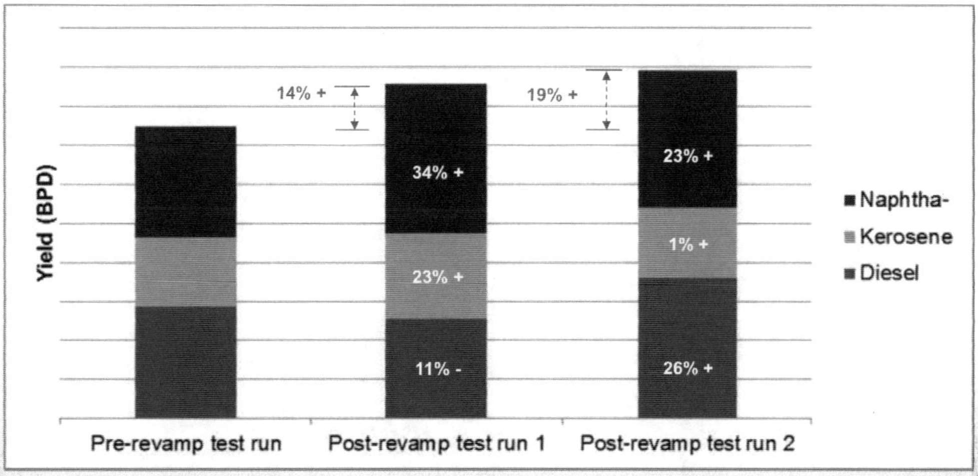

Case Study - Crude Distillation Unit
Pumparound Capacity Profile

GTC Technology – Engineered to Innovate

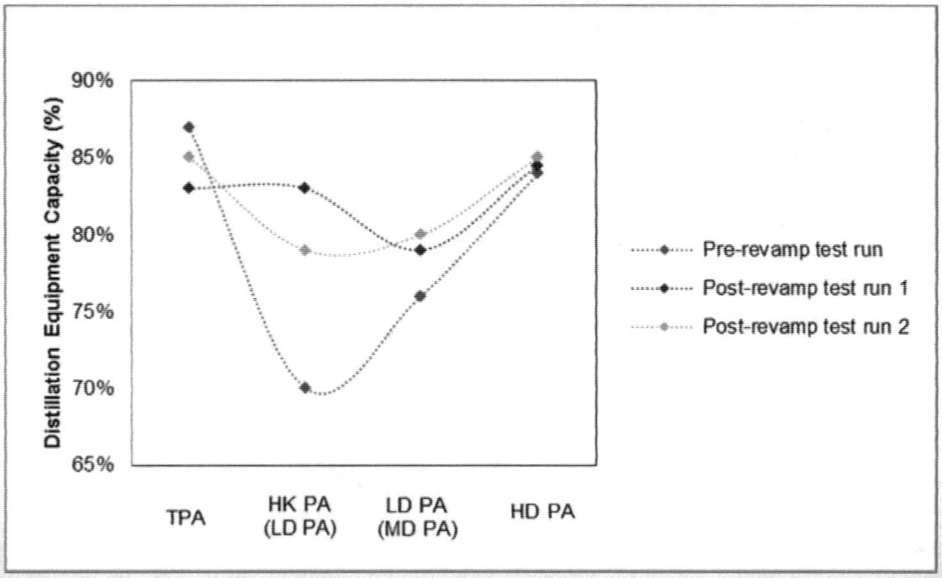

Case Study - Crude Distillation Unit Fractionation Performance- Naph./Kerosene

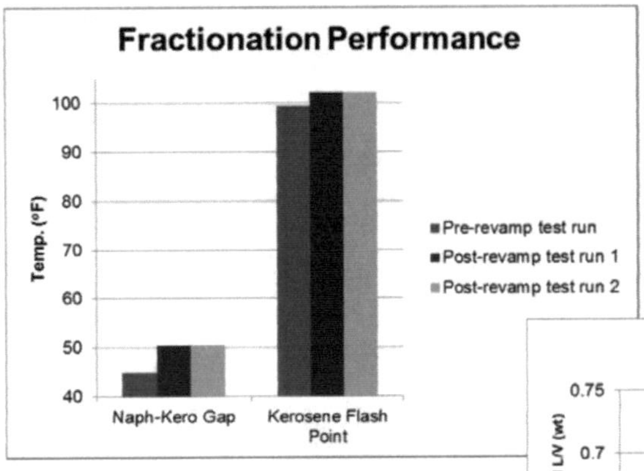

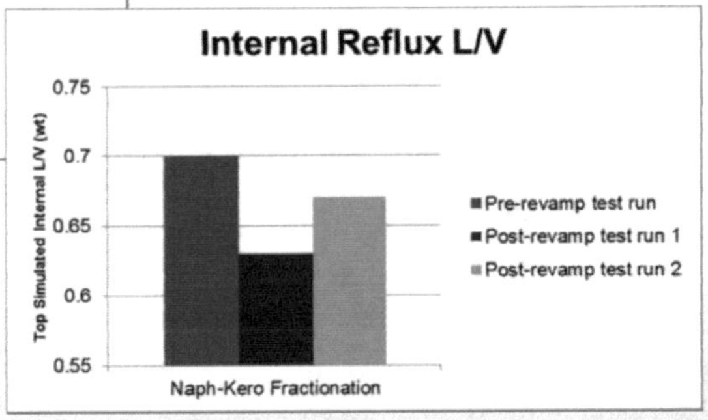

AIChE 2017 Spring Meeting / Kister Distillation Symposium

Page 15

Case Study - Crude Distillation Unit Fractionation Performance-Kerosene/LD

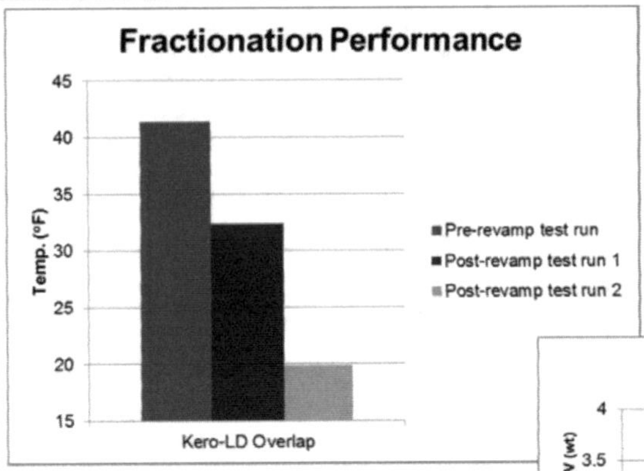

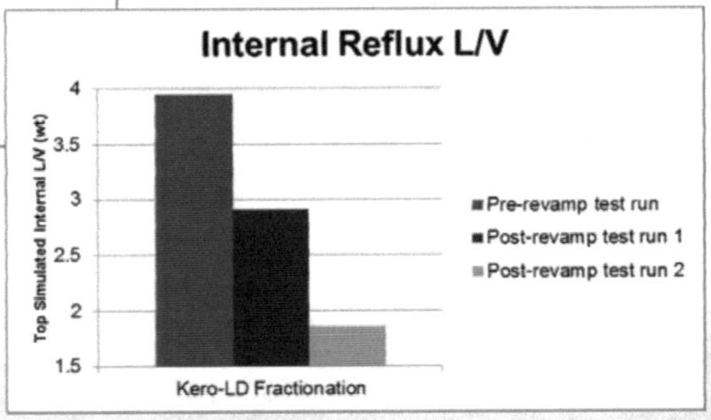

AIChE 2017 Spring Meeting / Kister Distillation Symposium

Page 16

Case Study - Crude Distillation Unit Fractionation Performance- Diesel / RC

GTC Technology – Engineered to Innovate

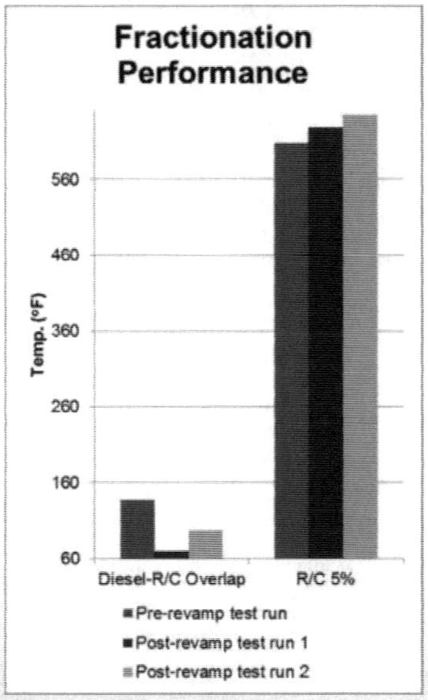

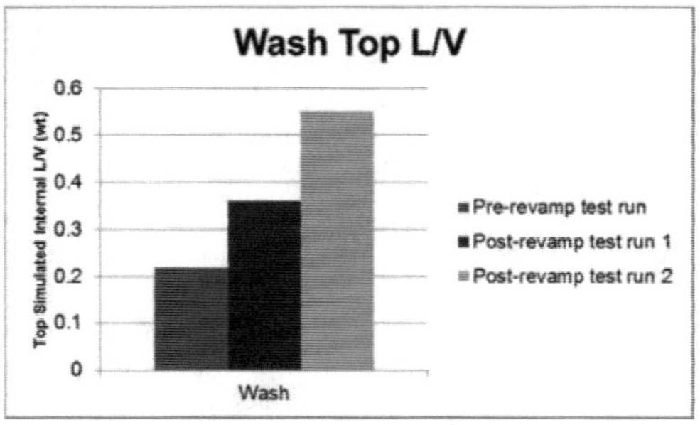

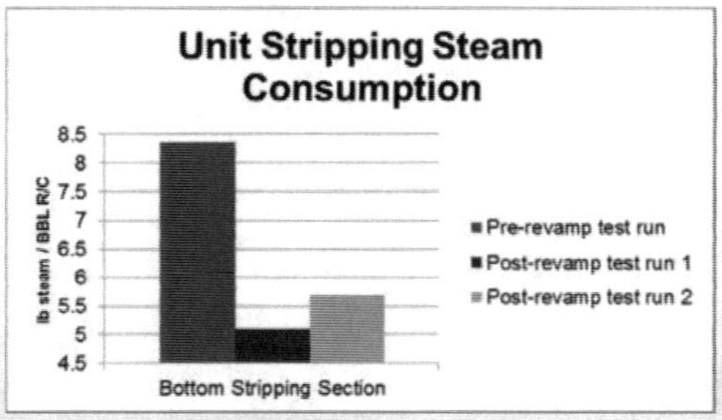

AIChE 2017 Spring Meeting / Kister Distillation Symposium

Page 17

Thank you

GTC Technology – Engineered to Innovate

Thank you

AIChE 2017 Spring Meeting / Kister Distillation Symposium

479525 Shell Consep Trays Improve Diesel Recovery and Naphtha Quality in a DHT Stripper

Abstract:

In 2013, Marathon was unable to make naphtha endpoint specifications due to being at 134% of FRI flood in the rectifying section of the low pressure stripper of a diesel Hydrotreater. Since the tower was flooded and the naphtha had such a high endpoint it could not be sent to the reformer, but it was instead recycled back to the crude unit as slop for cleanup. Shell ConSep trays were installed the rectifying section of the tower during the October 2013 Turnaround. Upon startup, naphtha endpoint met specification, which allowed the naphtha recycle to slop to be eliminated. Further optimization of the reflux up to flooding constraints on the tower allowed the naphtha endpoint specification to be decreased, thus increasing the diesel yield. The Shell ConSep trays allowed optimization of the unit with an increased profit of $6.5MM/yr.

481669 Rigorous Troubleshooting Identified Crude and Vacuum Distillation Column Constraints

Abstract:

Valero Ardmore refinery operates a 90 MBD Crude and Vacuum distillation unit, charging primarily mid-continent crude through the Cushing terminal. The heavy products from this unit are routed directly a gas oil hydrotreater (CFHT) or a blowstill tower for asphalt sales. Crude charge capacity became severely constrained 2-years into the operating cycle between turnarounds. Rigorous troubleshooting, including extensive isotope scanning, identified three separate and unrelated areas of limitation in the crude column and one area in the vacuum tower.

The Ardmore refinery crude tower is a typical atmospheric operation (630degF, 40 psig flash zone) with 42 trays, a packed gas oil wash section, and a shrouded stripping section. Overhead light naphtha is routed to a gas plant and atmospheric tower bottoms are routed partially to at CFHT, with the balance to the vacuum tower. Five side draw products from the crude tower are further processed within the refinery. Severe instability and tray flooding was noted throughout the tower, with specific constraints initiating in the top reflux trays, kerosene pumparound trays, and AGO wash bed. Because the limitations were extended through a very large area of the column, dissecting the data and identifying the true limits was a complicated task. The root cause analyses ultimately pointed to contaminants within the crude feed (phosphorous), inherent design issues with the tower internals, as well as operating practices.

The vacuum tower is operated dry, with a moderate flash zone severity (C-factor <0.4fps). The wash bed reliability is a typical run-limiting constraint to the unit, with wash oil controlled to maintain a slop wax draw rate (metered and pumped to CFHT) indicative of adequate overflash. Loss of slop wax draw occurred at a similar time to the atmospheric constraints. The troubleshooting identified an obstruction on the tray that created residence time and ultimately coke.

In this paper, the authors will describe the troubleshooting methodology, discuss an unconventional hot tap bypass that allowed the unit to remain in operation until a planned turnaround and outline design changes that were implemented during the outage. The authors will also present turnaround findings that prove the conclusions drawn from operational and troubleshooting data analysis were correct.

Correct Problem Diagnosis Debottlenecks a Gas Plant with No Hardware Changes

By

Henry Z. Kister, Fluor, Aliso Viejo, CA, USA

Presented at the Distillation Topical Conference, AIChE Spring Meeting, San Antonio, Texas, March, 2017

Summary

This paper describes a case study of a gas plant (designed by others) that was bottlenecked by the reboil system of a deethanizer stripper. The reboiler system that consisted of two parallel once-through thermosiphon circuits experienced two different steady states. When operated at low rates, an undesirable steady state was established with maldistribution between the two circuits, in which one was circulating and the other remained liquid-full. This steady state persisted when the tower feed rates were raised until the circulating reboiler reached its heat transfer limitation, still with the other circuit liquid-full.

Pressure drop calculations, close design reviews, examination of operating charts, and gamma scans, led to the correct diagnosis that enabled debottlenecking the reboiler circuits and the entire gas plant with no hardware modifications. The desired steady state was established by an operation that temporarily generated higher liquid circulation and pressure drop through the reboiler circuits.

This case teaches a number of very important key lessons:

- With a parallel reboiler system, maldistribution between parallel circuits does occur and can cause severe bottlenecks. To the best of our knowledge, this is the first time that such maldistribution is reported in the literature.
- Pressure balances need to be considered at the design not only for the maximum design rates, but also at the minimum expected operation rates.
- It is important to provide flexibility in once-through reboiler systems that will allow operating as a circulating thermosiphon systems, at least at low rates.
- Finally, good troubleshooting can provide low cost (in this case, no cost) debottleneck of an entire gas plant.

Background

This newly-built gas plant was bottlenecked by the deethanizer stripper. The tower bottleneck appeared in the reboiler circuit. In early April, 2014, Fluor, who had no involvement in the tower or reboiler design, but has extensive experience in distillation design and troubleshooting, was requested to assist with investigating the root cause of the bottleneck and in finding ways of overcoming it. A task force of operating company process and operation engineers, and Fluor's distillation specialists, was formed. The joint team identified the root cause. Based on the root cause, a solution was successfully implemented. After this the deethanizer stripper and the gas plant were in full operation achieving the design rates. This paper describes our troubleshooting investigation and the lessons learnt.

Process Description

The deethanizer stripper is a 15'-6" tower equipped with 4-pass conventional valve trays with moving round valves. The tower base contains a preferential baffle that divides the bottom sump into a separate reboiler draw compartment and a separate bottoms draw compartment. Liquid leaving the bottom valve tray is diverted to the reboiler draw compartment (Figure 1). From there it flows into a battery of four horizontal thermosiphon reboilers. The bottom tray liquid is diverted to the reboiler draw compartment by notched bottom seal pans. As shown in Figure 2, the 8" deep notches in the seal pans are located right above the reboiler draw compartment. There are no notches above the bottom draw compartment. The purpose is to feed all the bottom tray liquid to the reboilers.

The vapor-liquid mixture returning from the thermosiphon reboilers enters above the bottoms draw compartment via two opposite 24" nozzles. Each nozzle is equipped with a pipe that extends to near the centreline of the tower and has a 4'-6" cutout in the bottom quadrant that is directed towards the bottom draw sump (Figures 1 and 2). The purpose is to have all the reboiler return liquid enter the bottom draw compartment, and none enter the reboiler draw compartment. The vapor returning from the reboilers ascends to the bottom tray, while the liquid returning from the reboilers descends to the bottoms draw compartment. Any excess bottom tray liquid not boiled by the reboilers overflows the baffle into the bottoms-draw compartment.

This design generates a once-through[1] thermosiphon reboiler system. Such a system intends to capture all of the liquid leaving the bottom tray and feed it directly to the reboiler. The reboiler return liquid is entirely drawn as bottoms product. None of the reboiler return liquid is recycled back to the reboiler, hence the name once-through.

Once-through systems can achieve one full theoretical stage of separation, if the trapout tray and draw pan do not leak[1]. With the cold bottom tray liquid being fed to the reboiler, it can also maximize the reboiler temperature difference and capacity. Reboil vapor is limited to about 40% of the bottoms product rate, due to the normal limitation of 30% maximum vaporization (by weight) in thermosiphon reboilers[1]. This makes once-through thermosyphon systems appropriate only for low-boilup systems such as strippers. Many deethanizer-strippers use once-through thermosiphon reboiler systems.

Figure 1. Deethanizer Stripper Reboiler Circuits

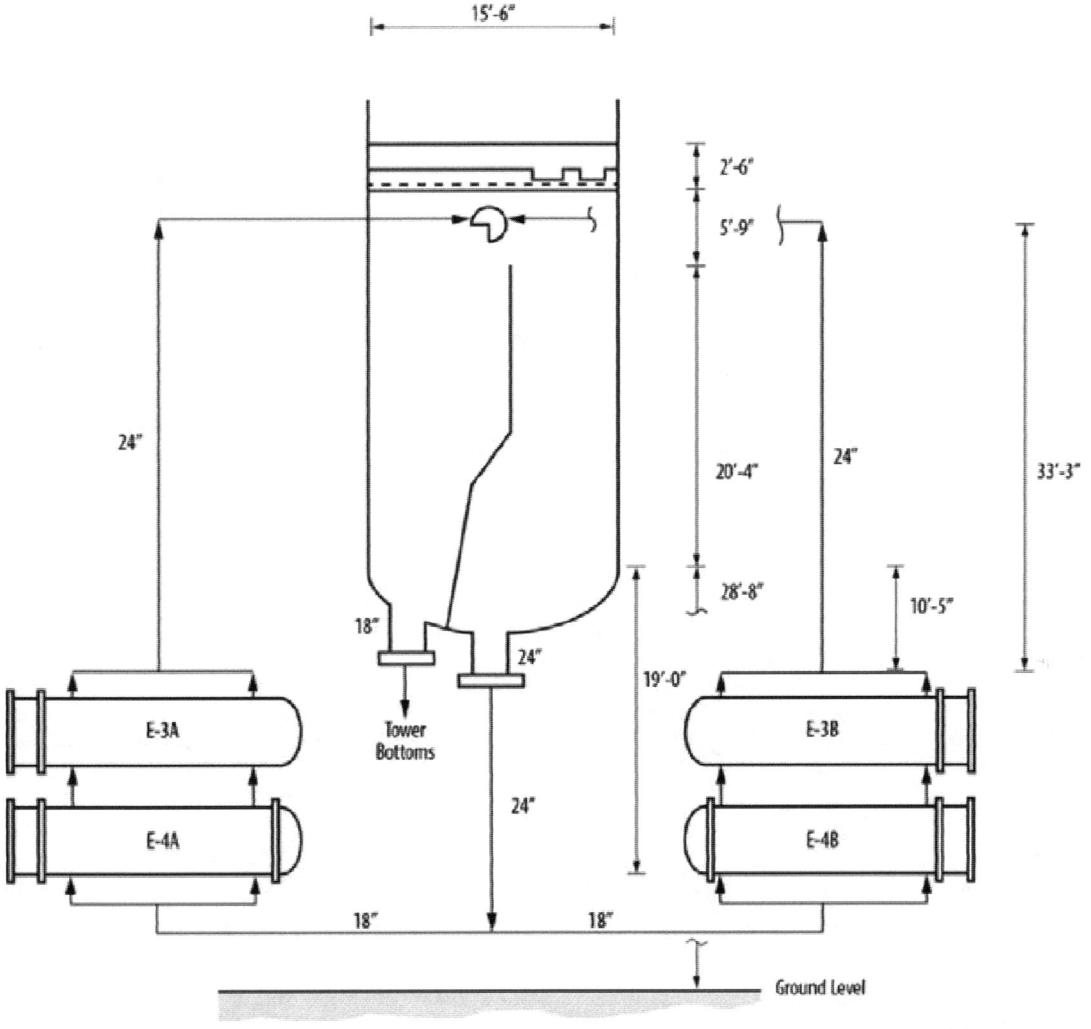

This deethanizer stripper has four horizontal reboilers, arranged in two parallel circuits, "A" and "B" (Figure 1). In each circuit, the lower reboiler (E-4A or E-4B) is heated by hot naphtha liquid, while the upper reboiler (E-3A or E-3B) is heated by high-pressure steam. In each circuit, the naphtha and steam reboilers are designed to provide duties of 32 MM Btu/h and 40 MM Btu/h, respectively. On the boiling side, liquid to the reboilers is provided from the reboiler draw sump via a 24" pipe which splits into two 18" pipes, each going to one circuit. The reboiler effluent from each circuit returns to the tower via a 24" line.

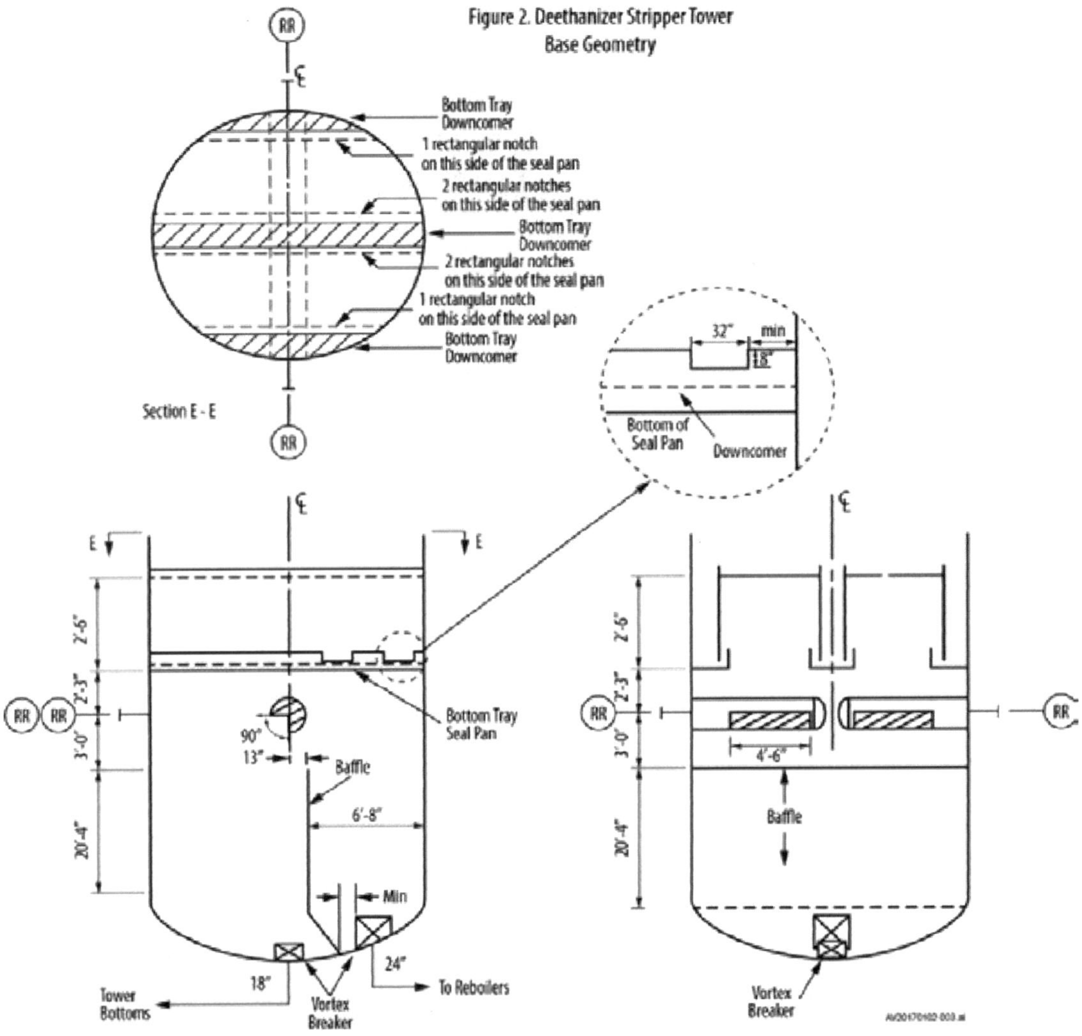

Figure 2. Deethanizer Stripper Tower Base Geometry

Initial Performance

During the start-up, the plant was at low rates and to improve tower stability it was decided to take the steam out of reboiler E-3A. This was successful at creating a steady steam flow through the E-3B reboiler circuit and provided tower stability but the plant has been unable to re-establish flow through the E-3A/E-4A set of reboilers ever since. For the next six months or so, the E-3B/E-4B circuit did all the reboiling, while the deethanizer bottoms liquid circulation has been difficult to establish through the E-4A/E-3A circuit.

Figure 3 shows typical operating data from 6 months after start up and compares them to the design values. The zero steam flow rate through E-3A and the absence of significant temperature difference across the heating side of E-4A indicate no boiling and no significant circulation through the A circuit. To maximize duty, steam to E-3B was raised to the maximum, well above the design steam rate for this exchanger, but still not high enough to make up the lost duty of E-3A. This restricted the feed flow rate through the tower and bottlenecked the entire gas plant. There was a need to get the stripper bottoms liquid to start flowing through the E4A/E-3A circuit.

Figure 3. Deethanizer Stripper
Operation vs. Design
Operation____ (Design____)

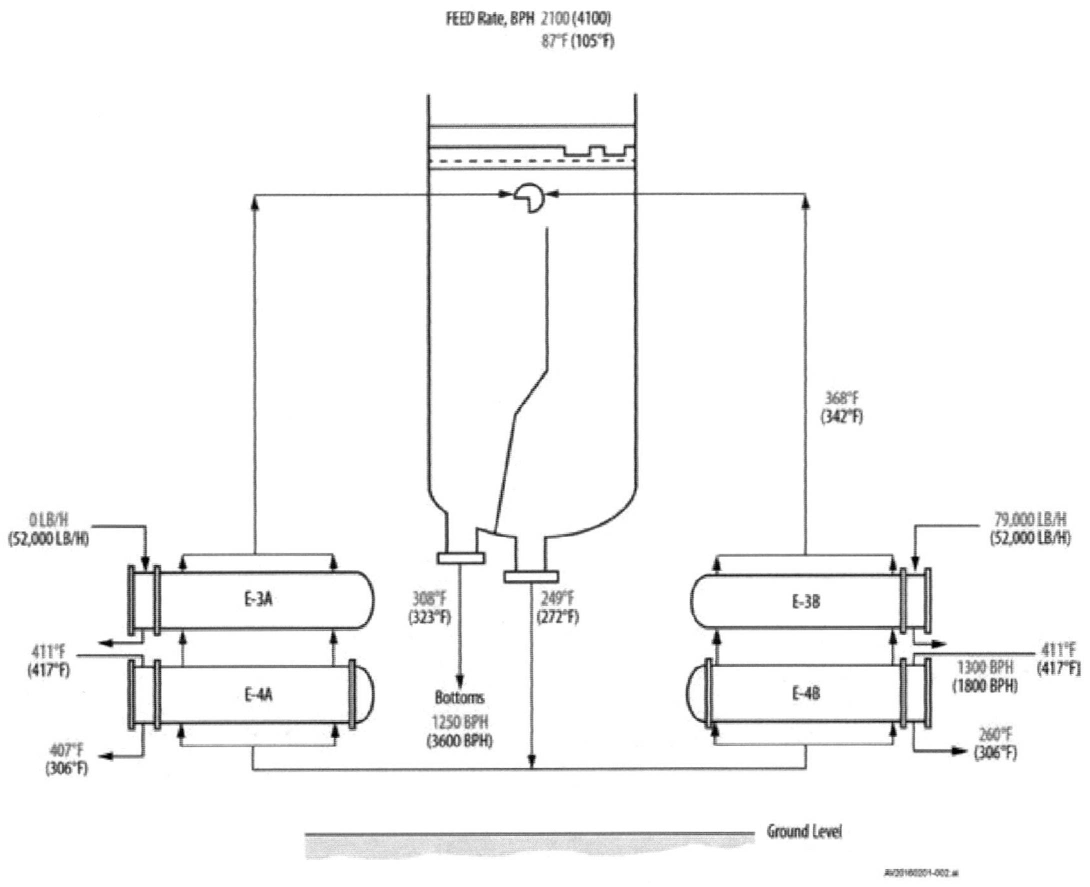

Troubleshooting

Initial troubleshooting endeavors focused on initiating flow through the A circuit by way of inducing a gas-lift. One possibility was that the lights were stripped out of the A-circuit reboilers, leaving behind heavy hydrocarbon liquid. Heating this heavy liquid may not have produced enough vapor to initiate reboiler action. Injecting a light gas like nitrogen could possibly gas-lift the heavy liquid and initiate thermosiphon circulation. This technique has been successful in the past for initiating circulation in liquid-full thermosiphon reboilers[2-4].

A source of nitrogen was connected to a bleeder valve immediately downstream of E-3A. Nitrogen was injected for 4-6 hours in an attempt to provide lift and jump-start the thermosiphon effect. This action was unsuccessful in initiating circulation through the A circuit.

This came as a surprise, as in other reboilers[2-4] a similar action had been successful. There were questions as to whether enough nitrogen was applied, whether it was applied at the correct point, and whether the liquid in the reboilers was just too heavy and needed draining and replacing by fresh liquid before the gas lift attempt. Addressing these problems would have required some painful modifications or operations with no assurance of success. The validity of the theory that heavy liquids in the A-circuit are the root cause was questioned. With these in mind it was decided to perform more troubleshooting before trying out a fix.

One useful reference point was process data from the very initial operation of the stripper. The operating charts from shortly after initial start up (Figure 4), at a stripper feed rate of about 1700 bph, show both the A and B circuits operating simultaneously and stably. Both E-3A and E-3B reboilers operated stably with a steam flow rate of about 32,000 lb/h to each. For both the E-4A and B reboilers, the temperatures of naphtha entering and leaving were about 400°F and 320°F, respectively. At 22:40 the stripper feed rate plunged, steadying at about 1200 bph forty minutes later. To improve stability, the plant manually reduced the steam flow rate to E-3A to about 8000 lb/h, with the steam rate to E-3B remaining at its previous value. The naphtha temperatures into and out of E-4A became the same, about 360°F, suggesting it stopped heating. At the same time the temperature outlet from E-4B decreased to about 260°F, suggesting that this reboiler was working harder than before. Both of these suggest that the A-circuit stopped circulating. Early on the next day, the feed rate to the stripper rose again, to 1500 bph, but this time the A circuit remained inactive. It was common practice to attempt to restart circulation by opening and closing the steam valve on the stalled set but circulation remained stalled.

Figure 4. Process Data for Initial Operations of Stripper Reboiler Circuits

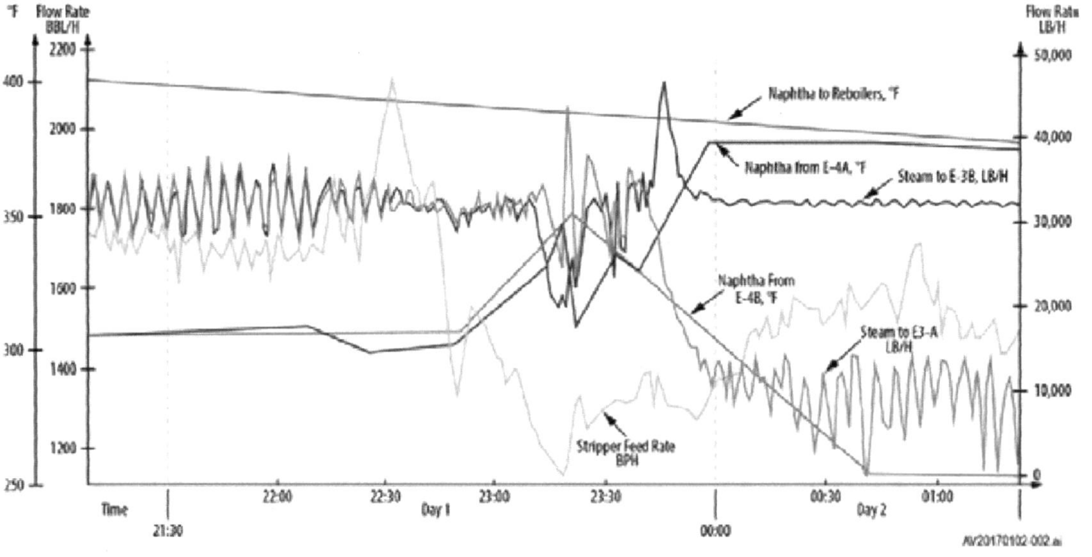

This sequence of events suggests the possibility of the existence of two hydraulic steady states, with an easy switchover from one to the other. This is likely to take place when the reboilers pressure drop on the boiling side is low.

Calculations performed during the design by the engineering contractor that designed the circuits showed a total pressure drop of 1 psi for the exchangers in each circuit at the design conditions. The contractor's calculated pressure drop were 0.4 psi and 1.3 psi for the reboiler liquid inlet line and the reboiler return lines, respectively. The total pressure drop is equivalent to 11 feet of liquid head.

Figure 1 shows the elevations of the reboilers circuits. The diagram shows that not much liquid head is required to overcome the pressure drop in the reboilers circuit. The liquid level in the reboiler draw compartment will therefore be near the bottom of the compartment, even at the design rates. At the low rates, the pressure drop will be lower still, and the liquid level will drop further, possibly dropping out of the reboiler draw compartment into the reboiler liquid inlet pipe. Further, at low boilup rates the bottom tray may weep, sending liquid directly into the bottom draw compartment, and causing the liquid level to be lower still.

At a low liquid head, fluids tend to form their favorite flow patterns. Two steady states can form: one, with operation of the two reboilers simultaneously, as intended. The other has all the boiling taking place in one circuit, and the other circuit filling with liquid to the same height as the liquid in the reboiler liquid inlet pipe. Once this steady state forms, the steam that is still going in (the 8000 lb/h going to E-3A after midnight on the second day in Figure 4) will strip the lights out of the near-stagnant liquid. Before long, heavy liquid will form in this circuit, and this liquid will be difficult to boil. This will stabilize this steady state and it will persist.

The proposed theory states that the liquid level in the reboiler compartment would be very low, possibly absent, and will not approach the overflow baffle. This can be easily tested using gamma scans. The tower base was gamma scanned both on the bottom compartment side and on the reboiler compartments side. In addition, neutron backscatter scans were performed to confirm or deny the gamma scan results.

The gamma and neutron scans showed the following:

- Nothing abnormal showed up on the inner active area scans for the bottom 6 trays. Other trays were not scanned.
- Both the gamma and neutron scans showed that the liquid level in the bottom draw compartment was 10-12 feet above the tangent line. This was the same as the level indicated by the level instrument.
- Both gamma and neutron scans had difficulty finding a liquid level on the reboiler draw compartment. There definitely was no liquid level above the bottom tangent line. The gamma scans showed some absorption at the tangent line, about the same absorption as the froth on the trays, but this could have been due to the welds at the tangent line. The neutron scans showed no liquid at all in the reboiler draw compartment.

The absence of liquid level in the reboiler draw compartment gave strong support to the two steady-states theory.

Solution

With the theory verified, the challenge was now shifting from the steady state with the stagnant "A" reboilers back to that of the two reboiler circuits operating simultaneously.

In a circulating thermosiphon reboiler (as distinct from once-through thermosiphon, like the stripper reboiler), increasing the liquid head in the tower generates more liquid circulation through the reboiler circuit and reduces the fractional vaporization[4]. Considering the pressure balance and assuming a constant heat duty, the higher liquid head supplied to the reboiler is matched by the additional friction head due to the enhanced liquid circulation rate plus the additional static head due to higher amount of liquid in the reboiler shell and outlet line. Extending this reasoning to the stripper reboiler, circulating liquid and increasing the liquid head in the reboiler compartment will raise the pressure drop through the active reboiler circuit. This higher pressure drop will induce some of the liquid to follow the easier path (less pressure drop) through the inactive reboiler. Once this liquid gets heated, it will start boiling, and the thermosiphon will be established in the previously-inactive circuit.

One way of generating circulation and increasing the level in the reboiler draw compartment is by raising the level in the bottom draw compartment until it overflows the baffle, and keep the bottom draw compartment full during the circulation. The risk is that the liquid level may ascend to the trays and cause flooding and possible damage there. It is important to avoid excessive buildup of liquid up the tower.

In principle, the level transmitter can be watched during the operation to avoid excessive liquid level. This was difficult in this deethanizer stripper. Since the reboiler return pipe is only two feet above the baffle (Figure 2) and the vapor from the reboiler return blows downwards causing a frothy dispersion, the frothiness may reduce the specific gravity and lead to optimistically low level measurement. Also, the high tap of the level transmitter is just less than two feet above the baffle, which may make it difficult to see when the level rises towards the trays. Fortunately, there is a differential pressure transmitter across the stripper that comes from the upper tap of the level transmitter and therefore can monitor any accumulation of liquid above this.

The plan was to raise the liquid level in the bottom draw compartment to the level of the reboiler return, an increase in height of about 20 feet. When the level was first raised, the level was controlled using the differential pressure transmitter. Subsequently a pair of electronic pressure gauges was added to the reboiler side of the column to monitor the level directly, an additional gauge was added to the product outlet line to monitor the level on the bottom draw side of the baffle. In order to avoid damage due to entrainment of liquid through the trays the dP across the bottom trays was closely monitored.

To raise level on the reboiler side of the baffle, the bottom draw side liquid was overflowed to the reboiler side by reducing the bottom product flowrate. The level rose until it reached the top of the baffle when it 'flat lined' as it overflowed.

After one hour overflowing the baffle, level had increased on the reboiler side by 13 feet. At that time, the following two process indications provided evidence that the "A" side reboilers were starting to circulate liquid:

(1) A rapid decrease in steam consumed by 3B reboiler suggested a decrease in liquid flow to the "B" side reboilers, and

(2) A rapid decrease in 4A hot side outlet temperature suggested an increase in duty in 4A reboiler.

Steam was quickly opened to 3A to ensure we maintained "A" side reboiler circulation. Almost immediately, even split and duties were confirmed between "A" and "B" side reboilers. Bottoms outlet flow was increased to stop overflowing the baffle, and bottoms level was returned to normal.

The thirteen foot level increase was probably not enough to get the level of weathered liquid (the liquid that was sitting in the A reboiler circuit for a lengthy period) up to the reboiler return nozzle on the tower. But the fresh liquid that had entered the reboiler as level was being raised provided enough vaporization to gas-lift the weathered naphtha into the tower and start circulation.

The column remains prone to losing reboiler duty. This is believed to be because a portion of the bottom tray liquid bypasses the reboiler side of the baffle. The bottom tray is an active tray. It remains unclear whether the tray weeps enough liquid to cause the problem or whether some of the liquid directed to the reboiler draw side of the baffle falls into the product side of the bottom. To fix this in the future the bottom tray will be replaced with a total trapout tray to ensure all bottoms liquid must pass through the reboilers.

Until the plant gets to an opportunity to install the chimney tray, the reboiler continues to stall periodically and raising the liquid level has become a standard procedure that has been executed several times since by operations to regain lost reboiler circulation

Lessons Learnt

The investigation highlights the importance of good troubleshooting and problem diagnosis for overcoming bottlenecks. In our case, pressure drop calculations, close design reviews and examination of operating charts, and gamma scans, led to the correct diagnosis that enabled debottlenecking an entire gas plant with no hardware modifications.

Maldistribution between parallel reboiler circuits has received little attention in the literature, with no published case studies picked up by Kister's survey[2]. To the best of our knowledge, this is the first time this type experience is reported. This maldistribution is especially severe at low rates, when the pressure drops in the reboiler circuits are low. Due to the low pressure drop, the fluids find their favourite paths and steady states, and can easily swing from one steady state to another, promoting instability. With parallel reboiler circuits, careful analysis is important of the pressure balance both at the normal and at a reduced rates operation, and should be performed at the design. This case study demonstrates that increasing the pressure drops in these circuits can help route the fluids to the preferred patterns and counter this maldistribution.

Baffled systems have their benefits[1], but can also generate or amplify hydraulic problems, and need to be designed carefully. In the case of the deethanizer stripper, the baffle was far taller than it needed to be. Had it been even a couple of feet shorter, it would have been easy to raise the liquid level in the bottom compartment above it and achieve better operational flexibility.

It is important to provide flexibility when designing once-through thermosiphon circuits. As pointed out[1,2,4], they can be quite troublesome, and simple features that can convert them temporarily to circulating thermosiphon circuits can go a long way towards easing startups and avoiding process bottlenecks.

This case teaches a number of very important key lessons:

- With a parallel reboiler system, maldistribution between parallel circuits does occur and can cause severe bottlenecks.
- The pressure balance both at high and at low rates needs to be considered at the design.
- It is important to provide flexibility in once-through reboiler systems that will allow operating as a circulating thermosiphon system, at least at low rates.
- Finally, good troubleshooting can provide low cost (in this case, no cost) debottleneck of an entire gas plant.

References

1. Fractionation Research Inc. Design Practices Committee "Reboiler Circuits for Tray Columns", Chemical Engineering, p.26, January 2011.
2. Kister, H. Z. , "Distillation Troubleshooting", Wiley-Interscience, NJ, 2006.
3. Kister, H. Z., T. C. Hower Jr., P. R. de M. Freitas, and J. Nery, "Problems and Solutions in Demethanizers with Interreboilers," Proc 8th Ethylene Producers Conf, New Orleans, Louisiana, 1996.
4. Kister, H. Z., "Distillation Operation", McGraw-Hill, NY, 1990.

Clark Solutions for AIChE's Kister Distillation Symposium 2017

On the Role of Proper Mist Elimination System for Distillation Tower Performance

Bruno B. Ferraro[1], Vitor A. Sturm[1], Michael D. Montani[1], João P. Torgal[1,2], Alexandre C. Bastida[1], Nelson P. Clark[1]

[1]*Clark Solutions, São Paulo, SP, Brazil*
[2]*University of São Paulo, São Paulo, SP, Brazil*

Abstract

Mist eliminators usually represent a low percentage of column internals overall cost, nonetheless these devices can increase tower capacity and separation efficiencies. A proper understanding of mist eliminators fundamentals, design and installation promotes better distillation operating conditions. This paper presents basic operating mechanisms of employed mist eliminators, basic design equations and installation techniques to ensure proper operation of these devices that also play a role in a fine separation process.

Keywords: Mist Elimination, Entrainment, Mist Formation

1 INTRODUCTION

Engineering will always find room for improvement as ideal solutions will never be met. Improvements can be cost-effectively achieved by means of design enhancements, technologies combination or innovations.

Mass transfer applied to distillation is recognized as a thoroughly developed field in regard to technologies and understanding – and its barriers are well known and studied. This article purpose is to advise on the understandings and limitations of one distillation process barrier – entrainment.

The studies for improving capacity or efficiency require the evaluation of basic mechanisms. In distillation, designers and operators research for internals that will produce finer gas bubbles, higher gas-liquid contact at the packing interface and thinner film interfaces, among others.

As a process consequence, entrainment is natural wherever gas and liquid phases are in close contact. As driving forces are further increased inside distillation columns – higher velocities, turbulence and contact, thinner films – to enhance production, an increased entrainment will be onset, and usually finer entrained particles will be generated.

Mist, fine liquid particles laden by a gas stream, is mainly generated by two major mechanisms: mechanical entrainment as a result of combined actions of dynamic shear stress due to forced contact between the liquid and gas phases and solid surfaces with liquid drag by the gas stream, generating droplets of micrometric diameters; and condensation due to thermodynamic changes in the system or chemical reactions in the gas stream, generating droplets of submicron diameters.

Mist eliminators in a properly designed and operated distillation plant will support operation and uptime. A proper selection and design of mist eliminators for an adequate removal of liquid carryover will increase purity and recovery lying in the fact that liquid entrained in process gas causes equilibrium disturbance on downstream stages, hindering proper mass transfer.

2 MECHANICAL ENTRAINMENT

Entrainment is the draw of liquid droplets of varying diameters alongside the gas flow. When two fluids flow together, their interface can become unstable depending on flow velocity and fluid properties such as densities, viscosities and surface tension. This instability may form waves and generate droplets in a wide range of sizes.

For absorption and stripping towers entrainment is the partial transition from countercurrent flow into concurrent flow for a liquid load fraction, partially affecting process efficiency, product composition and recovery.

Inter-tray entrainment is an often discussed subject mainly because it has a direct impact over tray efficiency. Between stages in a distillation there is a higher less-volatile components concentration at the bottom tray. The less-volatile and nonvolatile components can eventually be carried over as entrainment to the tray above and can consequently hinder mass-transfer and affect tray separation efficiency.

Several studies have been conducted over the past decades to determine which operational and design parameters are responsible for entrainment inception, quantity and effects. Colburn (1936) studied the effects of entrainment on tray efficiency. Fair (1961) correlation for crossflow sieve, valve and bubble-cap trays predicts entrainment flooding, including at froth regime for lower tray spacings. Kister and Haas (1987) correlations are good fits for collected data at spray regime for both sieve and valve trays. These studies and correlations have been developed to address not only process conditions but also design parameters such as orifice diameter, tray spacing and clear liquid height. These empirical models rely on large database and do not physically describe entrainment behavior.

After film rupture at the gas-liquid interface, a droplet is propelled at a given angle, rotation and velocity subject to drag, weight and buoyancy. The droplet vertical position will respect the following equation derived from applied position-dependent forces:

$$\rho_L V_L \frac{d^2\vec{x}}{dt^2} = \frac{\rho_G}{2}\left(\frac{d\vec{x}}{dt} - \vec{v_G}\right)^2 C_d A_{L,p} + \rho_G V_L \vec{g} - \rho_L V_L \vec{g} \; (1)$$

To numerically solve Equation (1) one must know the initial droplet vertical velocity $(\frac{d\vec{x}_0}{dt})$ at the gas-liquid interface $(\vec{x}_0 = \vec{0})$ and properly correlate drag coefficient (C_d). Gas and liquid densities (ρ_G and ρ_L), gas bulk velocity (v_G), droplet volume (V_L) and projected area $(A_{L,p})$ are known. The solution of this equation can show whether a droplet of given diameter (d) is ascending or descending depending on its travelled distance $(\vec{x})$. In other words, this equation determines whether or not a droplet reaches the downstream equipment (or tray above). Froth or liquid jets are not described by Equation (1), which addresses only spray regime.

Drag coefficient studies were reviewed by Aggarwal and Peng (1995).

Holbrook and Baker (1934) studied the effect of slot velocity for different tray spacing in a bubble-cap column. They observed that at lower tray spacing there is a carry-over increase for higher slot velocities whereas the inverse was observed at higher tray spacing. Their results can be partially explained by Equation (1). In both trays spacing at higher slot velocity droplets are assumed to be generated at an averaged higher diameter and quantity because of higher available kinetic energy. Independent to tray spacing, initial entrainment will be the same for the same slot velocity, but accordingly to Equation (1) droplets will tend to decelerate proportionally to diameter (d) as they travel upward $(\vec{x})$ until they either achieve an ascending terminal velocity $(\frac{d^2\vec{x}}{dt^2} = \vec{0})$ and are carried over or they fall back $(\frac{d\vec{x}}{dt} \cdot \vec{v_G} < 0)$, reducing entrainment proportionally to droplet distribution as gas rises. A finer distribution would reach higher than a coarser distribution, but a coarser distribution could still be carried over and reach the tray above if spacing is small enough as it was observed by Holbrook and Baker (1934). Therefore, Equation (1) partially explains entrainment stability and disengagement for any defined set of already suspended droplet distribution.

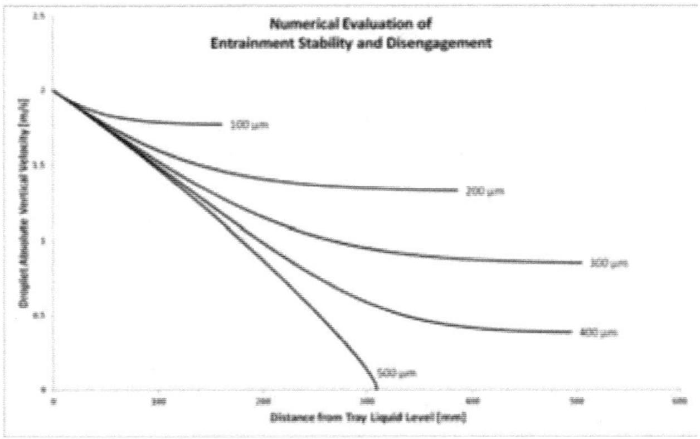

Figure 1. Numerical evaluation of entrainment stability and disengagement (air-water system at 2 m/s).

As vapor flows, gravitational, buoyancy and drag forces acting on droplets tend to equilibrium. The equilibrium point is represented by Equation (1) at $\frac{d^2\vec{x}}{dt^2} = \vec{0}$. A special case is when both $\frac{d^2\vec{x}}{dt^2}$ and $\frac{d\vec{x}}{dt}$ tend to zero.

The Souders-Brown equation (Souders and Brown, 1934) represents this position-dependent force balance special case. Souders-Brown was originally formulated to account for entrainment and capacity on design of fractionating columns. The equation was derived from the force equilibrium between drag ($\boldsymbol{F_d}$), weight ($\boldsymbol{F_w}$) and buoyancy ($\boldsymbol{F_b}$) applied on an averaged droplet diameter and drag coefficient to determine maximum allowed gas bulk velocity in a process:

$$\boldsymbol{F_d = F_w - F_b} \tag{2}$$

Or,

$$\frac{\rho_G}{2}(v_G)^2 C_d A_{L,p} = \rho_L V_L g - \rho_G V_L g \tag{3}$$

Souders-Brown Equation is then stated as:

$$v_G = \sqrt{\frac{4gd}{3C_d}} \cdot \sqrt{\frac{\rho_L - \rho_G}{\rho_G}} \tag{4}$$

And further simplified as:

$$v_G = K\sqrt{\frac{\rho_L - \rho_G}{\rho_G}} \tag{4}$$

The K-factor is often employed to set velocities for sizing a number of processes and internals, among then, trays and mist eliminators.

As a simplification for the de-entrainment problem, the K-factor is widely applied as a pressure dependent constant, relying in an averaged droplet diameter and drag coefficient which does not properly correlate for a wide range of processes and conditions. Notwithstanding its limitations, K-factor is a practical design parameter.

As continuation of Souders and Brown original work, the K-factors, remodeled as Flooding Capacity factors (C_{SB}), are frequently correlated for different operating conditions and design parameters. Equation (5) describes Kister and Haas (1987) correlation:

$$C_{SB} = 0.0277(d_o^2\sigma/\rho_L)^{0.125}(\rho_G/\rho_L)^{0.1}(TS/h_{ct})^{0.5} \qquad (5)$$

While Souders-Brown correlation gives a generalized approach, Kister and Haas (1987) empirical correlation also attempt to implicitly evaluate droplet generation with orifice diameter (d_o), surface tension (σ), liquid density (ρ_L) and vapor to liquid density ratio (ρ_G/ρ_L) as well as implicitly evaluate droplet distribution with tray spacing (TS) and clear liquid height at the froth-to-spray regime (h_{ct}).

Equation (1) only physically describes droplet distribution variation between trays for a given set of droplets.

None of these models properly describe droplet generation initial distribution neither coalescence nor atomization.

3 CONDENSATION

In a distillation column, pressure and temperature gradients applied over a gas stream may cause condensation with respect to composition. Although condensed liquid is consequently a form of liquid entrainment with a different composition compared to the gas stream composition, the gas-condensed liquid stream overall composition that reaches the next stage tray will be the same as the gas stream composition that left the previous stage tray unless there is disengagement. The condensed liquid entrained in the gas stream is composed mainly of less volatile components and would already have been condensed and separated on the next stage tray.

This entrainment is initially made of sub-micron droplets, differently from large droplets observed from liquid carryover. These droplets tend to increase in size depending on pressure and temperature gradient, gas and liquid composition and presence of condensation surface such as liquid flow, existing droplets or vessel wall. High levels of condensation cause internal reflux.

4 MIST ELIMINATION DEVICES

Process engineers currently have a wide range of mist elimination equipment to select depending on generated mist particle size distribution.

The simplest form of current mist eliminators employed in industry are sets of crimped parallel plates which cause gas direction change along channels, capturing particles by inertial impaction at every deflection. Chevron or vane mist eliminators parallel plates are separated by distances that typically vary from 10 to 50 mm with 3 to 6 passes.

Chevron mist eliminators bring lower efficiency than other mist eliminators, but also lower pressure drop and higher liquid draining capacity. Chevrons are usually the best option for high fowling operation as blades support higher drainage and allow simple washing when plugged.

Figure 2. MaxiChevron® mist eliminator

Clark Solutions MaxiChevron® Double-Pocket™ is similar to a conventional chevron. However, a set of pair of blades with draining slots replaces conventional blades, promoting draining. Therefore, this equipment is able to operate at even higher vapor capacities than those achieved by conventional chevron in the same service at the expense of higher pressure drop and lower fowling capacity.

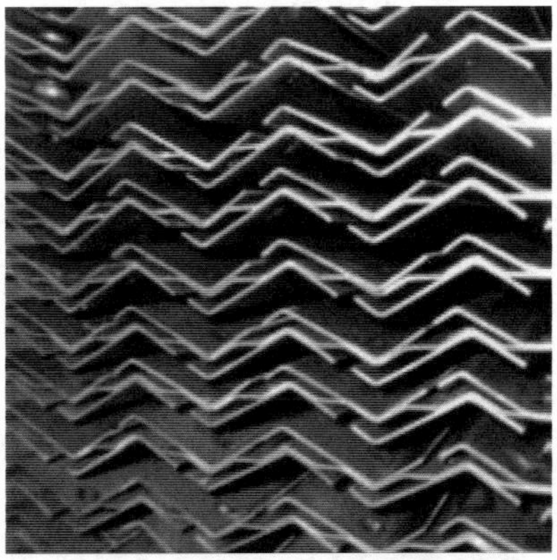

Figure 3. Clark Solutions MaxiChevron® Double-Pocket™

For higher efficiency, fine metal wire provides much smaller and difficult to avoid obstacles than large metal sheet plates. Wire mesh pads are sets of knitted wire mesh, arranged in particular geometries to capture small particles.

Figure 4. MaxiMesh® mist eliminator

Originally developed in the early 1950's with few different mesh densities, wire mesh pads have developed substantially in the last 70 years. Today, they can be offered in a wide variety of geometries, wire sizes and knitting with or without polymers which will allow capture of finer particles in the micrometric range.

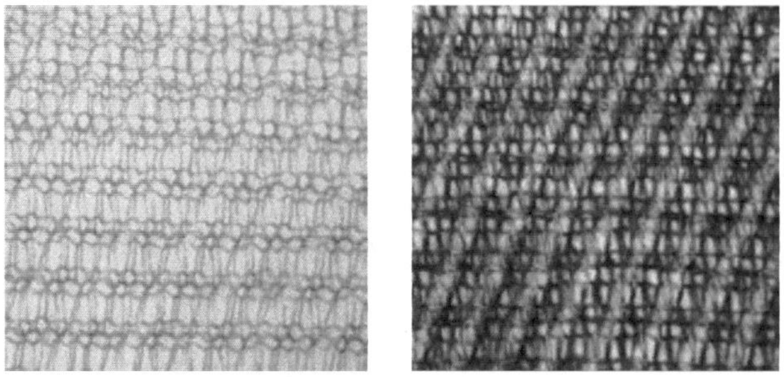

Figure 5. Knitted and Co-knitted mesh

Plain metal wire meshes are designed to achieve capture of particles with few micrometers. More efficient designs will generally require a co-knit mesh, but will also increase pressure drop. Special applications may require special alloys or plastics for the wire mesh structure.

There are other types of mist eliminators such as candle filters and swirl flow cyclones. However, chevron and wire mesh mist eliminators represent most applications in distillation. This paper discusses wire mesh mist eliminators because of its broadest use.

5 DE-ENTRAINMENT THEORY

De-entrainment is the disengagement of suspended droplets within gas flow. Equation (1) describes de-entrainment as an inflexion of the droplet vertical velocity vector in respect to gas flow ($\frac{d\vec{x}}{dt} \cdot \overrightarrow{v_G} < 0$). But de-entrainment can be further increased by mist elimination collection mechanisms.

The drawback of mechanically achieving de-entrainment is pressure drop. Studies have been conducted to describe and correlate collection mechanism efficiency, overall efficiency and pressure drop from dry to flood operations.

Different devices exist in order to separate the entrained liquid from the gas phase, making use of three major mechanisms to separate and coalesce droplets.

Inertial impaction occurs when a gas stream carrying liquid particles passes through an obstacle. The gas will avoid the obstacle, but particles with sufficient inertia will be projected from the gas flow and impact on the obstacle. Inertial Impaction will be as pronounced as the gas velocity raises, as liquid to gas density ratio increases and as target gets smaller (*i.e.* as steep is the gas trajectory change).

Direct interception occurs when particle diameter is small enough to follow gas streamlines, but big enough to collide on obstacles' edges. This effect will increase as the target gets smaller and the mist eliminator void fraction is reduced.

Brownian diffusion occurs with extremely small particles which acquire a Brownian movement caused by energy transfer on collision with randomly moving gas molecules. This movement will drive these particles to eventually collide with the mist eliminator targets. Brownian diffusion capture will be favorably affected by the reduction in target diameter and bed void fraction. Lower gas velocities will increase particle residence time within the mist eliminator bed, resulting in a higher probability collection, increasing the overall efficiency of the mist eliminator designed for this latter capture mechanism

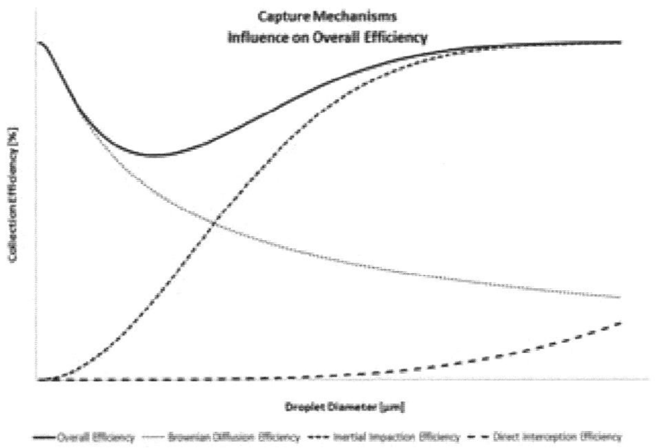

Figure 6. Collection efficiency for different capture mechanisms

(Co-)Knitted wire mesh pads mist eliminator devices collects droplets primarily via inertial impaction, thus, the droplet size range collected is specific. Since entrainment raised from mechanically dragged droplets in distillation towers contains size distributions fit for this equipment efficiency, wire mesh pads are the first de-entrainment devices to be considered – unless fowling is a primary concern.

The efficiency of a filtering media can be expressed as the fraction of the particle distribution that is collected. One way to account for efficiency is to sum contributions of individual wires or fibers, assuming that they are equal in size and provide the same individual collection efficiency for the uniformly packed bed. This can be expressed as:

$$\boldsymbol{E = 1 - exp\left[-f \sum \eta\right]} \qquad \textbf{(6)}$$

The above equation is derived originally from the Kleinschmidt equation (Kleinschmidt, 1939) designated to calculate particle collection efficiency of Pease-Anthony cyclonic scrubbers, and was later used for particle collection on fiber filters, where the exponential term is also referred as Penetration (Pan and Zhong, 2006). The equation was formulated to account for single filament efficiency and generalized for the equipment extension.

This equation (6) can be used for wire mesh mist eliminators, with f representing the effective contact area expressed as $\frac{-A_e LF}{\pi}$. Mesh pad thickness (L), specific surface area (A_e) and fraction of wires in an adequate collection position (F) are usually known.

The efficiency sum in equation (6) can be stated as the contribution of inertial impaction (η_{II}), direct interception (η_{DI}) and Brownian diffusion (η_{BD}) minus the interfering efficiency of interacting mechanisms (η_{IM}):

$$\boldsymbol{\sum \eta = \eta_{II} + \eta_{DI} + \eta_{BD} + \eta_{IM}} \qquad \textbf{(7)}$$

For wire mesh pads, the most substantial contribution for equation (7) is from inertial impaction.

This approach, although a good method, may sometimes be misleading for some mesh-type mist eliminator devices since specific geometric parameters (*e.g.* layering of the knitted meshes and crimps) are not taken in consideration, oversimplifying the mesh pads into a bed of wires.

Therefore efficiency models must be evaluated before applying on wire mesh mist eliminators, since some equation premises are not always met. Mead-Hunter et al. (2014) reviewed several efficiency models from literature. One widely used single target efficiency model was developed by Langmuir and Blodgett (1946).

More specific equations were developed for wire mesh mist eliminators overall equipment efficiency, such as the semi-empirical model of Carpenter and Othmer (1955) and the mechanistic models of Brunazzi and Paglianti (1998 and 2000). These models rely on single target efficiency models such as Langmuir and Blodgett (1946) to predict overall equipment efficiency along all wire mesh layers. Brownian diffusion and direct interception can be neglected for the evaluation of wire mesh mist eliminators efficiency as shown by Holmes and Chen (1984).

Both Carpenter and Othmer (1955) semi-empirical model and Brunazzi and Paglianti (1998) assume no re-entrainment and no liquid hold-up. While Carpenter and Othmer (1955) consider droplets remixing along wire mesh layers, Brunazzi and Paglianti (1998) consider no mixing in their multi-cell efficiency model.

Operating outside creeping flow droplets may not experience all cells as assumed by Brunazzi and Paglianti (1998). Droplets may be disturbed and eventually follow interstitial paths between cells, which can explain Brunazzi and Paglianti (1998) overestimated efficiency.

Most correlations for single fiber efficiency through different mechanisms assumes specific flow fields that exist when diameter-based Reynolds numbers are very low, such as Stokes or Kuwabara Flow Fields (Kuwabara, 1959). For typical vapor velocities the flow field around mesh wires may differ significantly from creeping flow. For these reasons, many wire mesh applications may be misrepresented.

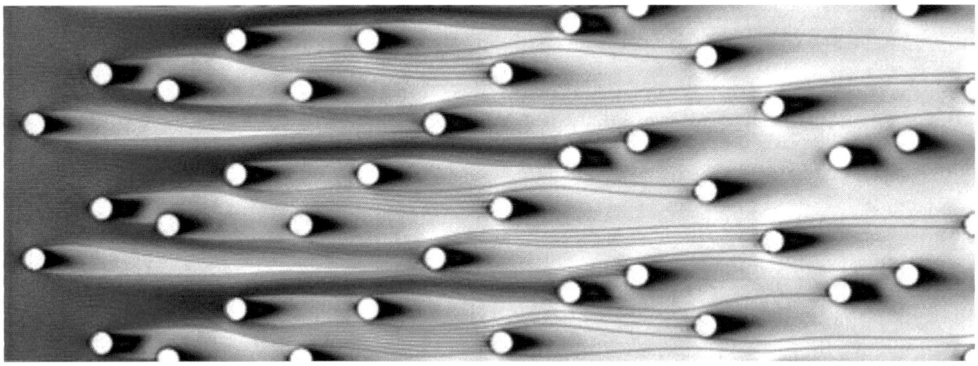

Figure 7. Computational Fluid Dynamics for wire arrays (approximation velocity = 2 m/s, gas density = 1.2 kg/m^3, particle density = 1.000 kg/m^3, wire diameters = 0.3 mm, particle diameter = 10 μm, void fraction =0.96) – layers are not always aligned, but wire spacing inside layers are the same

When designing wire mesh mist eliminators, efficiency and pressure drop balance is sought. Various factors influence equipment pressure drop (*e.g.* operational conditions, mesh thickness, mesh geometry, wire diameters and void fraction). Generally, higher efficiencies bring higher pressure drop. Co-knitting allows for a higher increase in efficiency to pressure drop ratio. Multi-zones allow increase in capacity as lower zones can be designed with lower density mesh and upper zones with co-knitted higher density mesh.

Figure 8. Clark Solutions MaxiMesh® with dual zones (plain metal bottom drainage zone and fiberglass co-knitted high efficiency zone)

Efficiency and pressure drop are often correlated under specific conditions, processes and equipment styles. Many of these correlations are proprietary models.

El-Dessouky et al. (1999) presents a set of correlations for capacity, efficiency and pressure drop built from a series of experiments in typical mesh pads specifically used in multi-stage flash desalination units. Al-Dughaither (2010) studies the behavior of mesh pads on low velocity bubble columns.

6 PRACTICAL DESIGN

Although technology develops and more products and tools become available, many designs still follow previous standards. Souders-Brown equation and empirically determined K-factors are a dominant design technique.

On recurrent cases, and when pressures are near atmospheric, simple standard equipment such as plain metal wire mesh pads and single pocket chevrons can be calculated using Souders-Brown equation with known K-factor values.

As an example, a Maximesh® Style 431 wire mesh pad for an air and water system with an actual gas flow of 10,000 m³/h, operating at 1 bar and 25°C, can have its ideal area calculated following Souders-Brown equation as:

$$K_{431,1\,bar} = 0.106[m/s]$$

$$V \cong 0.106[m/s]\sqrt{\frac{997[kg/m^3] - 1.2[kg/m^3]}{1.2[kg/m^3]}} \cong 3.0[m/s]$$

$$A \cong \frac{10{,}000[m^3/s]}{3.0[m/s]} \cong 0.9[m^2]$$

Clearly, K-factors should be corrected for high pressures and more precise correlations could retrieve a more compact equipment area. Pressure changes, substantial gas or liquid density variations, excessive liquid loads and other factors can lead to a different choice of mist eliminator design and area.

This ideally calculated area may be modified through more detailed calculations to achieve desired droplet diameter range removal efficiency. Nonetheless, this is a valuable widely used method for estimating mist eliminator sizes.

7 INSTALLATION

Proper equipment installation is a major requirement for mist eliminator system performance.

Wire mesh and chevron mist eliminators are designed to fit normal sized manholes of 600 mm (24"). Also, sections should not weight more than 20-25 kg to allow for a single man to hold and manipulate the individual sections. As a general rule, sections width will be around 350-500 mm.

Depending on tower diameter the number of sections will change. They can be full length, half length or even cut in thirds, fourths or sixties, depending on sections weight and tower diameter, similarly to structured packing and trays.

Mist eliminator arrangement and flow direction is chosen depending on calculate area, space availability and geometry as well as operating conditions.

Inside distillation, absorption or stripping towers, mist eliminators installation are mainly horizontal as gas vertically flows.

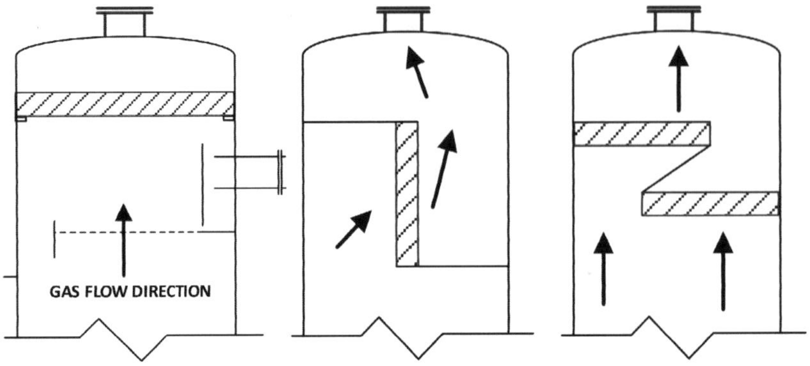

Figure 9. Mist eliminators possible arrangements

However, other arrangements are possible, such as vertical installation with horizontal flow and Z-shapes, depending on design requirements. Computational Fluid Dynamics can assist on design evaluation and improvement of mist eliminators geometry and installation, assessing mist eliminators effective area in respect to flow velocity fields.

For applications where operating pressures are close to atmospheric, some rules of thumb can be applied regarding mist eliminator location.

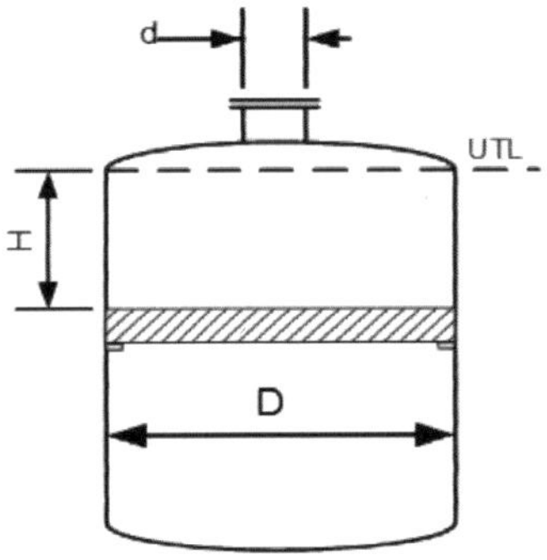

$$H \geq \frac{D - d}{2}$$

$$H \geq \frac{D + d}{2}$$

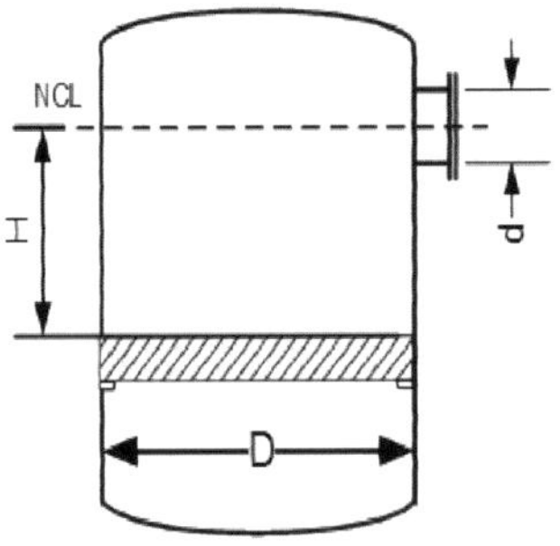

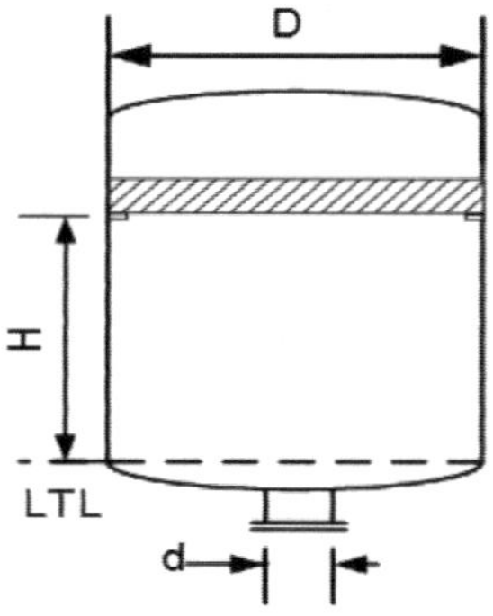

$$H \geq \frac{D}{2} - \frac{d}{3}$$

$$H \geq \frac{D + d}{2}$$

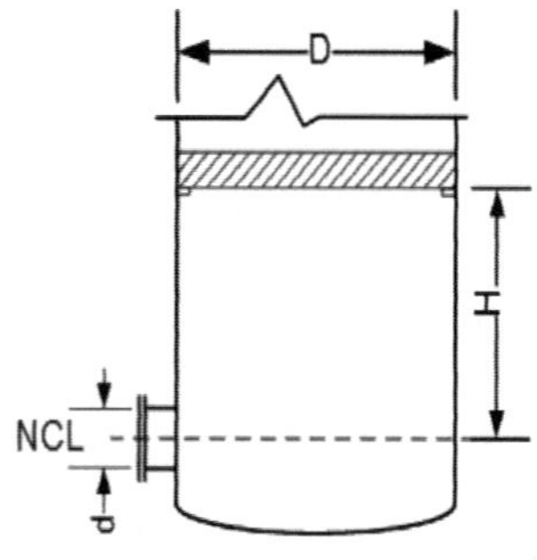

Figure 10. Mist eliminators positioning practical standards

Simulations can be made to evaluate and achieve more compact vessels. Proper installation is sought to enhance gas distribution across mist eliminator surface as uneven flow distribution will result in uneven velocity distribution and compromise performance.

There is a wide range of fastening options to install mesh pads. Tie wires fasten mist eliminator sections to a support ring welded or attached to the vessel wall. The ring contains holes that are usually spaced by 200-250 mm, depending on mist eliminators sections width.

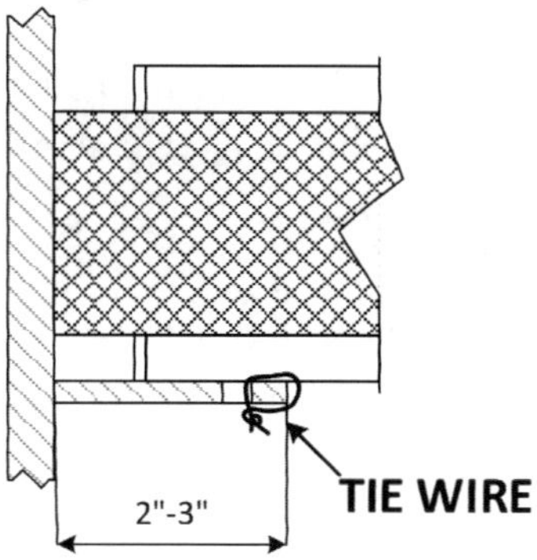

Figure 11. Tie wire fastening

Tie wires can be applied provided a non-corrosive and low pressure service. High pressure systems may submit tie wires to stresses that eventually result one or more sections displacement.

On corrosive or more demanding applications, J-Bolts are the preferred choice. J-Bolts offer a more resistant and stronger fastening between the mist eliminator and the support ring.

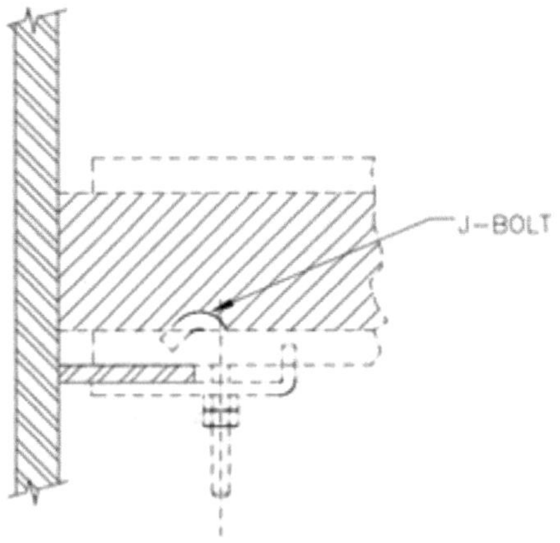

Figure 12. J-bolt fastening

If bottom access is difficult, a third option is the L-Bolt. Similarly to a key, once the section is in place, turning the bolt to the proper position will firmly fasten the section.

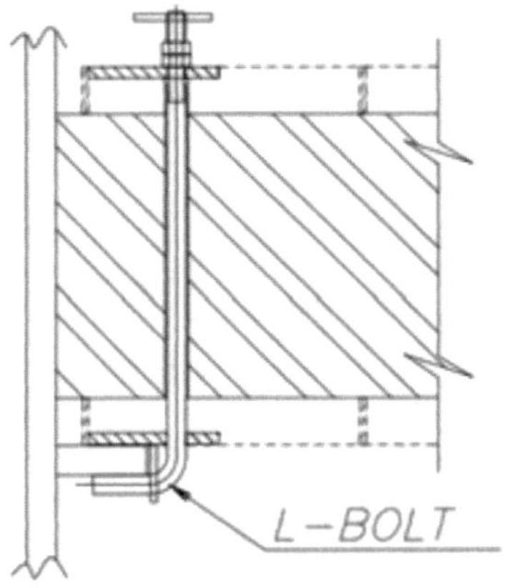

Figure 13. L-bolt fastening

There are other ways to install mist eliminators (*e.g.* between rings with removable sections or using upper support bars). Selection is usually made based on operating conditions and maintenance comfort.

8 APPLICATIONS

Typically the rectifying section is above the feed tray, where top vapor is corrected to achieve desired composition. On this section both foaming and entrainment are less significant due to lower vapor density and velocity. As such, tray spacing can be reduced at the rectifying section with lower compromise of system capacity.

The stripping section, on the other hand, experiences higher vapor density and velocity and as a consequence an increased foaming and entrainment. Solids, when present, are also mainly observed on the stripping section. Therefore, higher tray spacing may be used to guarantee system capacity and avoid mixing between trays. An alternative solution is to provide entrainment removal between trays. Some applications, such as a whiskey still, utilizes an entrainment removal section only between the stripping and rectifying sections, consisting solely of a plate.

But for some operations the feed tray is above the rectifying section, such as the wine feed on sugarcane alcohol distillation. Therefore both liquid and solid entrainment must be removed from the phlegm lateral removal below feed. In some cases a wave-plate mist eliminator is positioned between the distillation column and the rectification column in order to prevent entrainment to reach the rectification column feed.

There are plenty of other well proven typical mist eliminators applications such as wire mesh pads installed at the gas exit of overhead condensers to protect downstream compressors and increase condenser's throughput; chevron or fast draining wire mesh pads such as Maximesh® Style 353 installed on vacuum tower's steam stripper section; and very high efficiency, co-knitted wire mesh pads installed on glycol, amine and solvent drying and absorption towers to reduce carryover losses.

As a practical example one can consider a Jet-A side stream stripper producing 15,000 BPD of Jet-A. Assuming a 0.5% carryover from the stripper the amount of Jet-A which was carried to the Naphtha section would be 75 BPD of Jet-A, roughly $ 125,000 in losses per year considering a $ 5.00 price difference per barrel between Jet-A and Naphtha. A wire mesh pad of 5 square meters for this service will cost around $ 5,000. Payback will therefore be less than 30 days.

This same approach can be applied for other applications. While the economics may not be as favorable, payback is typically lower than 3 to 4 months.

9 CONCLUSION

The growth in mist elimination science reflects the perceived interest in attainable product recovery and purity with these low investment equipment. Proper selection and design of mist eliminators are a cost-effective approach to increase operational revenue and may support capacity increase.

Although entrainment is known as a distillation barrier for nearly a century, mist elimination is still a developing science. New studies and modern computational tools offer new perspectives in the understanding of mist eliminators as new products and developments are established.

10 REFERENCES

Aggarwal, S. K. and Peng, F. 1995. A Review of Droplet Dynamics and Vaporization Modeling for Engineering Calculations. Journal of Engineering for Gas Turbines and Power 117: 453-461.

Al-Dughaither, A. S.; Ibrahim, A. A. and Al-Masry W. A. 2010. Investigating droplet separation efficiency in wire-mesh mist eliminators in bubble column. Journal of Saudi Chemical Society 14(4): 331-339.

Al-Sheikh, J. N. 1959. A Study of Entrainment in Distillation Column for Sieve Plate. Master Thesis, Rice University, Houston. 59 p.

Bujake, J. E. 2007. Beverage Spirits, Distilled. Kirk-Othmer Encyclopedia of Chemical Technology.

Brunazzi, E. and Paglianti, A. 1998. Design of wire mesh mist eliminators. American Institute of Chemical Engineers Journal 44(3): 505.

Brunazzi, E. and Paglianti, A. 2000. Design of complex wire-mesh mist eliminators. American Institute of Chemical Engineers Journal 46(6): 1131-1137.

Carpenter, C. L.R. and Othmer, D. F. 1955. Entrainment removal by a wire-mesh separator. American Institute of Chemical Engineers Journal 1(4): 549-557.

Colburn, A. P. 1936. Effect of Entrainment on Plate Efficiency in Distillation. Industrial and Engineering Chemistry. 28(5): 526-530.

Cristini, V. and Renardy, Y. 2006. Scalings for Droplet Sizes in Shear-Driven Breakup: Non-Microfluidic Ways to Monodisperse Emulsions. FDMP 2(2): 77-93.

El-Dessouky, H. T.; Alatiqi, I. M.; Ettouney. H. M. and Al-Deffeeri, N. S. 2000. Performance of wire mesh mist eliminator. Chemical Engineering and Processing: Process Intensification 39(2): 129-139.

Fair, J. R. 1961. How to predict sieve tray entrainment and flooding. Petro/Chem. Eng. 33(10): 45-52.

Holbrook, G. E. and Baker, E. M. 1934. Entrainment in a Bubble-Cap Distillation Column. Industrial and Engineering Chemistry 26(10): 1063-1067.

Holmes, T. L. and Chen, G. K. 1984. Design and selection of spray/mist elimination equipment. Chemical Engineering 91(21): 82-89.

Ishii, M. and Grolmes, M. A. 1975. Inception Criteria for Droplet Entrainment Two-Phase Concurrent Film Flow. AIChE Journal 21(2): 308-318.

Kister, H. Z. and Haas, J. R. 1987. Sieve tray entrainment prediction in the spray regime. Institution of Chemical Engineers Symposium Series No. 104.

Kister, H. Z. and Haas, J. R. 1990. Predict Entrainment Flooding on Sieve and Valve Trays. Chemical Engineering Progress 86(9): 63-69.

Kleinschmidt, R. V. 1939. Factors in spray scrubber design. Chem. Met. Eng. 46: 487.

Kuwabara, S. 1959. The forces experienced by randomly distributed parallel circular cylinders or spheres in a viscous flow at small Reynolds numbers. Journal of the Physical Society of Japan 14(4): 527-532.

Langmuir, I. and Blodgett, K. 1946. A mathematical investigation of water droplets trajectories. Collected works of Irving Langmuir. Pergamon Press 10: 348-393.

Pan, N. and Zhong, W. 2006. Fluid transport phenomena in fibrous materials. Textile Progress 38(2): 1-93.

Pyott, W. T.; Jackson, C. A. and Huntington, R. L. 1935. Factors Affecting Entrainment in Bubble-Cap Columns. Industrial and Engineering Chemistry 27(7): 821-825.

Souders Jr., M. and Brown, G. G. 1934. Design of Fractionating Columns: I. Entrainment and Capacity. Industrial & Engineering Chemistry 26(1): 98-103.

Ujile, A. A. and Iminabo, J. 2014. Evaluating Sieve Tray Flooding in a Distillation Column Using Kister and Haas; and Fair's Correlations. Chemical and Progress Engineering Research 25: 16-23.

485321 Troubleshooting Tower Carryover

Abstract:

For most traditional TEG contactors, high capacity structured packing is the ideal mass transfer device due to the low liquid rate in this service. However, structured packing should only be used in a fairly clean system as it is susceptible to plugging. For dirty systems, bubble cap trays have been traditionally applied. There are a few high capacity tray technologies available that can be used to achieve a capacity boost over bubble cap trays albeit with higher glycol carryover. These new high capacity mass transfer devices exceed the capacity of most existing de-entrainment equipment, such as wire mesh and vane packs. The capacity of demisting devices is further reduced with increased entrained liquid loading as the high capacity devices push to higher superficial velocities. Thus, TEG carryover can ultimately dictate the capacity of the TEG contactor if the de-entrainment internals are not specified correctly or not modified during a revamp to handle the additional TEG carryover. High concentrations of H_2S and CO_2 and even small amounts of contaminants such as heavy hydrocarbon liquids, iron sulfides, salts and other organics in the feed gas can also present operational difficulties. These contaminants can be absorbed in the TEG, accumulate, foul, foam and react to form other byproducts. Therefore, the cleanliness of the system must be considered when selecting internals.

This article summarizes the lessons learned in debottlenecking a "dirty" TEG contactor at our LaBarge Black Canyon Facility in Wyoming US. The facility modifications included replacement of all bubble cap trays with fixed valve trays, modifications to the downcomer inlet sump and seal pan, addition of a new Inlet Vane Diffuser at the gas inlet to the tower, a new full stream carbon filtration system, and optimization of the mist eliminator section at the top of the tower. This work was performed in progressive steps between 2001 and 2004. After completion of this work, the following conclusions can be drawn:

- A clean TEG system will minimize foaming and increase capacity as well as decrease maintenance, operating expense and downtime. As demonstrated, the full stream carbon filtration system reduced TEG liquid carryover by 65-70%.
- Although vendor data was available on demisting cyclones with glycol at elevated pressure and good experience was also available with cyclones in hydrocarbon service at high pressure, this application was outside the data range. The entrained liquid load is also an important parameter that often is ignored or is unknown.
- The combination of the high capacity trays and mist elimination equipment, along with the new solvent filtration system, allowed the gas capacity of the Black Canyon Facility to be almost doubled.

479973 The Ergun Equation: Right or Wrong?

Abstract:

When it comes to fluid flow through packed beds, the Ergun equation is the most popular model for relating fluid pressure drop to flowing fluid Reynolds number. The popularity of this equation results from it being

- presented in most engineering texts concerned with such flow;
- taught as "the gospel" in most top-tier engineering schools.

Thus, future engineers leave university with the Ergun equation well-represented in their intellectual toolkit.

Unfortunately, the Ergun equation rests upon shallow foundations laid in shifting sands. Basically, the Ergun equation is nothing more than the addition of the Blake-Kozeny and Burke-Plummer equations, both of which contain a plethora of questionable assumptions.

This presentation reports an experimental test of the Ergun equation and discusses the theoretical foundations of the Ergun equation.

This presentation also contains a "new look" at fluid flow through packed beds and demonstrates how to scale packed beds from laboratory-sized equipment to commercial-sized equipment.

480821 Product Development Challenges for Small or Start-up Companies

Abstract:

Product development challenges for small or start-up companies

Pratik C Pednekar, Research Chemical Engineer, Mid-Atlantic Technology Research and Innovation Center
pratik.pednekar@matricinnovates.com

John Dever, Executive Vice President, Mid-Atlantic Technology Research and Innovation Center
jack.dever@matricinnovates.com

Development is a key stage towards the commercialization of a product or a process. For a startup company or a department that is still in the process of looking for seed money, and is currently working with limited resources, this step becomes even more crucial. In order to engage potential investors, a small scale demonstration of the technology or a scale up of the product to send samples becomes a limiting step due to the investment required to perform this. Setting up a facility from the ground up may not be an option for many such companies due to cost or risk reasons. Other hurdles include the need to hire specialized vs flexible manpower, to have continuous access to expensive analytical tools and to be in touch with market experts who have worked with products that may put the company through a slow growth curve.

Another key aspect of this stage is the flexibility to handle growth of the project. For a company developing a candidate product, the goals for a project may change very rapidly from producing a defined amount of product to optimizing the product properties to being able to demonstrate continuous production or sustained quality. These changes can make some of the initially obtained equipment or workforce obsolete through the course of development and these are added costs towards a necessary path of progress for the company. Additionally after all this effort, it is quite possible that the product or technology may not be cost effective or desirable at this point in time. To know this information as early as possible through the development stage can make a huge difference to the survival of a start-up company.

Getting in touch with a facility that has most of the infrastructure already in place can be vital to help the project get off the ground quicker and cost effectively. This could save on costs for setting up a facility that is compliant for production, procuring the right equipment quickly and obtaining the manpower to build, operate and guide the development process. Through university or company collaboration or through access to the consultation entities, the development stage can be streamlined to the benefit of the start-up company. However, each of these options differ in several ways such as their access to resources, industrial contacts, intellectual property agreements, ability to adapt the manpower according to the growth of the project etc. Through working with start-up companies previously, MATRIC would like to present a discussion on some of the challenges that such companies face and how technology development companies with such facilities can help make the product development phase cost effective and efficient. The discussion is intended to demonstrate some of the cost saving means that can be employed through the development stage to reduce risk and at the same time provide access to a flexible workforce that can be adjusted based on the growth of the project.

479178 Scaling Packed Bed Columns and Reactors

Abstract:

In 1952, Sabri Ergun published an article reconciling all the data pertaining to flow through packed columns at that time. The equation presented in that paper has since become known as the Ergun equation. It has been used with varying degrees of success since its publication. The Ergun equation has various problems. First, it uses a Reynolds number based on particle diameter, not fluid conduit diameter. Thus, the Reynolds number used in the Ergun equation does not describe fluid flow. Second, the Ergun equation was developed for spheres. However, myriad odd-shaped packing exists; thus, the Ergun equation must be modified when using non-spherical packing. That modification is generally encapsulated in a dimensionless "sphericity factor" ... a cheat factor, in other words. Third, the Ergun equation provides no ready mechanism for upscaling or downscaling a process.

In this paper, we address these problems via Dimensional Analysis. We use Dimensional Analysis to formulate a new method for characterizing flow through packed columns. We also propose a new method for characterizing the packing material. Since our results obey the laws of similarity, our analysis can be used for upscaling and downscaling packed columns, which we will demonstrate in this presentation.

Best Practices for Writing Operating Procedures and Trouble-Shooting Guides

William G. Bridges, President
PROCESS IMPROVEMENT INSTITUTE, INC. (PII)
1321 Waterside Lane, Knoxville, TN 37922
Phone: (865) 675-3458
Fax: (865) 622-6800
e-mail: wbridges@piii.com

Revonda Tew, Principal Engineer & Instructor
PROCESS IMPROVEMENT INSTITUTE, INC. (PII)
1321 Waterside Lane, Knoxville, TN 37922
e-mail: rtew@piii.com

Prepared for Presentation at
13th Global Congress on Process Safety
San Antonio, TX
March 27-29, 2017

UNPUBLISHED

13TH GLOBAL CONGRESS ON PROCESS SAFETY

Best Practices for Writing Operating Procedures and Trouble-Shooting Guides

William G. Bridges, President
PROCESS IMPROVEMENT INSTITUTE, INC. (PII)

Revonda Tew, Principal Engineer & Instructor
PROCESS IMPROVEMENT INSTITUTE, INC. (PII)

Keywords: Procedures, Human Factors, Risk control, process safety management, PSM

Abstract

There is no complete, best practice guideline or textbook for writing operating procedures and trouble-shooting guides. This paper presents the proven, best approach for developing accurate operating procedures and for ensuring the page formatting and step writing are optimized to reduce human error rates. This approach and the 32 rules established in this paper are based on the foundations set by Swain and others (in 1970) for control of human error rates, but uses experiences from more than 100 sites where this approach has been successfully followed. The approach and rules for developing operational troubleshooting guides (procedures for responding to process deviations such as those needed for Human IPL) are again the best approaches found and have been applied successfully since the early 1990s. Several case studies are provided that show the gains from following this approach. The guidelines in this paper build upon ones presented in 1999 at CCPS and 2016 at GCPS.

Background

Operating procedures have always been crucial to the safety, quality, and productivity of process systems. With the advent of new safety and quality standards such as OSHA's process safety management (PSM) regulation, OSHA's personal protective equipment (PPE) regulation, EPA's risk management program (RMP) regulation, and the ISO 9000 quality standard, many companies are facing the daunting task of developing or upgrading their procedures to satisfy varied and sometimes complex and conflicting requirements.

Although there are overlapping characteristics that the various regulations and standards share, each approaches the procedure-writing process from a somewhat different perspective (e.g., quality or safety, protection of workers or protection of the public/environment), and the required level of detail for documenting the procedures differs greatly between these regulations and standards. For instance, the ISO 9000 quality standard advocates documenting procedures and work instructions that impact quality and suggests keeping the procedures as simple as possible, while the OSHA PSM regulation requires detailed procedures that address: all operating modes (startup, shutdown, etc.), operating limits, consequences of deviations, means to avoid hazards, safety and health considerations, and safety systems and their functions. How can companies develop and maintain procedures that ensure productivity and simultaneously satisfy the different regulations and standards? The key is to remember the ultimate goal of the regulations and standards: to reduce human errors that can impact quality, productivity, and/or safety.

Procedure-related errors are errors that occur because some characteristic of the procedure caused task performance to fail. ***This is currently the most critical human factor at most sites since 90% of accidents have at least one root cause related to mistakes within procedures.*** Reducing these procedure deficiencies can reduce human error rates by a factor of 2 to 10, normally. [1, 2]

The concepts and rules presented in this paper are applicable to all industrial settings. Most importantly, this approach **is effective in satisfying both safety and quality concerns.** With this approach, procedures can and have been developed that:

- Provide a sound basis for training
- Provide the users and supervisors with an understanding of what can go wrong and how it can go wrong
- Provides management with an understanding of the inherent risk (to quality, productivity, safety, etc.) of an activity (i.e., the activity addressed by the procedure) and an appreciation of the safeguards that are being relied upon to mitigate this risk.

Three (3) case studies are presented (with data from other sites as well), which illustrate this approach, along with supporting the benefits of the accurate and clear procedures. These case studies' results are representative of the benefits seen at more than 60 sites where the same approach has been followed.

Fundamentals of Writing Effective Procedures

When developing procedures, or any written material, always remember that the burden of written communication is on the writer, not on the reader. It should never be up to the reader to decipher the meaning of a message.

- Engineers who write procedure manuals "tend to include information that is more relevant to them than it is to the operators actually running the unit" [Sutton, 1992].[3]
- When writing procedures, your guiding principle should always be: if the written procedures are difficult to read, difficult to locate (e.g., if titles do not have a clear meaning), or inconvenient to use, they will seldom be used [Swain, 1983].[1]

Deficient Procedures are the most prevalent problem in process industries since procedures have not traditionally been developed from the perspective of optimizing human factors; instead procedures have been traditionally developed to meet a compliance requirement. Examples of procedure deficiencies (inaccuracies) include:

- Incorrect/incomplete/nonexistent (most procedures we have audited have been only 70-75% accurate – the inaccuracies include missing critical steps, steps as written are not what needs to be done, or the steps are out of sequence)
- No/misplaced/incorrect information in warnings (for example, a warning should **never contain** the action to take; it should instead **emphasize** the action to take) and warnings should be located BEFORE the step applicable to the warning
- Poor format and presentation rules

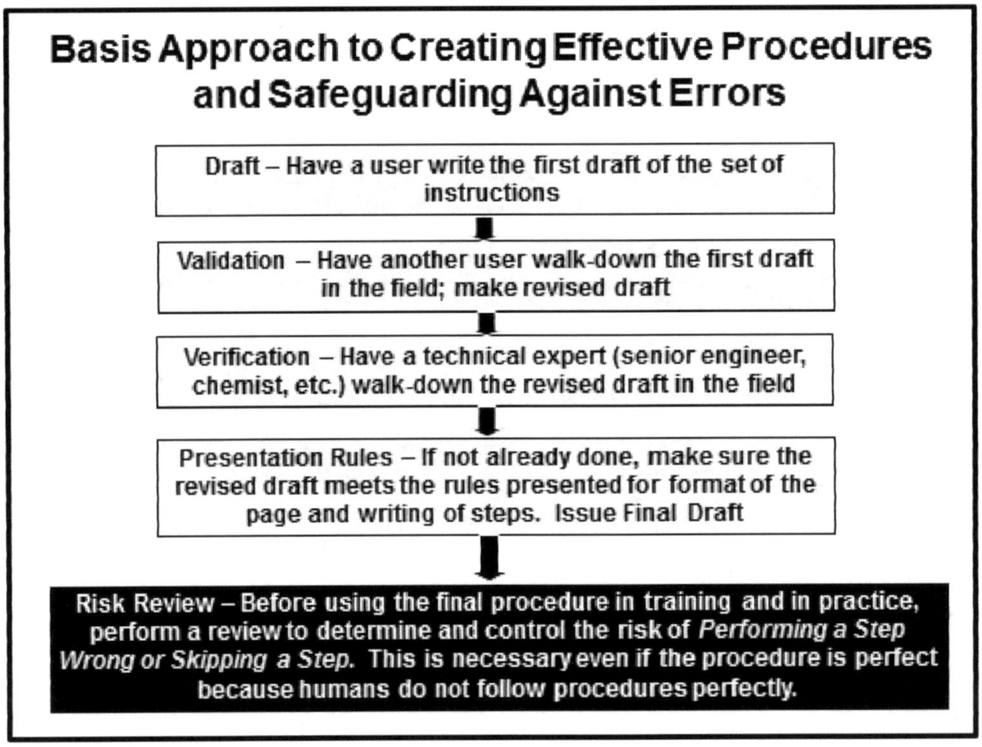

Figure 1: Best Approach to Developing written Procedures – courtesy of Process Improvement Institute, Inc. (PII), copyright 2003-2017[4]

The flowchart above describes the basic sequence of steps for writing effective procedures. Note that some basic rules are contained in the flowchart, such as "operators write operating procedures," "maintenance craft-persons write maintenance procedures," and "Lab technicians write lab procedures." Procedures are NOT written by engineers or department superintendents. This is a critical to ensure procedures are written in the user's common language and are written at the right level of detail.

Universal Purposes of a Procedure[4]

1. A written guide to accomplish initial training of employees
2. A quick reference manual for experienced (trained) employees

The Right Information to Write – Content is KEY[4]

The procedure procedural content controls the human error related to "procedure-based" error – if content is wrong, the trainee is very likely to learn the step wrong (trainer uses the procedure as a basis for training or the trainee refers to the procedure step that is wrong when executing the procedure).

Even if the content is correct, not following page formatting best practices can still increase human error rates.

If the content of the step is wrong it does not matter if the presentation is clear. If the content is perfect, you can further reduce error rates by 50% to 70% by presenting the information properly.

Ask experienced mechanics or operators to describe a simple task that they perform. You will quickly see that even simple tasks can involve many steps. You should consider the following questions when writing each procedure step:

- How is the step performed?
    - Does the action involve interacting with a computer terminal, an automatic controller, or simple devices (gauges, valves, micrometers)?
- Can the actions be performed as written and in the sequence written?
    - Can the equipment be operated as specified?
    - Can the steps be physically performed?
    - Does the user have the training or experience to understand and carry out the action using the information available, or is additional information needed?
- Does the user need to be alerted of any potential hazards (Cautions or Warnings) or need any supporting information (Notes) before performing the action?
- Does the user need to know specific operating ranges or limits to:

  - Perform this action
  - Recognize the successful completion of the action
  - Recognize an actual or potential problem
  - Make a decision

For example, does the user have to verify temperature or pressure visually? Is the user waiting for a computer command or verification? How should the process react? Will exceeding a limit result in an accidental chemical release, personal injury, or quality concern? What is the chance of not operating within safe operating limits, and what would be the consequences? If operation of the component is critical and a deviation will probably cause a failure, state the corrective actions. If failure is unlikely, keep going.

- Is needed information found on an instrument, panel, or monitor or is it in the procedure or another source such as a graph, table, drawing, or specification sheet? Should this information be included in the procedure or be referenced?
- What is the next logical step? How is the next step affected by what is performed in the current step?
- What are the results of improper task performance? If an action is critical, spell it out in detail. For example, when starting a special pump that may bind or cavitate if not vented and primed properly, the procedure may require more information than simply stating "start the pump."
- Is the action frequently performed? Is it an action that is easily overlooked? Is this a complex piece of critical equipment that has a low frequency of use? These questions can help determine the level of detail in your procedure.
- Is the action performed so infrequently or is it so complicated that the user is not sure how to do it? Is the action so complicated that nobody is ever certain it's done right the first time?
- Is the decision point clearly defined if a decision is required? Unclear decision points can cause arguments and delays in performing actions.

To address the concerns listed above, a few simple rules are needed for developing content:

- The procedures should be written at the level of someone who has just completed the basic training for that task. Do not write the procedure for someone just hired or for the 10+ veteran.
- Clearly identify ahead of time what activities need procedures to help ensure error rates are controlled low enough, and identify which activities are considered common "skills" of all of the staff.

  Example: Starting a pump is typically considered a "skill." This means a procedure step can simply be "Start the benzene recycle pump (P-119)." The procedure does not need to explain how to start the pump. But, starting a pump is a skill that nonetheless must be learned; it requires doing several sub-steps related to positioning of intake and discharge valves, how the pump is throttled at startup, checking local pressure gauge, etc. So, a training module is needed for starting pumps of various types, and the operator needs to learn this necessary skill. Once the skill is learned, it will be applied to so many various cases and so often, that a procedure is no longer needed as a "refresher" on how to "start the pump."

- The first draft of the procedure must be walked-down in the field by another user. Simply reading through the procedure does not catch enough of the mistakes.
  The revised draft should be reviewed in the field by a technical staff person, such as a process engineer (for operating procedures).
- The procedure needs to be checked to ensure it follow the page format and step writing rules described in the next section. This ensures the accurate steps are clearly presented.

The target is to reach an accuracy of 95% or better (so, no more than one wrong or missing step out of 20 steps). The following have been observed in the field by PII staff by direct data collection at more than 110 sites/plants around the world, based on walk-down review of several to dozens of procedures at each unit/plant:

- At 95% accuracy or better and when the same procedures follow 80% of the rules for procedure clarity (presented next), then most users will follow the written procedures and will try to keep the procedures up-to-date
- At about 85% accuracy or less, about half of the users stop using the procedures
- At about 70% accuracy or less, less than 10% of the users will refer to the procedure or will try to keep it up-to-date. ***So, the written procedures are useless***
- Unfortunately, the typical operating procedure walked down by PII staff (accompanied by senior operators, a process engineer, and a shift supervisor) is about 75% accurate (so one step in four is missing or wrong). Usually these inaccurate procedures also follow less than half of the best practices for procedure format, presented next.

PHA/HAZOP of Procedures

Just as critical as getting the content accurate and presenting the steps clearly, is performing a PHA/HAZOP of the written steps to determine if you have enough independent protection layers (IPLs) against errors that the "imperfect" operators will eventually make when using the "near perfect" written procedures. This analysis is described in detail in Chapter 9 of the *Guidelines for Hazard Evaluation Procedures*, 3rd edition, 2008[5]. **For an update on the methods for such analysis and for a discussion of the extra emphasis on PHA of procedures by US OSHA, refer to the more recent paper (Bridges, Marshall; 2015).**[6, 7, 8]

Procedure Clarity

The importance of the accuracy of the procedure steps were discussed above. The clarity of the procedure steps (how they are written and how the page is formatted) is also important. Following best practices for step and page format reduces human errors by a multiplying factor of 3 to 5.[1, 9]

The best practice rules for writing and validating procedures have been published for many years *(see Bridges & Williams, 1997; Madden & Bridges).*[11, 12] These rules include best practices for formatting of the pages and steps. These have been gradually improved over the past decades and now are incorporated into PII's training materials.[4]

Table 1 below is a checklist (in auditing format) based on the current set of best practice rules for developing accurate and clear operating, maintenance, and other work-instructions/procedures.

TABLE 1 PROCEDURE QUALITY CHECKLIST *(courtesy PII, copyright 2003-2017)*[4]

#	Issue	Response
	Procedure Content Checklist	
1	Is the procedure drafted by a future user of the written procedure? (Engineers should not author procedures to be used by operators or maintenance staff.)	
2	Is the procedure validated by a walk-down in the field by another future user of the procedures?	
3	Is the procedure reviewed and commented on by technical staff (engineers or vendors)?	
4	Is the procedure checked versus the Page and Step format rules below?	
5	Is a hazard review of step-by-step procedures performed to make sure there are sufficient safeguards (IPLs) against the errors that will occur eventually (when a step is skipped or performed wrong)?	
6	Is the content measured using "newly trained operators" to judge the % of steps that are missing, steps that are confusing or wrong, and steps that are out-of-sequence? (A score of 95% accuracy of content is good; 98% should be the targeted average.)	
	Page Format Checklist	
1	Is the title of the procedure the largest item on the page?	
2	Is the procedure title clear and consistent with other titles, and does it uniquely describes the topic?	
3	Are the document control features the smallest items on the page?	
4	Are temporary procedures clearly identified?	
5	Is white space used effectively? • Is there one blank line between each step? • Does indentation help the user keep their place? • Are the margins large enough to reduce page congestion?	
6	Is type size is 12 pt font or larger?	
7	Is mixed case used for words of steps, with ALL CAPS used only for special cases (such as IF, THEN, AUTO, and WARNING)?	
8	Is the step number very simple (single level of number used)? Only an integer?	
9	Have sections or information not necessary to performing the steps been moved to the back or to another part of the manual or training guide?	
10	Are section titles bold or larger than the text font? Do sections have clear endings?	
11	Is the decision on electronic presentation versus hard copy correct? Are electronic linkages to procedures clear and accurate and easy to use? If paper is chosen, is the color of the paper appropriate?	
12	Is the overall page format (such as Outline format or Two-Column [T-Bar] format) appropriate to the use of the procedure?	
13	Are play script features added for tasks that must be coordinated between two or more users? • Play script is normally used when there are two or more hand-offs of responsibility for steps.	
14	Are rules followed for formatting of Warnings, Cautions, and Notes? (See annotated rules, such as Warnings are for worker safety and Warnings must clearly standout from rest of page.)	
	Step Writing Checklist	
1	Is each step written as a command?	
2	Is the proper level of detail used throughout? This is judged based on:	

	• *Who* will use the procedures • Same level of detail used in similar procedure steps	
3	On average, is there only one implied action per instruction? Best practice is to average 1.2.	
4	Does the procedure indicate when sequence is important? • If sequence matters, each step should be numbered (with an integer or letter) • If sequence does not matter, bullet lists should be used	
5	Are only common words used? Apply "education" level test (5 grade reading level is best)	
6	Do all acronyms, abbreviations, and jargon aid understanding? • Develop a list of such terms for use in procedures *and* communication. • Use terms that users use (within reason)	
7	Is each step specific enough? No room left to guess/interpret: • The meaning of a word or phrase (Check vs. Make sure) • The intent of a step or series of steps • A desired quantity or value • To what equipment the step applies	
8	Is the procedure free of steps that require in-your-head calculations? • Values expressed as ranges rather than targets with error bands • Conversion tables, worksheets, or graphs provided where needed	
9	Are graphics to the user's advantage? • No explanatory paragraphs or lengthy instructions that could be replaced by a picture • No impressive graphics that provide no real advantage	
10	Are references to the user's advantage? • No lengthy explanations or instructions that could be replaced by branching to a reference • No references to a procedure that references still another • No gaps or overlaps between this procedure and a referenced document • If branching, must branch to a procedure, not to a specific step in a procedure	
11	Are rules followed for writing warnings, cautions, and conditional steps? • No more than 2 per page • No actions within a warning or caution (actions must always be numbered steps) • Warnings and Cautions contain descriptions of potential consequences	

Each of these rules is explained in detail in the course *Writing Effective Operating and Maintenance Procedures*, by PII[4] (visit www.piii.com for more details). Each rule is backed by experimental or observed data that indicate there is a clear advantage in error reduction by using the rule. Below are a couple of examples from others on procedure writing rules.

> ***Proper Page Format:*** *In one study, Haney took a procedure presented in narrative format and converted it to column format. He then asked experienced technicians who had been completing the procedure using the narrative format to complete it as presented in the column format. He found that when the technicians used the column format, their errors were reduced by a factor of 3 compared with the narrative format [Haney, 1969].*[10]

> ***Including Supplemental Checklists:*** *Checklists can be effective aids to performance and often should be used to supplement procedures presented in any format. Checklists can be particularly effective in terms of preventing errors of omission. Some of the literature supports using checklists only for simple procedures that do not need to be performed under time or other constraints [Sutton, 1992].*[3] *However, there is also support for using checklists for emergency procedures in which a core set of steps is used to respond to emergency situations, particularly when the actions follow a simple sequence of steps for shutting down a process [Connelly, 1992].*[13] *Overall, experimental data have shown that the probability of errors of omission and commission can be reduced by a factor of 3 to 10 when checklists are*

used as reminders of key steps in a work instruction [CMA's A Managers Guide to Reducing Human Error, 1990].[9]

For more data on the history and development of these rules, see Bridges & Williams (1997)[11] and Madden & Bridges (2016)[12] especially refer to originating data and discussion for most of the rules (Swain, 1983).[1] **The procedure on the next page follows the rules** and is an example of the T-Bar format. Several other page formats are also valid.

Trouble-Shooting Guides (Procedures for Responses to Critical Deviation Alarms)

Trouble-shooting guides are a special form of operating procedure. They are written for the actions we want the operators to take to recover from a process deviation, *before an emergency situation occurs.* They are called guides since rarely can we predict the process conditions at the time the action is required. Trouble-shooting guides (and necessary training and drills) are required for any action that is considered a Human IPL. The Action Limit is what we show as the Min or Max in a Trouble-Shooting Guide. The action limit triggers the demand to use the trouble-shooting guide.

The starting criteria should be true before going to the effort of developing a procedure for response (before developing a trouble-shooting guide):

- The response is typically still possible, but it is time dependent. The time available is called the process safety time (PST). The operator must complete the diagnosis, make the necessary change(s), and make sure they are out of harm's way by the end of the Maximum Allowable Response Time (MART). [14]
- We usually set an alarm or a pre-alarm to trigger this action. This is usually before the shutdown triggers (ESD occur automatically) or release points (PSV set points) are reached
- The Min and Max shown in the Trouble-shooting Guide table are not the absolute safety limits for a system, but are instead some values that leave us some time to take action to prevent from reaching the absolute limits.
- There is still time to prevent or avoid the final consequence that could occur if we reach the ultimate limits of the process. Usually, we want the MART to be ½ or less of the PST, and we want MART > 10 minutes for trouble shooting in the field/plant and MART > 5 min for trouble-shooting only from the control room. [14]

Figure 2: Example of Procedure that Follows Most of the Rules in Table 1

PII CHEMICAL COMPANY — Page 1 of 2
Unloading Monomer from Tanker to Storage — Revised 11/4/02, Printed 2/21/16

Unloading Monomer from Tanker to Storage

STEPS	DETAILS
1. Wear standard PPE, plus rubber gloves and full-face organic respirator.	Standard PPE includes hard hat, safety glasses with side shields, and steel-toe shoes.
2. Weigh in the tanker.	Record weight on GROSS line of form.
3. Check bill of lading.	... to verify correct type of material is in tanker.
4. Sign in tanker driver.	Driver must sign in as a visitor and be escorted at all times.
5. Take Certificate of Analysis to QC.	***CAUTION:*** *To avoid contamination,* ***DO NOT*** *unload until you receive approval from QC.*
6. Spot tanker.	A) Direct driver to location. B) Verify brake is set. C) Chock the wheels on at least one side between the two rear axles. D) Ground the tanker (attach grounding strap to an unpainted metal surface).
7. Have the storage area operator make sure the storage tank can hold contents of tanker.	***WARNING:*** *Failure to perform this step can result in an overflow and spill of hazardous Monomer.* Perform **Tank Gauging Procedure (SOP-01-804)** or check the storage tank load cell readout. Storage tank operating limits are stenciled on the control panel.
8. Place 5-psig nitrogen pad on tanker.	A) Connect from local 5-psig nitrogen drop to aft bottom nitrogen connection on tanker. B) Open nitrogen supply valve. C) Then open valve at tanker.
9. Connect Monomer unloading hoses.	A) Verify hoses and gaskets are in good condition. (Hoses are normally stored on hose rack at unloading spot.) B) Connect from tanker to air pump. (Use aft bottom connection on tanker. Remove cap.) C) Connect from air pump to storage tank. (Connect to storage tank line labeled "from TW.")

If the unit has a good PHA/HAZOP, then it is best to extract information from the HAZOP (or What-If) analysis tables to start the development of each guide. The guide is then finished with input from the process experts.[4]

Table 2 Examples of Creating a Trouble-Shooting Guide from a HAZOP Table

HAZOP Table Entry	**Trouble-Shooting Guide Entry**
Cause: *Bypass valve is open or passing*	Make sure the bypass is tightly closed
Safeguard: *Isolation valves for the vessel*	Isolate the vessel, if necessary
Safeguard: *Relief valve*	Make sure the relief valve block valves/relief path are open

The key categories of information needed in a trouble-shooting guide are:[4]

- IMMEDIATE ACTION (by system or by operator)
- DECIDE IF ALARM is REAL
- FINDING and FIXING the CAUSE
- FIX or BYPASS PROBLEM

The example below is of a trouble shooting guide for one critical alarm/action.

Figure 3: Example of Trouble-Shooting Guide that Follows Best Practices

Trouble-Shooting Guide

Alarm or Indicator:	**PAL 4446 – Low Pressure Alarm for Suction of Organic Feed Pump 40-PM-18.445**

Action Limit:	5 kPa		
Consequence:	Possible pump seal failure, releasing or spraying organic waste into the berm.		
Process Area:	FB&D Incinerator; Liquid Organic Liquid Feed	**Oper. Mode:**	Normal
Drawing #s:	D-400-PI-013		

IMMEDIATE ACTION (by system or by operator)

- DCS should shut down the organic feed pump (40-PM-18.445).
- From the DCS display, MAKE SURE the organic feed pump is shutdown.
- HAVE the field operator check for leaks near the organic feed pump.
- IF there is a large leak/release, THEN use the ESD switch to shutdown the unit and then follow/complete the shutdown and isolation procedure, OPS-ESD-117.
- IF there is a minor leak or no leak, THEN:
    - COMPLETE the rest of the trouble-shooting,
    - and DECIDE how to contain the leak for now,

DECIDE IF ALARM is REAL

- From the DCS, CHECK the pressure and feed tank level trends. IF the trends indicate the alarm if valid, THEN continue with finding the cause or fixing or bypassing the problem.

FINDING and FIXING the CAUSE

- CHECK valves upstream of the organic feed pump to see if any are closed too far, including checking ESD valves.
- CHECK, by feel with hand, if the heat tracing is on; IF Not, then TURN ON or open heat trace valves
- MAKE SURE nitrogen to the pump seal is at the normal operating pressure.
- CHECK if the line if plugged or frozen (skill)

ABC Chemical Company Prepared by: *Printed copy of this procedure is good for one job task duration.*	OPS-76-TSG-233	Incinerator Unit Revised 8/24/2015 Printed 2/21/16 Page 1 of 2

Figure 3: Example of Trouble-Shooting Guide that Follows Best Practices (continued)

Trouble-Shooting Guide

Alarm or Indicator:	**PAL 4446 – Low Pressure Alarm for Suction of Organic Feed Pump 40-PM-18.445**

- CHECK if the line if plugged or frozen (skill)
- CHECK if the level is actually low, use the Organic Feed Tank.
- IF the cause is low level in the feed tank, THEN resolve the problem if necessary based on cause that is found (skill)

FIX or BYPASS PROBLEM

- IF necessary, SHUT DOWN the Unit to allow fixing of the problem.
- FOLLOW the proper procedure to resolve problems (repair procedure, line clearing procedure, etc.)
- IF the decision is made to continuing operation without all equipment in normal condition, THEN:
    - FOLLOW Temporary Operating mode, if there is a temporary procedure already written for this possible problem/condition
    - FOLLOW MOC procedures to obtain approval for any non-standard temporary operating procedure or mode

END

Optimal Presentation of Trouble-Shooting Information

Ideally trouble-shooting information should be imbedded in the basic process control system (DCS) so that the operators can access the information on demand, with the click of mouse or key. Using the DCS for display of the steps for response to alarms (trouble-shooting), to be displayed "on demand" is becoming more of the norm each year.

Comments on Current Industry Consensus Guides for Writing Procedures

None of the consensus textbooks available for procedure writing cover all of the major points of this paper. The best "guidelines" for writing procedures currently exist in the course notebooks used by PII (and likely in the course notebook used by ABS Consulting, as they had similar origins).

Notably, the AIChE/CCPS textbook, *Guidelines for Writing Effective Operating and Maintenance Procedures*, 1996[15]:

- Contains NO description at all on the development, format and use of Trouble-Shooting Guides
- Mentions only about half of the rules listed in this paper (and the related PII course notebook) for formatting of pages and steps
- Emphasizes technical staff or professional procedure writers writing procedures, rather than the Users themselves writing the procedures (it violates the first rule of procedure writing)
- Fails to cover the competency needed by procedure writers
- Recommends and emphasizes having subsections in each work instruction/procedure (between the Title of the task and Step 1) rather than relegating such information either to other sections of the operating manual or rather than turning this information into steps
- Fails to mention the need for a PHA of procedures that focuses on hazard review of deviations from the steps; this is required by best practices and by US OSHA for PHAs under PSM

A better example, though still missing key best practices, is the ASM Consortium *Guidelines for Effective Procedural Practices,* 2010[16]:

- Includes a good discussion on the need and thought processes related to handling abnormal situations (trouble-shooting or response to critical deviations), but there is little description on the format and detailed content of the trouble-shooting guides. The guideline fails to differentiate between routine procedures for startup, shutdown, and online maintenance, and those used for response to critical alarms (trouble-shooting)
- Mentions only about half of the rules listed in this paper (and the related PII course notebook) for formatting of pages and steps
- Mentions that experience users themselves should write their own procedures, but it does not emphasize this rule and leaves the authorship open to engineers or a committee. It specifically mentions using retired senior operators for writing operating procedures. While this is one good way to retain institutional knowledge, we have found that the current users do a better job overall, once they have achieved competency in how to write procedures
- Mentions the competency needed for procedure writers in general
- Mentions the need for a PHA of procedures, but does not reference the method documented in Chapter 9 of *Guidelines for Hazard Evaluation Procedures*, CCPS/AIChE, 3rd edition, 2008,[5] or other related papers on the topic. Also, the description of the PHA is misleading because it focuses on how PHAs should be used in writing procedure content (which is fine for trouble-shooting guides) but fails to emphasize that a step-by-step analysis of deviations from the steps are needed to ensure that the process has enough IPLs against mistakes that can occur when imperfect humans use the procedure.

Though this guideline has deficiencies, it is worth the investment and provides useful insight not found elsewhere. Just be sure to address the deficiencies noted above.

Implementation of This Approach

PII staff has helped more than 60 sites implement the approach above. The typical approach for successful implementation is:

1. Perform a gap analysis to determine the average accuracy of procedure content, average clarity (as judged versus the procedure format rules), and identify any examples of excellent procedure writing.
2. Convince management of the problem and get their buy-in on the solution.
3. Train (1.5 to 2 days for operators; 1 day for lab or maintenance staff) and coach the users of procedures (normally for 3 to 8 days, to have them achieve full competency) so they can be effective authors. Normally, 20 to 25% of the users (hands-on operators, hands-on maintenance and inspection staff, lab staff) need to be trained as authors. Not all of those trained will make good writers. The organization needs to judge who does the best jobs. About 15 to 20% of the staff should end up being authors.
4. Get organized using project management principles for tracking and scheduling the effort. Use master spreadsheets such as those used at REC Silicon (case study presented later).
5. Use a Traveler form (described later) to make sure each new procedure gets a proper amount of review by other users and by technical staff, such as engineers.
6. Reward procedure writers for their efforts. Especially ensure there are no disincentives for taking the extra effort to write procedures.
7. Track the progress on the effort; fix problems as they arise; enforce schedules (to show management cares).
8. Perhaps use specialized software to make it easy to output the content of procedures into different user-friendly formats such as training tools, training competency records, and field checklists. WordPerfect (formerly from Corel) made this relatively easy with the use of Primary and Secondary merge files; MS Word makes this more difficult. As a result, some specialized software have been developed to help keep formats consistent and to help reformat the content, on demand.
9. Issue procedures through a document control system that does not add extra workload. This will of course make Management of Change (MOC) to procedures easier to track through the steps of change request, to risk review and technical review, and finally to issuing (and replacement of prior versions).
10. Of course, ensure there is JSA performed for procedures that have "personal injury" potential and perform a PHA (or mini-PHA for an MOC) of the steps in the procedures whenever there is "process safety" potential.
11. Train all affected workers in the new procedures.
12. Identify the best writers to become the trainers of new authors, over the long run.

CASE STUDY 1 – REC Silicon (note, the general approach used for the REC Silicon effort is the same used in nearly all of PII's procedure writing projects, including those described in CASE STUDY 2 and 3; the forms and general approach described in detailed only in CASE STUDY 1 to allowing presenting the approach in the context of actual plant work).

REC Silicon is one of the world's largest manufacturer and suppliers of Silane Gas and specialty silicon gases, Granular Polysilicon from Fluid Bed Reactor technology for solar photovoltaic applications and ultra-pure float zone Polysilicon for the electronic industry.

SOP Improvement Initiative

REC production facilities in the USA have been actively implementing process safety for many years, with a renewed interest in 2013 to exceed OSHA PSM requirements by meeting industry best practices. REC commissioned gap analyses at both the Butte and Moses Lake production sites to identify areas for PSM best practice implementation. One common finding was that the operating procedures and maintenance procedures needed improvement with respect to accuracy of the procedural steps and presentation clarity (page format and step format). The following are audit results pertaining to procedures:

- Procedure Inaccuracy
- Technical Staff were viewed as owners, not shop floor
- Annual review process ineffective
- Procedure formats needed improvement for reducing human error
- Inconsistent use of Warning, Caution, and Note
- Procedure Naming – not consistent
- Modes of operation contained in one procedure title
- Procedure accessibility was difficult in the legacy database; multiple searches needed to find tool for job
- Procedure Safety Reviews – did not utilize recognized methodologies for procedure safety reviews; i.e., there were no PHA of procedures with process hazards or JSA of tasks with personal hazards

Step 1 - Procedure Author Training

In early 2014, REC Silicon adopted Process Improvement Institute's (PII's) procedure author training module (*Writing Effective Operating and Maintenance Procedures*[4]) with minor customizations. The training consists of a 2-day class room lecture and about 6 exercises, followed by practical procedure development and field validation steps in the 2-day coaching phase. The classroom course covered all aspects of procedure writing with the focus on techniques that will lower human error rates when using the procedures. Topics included:

- human factors overview
- getting the content right

- page format rules and selection
- step format rules
- addressing operating limits (including creating trouble-shooting guides
- nuances of emergency shutdown and emergency operating procedures
- implementation guides and high level procedures for the task of getting all of the work instructions written
- examples/exercises on each topic

Step 2 - SOP Management System Revisions

REC Silicon revised its management systems to require SOP writing competency for all personnel developing or changing procedures and to reach full operator certification. The unit trainers have taken over leading the procedure author training. Operators took over procedural ownership from technical staff since these are the people performing the work in the field. Additionally:

- The quality and accuracy of procedures improved and subsequently the quality of operator training and certification program due to developing procedure writing competency
- The quality of the annual reviews and audits improved tremendously with the rewritten procedural format and content changes.

Step 3 - Developing Troubleshooting Guides Engineering Standard

REC Silicon developed an engineering standard governing the content and format for troubleshooting guides:

- *CCPS Guidelines for Engineering Design for Process Safety*[17] is the basis for determining the safe operating limits.
- Production engineers are the owners of these procedures.
- Engineering competency is required to develop the safe operating limits table based on the available process safety information.
- A consequence of deviation is then developed by the Process Hazard Analysis team.
- Involvement of certified operators is required to develop steps to correct or avoid exceeding the safe limit.
- REC silicon risk ranked its process areas and focused on developing safety troubleshooting guides for high risk process first.
- Over thirty six (36) guides were developed for the Moses Lake Facility.
- These have been developed in conjunction with PHA updates to the safe operating limits.
- Formal training and competency testing of these safety procedures.

Step 4 – Beginning and Managing The Re-Write Effort:

- Each operating department listed all existing operating procedures into the procedure development tracking form (also called a Traveler, because the form travels with the draft procedure until the procedure is final; see Figure 4 on the next page).

- Each department also developed a list of "skills" that are used across multiple procedures and that operators or maintenance staff are required to attain before using complex procedures.
- Procedures were screened for safety, equipment damage, or quality risk to determine whether 2, 3 or 4 peer field walk-downs were required (see validation traveler workflow in Figure 4).
- The procedure re-write assignments were divided up between all four rotating shifts and the superintendent assigned certified authors (e.g., operators trained/coached as procedure writers) from their pool.
- Each department manages their pool of qualified authors. Cross shift validation was utilized to ensure all operating approaches were considered. The assigned author was responsible for consolidating red line comments.
- A **Validation Traveler** was developed and used to ensure all steps in the SOP re-writing workflow were completed.

Figure 4 - A **Traveler form** shown to the right accompanies each new or revised procedure through the writing and editing/review process. When complete, the traveler form is then kept in the QA file for the procedure to validate long-term how the initial procedure was developed.

Procedure Development Traveler - Moses Lake			
New Document Number:		Area:	
Procedure Title:			
Procedure Type:			
Author:			
Initial Draft Completion Date:			
Author Walkdown Date:			
Author Revision 1 Completion Date:			
Peer 1:		Shift:	
Peer 1 Walkdown Date:			
Optional Peer 2:		Shift:	
Peer 2 Walkdown Date:			
Optional Peer 3:		Shift:	
Peer 3 Walkdown Date:			
Optional Peer 4:		Shift:	
Peer 4 Walkdown Date:			
Author Revision 2 Date:			
Engineer/Verifier:			
Verification Completion Date:			
Author Revision 3 Completion Date:			
QA Person:			
Author Revision 4 Completion Date:			

Images and layouts copyrighted by PII, 2008-2017[4]

- A color coding process was used to flag SOP content that would be deleted, or moved into the unit training manual.
- A total of 184 Moses Lake procedures were tracked through the development, review, and revision process.
- **The overall tracking form below was used to develop KPIs to show accomplishments and to meet set completion dates.**

Figure 5 Procedure Development Project Status Table
Images and layouts copyrighted by PII, 2008-2017[4]

Procedure Development Tracking Form - Moses Lake

							Initial Development			Validation				
Item	New Doc #	Old Doc #	Area	Procedure Title	Proc Type	Author(s)	Initial Draft	Author Walkdown	Author Rev 1	Peer 1 Walkdown	*Opt. Peer 2 Walkdown*	*Opt. Peer 3 Walkdown*	*Opt. Peer 4 Walkdown*	Author Rev 2
1	FBP.3.xxx	FBP.3.160	FBR	Hydrogen Booster Compressor Cleanup	Cleanup	LT/FG	3/19/14	3/19/14	3/20/14	3/21/14	3/21/14			
2	FBP.3.xxx	FBP.3.160	FBR	Hydrogen Booster Compressor Operation and Op Restrictions	Operation	LT/FG	3/20/14	3/20/14	3/20/14	3/21/14	3/21/14			
3	FBP.3.xxx	FBP.3.160	FBR	Hydrogen Booster Compressor LOTO	LOTO	BCA/OL	3/19/14	3/19/14	3/21/14					
4	FBP.3.xxx	FBP.3.160	FBR	Hydrogen Booster Compressor Standby	Standby	BCA/OL	3/20/14	3/20/14	3/21/14					
5			4.0	STC/TCS from Receiving Tank to STC/TCS Storage Transfer Through Metering Line	Transfer	ME/RR	3/19/14	3/19/14	3/20/14	3/20/14	3/21/14	3/21/14	3/21/14	
6			4.0	STC/TCS from Receiving Tank to STC/TCS Storage Transfer Through Bulk Line	Transfer	ME/RR	3/19/14	3/19/14	3/20/14	3/20/14	3/21/14	3/21/14	3/21/14	
7			4.0	H2 Recycle Compressor Switching to Standby	Standby	ME/RR	3/20/14	3/20/14	3/20/14	3/20/14	3/21/14	3/21/14	3/21/14	
8			4.0	H2 Recycle Compressor Cleanup	Cleanup	ME/RR	3/20/14	3/20/14	3/20/14	3/20/14	3/21/14	3/21/14	3/21/14	

- From PII's procedure author training,[4] the Quality and Content checklist was modified and used as quick reference guide to make sure that all the content development and formatting rules are being followed during the re-write process. We used Microsoft Word and made templates for the T-bar, Flat and Consequence of Deviations.

Step 5 - Procedure Management of Change

REC Silicon adopted formal safety review methods for procedures and were included in a new procedure management of change procedure. The REC Silicon IT department developed a custom tool to execute these requirements in an electronic workflow. Benefits included improving consistency and quality of MOC procedural safety and health reviews and improving the overall quality of the SOPs with respect to clarity and content; and helped identify human factor deficiencies such as with human machine interface (equipment accessibility, operator feedback, labeling, missing or deficient instruments/controls).

Overall Summary of the procedure writing initiative:

For the Moses Lake facility, 398 procedures (operating, maintenance, and trouble-shooting guides) have been developed through this initiative (new and rewrites combined.) REC Silicon has seen great improvements based on the following feedback from workers involved in the initiative:

- Operators are receiving better work instructions
- Sequence of steps is well understood
- Much easier to audit operator compliance to procedure
- **Standardizes how we do work – protects our equipment from damage, protects our people, protects the environment**
- No longer need 20 year operator tribal knowledge and interpretation to complete procedure
- Procedures are not cluttered anymore, they just have the necessary information
- Procedure element now requires involvement of many shop floor workers

- JSA technique is alive on the shop floor
- Procedure can be trusted
- Technical staff no longer doing all procedure work. Freed up to optimize processes and realize other cost saving benefits.

REC Silicon's procedural initiative's success was due to forming a cross-functional project team of key decision makers from all the correct disciplines and having a project sponsor who could remove barriers. REC Silicon held weekly meetings, set short term goals, and tracked KPIs to meet those goals. Leadership was key to support the initiative and the related culture change.

CASE STUDY 2

As with REC Silicon, PII worked with ROQUETTE in Keokuk, IA (USA) to develop operating procedures for a process that a food additive. There were many procedures for this process and the detailed procedures for each task had never been written. Many of the older staff were being replaced with younger staff and the company decided to take this opportunity to develop detailed procedures to help ensure the consistency of training and coaching of the new workers.

Since there were limited experienced staff, the company trained the newly hired staff (many of whom had years of experience at other companies) in how to write and validate operating procedures. In addition, since ROQUETTE could not spare sufficient workers for the writing effort, PII provided 2 senior staff onsite for 2 weeks. Five (5) workers were trained and coached in procedure writing. These PII senior staff and the newly trained writers (who were also new operators) then worked with the experienced operators to outline the procedure for each task and then develop the best sequence of steps for each task. About 50 tasks were identified that needed procedures. 24 were drafted in the first week onsite and a similar number on the second week onsite. The forms shown in Figure 4 and 5 above were used extensively in this project.

In between the onsite visits by PII, then newly drafted procedures were walked down by the senior operator and by the other new operators. The revised procedures were finally issued for use. Even though the newly hired operators were inexperienced in the process operation, the ones that were trained as procedure writers developed several procedures (with the senior operator input) without help from the PII writers.

The company is considering expanding this effort to the many other process in Keokuk and at their other sites around the world.

CASE STUDY 3

As with REC Silicon and ROQUETTE in the USA, PII worked with Bahrain National Gas Company (BANAGAS) in Bahrain to develop operating procedures for their two gas process plants and for the gas compression stations upstream of the gas plants. Although the native language of most of the workers is Arabic or Hindi, the official language of the company is English.

This effort was part of a larger project to full implement PSM (and ongoing project that began in 2015 and is scheduled for completion in 2018). There were hundreds of operating procedures for these processes and the detailed procedures for each task had never been written. Similar more than 50 procedures were identified for maintenance activities related to repair, inspection, test, PM, calibration, etc.

The company has many very experience operators and maintenance technicians and many new operators. PII trained 12 operators and 7 maintenance technicians how to write and validate operating procedures. PII provided 2 senior staff onsite for 3 days of coaching following the 2-day course. These PII senior staff and the newly trained writers then worked with the other experienced operators to develop the best sequence of steps for each task and to get the procedures properly formatted. More than a dozen procedures (a combination of operating procedures and maintenance procedures) were written while PII was onsite; the main function of PII staff was to coach newly trained writers as they developed procedures for their departments. About 12 of the 18 BANAGAS writers that were trained and coached reached reasonable competency by the end of the first week (2 days of training and 3 days of coaching). During subsequent visits to the site by PII staff and between site visits via e-mail, PII continue to critique and coach the new minted procedure writers at BANAGAS.

As in the other two companies listed in these case studies, the forms shown in Figure 4 and 5 above were used extensively in this project, and continue to be used as the effort is ongoing. In addition, PII developed a flowchart for the procedure development process (see Figure 6) and developed a procedure writing Guide for use by the BANAGAS procedure writers.

The procedure writing effort is nearing completion, with very minimal input required from PII staff.

OTHER CASE STUDIES

PII has used the approach above since 2003 at more than 60 plants/sites in North America, South America, Australia, Europe, and the Middle East. The approach has consistently yielded great results; accurate procedures and clear formats. In these cases, the local operators and maintenance staff (the trained and now competent writers) put in 95% to 99% of the procedure writing effort.

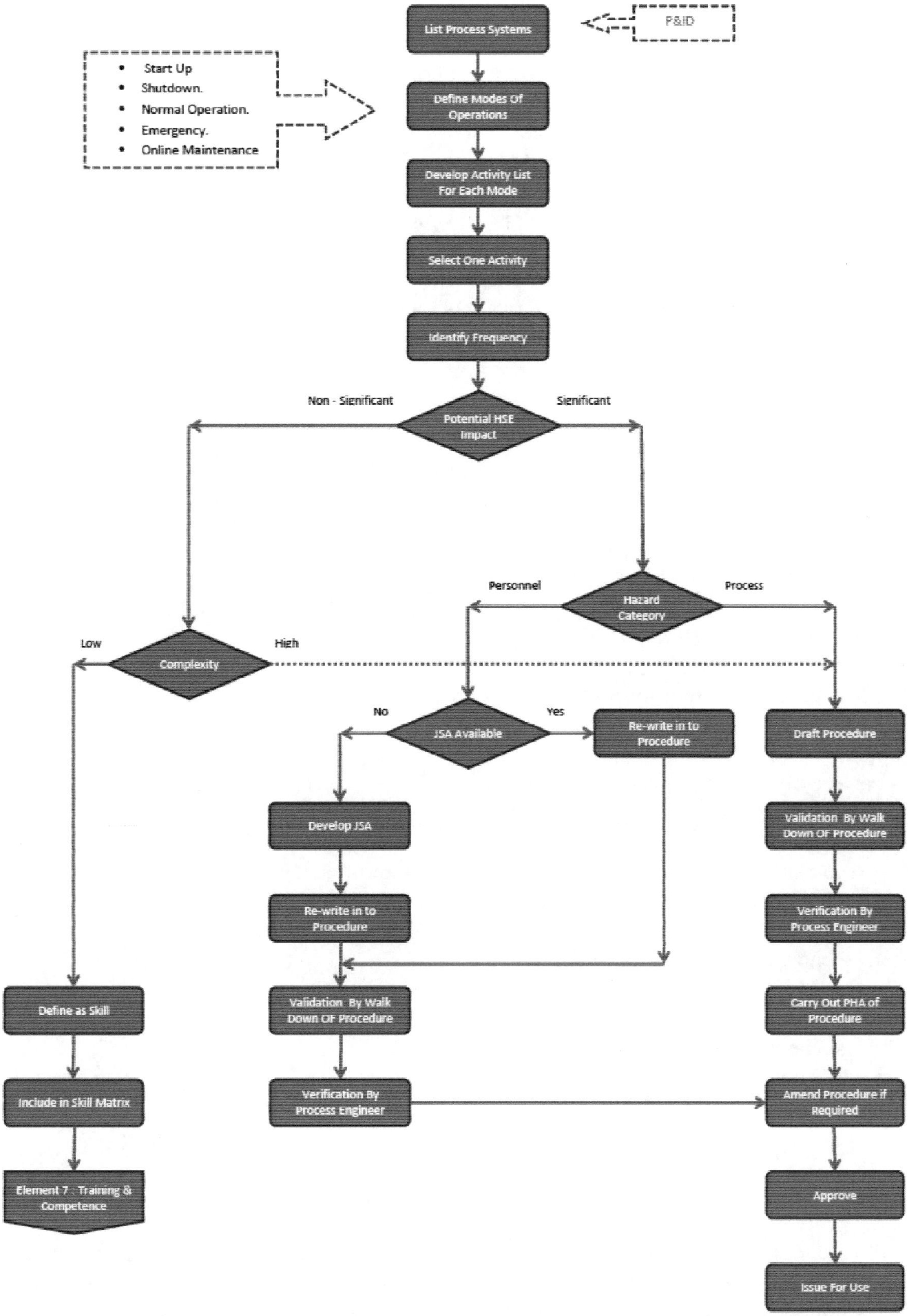

Figure 6 Procedure Development Flowchart (copyright PII, 2015-17)

Conclusions

Having clear and accurate procedures is a must. Poor procedures show up as one of the root causes in 90% of accidents. The approach described in this paper has worked for more than 35 years for one of the authors, and continues to work today. Controlling risk is mostly about controlling human error rates, and having accurate and clear procedures is one key to this control. One other gain from the approach described in this paper is that the users become the owners, and therefore they have a vested interest to keep the procedures up to date and clear.

No doubt, writing all of the procedures for a facility is a large task. But, it can be accomplished by site staff, with support from company process safety or quality control leadership, after enough users become competent authors. Attaining this competency takes less than 2 weeks of work, on average.

Acronyms Used

AIChE– American Institute of Chemical Engineers
ASM – Abnormal Situation Management consortium
BANAGAS – Bahrain National Gas Company
CCPS – Center for Chemical Process Safety (a division of AIChE)
CMA – Chemical Manufacturer 's Association, now American Chemical Council (ACC)
FBR – Fluidized Bed Reactor
HAZOP – Hazard and Operability Analysis
IPL - Independent Protection Layer
JSA – Job Safety Analysis
LOPA – Layer of Protection Analysis
MART – Maximum Allowable Response Time
MOC – Management of Change
NUREG – Nuclear Regulation, US Nuclear Regulatory Commission
OSHA – Occupational Safety and Health Administration, US Department of Labor
PHA – Process Hazard Analysis
PII – Process Improvement Institute, Inc.
P&ID – Piping & Instrumentation Diagram
PSI – Process Safety Information
PSM – Process Safety Management
PST – Process Safety Time
SOP – Standard Operating Procedure

References

1. Swain, A. D., Guttmann, H. E., *The Human Reliability Handbook - with Emphasis on Nuclear Power Plant Applications*; NUREG/CR-1278, U.S. Nuclear Regulatory Commission, Division of Facility Operations, 1983.

2. Gertman, D., Blackman, H., Marble, J., Byers, J. and Smith, C., *The SPAR-H Human Reliability Analysis Method,* NUREG/CR-6883, U.S. Nuclear Regulatory Commission, Office of Nuclear Regulatory Research, 2005.
3. Sutton, Ian S., "Writing Operating Procedures," *Proceedings of the International Conference on Hazard Identification and Risk Analysis, Human Factors and Human Reliability in Process Safety*, AIChE, Center for Chemical Process Safety, New York, NY, 1992.
4. *Writing Effective Operating and Maintenance Procedures*, Course Notebook, Process Improvement Institute, Inc., 2003-2017.
5. *Guidelines for Hazard Evaluation Procedures*, 3rd Edition, 2008, CCPS/AIChE.
6. "Necessity of Performing Hazard Evaluations (PHAs) of Non-normal Modes of Operation (Startup, Shutdown, & Online Maintenance)", W. Bridges and Mike Marshall (US OSHA), *18th Annual International Symposium, Mary Kay-O-Connor Process Safety Center, College Station, TX, October 2015. (also updated at 12th GCPS, 2016)*
7. "How to Efficiently Perform the Hazard Evaluation (PHA) Required for Non-Routine Modes of Operation (Startup, Shutdown, Online Maintenance)," W. Bridges and T. Clark, *7th Global Congress on Process Safety*, Chicago, AIChE, March 2011.
8. Bridges, W.G., et. al., "Addressing Human Error During Process Hazard Analyses," *Chemical Engineering Progress*, May 1994
9. *A Manager's Guide to Reducing Human Error*, 1990, Chemical Manufacturers Association (CMA)
10. Haney, R. W., "The Effects of Instructional Format on Functional Testing Performance," *Human Factors*, 11, 1969, pp. 181-187.
11. "Effective Procedure Writing – A Two-Part Series (Part 1: Reducing Risk of Worker Errors; Part 2: Addressing Process Safety Management Requirements for Operating Procedures)," *Chemical Engineering Progress*, 1997, W. Bridges with T. R. Williams (both of JBFA).
12. Madden, L., and Bridges, W., Best Practices for Writing Operating Procedures and Trouble-Shooting Guides, L. Madden and W. Bridges, *12th Global Congress on Process Safety*, Houston, TX, AIChE, March 2016.
13. Connelly, C. S., "Emergency Operating Procedure Writing for Oil Refinery Process Units," *Proceedings of the International Conference on Hazard Identification and Risk Analysis, Human Factors and Human Reliability in Process Safety*, AIChE, Center for Chemical Process Safety, New York, NY, 1992.
14. *Guidelines for Initiating Events and Independent Protection Layers*, 2015, CCPS/AIChE.
15. *Guidelines for Writing Effective Operating and Maintenance Procedures*, 1996, CCPS/AIChE.
16. *ASM Consortium Guidelines – Effective Procedural Practices*, 2010, ASM Joint R&D Consortium.
17. *Guidelines for Engineering Design and Process Safety*, 2nd Edition, 2012, CCPS/AIChE

478864 Process Safety Management: It's Not Just for Chemcal Plants!

Abstract:

Process Safety: It's Not Just for Chemical Plants & Refineries!

Presented By

Richard W. Sarnie, CSP, P.E., ARM-E

Director, CCPS Projects

Center for Chemical Process Safety

American Institute of Chemical Engineers

In this presentation Mr. Sarnie will discuss lessons learned and relearned over decades in manufacturing and cold storage facilities on preventing and mitigating process safety incidents (LOSS OF CONTAINMENT EVENTS) and how these lessons learned can and should be applied across many types of manufacturing and cold storage operations such as the food, beverage, pharmaceutical and nutraceutical industry. He will:

- Present a rationale as to why a **risk-based** rather than **regulatory-based** approach to managing process safety and other risk exposures is prudent
- Discuss the continuing movement by many organizations to more formal and comprehensive enterprise risk management systems
- A review of the typical steps or phases an organization usually takes in developing and improving management system processes
- Provide an overview of chemical industry best practices for managing process safety (with a focus on AIChE CCPS recommendations as described in the **CCPS Guidelines for Risk Based Process Safety** and many other CCPS books)
- Describe numerous process safety management system deficiencies (root causes) associated with serious chemical process safety incidents that have repeatedly occurred during the last 50+ years
- Discuss how proven process safety management practices and use of multiple layers of protection can help better manage risks in manufacturing and cold storage facilities.

Hit the Ground Running (Safely):

Process Safety for the Early Career Engineer

Jerry J. Forest, CCPSC
Celanese
225 E. John Carpenter Freeway Suite 1200, Dallas, TX 75234
Jerry.Forest@Celanese.com

Don Connolley, CSP
BP Corporation
Don.Connolley@bp.com

Amy Gray, Ph.D., PE, CFEI
ESI
aegray@engsys.com

Prepared for Presentation at
American Institute of Chemical Engineers
2017 Spring Meeting and 13th Global Congress on Process Safety
San Antonio, TX
March 26 – 29, 2017

Hit the Ground Running (Safely):

Process Safety for the Early Career Engineer

Jerry J. Forest, CCPSC
Celanese
225 E. John Carpenter Freeway Suite 1200, Dallas, TX 75234
Jerry.Forest@Celanese.com

Don Connolley, CSP
BP Corporation
Don.Connolley@bp.com

Amy Gray, Ph.D., PE, CFEI
ESI
aegray@engsys.com

Keywords: New Hire, Training, Competency, Mentoring

Abstract

Engineers entering the workforce are frequently expected to participate in process safety elements such as PHA, MOC, incident investigation, mechanical integrity programs and more. Although there is a recent ABET requirement to include process safety in engineering curricula, many schools do not help students define what to expect in their first year.

This paper will discuss common process safety duties, activities, and responsibilities expected of new engineers. The paper will follow the recently published CCPS book *Process Safety for Undergraduates and Engineers*. It will also introduce the AIChE's Health and Safety Division's Professional Outreach concept on mentoring. The mentoring project will provide a platform to connect experienced process safety engineers with student and local chapters, and individuals who are seeking assistance with specific process safety topics.

1. Introduction

When engineers enter industry for the first time they are often surprised that one of their first assignments includes an onboarding process that can include several years of training. The material in these professional development courses is usually not covered in school. Companies may use different onboarding tools and the early career engineer must learn the company specific methodologies for process safety elements.

This paper discusses an overview of how new engineers can hit the ground running with respect to process safety when first entering industry. It includes a brief overview of the Accreditation

Board for Engineering and Technology Inc. (ABET) requirements for university chemical engineering students, typical onboarding training in the first few years of employment, and the type of process safety duties a new hire engineer can expect. The paper also discusses a new pilot process safety mentoring program from the American Institute of Chemical Engineers (AIChE) Safety and Health Division designed to help new hire engineers hit the ground running safely by drawing on those who have years of industrial process safety experience.

2. T2 Laboratories and ABET Requirements

In 2007, an explosion occurred at T2 Laboratories in Florida, killing 4 employees and injuring 32 including members of the surrounding community. The U.S. Chemical Safety Board (CSB) investigated the incident and found that the primary cause of the incident was a lack of understanding of the reactive hazards involved with the process. A key recommendation from the incident is addressed to the AIChE:

"Work with the Accreditation Board for Engineering and Technology Inc. to add reactive hazard awareness to baccalaureate chemical engineering curricula requirements." [1]

AIChE responded to this recommendation by working with ABET to require not just reactive chemical hazards in chemical engineering curricula, but all chemical process hazards. [2] Given that ABET accredited schools must address all chemical process hazards in their undergraduate curricula, the first way early career engineers can hit the ground running is by understanding the process safety topics presented in their curricula.

The Center for Chemical Process Safety (CCPS) has gone beyond the original intent of the CSB recommendation with two additional initiatives. The first is publication of *Introduction to Process Safety for Undergraduates and Engineers* (hereafter referred to as *Introduction to Process Safety*) and the second is the undergraduate process safety education initiative.

Introduction to Process Safety supplements Crowl and Louvar's *Chemical Process Safety: Fundamentals with Applications* [3] by introducing the student to Risk Based Process Safety. The book is designed to be used as a stand-alone course book or as supplemental material for several chemical engineering courses. *Introduction to Process Safety* was published in 3rd quarter of 2016 and early adopters include Louisiana State University and Kansas State University. Celanese is presently evaluating making the book required reading for all new hire engineers and is considering an onboarding course from the material.

The second initiative in progress from CCPS that goes beyond the original intent of the T2 Laboratories CSB recommendation is an undergraduate process safety initiative. This initiative builds on AIChEs existing Safety and Chemical Engineering Education (SaChE) program and *Introduction to Process Safety* with four key elements:

- Curriculum – enhancing SaChE e-learning modules and translating to multiple languages
- Teaching – Educate professors about process safety with workshops at industrial sites
- Addressing gaps – with industry led "Boot-Camps" for students on campus and
- Maintaining focus – funding key research in process safety [4].

The early career chemical engineer has an opportunity to hit the ground running safely by taking advantage of the material presented by their undergraduate curriculum and by utilizing the resources available by the CCPS. These include the aforementioned SaChE modules, "Boot-Camps", and the resource books *Introduction to Process Safety for Undergraduates and Engineers* and *Chemical Process Safety: Fundamentals with Applications*.

3. Onboarding & Training for New Hires

Unfortunately for the new hire engineer, each company will address the various process safety elements in different ways and with differing methodologies. For example, while management of change (MOC) processes may have common elements of risk identification and mitigation, approval, and action item identification and closure, the methodologies used for these elements may differ among companies. While the fundamental principles will look familiar to what was presented in the undergraduate curriculum, the specifics particular to each company may be unfamiliar. Therefore, it is very common for an early career engineer to undergo an onboarding process that might include several years of professional development training.

The final onboarding content is a function of the hiring company and the job position of the early career engineer. The process safety courses given can be mandatory or optional and the material can be delivered by an internal training department or offered by third party vendors. Whatever the case, the onboarding process safety training represents an excellent opportunity for the early career engineer to hit the ground running.

Early career engineers typically are immediately immersed in the following risk based process safety elements: process safety information (PSI), process hazard analysis (PHA), management of change (MOC), pre-startup safety reviews (PSSR), conduct of operations, mechanical integrity (MI), standard operating procedures (SOP), training, metrics, and incident investigation. The extent to which early engineers contribute is usually based on training and experience. Therefore, it is in the engineer's best interest to pursue at a minimum, initial training on these topics.

4. What to Expect in the First Year

Although this is a paper on how to hit the ground running with regard to process safety, we would be remiss if we did not mention occupational safety expectations for the early career engineer. Note the majority of this section is taken from *Introduction to Process Safety* [5]. It is safe to say that most U.S. chemical manufacturers place an enormous emphasis on occupational safety – to the extent that they have been criticized by OSHA for placing an emphasis on occupational over process safety. Nevertheless, all engineers should understand their company's expectations for occupational safety and become a model leader by demonstrating behaviors consistent with standards and regulations and exceeding expectations set by the company leadership in this area. Activities that early career engineers can expect include:

- Leading and participating in safety meetings and shift safety meetings
- Unit inspections
- Life critical procedure audits
- 8 – 24 hours of occupational safety training per year

- And more, depending upon an individual company's requirements

4.1 Social Interactions

Engineers entering the manufacturing environment are thrust into social interactions for which many are ill prepared. Operators and maintenance crafts have often been part of the same team for years and have unique understanding of their units even if they do not have the vocabulary to express the physical processes in the same way as the new engineers. The main thing for an early career engineer to understand is to approach these groups with great respect. They have a deep understanding of process that is only gained by experience. To become a mentee of these experienced individuals, one must first respect their knowledge, abilities, and attitudes gained by experience.

This section comes with a warning. The early career engineer who adopts an attitude of knowing more than those hourly employees that have run the manufacturing units for years will most certainly fail. The reason simply is the early career engineer does not know as much as these experienced hands. The engineer may possess an understanding of first principals, but must also gain an understanding of the physical process in order to apply first principals.

Learning from operators and crafts will accelerate the learning curve for a new engineer and provide a way to apply technical skills learned at university in a practical way that helps the company achieve results. Interacting with the operators and crafts is also a way to eventually gain their respect and trust, and learn how to function well as a team.

4.2 New Skills

Many assignments in the university curriculum include group projects, formal reports, and working with teams. They are designed to prepare students in doing the same activities in their new engineering jobs. New skills in industry related to process safety include assembling process safety information, leading project teams, training operators, writing standard operating procedures, scribing or leading hazard reviews and management of change hazard reviews, delivering presentations and writing formal technical reports (just to name a few skills).

One of the most important skills to acquire is the ability to concisely document reasoning and thinking related to solution of items. Findings and action items are generated from process hazard analysis, management of change, incident investigation, audits, and other team meeting activities. The early career engineer is often assigned much of this important workload and must find solutions to the problems and document with proof that concern is addressed. In many cases, this documentation is regulatory requirement and becomes part of the operating record for the life of the process.

4.3 Conduct of Operations & Operating Discipline

One process safety element that has been overlooked in university engineering curricula is conduct of operations. Conduct of operations is a term that "describes human factors and tools that are necessary to produce repeatability in performance and consistency in results" [5]. While much of process safety is concerned with maintaining the integrity of engineered controls, conduct of operations is concerned with maintaining the integrity of administrative and human controls.

The day in a life of an operator can be described by discrete activities performed. These include: reporting to work outfitted, making relief in the field, attending shift change meetings, informal and formal evaluation rounds, issuing permits, preparing equipment for maintenance or operation, operating equipment, and documenting each activity. Early career engineers are often asked to participate in several or all of these aspects of operation.

For example, operators usually document their activities in a log book. Engineers should monitor the operator logs to gain a deeper understanding of unit activities. When problems are documented, engineers should take action. Failure to do so leads to an acceptance of problems and a normalization of deviance. Furthermore, engineers might be asked to create logs and periodically train operators on how to use them.

Other examples of an engineer's role in conduct of operation include participation in equipment evaluation. An engineer might create evaluation sheets by mapping out how an operator should travel through the unit, and define exactly what equipment should be evaluated and how it is to be evaluated. Then, by periodic monitoring of the evaluation rounds, engineers better understand normal operation and are better equipped to solve problems.

Early career engineers are often called to participate in all aspects of conduct of operations and should be familiar with the operating discipline tools to be effective. This familiarity is only gained by experience as it is rarely formally taught in training.

4.4 Engineering Discipline

Engineering discipline is closely related to operating discipline in that it describes activities that engineers and technical personnel do to monitor key operating parameters in the plant to keep the unit running at peak efficiency and normal safe operation. This new skill involves the selection and monitoring of key performance indicators and determination of, monitoring, and evaluating safe operating limits. Engineers monitor trends and take action to prevent process safety incidents based on their understanding of the hazards and risks gained from process safety information and hazard identification and risk assessment processes of their company.

4.5 Process Safety Duties

Early career engineers will be active in most of the 20 elements of risk based process safety. In addition to the company specific training described earlier, participation in these elements will be required. A few common areas are shown in table 1, summarized from *Introduction to Process Safety* [5].

Table 1: Process Safety Activities for New Engineers

Element	Activity
Process Knowledge Management	Ensure accuracy of PSI Be familiar with Safety Data Sheets Develop reactivity matrix Develop and update design basis calculations Develop and update P&IDs Verify electrical classification Maintain equipment files
Hazard Identification and Risk Assessment	Participate in HIRA, & PHA Assemble required PSI Review applicable RAGAGEPs Compile industry incidents Scribe, document reviews Resolve findings
Operating Procedures	Develop and update operating procedures
Safe Work Practices	Write and approve permits
Asset Integrity and Reliability	Identify equipment covered by inspection, testing, and preventative maintenance (ITPM) programs Write ITPM procedures Analyze inspection results & maintain records
Contractor Management	Approve and train contractors Write contractor procedures
Management of Change	Identify changes Participate in MOCs Assemble required PSI Resolve items
Operational Readiness	Participate in readiness reviews Resolve items
Emergency Management	Learn emergency response plans Participate in drills
Incident Management	Participate in investigations Resolve items
Metrics	Record and maintain data Analyze data
Audits	Assemble requested information for audits Become a certified auditor

The main point of this section is that there is much more to process safety for early career engineers than is generally presented in the university curricula even with the new ABET requirement. While many companies will have onboarding and training to prepare the engineers, many of the skills illustrated above are gained by experience. This is just a sampling of what early career engineers can expect in their first year.

To help engineers prepare for these activities in their first few years, the Safety and Health Division of AIChE is piloting a mentoring program.

5. AIChEs Safety and Health Division Pilot Mentoring Program

The American Institute of Chemical Engineers (AIChE) Safety and Health Division's mission is:

"To share, promote, and shepherd process and personal safety for individuals in the field of process safety."

In support of the mission the Division has formed several committees including Professional Outreach, Student Outreach, Communication, and Membership. The Professional Outreach Committee has identified a potential avenue to help the early career engineer hit the ground running safely by creation of a process safety mentoring program.

The objective of the mentoring program is to connect AIChE members experienced in process safety with members new to the area of process safety for the purpose of providing guidance and a resource point. It is envisioned that eligible mentors will be engineers who have at least 10 years of experience in process safety. This experience could have been gained through a variety of pathways. Reviews of the volunteer mentors' experience will be performed by the Safety and Health Division Mentoring Committee.

Eligible mentees will be early or transitional career professionals with less than 5 years of experience in the area of process safety. This career can be in a variety of industries. Reviews of potential mentees' eligibility will be performed by the Mentoring Committee.

The mentoring program is in a pilot stage with plans to focus on Louisiana to solicit mentors and mentees. It is expected that applications for mentors and mentees will be available on the Safety and Health Division website and will include information on where to return information. The mentor application will include name and company info, time commitment desires, areas of expertise, and a brief statement on why the applicant wants to be a part of the Mentoring Program. The mentor should attach an updated resume and provide 3 references. Mentee applications will require similar information and can be submitted by individuals or engineering supervisors interested in pairing mentors with groups of young professionals.

The Safety and Health Division Mentoring Committee will review all Mentor applications and approve or reject based on experience, reputation in industry, and referral comments. Mentors and Mentees will be paired by the Safety & Health Division Mentoring Committee. Decisions will be based on location, industry, and desired commitment level.

At the time of writing this paper, the pilot mentoring program is in its infancy. Those interested in being either a Mentor or Mentee should contact a member of AIChEs Safety and Health Division Executive Committee.

When it is fully established, we anticipate that this Safety and Health Division Mentoring Program will provide an excellent opportunity for the early career engineer to hit the ground running.

6. Conclusion

The early career Chemical Engineer quickly learns that education and competency are an integral part of job duties and in fact, will be an ongoing part of the job during the course of a career. A large portion of this continuous learning is gaining expertise in occupational and process safety; skills that often are not covered in depth at many universities.

It is in the early career engineers interest to gain competency in process safety as the elements of process safety are weaved throughout all aspects of an engineer's job. The CCPS has published many resources that provide excellent background in process safety. Most companies have an onboarding process that details several years of process safety training. Engineers will also gain "hand-on" experience through the early assignments given in common process safety elements such as MOC and PHA. And if successful with the pilot Mentoring Program, the AIChE Safety and Health Division may have an additional tool to help the engineer hit the ground running (safely).

7. References

[1] U.S. Chemical Safety and Hazard Investigation Board. *Investigation Report T2 Laboratories, Inc. Runaway Reaction (Four Killed, 32 Injured)*, CSB Report No. 2008-3-I-FL, 2009

[2] U.S. Chemical Safety Board, http://www.csb.gov/csb-applauds-aiche-s-response-to-including-reactive-hazard-awareness-in-college-chemical-engineering-curriculum/ (accessed January 29, 2017)

[3] Crowl, Daniel, A. and Louvar, J. F., *Chemical Process Safety: Fundamentals with Applications*, 3rd ed., Prentice Hall, Englewood Cliffs, NJ (2011)

[4] Center for Chemical Process Safety, http://www.aiche.org/ccps/community/undergraduate-process-safety-learning-initiative (accessed January 29, 2017)

[5] Center for Chemical Process Safety, Introduction to Process Safety for Undergraduates and Engineers, 1st ed., John Wiley & Sons, Inc., 2016

480957 Design and Development of a Supercritical Fluid Processing System for Conversion of Used Cooking Oil to Biodiesel

Abstract:

The design project involves collection of waste cooking oil from the University of South Florida and vicinity, conversion of waste cooking oil using a novel catalyst-free supercritical fluid processing facility, making novelty soap from the by-product glycerin, and utilizing the biodiesel in powering the University bus system that utilizes B20. The processing time is 15 minutes and allows a mobile solar powered unit. The design involves scale-up, heat integration, safety and environmental impact analysis.

2017 AIChE®
SPRING MEETING &
13TH
GLOBAL CONGRESS
ON PROCESS SAFETY

March 26–30, 2017
Henry B. Gonzalez Convention Center
San Antonio, TX

17 AIChE Spring Meeting, San Antonio, TX

13TH GLOBAL CONGRESS ON PROCESS SAFETY

17 AIChE Spring Meeting, San Antonio, TX

13TH GLOBAL CONGRESS ON PROCESS SAFETY

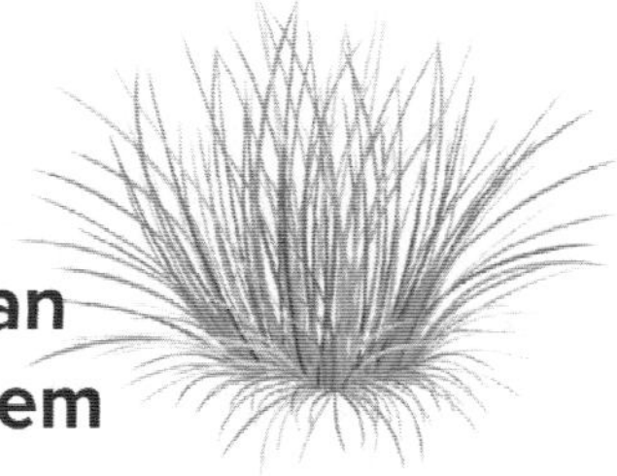

Carbon dioxide recapture utilizing an integrated vertical aquaponics system

Brittany Foley, Andrew Horowitz
Joe Laviano, Katy Wardzala
Northeastern University
Boston, MA

17 AIChE Spring Meeting, San Antonio, TX

13TH GLOBAL CONGRESS ON PROCESS SAFETY

Increasing food demand exists worldwide

- Global food production will require a **50-60% increase** over the next 30 years (*LA Times*)
- It is estimated that ⅔ **of the world's population** will live in cities by 2050 *("Growing Food for Growing Cities")*
- **50-70% of the U.S.'s food supply** is consumed by people living in **urban communities** *(LA Times)*
- Food supply for **urban areas** is largely **unsustainable**
- Food today is often **shipped thousands of miles**

17 AIChE Spring Meeting, San Antonio, TX

13TH GLOBAL CONGRESS ON PROCESS SAFETY

"The threat is, will there be enough food produced for those people living in cities?"

Dan Glickman, former U.S. Secretary of Agriculture

17AIChE Spring Meeting, San Antonio, TX — 13TH GLOBAL CONGRESS ON PROCESS SAFETY

Hydroponics

CityFARM. MIT, n.d. Web.

- No need for **pesticides**, **fungicides**
- **Year-round** crop cultivation
- Not weather-dependent
- Drastically reduces agricultural waste
- **Any unused space** can be turned into a vertical hydroponic farm
- ***Challenge***: systems require fertilizer to be added on a daily basis

17AIChE Spring Meeting, San Antonio, TX — 13TH GLOBAL CONGRESS ON PROCESS SAFETY

Aquaponics

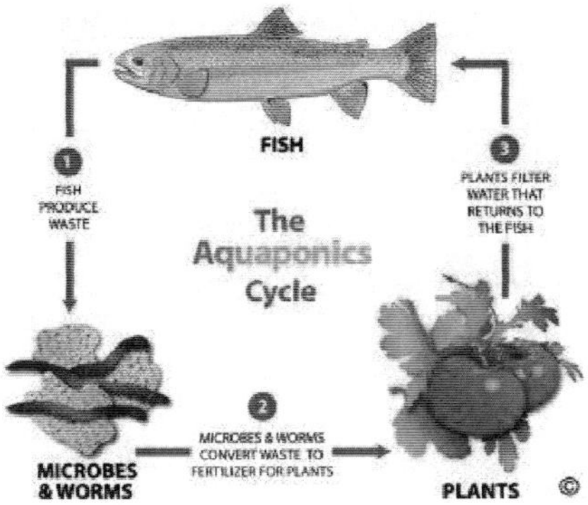

The Aquaponics Source. (n.d.). https://www.theaquaponicsource.com/what-is-aquaponics/

- A form of **hydroponics** that utilizes **waste** produced by fish to fertilize plant growth
- Farming of both **fish** and **plants**
- Fish may be ornamental (koi, etc.) or grown for food (i.e. tilapia, trout, etc.)
- **Fully continuous system**
- **Two inputs:** light and nutrients for fish

17 AIChE Spring Meeting, San Antonio, TX — 13TH GLOBAL CONGRESS ON PROCESS SAFETY

"In the last 20 to 30 years, the amount of food being transported from rural to urban areas has increased 300% in India, 800% in Africa and 1,000% in Southeast Asia."

Thomas A. Reardon, Chicago Council on Global Affairs, for *L.A. Times*

17 AIChE Spring Meeting, San Antonio, TX — 13TH GLOBAL CONGRESS ON PROCESS SAFETY

Carbon dioxide recapture utilizing an integrated vertical aquaponics system

Our system: A commercial vertical aquaponics system with integrated controls, using carbon dioxide produced from fermentation and heat recapture from LED grow lights.

Recycled Carbon Dioxide

- Enhancing growth with CO_2 but with reduced energy consumption relative to competitors who burn propane for CO_2
- Supply can be sourced from other systems' waste (i.e. breweries)

Heat Recapture

- Heat released by LED growth lights used to control water tank temperature at 75° ± 3°F

Demonstrative model

Fish tank for tilapia

Sump tank to maintain water level

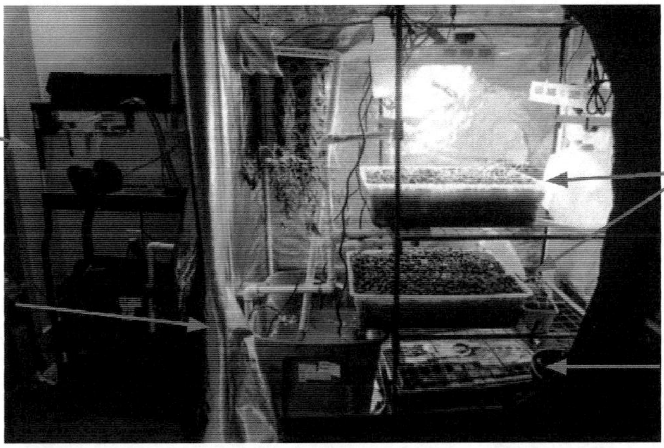

Two growth beds with expanded clay media

Seed starters in rockwool and humidity dome

Setting & Market

Target Market

- Small-scale breweries
- Alleviate waste CO_2 expenses
  - Provide incentives
  - Offer alternative income
  - Market popularity = expansive growth

Plant Location

- Greater Boston Area

17 AIChE Spring Meeting, San Antonio, TX — 13TH GLOBAL CONGRESS ON PROCESS SAFETY

Design of a 1,000 ft^2 scalable model

Model Assumptions

- 8,000 barrel/year production capacity
- Assumes beer brewed at 14 → 2 degrees plato
- Assumes ⅔ CO_2 recapture

Product Basis

- Cultivating 50:50 lettuce and basil
- Crop densities planted at $4/ft^2$ - $12/ft^2$
- Short life cycles (8 weeks)
- Plant consumes ½ of its weight in CO_2 during its lifetime

Specifications

- Four quad-tier growth beds (125 ft^2/bed)
- Four 1,200 gallon fish tanks (50 ft^2/tank)
- 1000 ft^2 total occupied floor space

17 AIChE Spring Meeting, San Antonio, TX — 13TH GLOBAL CONGRESS ON PROCESS SAFETY

Major equipment specifications

- Media Beds
  - Food-grade containers
- Fish & Sump Tank
  - 1,200 gallon receptacles
- Heat Exchanger
  - Shell-and-tube design
- Pumps
  - Centrifugal
  - 1000 gph rated (assumes 75% packing)

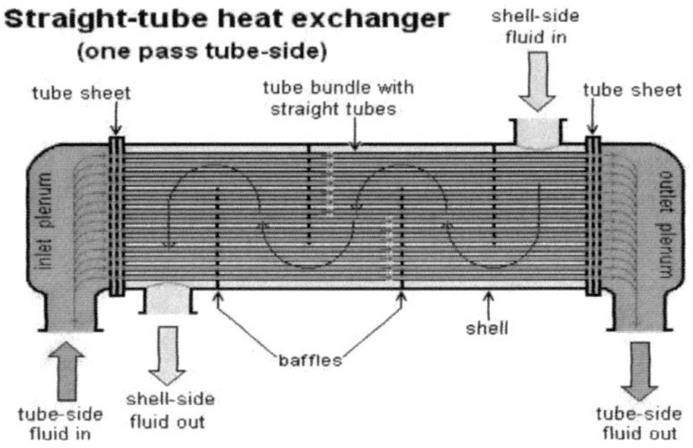

Control philosophy

Automated Control

- Pumps and Valves
- Temperature Control
- Lighting

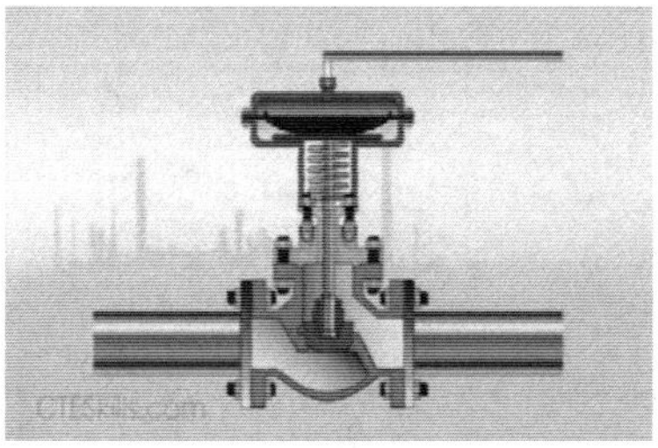

Manual Control

- Seeding
- Sampling
- Harvesting
- Water Quality Control

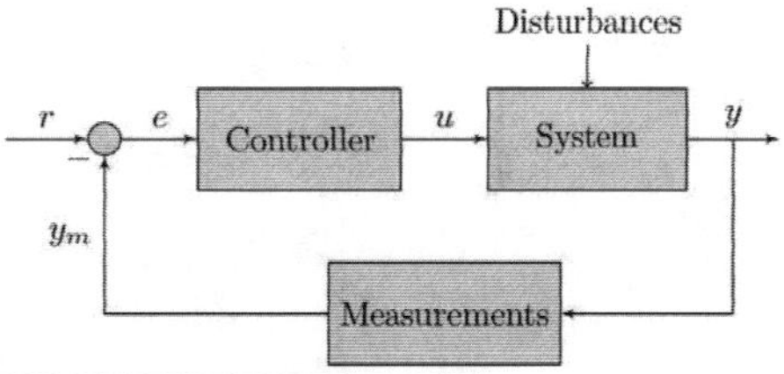

Automated controls

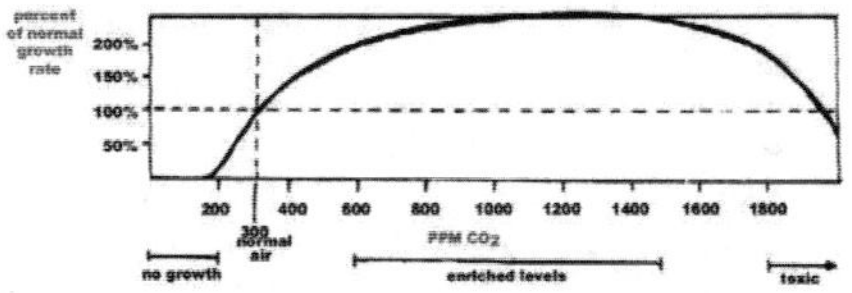

Temperature

- Logged once per second
- Alarms outside of 72-78°F
- Distributed Control System (DCS)

Carbon Dioxide

- Gas analyzer
- High alarm at 2,500 ppm
- Low alarm at 1,200 ppm

Flow

- Valves and pumps monitored by DCS

Level

Manual controls

Water quality will be continuously monitored, only requiring operator intervention when necessary.

Harvesting/Planting

- Plants: 4-8 weeks
- Tilapia: 10 months
- Staggered life cycles throughout

Daily Sampling

- Plant growth
- Fish health

Overall process flow

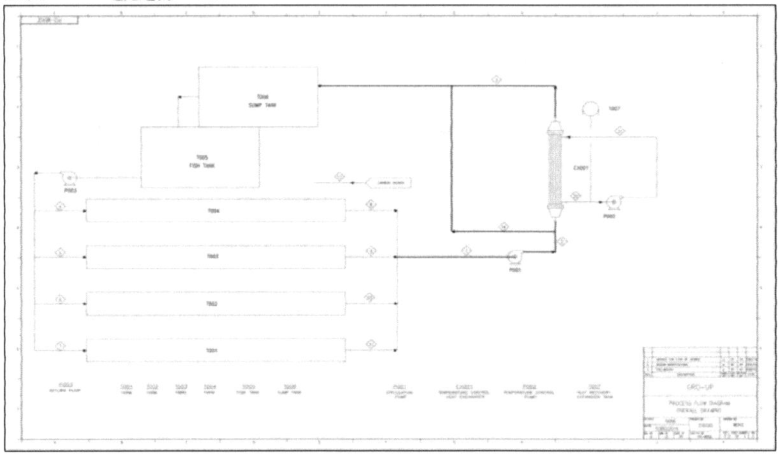

17 AIChE Spring Meeting, San Antonio, TX — 13TH GLOBAL CONGRESS ON PROCESS SAFETY

Installed capital costs

(for client)

	Units	Units/Unit product	Quantity	Cost ($)/unit	Cost ($)
High-powered 300W LED grow lights	ea	1	16	109.00	1,744.00
Fish filter pump	ea	1	4	110.00	440.00
HDPE tubing 2"	ft	500	2	573.00	1,146.00
Metal supports (shelving)	ea	1	4	510.00	2,040.00
Water pumps for circulation system	ea	1	8	71.05	568.40
Air pump for oxygenation	ea	1	4	94.24	376.96
Air hoses	ft	1	80	2.00	160.00
Ammonium chloride	g	500	1	14.06	14.06
Fish and sump tanks	ea	1	8	724.00	5,792.00
Temperature transmitters	ea	1	16	30.35	485.60
Flow meter	ea	1	8	79.99	639.92
Composition analyzers	ea	1	12	185.00	2,220.00
Automated on/off valves	ea	1	36	46.65	1,679.40
Control valves	ea	1	12	53.42	641.04
Food-grade plastic containers for grow beds	ea	1	16	625.00	10,000.00
CO_2 Analyzer Thermo 41H	ea	1	1	600.00	600.00
TAC I/A Series: MNB-300 BACnet	ea	1	1	373.00	373.00
TAC I/A Series: MNB-V1-2	ea	1	1	216.30	216.30
TAC I/A Series: MNB-V2-2	ea	1	1	231.70	231.70
TAC I/A Series: MNB-1000	ea	1	1	1,080.45	1,080.45
FoxView & FoxDraw	ea	1	1	0.0	0.00
Computer	ea	1	2	499.0	998.00
(Table 3 cont.)	Units	Units/Unit product	Quantity	Cost ($)/unit	Cost ($)
Heat exchanger	ea	1	1	834.0	834.00
Air sealed tent, 1000 sq. ft.	ea	1	1	550.0	550.00
Total ISBL Costs					**$32,830.83**

- Economic model is based on brewery with an **8,000 barrel/year** production rate
- **Four** total cultivation units per system
- Number of units depends on:
  - Brewery size
  - Desired CO_2 recapture rate

17 AIChE Spring Meeting, San Antonio, TX — 13TH GLOBAL CONGRESS ON PROCESS SAFETY

Fixed cost of production

(for client)

- 2 **Full-time operators** required
- Additional seasonal help as necessary (Internship opportunities)
- Size of project allows management to be absorbed by parent company

	Units	Units/year	Cost	Cost/year
Fish Care	kg	1,000.00	$0.50	$500.00
Clay media	ton	72.70	$60.00	$4,362.00
Seeds	2ea	104,000.00	$0.01	$1,308.32
Fish	ea	1,050.00*	$0.99	$1,039.50
Total Raw Materials (RM)				**$7,209.82**

			Cost/year
Labor			
	2.0	Operators per Shift Position	
# Shift positions	1.0	salaried at $36,800 ea.	$73,600.00
Maintenance			
		3% of ISBL Investment	$984.92
		GroUp Yearly Maintenance	$2,000.0
Overhead Expense			
Plant Overhead		40% of Labor & Maintenance	$29,833.97
Tax & Insurance		2% of Fixed Investment	$3,324.32
		Fixed Cost of Production (FCOP)	**$109,743.22**

Revenue streams
(for client)

- **Market value** ranges depending on consumer
  - $1.25 - $5.00 per unit
- **Three primary sources of distribution**
  - Wholesale distributors
  - Direct-to-consumer
  - Local restaurants

Key Products	Units	Units/unit product	Units /year	Cost/unit	Cost/year
Lettuce (Wholesale)	26,000	1	13,000	$1.25	$16,250.00
Lettuce (Direct)		1	6,500	$3.00	$19,500.00
Lettuce (Restaurants)		1	6,500	$2.50	$16,250.00
Basil (Wholesale)	78,000	2	78,000	$2.50	$195,000.00
Basil (Direct)		2	39,000	$5.00	$195,000.00
Basil (Restaurants)		2	39,000	$3.00	$117,000.00
Fish (Wholesale)		1	1,000	$3.00	$3,000.00
Total Key Product Revenues (REV)					**$562,000.00**

Projected growth
(for client)

Total Investment sums to **$165,672**

- Simple payback period of **0.57 years**
- 224.83% ROI after 10 years
- Engineering costs include installation, maintenance, training, and consulting fees
- Areas for sensitivity
  - Plant type
  - Price points

	Cost
ISBL Capital Cost	$32,482.03
OSBL Capital Cost	$6,496.41
Engineering Costs	$118,897.84
Contingency	$7,795.69
Total Fixed Capital Cost	**$165,671.97**
Working Capital	$16,567.20

17 AIChE Spring Meeting, San Antonio, TX — 13TH GLOBAL CONGRESS ON PROCESS SAFETY

Economic incentives

(for client)

Direct Incentives:

- Tax Credits
  - Fish and Agriculture Income Tax Credit (Mass DOR) - 3% of production cost, including property
  - Lighting controls (MassSave) - $0.50/ft^2$
- Emissions Breaks - CO_2 recapture

Indirect Value Added:

- **Increased market value** for products grown hydroponically, locally, and pesticide free
- **Tourism/Novelty** of system

17 AIChE Spring Meeting, San Antonio, TX — 13TH GLOBAL CONGRESS ON PROCESS SAFETY

Revenue breakdown

(GroUp)

Key Products/Services	Units	Units/yr	Price $/unit	$/yr
Plant Scale-Up/Scale-Down Spec	free	15	-	$0.00
Scale Quote	free	8	0	$0.00
Installation	fixed	4	$100,000.00	$400,000.00
Annual Service/Maintenance Charges	$/yr	4	$2,000.00	$8,000.00
Training Services	fixed	4	$10,000.00	$40,000.00
Crop and aquaculture consultation	fixed	4*	$2,500.00	$10,000.00
Design modifications	fixed	1**	$5,000.00	$5,000.00
Additional maintenance charges	fixed	2**	$5,000.00	$10,000.00
Total Key Product Revenues		42.0		**$473,000.00**

*Assuming all opt
**Estimated occurrence in a one-year period

Business model
(GroUp)

Location: Charlestown, MA

Team: Four full-time chemical engineers employed at salaries of $65,000 annually

CoP Costs	$/yr
Variable Cost of Production	$33,926.40
Fixed Cost of Production	$365,801.86
Cash Cost of Production	$399,728.26
Gross Profit	**$73,271.74**

GroUp Costs	$
ISBL Capital Cost	$25,450.00
OSBL Capital Cost	$5,090.00
Engineering Costs	$3,054.00
Contingency	$3,054.00
Total Fixed Capital Cost	$36,648.00
Working Capital	$3,664.80

Acknowledgements

We would like to offer thanks to...

AIChE & the 2017 Spring Meeting team for welcoming us to this year's meeting in San Antonio and inviting us to present.

Our advisors Thomas Webster, Devin Hersey, Don Wood, and Amit Roy for their leadership and guidance in our Capstone course.

Northeastern University and the Northeastern Chemical Engineering Department for making it possible for us to present today.

Nightshift Brewing Company, Downeast Cider House, and Harpoon Brewery for their interest, information, and support.

17 AIChE Spring Meeting, San Antonio, TX

13TH GLOBAL CONGRESS ON PROCESS SAFETY

Questions?

2017 AIChE® SPRING MEETING & 13TH GLOBAL CONGRESS ON PROCESS SAFETY 25

479802 Economic Optimization and Hazard Analysis of a Fluidized Catalytic Cracker

Abstract:

A fluidized catalytic cracking process was modeled along with regenerator and subsequent fractionator. The system was modeled using the process simulator ChemCAD and priced using cost equations from Turton, Bailie, Whiting and Shaeiwitz (2012). Decisions of representative feed and product compositions along with reaction pathways were made based on literature and requirement of model fidelity for economic analysis. Kinetics were regressed from industrially representative literature data. Both topological and parametric optimizations were considered. Process hazards will be identified and suggestions to mitigate hazards will be presented.

480893 Miniaturized Biomass Gasification System for Onboard Applications

Abstract:

Existing biomass gasification systems used for transportation are bulky, slow to start, and not much more efficient than traditional gasoline engines. Grassoline utilizes a small, efficient, and rapid-heating process to generate carbon-neutral electrical power for onboard applications. The design includes a packed-bed gasification reactor, a solid oxide fuel cell, and systems for fuel handling and electrical processing. The performance of a 1:18 scale model of the system is simulated in Aspen Plus.

480979 Parameters for Properly Designed and Operated Ground Flares: 2017

Abstract:

Multi-point Ground Flares (MPGF) are the preferred technology for pressure relief in many ethylene plants. Like elevated flares they provide a safe means of depressurizing the plant should an upset occur. Unlike elevated flares proper design criteria have not been widely disseminated or discussed.

Based on a series of performance studies conducted at the UOP Callidus multi-point flare test facility, engineering calculations, and computational fluid dynamics (CFD) models greater insight into the design tradespace and the resulting safe and environmentally sound operation can be gained.

Various design inputs for these flares results are discussed through physical test data and calculation including burner head design, plot space requirements, fence height requirements, and number of stages required for correct operation. The resulting impact on outputs, capital equipment cost, operating cost, environmental impact, and safety are also highlighted.

NOISE TESTING AND PREDICTION METHODS FOR MULTI-POINT GROUND FLARES

By

Scot Smith
Zeeco, Inc.
Broken Arrow, OK, U.S.A.

Brian Duck
Zeeco, Inc.
Broken Arrow, OK, U.S.A.

Justin Roberts
Zeeco, Inc.
Broken Arrow, OK, U.S.A.

Spencer Randolph
Zeeco, Inc.
Broken Arrow, OK, U.S.A.

Reginald Keith
Hoover & Keith
Houston, TX, U.S.A.

Garrett Porter
Hoover & Keith
Houston, TX, U.S.A.

Mr. Roberts joined Zeeco in 2010. As a graduate of Oklahoma State University, Mr. Roberts received a Bachelor of Science in Mechanical Engineering as well as a Bachelor of Science in Aerospace Engineering from Oklahoma State University. He currently serves as a Flare Applications Engineer.

Abstract

As demand for Multi-Point Ground Flares increases both domestically and internationally, noise generation from these flares has become a focal point of discussion. Currently, there are no set industry standards for noise predictions and theoretical noise values, which often vary between flare suppliers.

Over the years, Zeeco has conducted numerous noise measurement tests on Multi-Point Ground Flares. This paper will cover the testing and test results obtained in conjunction with one of the foremost noise consultants in the industry. In addition, this paper will address in detail noise generation from various gases at multiple sonic and sub-sonic flow rates. Further information and analysis will be provided to discuss the impact of jet noise vs. combustion noise, as well as calculating noise at various distances.

Zeeco hopes to set standards within the combustion industry in regards to predicting noise for Multi-Point Ground Flares based on the relationships correlated from the test data.

Presented at the

2017 Spring Meeting and 13th Global Congress on Process Safety
San Antonio, Texas – March 26-30, 2017

Overview of Multi-Point Ground Flares

Developed in the 1970s, Multi-Point Ground Flares (MPGF) derive their name from their physical layout. Instead of locating the flare flame on an elevated structure, the flame is spread out in a grade-mounted field of multiple pressure assisted flare tips. The tips are then arranged in stages that open as the upstream pressure and gas flow increases and close as pressure and flow decreases.

MPGF are often selected for heavy hydrocarbon service with high available pressure; however, they can be used for a wide range of gas compositions. High pressure assists in obtaining full smokeless operation, which can be difficult to do with other assist mediums. Each tip in a MPGF has unobstructed air access, allowing the momentum from the high exit velocity of the flare gas to entrain the necessary air for full combustion. MPGF are designed to provide maximum smokeless performance, while minimizing radiation impacts and the need for a large sterile area around the flare. Installing a fence around the field can block the visibility of the flare flame, serving a dual purpose of reducing radiation and the likelihood that flaring operations will be a nuisance to the public. Figure 1 is an example of a typical MPGF installation.

FIGURE 1: TYPICAL MPGF INSTALLATION

Unlike elevated flares, MPGF provide easy access for maintenance. Typically, all staging equipment is located at grade and outside the fence. As a result, workers can safely perform normal operations near the flare without being affected by a flaring event. Figure 2 illustrates a typical fence installed with a MPGF.

FIGURE 2: TYPICAL MPGF FENCE

Noise

Noise can be defined as "excessive or unwanted sound. In general, any sound that is annoying, interferes with speech, damages the hearing, or reduces concentration or work efficiency may be considered noise. In air, sound is radiated spherically from its source as a compressional wave, being partly reflected, absorbed or transmitted by hitting an obstacle. Noise is usually a non-periodic sound wave, as opposed to a periodic pure musical tone or sine –wave combination. It is characterized by its intensity (measured in decibels), frequency, and spatial variation." (Darling, 2016)

A decibel (dB) is a log base scale developed to quantify sound. There are two common uses of decibel levels. One is sound power (PWL) and the other is sound pressure (SPL).

Sound power (PWL) or acoustic power is the rate at which sound energy is emitted, reflected, transmitted or received, per unit time. Sound pressure or acoustic pressure is the local pressure deviation from the ambient atmospheric pressure, caused by a sound wave. The sound pressure scale usually ranges from 0 to 140 dB. The 0 value of the scale occurs when sound pressure equals the threshold of human hearing. Figure 3 shows a comparison of sound pressure levels and sound pressures for a variety of different common objects.

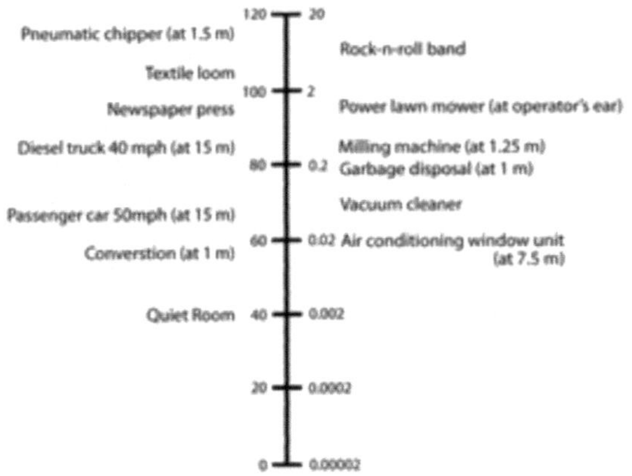

(Breysse, 2006)

FIGURE 3: SOUND PRESSURE LEVEL (SPL) AND SOUND PRESSURE (PA) OF COMMON ITEMS

Even though a decibel has a unit of dB, noise valves are often given in dBA. The "A" designation means the noise level has been modified (i.e. A-Weighted) to de-emphasize the low and very high frequencies which pose less of a risk to hearing. In this paper, all noise values will be shown as unweighted unless stated otherwise.

Another important item when looking at noise data is Leq. When a noise varies over time, the Leq is the equivalent continuous sound which would contain the same sound energy as the time varying sound. In essence, this is the average measurement over a duration of time.

Test Setup

Testing was conducted at Zeeco's test facility in Broken Arrow, Oklahoma. One MPGF flare tip was installed on the test stand which can be seen in Figure 4.

FIGURE 4: MPGF TEST SETUP

With this test setup, gas flow was measured using a 4-inch orifice run. The upstream pressure in the orifice run was measured using a digital pressure transmitter and the temperature was measured using a thermocouple. The differential pressure across the orifice was measured using a differential pressure transmitter.

Tip pressure and gas temperature were also recorded for secondary flow measurement verification. Tip pressure was recorded using a digital pressure transmitter. Gas temperature was recorded using a thermocouple. All aforementioned data was recorded simultaneously using a data acquisition system (DAQ). A weather station was also connected to the DAQ that measured wind speed, wind direction, ambient temperature, barometric pressure, and relative humidity throughout the entire test, which allowed for accurate accounting of atmospheric attenuation in the analysis.

The test fuels used were Tulsa Natural Gas (TNG) and Propane.

Noise measurements were recorded at distances of 100′-0″ and 200′-0″ to the East of the flare tip using two Norsonics NOR140 Type I noise meters. One meter was placed at each distance to measure simultaneously during the test points. Each measurement point lasted 60 seconds.

In order to minimize the amount of background noise, testing was conducted at night with all non-essential equipment (compressors, forklifts, etc.) shut off to avoid contamination of the noise results.

To aide in the testing, Zeeco partnered with Hoover & Keith, a well renowned noise consultant located in Houston, Texas. A consultant from Hoover & Keith was present during all testing and was involved in data analysis.

Results

Background Noise

To ensure accuracy of the test data, background noise points were taken before and after testing. The average ambient sound pressure level was approximately 64 dB. As observed in Figure 5, ambient noise was dominated by low frequencies. While every action was taken to reduce ambient noise, proximity to city streets and highways were uncontrollable factors that likely led to the slightly elevated levels of low frequency sound.

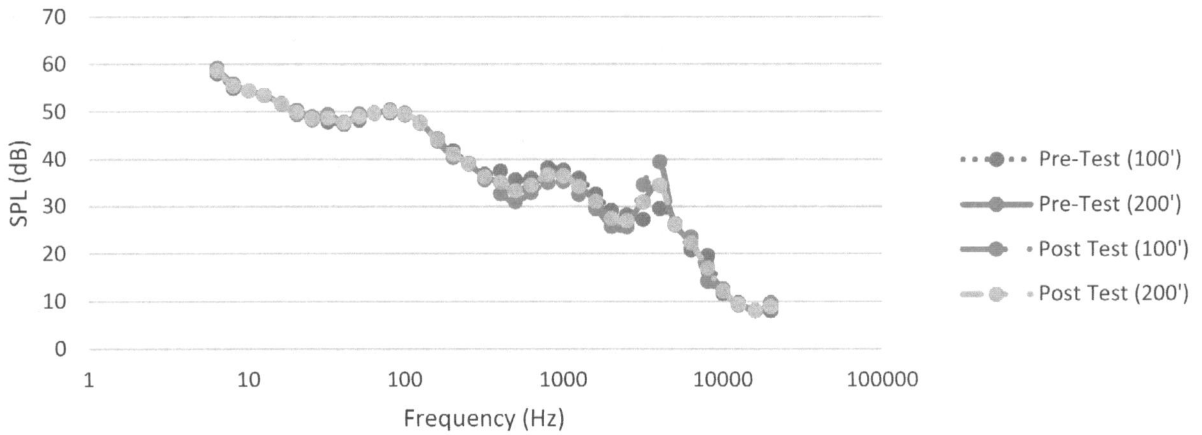

FIGURE 5: AMBIENT NOISE AS SOUND PRESSURE LEVEL IN 1/3 OCTAVE BANDS

Sound Power Levels as a Function of Sonic and Subsonic Flows

All resulting noise data was analyzed on an unweighted basis as sound pressure levels in 1/3 octave bands. All noise measurements were recorded over a 60 second measurement duration. During the measurement duration, a 1/3 octave band spectrum was recorded at 1-second intervals. The 60 measured 1-second 1/3 octave band spectra were then used to compute a single 60-second Leq 1/3 octave band spectrum. The data presented in this paper are the computed 60-second Leq 1/3 octave band spectra or the overall (or total) level derived from these 1/3 octave band spectra.

Shown in Figure 6 is the sound pressure level (SPL) in dB versus fuel flow rate in pounds mass per hour.

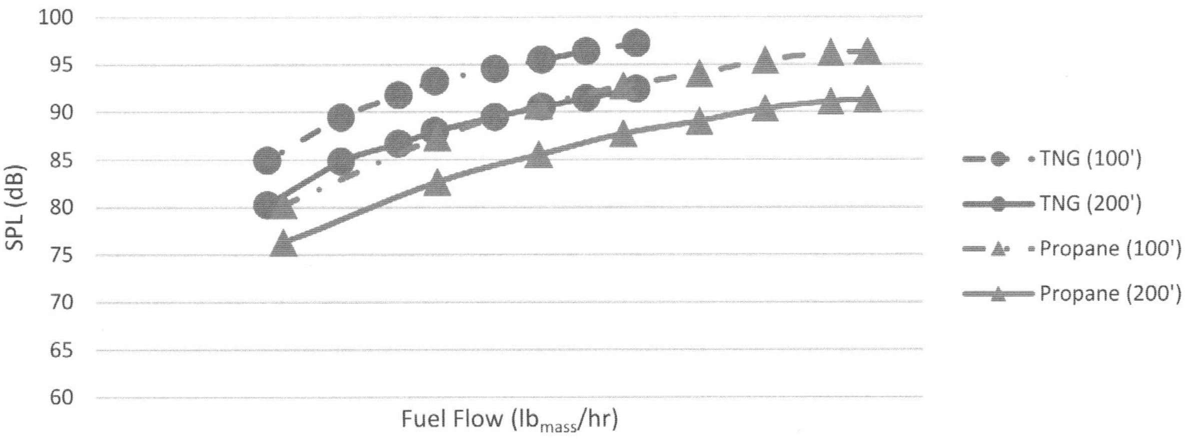

FIGURE 6: SOUND PRESSURE LEVEL (SPL) AS A FUNCTION OF FUEL MASS FLOW RATE

The sound pressure level for each test point was converted to a sound power level using Eq. (1).

$$PWL = SPL + 20Log(r) + 0.5 \quad \text{Eq. (1) (Cunha-Leite, 1988)}$$

Variable "r" is the direct distance in feet from the noise source to the noise measurement location. When converting to sound power level, atmospheric attenuation was taken into account using onsite meteorological data. Upon analysis of the calculated sound power levels, as shown in Table 1 and Table 2, the largest percent difference between the two measurement points was 0.6%, with an average percent difference of 0.4% for TNG. A maximum percent difference of 1.5% was observed for propane with an average percent difference of 0.4%. Each type I meter has an accuracy range of ± 1 dB. When comparing both meters this yields a ± 2 dB range. Therefore, the percent differences in the measurements between the two noise meters fall within the combined accuracy range of the meters. The following results are analyzed as a sound power level. Figure 7 shows a graph of the data from Tables 1 and 2.

TABLE 1: COMPUTED SOUND POWER LEVEL (PWL) – TNG

TP#	PWL - 60 Second Log Average			
	100'	200'	Δ dB	% Diff
1	138.7	139.3	0.6	0.4
2	137.9	138.4	0.5	0.4
3	137.0	137.4	0.5	0.3
4	136.0	136.4	0.4	0.3
5	134.7	134.9	0.1	0.1
6	133.2	133.5	0.3	0.2
7	130.9	131.7	0.8	0.6
8	126.3	127.1	0.8	0.6
Average % Difference			0.5	0.4

TABLE 2: COMPUTED SOUND POWER LEVEL (PWL) - PROPANE

TP#	PWL - 60 Second Log Average			
	100'	200'	Δ dB	% Diff
9	137.8	138.2	0.4	0.3
10	137.7	138.0	0.3	0.2
11	136.9	137.2	0.3	0.2
12	135.5	135.9	0.4	0.3
13	134.2	134.6	0.4	0.3
14	132.1	132.4	0.3	0.2
15	128.7	129.5	0.8	0.6
16	121.7	123.1	1.5	1.2
Average % Difference			0.5	0.4

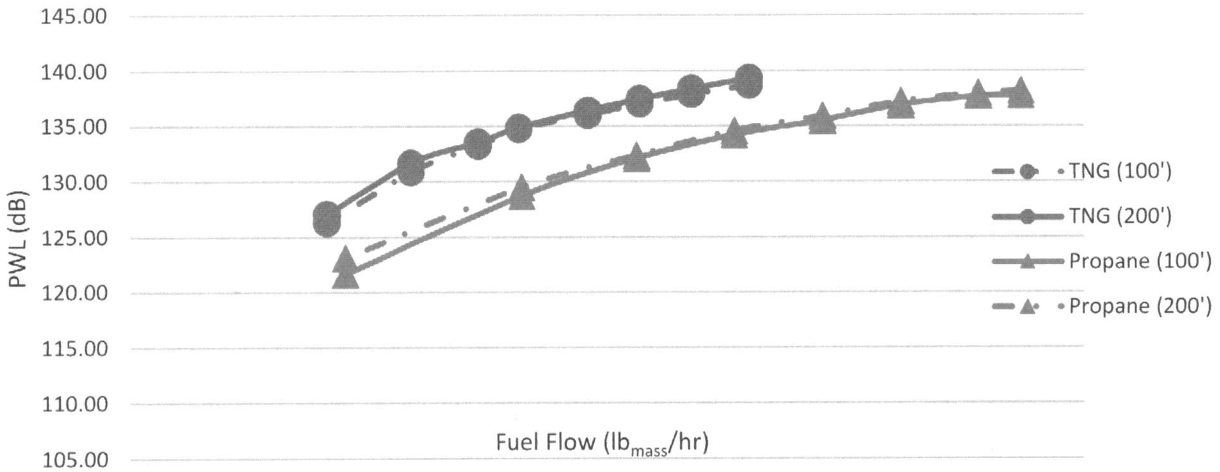

FIGURE 7: SOUND POWER LEVEL (PWL) AS A FUNCTION OF FUEL MASS FLOW RATE

Shown in Figure 8, the sound power levels versus tip static pressure. Also noted are the calculated critical pressures for both TNG and Propane. The critical pressures were calculated from the specific heat ratio of the fuel gas at their respective flowing temperatures per Eq. (2) and correspond to the point at which the fuel gas reaches sonic velocity.

$$p_c = p_2 \left(\frac{k+1}{2}\right)^{\frac{k}{k-1}} \qquad \text{Eq. (2)}$$

$P_c = Critical\ Pressure, psia$

$P_2 = Ambient\ pressure. psia$

$k = Ratio\ of\ Specific\ Heat$ (Reed, 2001)

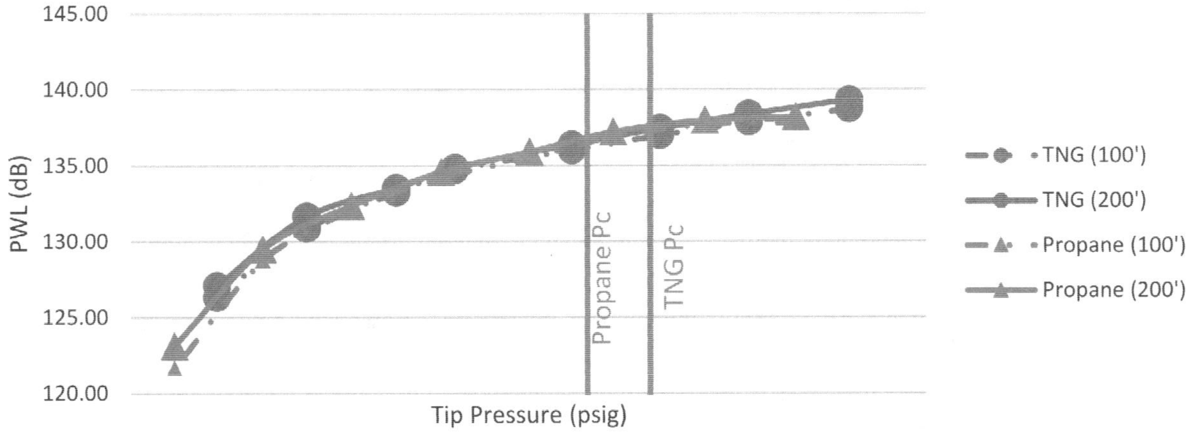

FIGURE 8: SOUND POWER LEVEL (SPL) AS A FUNCTION OF TIP STATIC PRESSURE

Acoustical Efficiency

Using the empirical sound power level, fuel flow rate, and fuel composition, the acoustical efficiency was calculated for each test point. Figure 9 shows the observed trend such that as the fuel flow rate increases, the acoustical efficiency increases for a constant exit area.

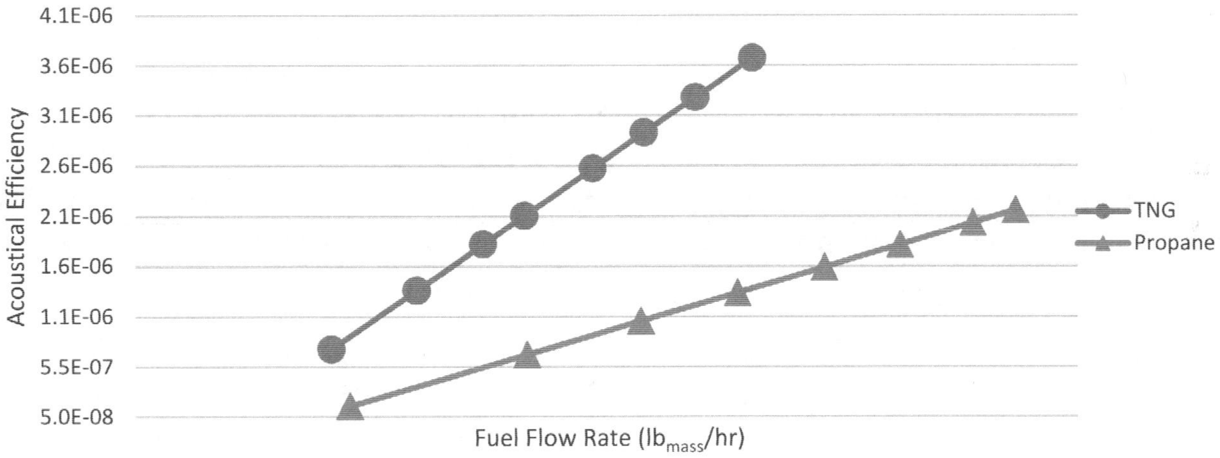

FIGURE 9: ACOUSTICAL EFFICIENCY TREND AS A FUNCTION OF FUEL FLOW RATE

Combustion vs. Venting

Two sets of data points were tested to compare venting jet noise to combustion noise. Shown in Table 3, venting without combustion produced an average 20 dB decrease in sound power level of TNG and an average 23 dB decrease for propane.

TABLE 3: SONIC VS SUBSONIC SOUND POWER LEVEL

Test Point	PWL (Leq dB) at Respective Meter Locations		
	100' (dB)	200' (dB)	Average Δ (dB)
TNG Sonic - Combustion	139	139	20
TNG Sonic - Venting	119	119	
TNG Subsonic - Combustion	136	136	20
TNG Subsonic - Venting	116	116	
Propane Sonic - Combustion	138	138	23
Propane Sonic - Venting	114	117	
Propane Subsonic - Combustion	136	136	23
Propane Subsonic - Venting	112	114	

A further analysis of individual 1/3 octave bands indicates that combustion noise predominantly occurs at frequencies below approximately 500 Hz and jet noise predominantly occurs at frequencies above approximately 2500 Hz. It is important to note that the venting test point low frequency noise could be influenced by the proximity to city streets and highways. The higher trends of the venting case for low frequencies less than 1000 Hz does not appear to correlate to noise mechanisms of the flare, but when compared to background noise frequency appears to be originating from test site surroundings. At these lower frequencies for the venting case, the ambient noise pressure level is higher than that of the flare tip. When analyzing the data as sound power level, the calculation is not applicable to these lower frequencies due to the measured sound pressure level being from ambient surroundings and not the venting point source. When the sound power level is calculated from sound pressure level of ambient noise, the ambient noise is arbitrarily calculated as though it is a point source at 100′ to 200′ from the noise measurement location. This arbitrarily amplifies the sound power level, but is necessary to be applied over the full noise spectrum for the analysis of jet noise in frequencies above 2500 Hz. Figures 10 and 11 show the sound power levels plotted for each 1/3 octave band for both TNG and propane at sonic and subsonic flows.

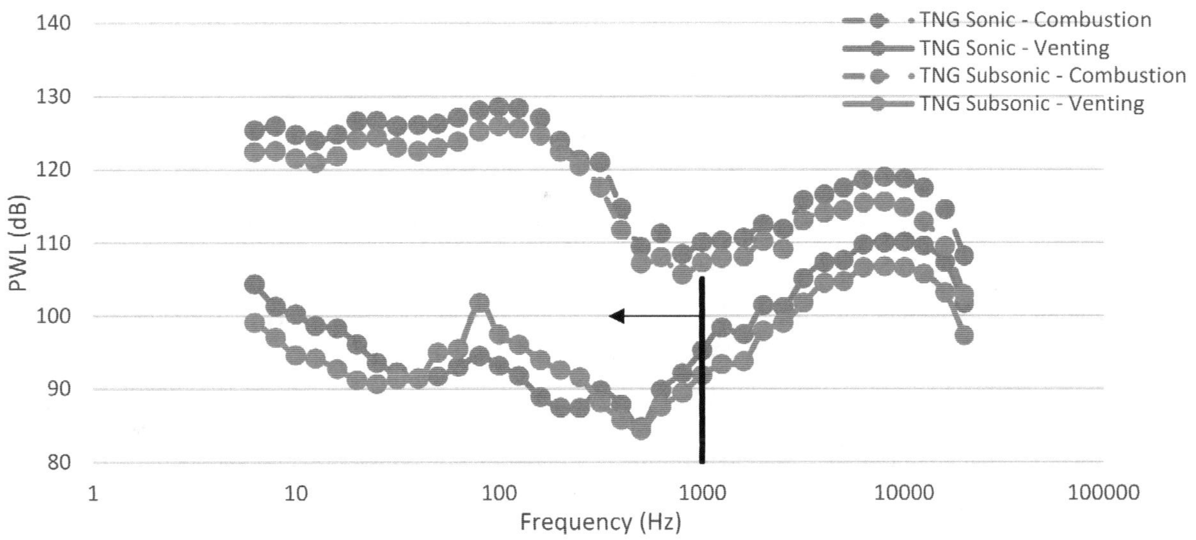

FIGURE 10: SOUND POWER LEVEL (PWL) AS A FUNCTION OF FREQUENCY - TNG

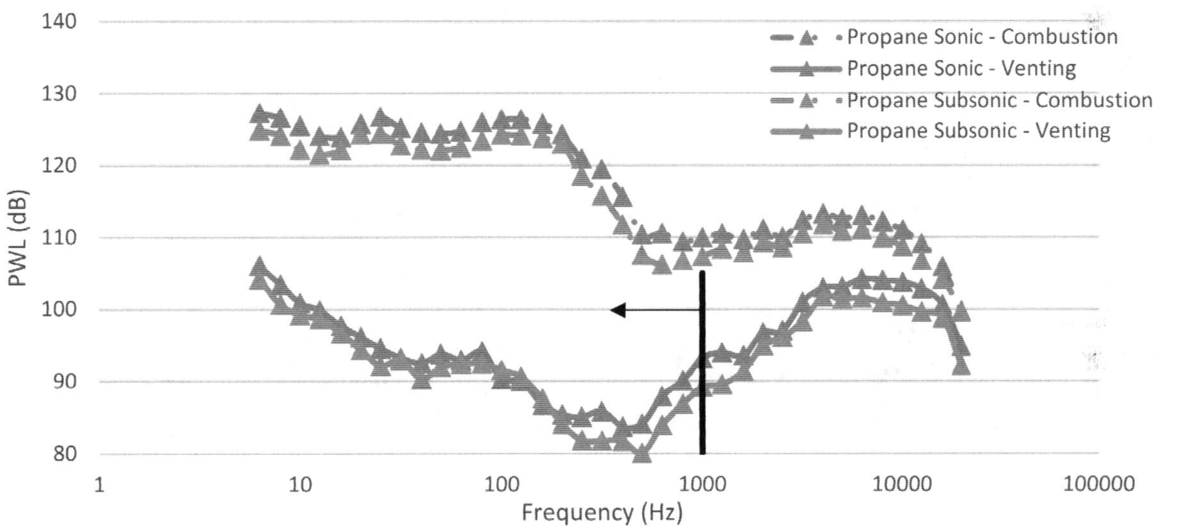

FIGURE 11: SOUND POWER LEVEL (PWL) AS A FUNCTION OF FREQUENCY - PROPANE

In Figure 12 and Figure 13, the sound power level versus 1/3 octave band frequencies have been plotted for each test point, again note the noise spectrum is dominated by below 500 Hz and above 2500 Hz. Individual test point information in these figures are of less concern. The main observable trend is the sound power level increase for low (500 Hz and below) and high (2500 Hz and above) frequencies. This trend is observed in both propane and TNG testing.

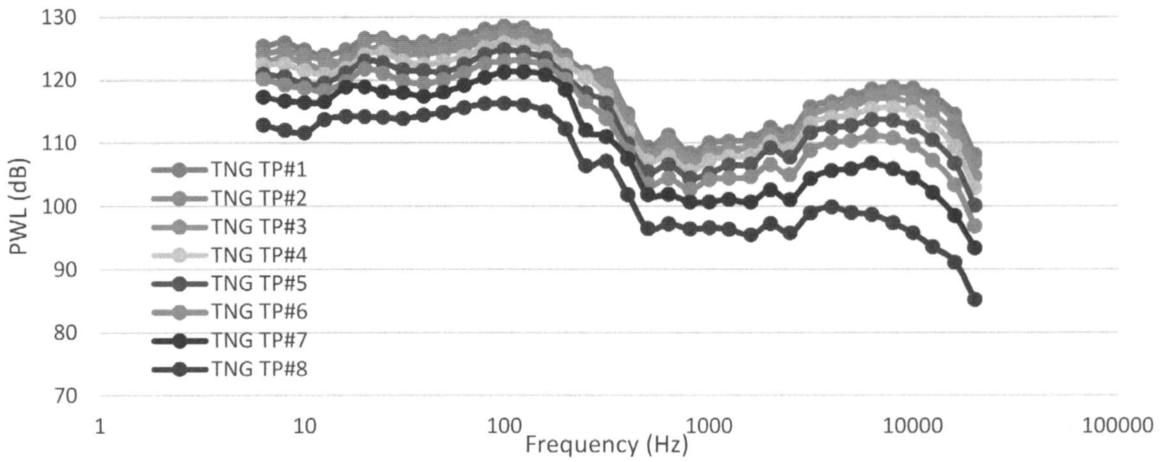

FIGURE 12: SOUND POWER LEVEL (PWL) AS A FUNCTION OF FREQUENCY AND FUEL FLOW RATE – TNG

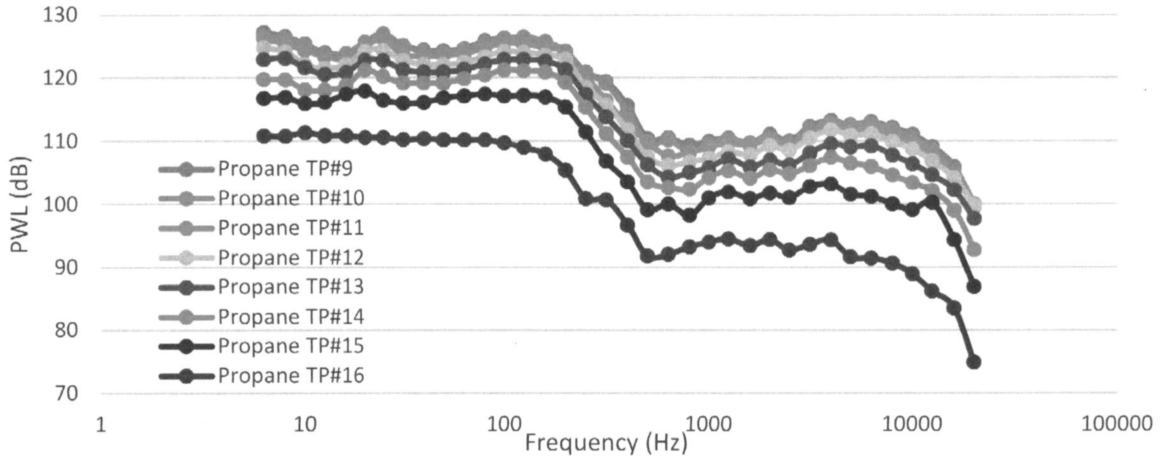

FIGURE 13: SOUND POWER LEVEL (PWL) AS A FUNCTION OF FREQUENCY AND FUEL FLOW RATE - PROPANE

Conclusions

10Log vs 20Log Analysis

Previous information debates using a 10Log versus 20Log relationship to calculate the overall PWL. For each case, a reference fuel mass flow rate and corresponding power level is used to determine a sound power level over a range of fuel mass flow rates. In equations 3 and 4, G_0 references the known fuel flow rate corresponding to a known sound power level. G, is the fuel mass flow rate at which a sound power level is to be determined.

$$PWL_{calculated} = PWL_{reference} + 10Log\left(\frac{G}{Go}\right)$$ EQ. (3) (NARASIMHAN, 1986)

$$PWL_{calculated} = PWL_{reference} + 20Log\left(\frac{G}{Go}\right)$$ EQ. (4) (NARASIMHAN, 1986)

$$PWL_{Calculated} = Sound\ Power\ Level\ at\ Desired\ Flow\ Rate, dB$$

$$PWL_{reference} = Sound\ Power\ Level\ at\ Empirical\ Reference\ Point, dB$$

$$G = Desired\ Fuel\ Flow\ Rate, \frac{lb_{mass}}{hr}$$

$$G_0 = Fuel\ Flow\ Rate\ at\ Empirical\ Reference\ Point, \frac{lb_{mass}}{hr}$$

Figures 14, 15, 16, and 17 show the analysis of the 10Log vs 20Log trends. The actual test data was graphed as well as the data calculated based on G_0 being a low flow point, mid flow point, or a high flow point. These low, mid, and high flow points are all empirical data collected during the testing. The low flow point references the lowest fuel flow point in the data set for that fuel gas, mid flow references a mid-range point, and high flow references the highest fuel flow of the data set for each fuel.

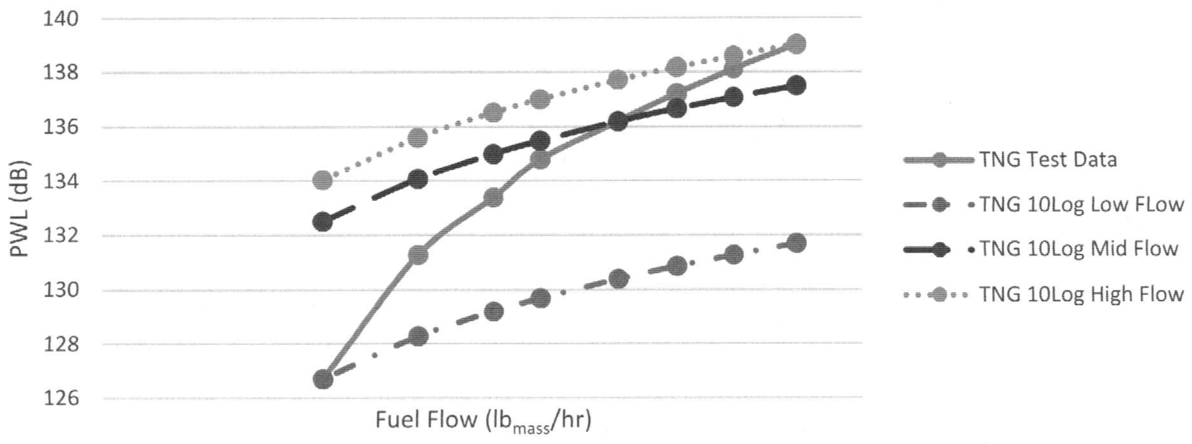

FIGURE 14: 10LOG TREND ANALYSIS - TNG

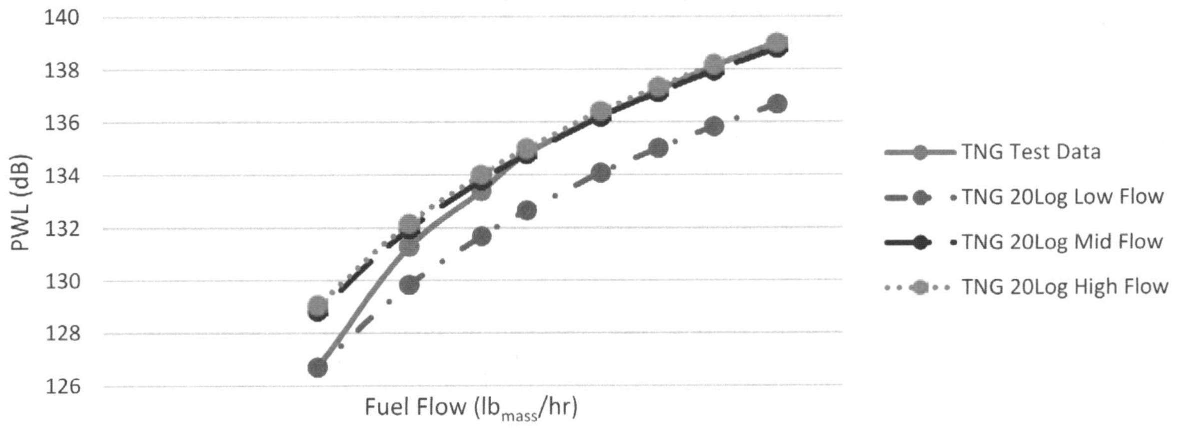

FIGURE 15: 20LOG TREND ANALYSIS – TNG

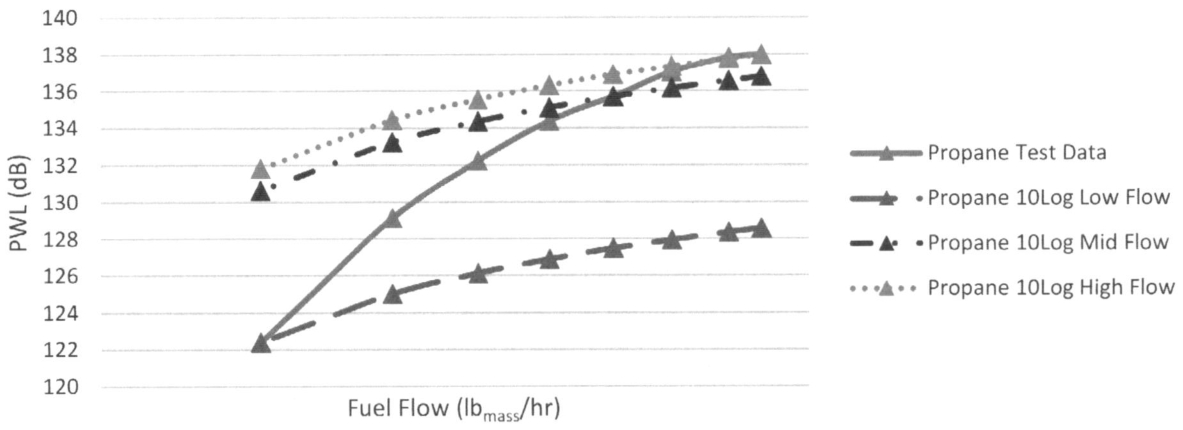

FIGURE 16: 10LOG TREND ANALYSIS – PROPANE

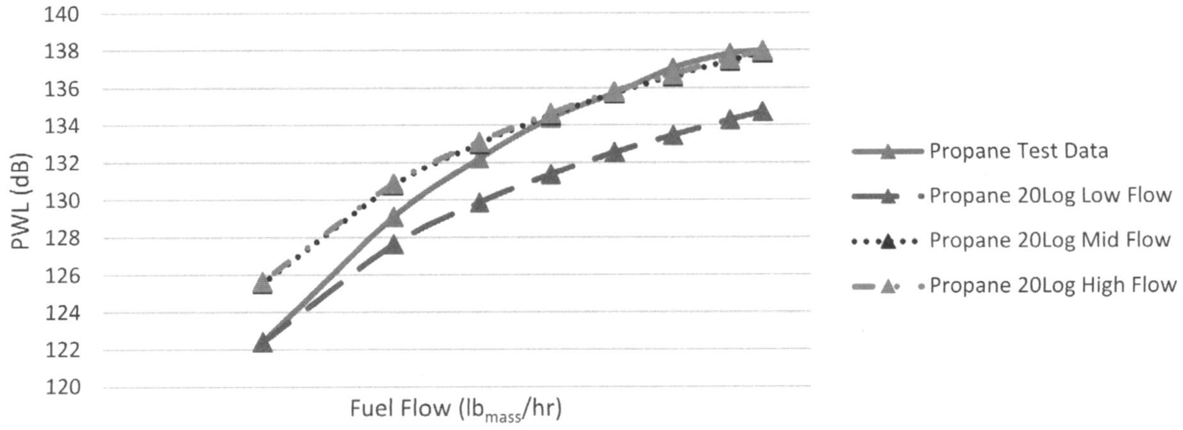

FIGURE 17: 20LOG TREND ANALYSIS - PROPANE

As shown in the figures above, for both propane and TNG, the 20Log trend fit resulted in a smaller percent error. Also observed in the figures above, the referenced empirical data point impacts the percent error. When analyzing a data point corresponding to a smaller mass flow rate, the trend does not accurately represent empirical data.

TABLE 4: PERCENT ERROR IN 10LOG AND 20LOG TRENDS

		Low Flow		Mid Flow		High Flow	
		Avg % Error	Max % Error	Avg % Error	Max % Error	Avg % Error	Max % Error
TNG	10Log	-3.5	-5.3	0.8	4.6	1.9	5.8
	20Log	-1.3	-1.7	0.3	1.7	0.4	1.8
Propane	10Log	-5.0	-6.9	1.2	6.7	2.1	7.7
	20Log	-1.9	-2.6	0.5	2.6	0.6	2.7

As shown in Table 4, the most accurate trend while yielding conservative results, is a 20Log trend based on a data point corresponding to a mid-range mass flow rate. It is important to note that as the calculated sound power level and associated mass flow rate deviate from the referenced point, the percent error increases. This increase is greater at calculated sound power levels associated with low mass flow rates.

Acoustical Efficiency

Multiple publicly available reference articles include discussions on the topic of acoustical efficiencies. Several discrepancies exist between these reference articles as shown in Table 5. The empirical data collected for this testing was investigated for acoustical efficiency trends. Using Eq. (5) and Eq. (6), acoustical efficiencies for each test point were calculated.

$$\boldsymbol{PWL = 10Log(\frac{W}{W_0})} \qquad \text{EQ. (5)}$$

$$\boldsymbol{W = e\left(0.293\frac{hr\cdot W}{Btu}\right) G \cdot H_v} \qquad \text{EQ. (6)}$$

$$e = Acoustical\ Efficiency$$

$$W = Sound\ Power, Watts$$

$$W_0 = Reference\ Sound\ Power, Watts$$

$$G = Fuel\ Flowrate, \frac{lb}{hr}$$

$$H_v = Heat\ Value, Btu/lb$$

TABLE 5: ACOUSTICAL EFFICIENCIES FOR A TYPICAL HYDROCARBON

Reference Article	Acoustical Efficiency for a Typical Hydrocarbon
"Predict Flare Noise and Spectrum" (Cunha-Leite, 1988)	$5(10^{-8})$
"Predict Flare Noise" (Narasimhan, 1986)	$1(10^{-6})$
"Noise Generation by Open Turbulent Flames" (Smith, 1963)	$1.23(10^{-8}) - 8.20(10^{-8})$
"Ecological Aspects of Combustion Devices (with Reference to Hydrocarbon Flaring)" (Swithenbank, 1972)	$1(10^{-7}) - 1(10^{-9})$

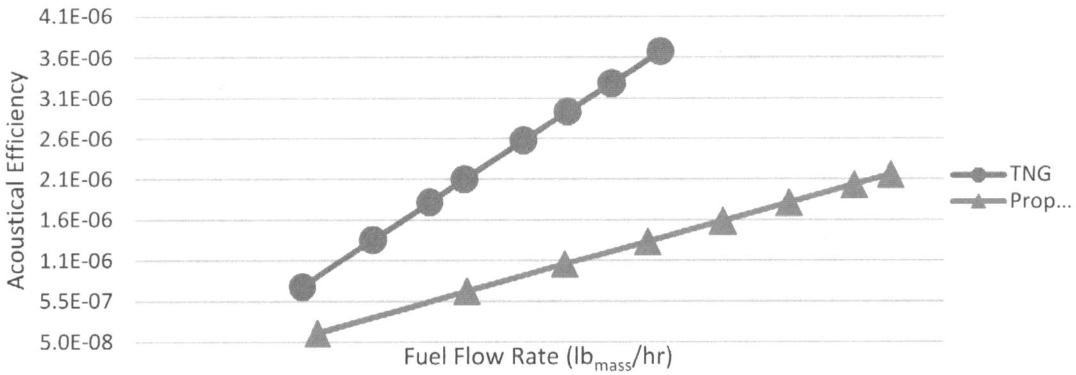

FIGURE 18: ACOUSTICAL EFFICIENCY (E) AS A FUNCTION OF FUEL MASS FLOW RATE

As shown in Figure 18, it was observed that the acoustical efficiency increases with increasing fuel mass flow rate for a constant exit area. Trends of both TNG and Propane were similar with differing rates of change.

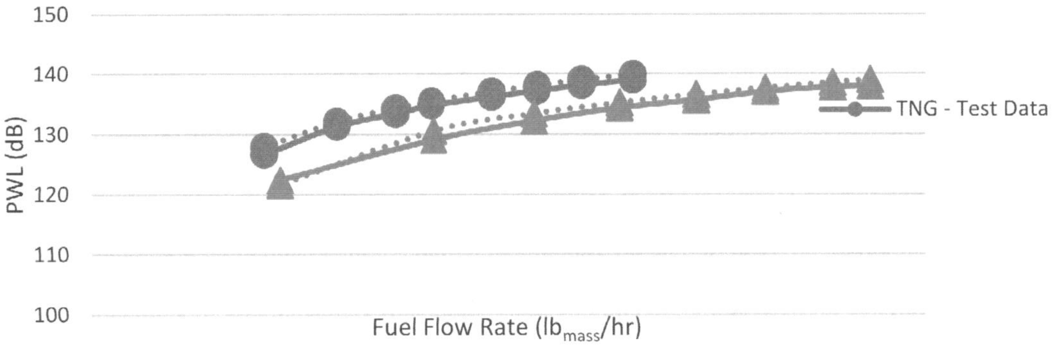

FIGURE 19: SOUND POWER LEVEL (PWL) AS A FUNCTION OF FUEL MASS FLOW RATE

In conclusion, it was observed that as the fuel mass flow rate increased, the acoustical efficiency increased for a constant exit area. When applying the changing acoustical efficiency as a function of fuel composition and flowrate, the prediction model shows a much stronger correlation, as shown in Figure 19. Zeeco is continuing work with Hoover & Keith to determine the impact of these acoustical efficiency trends.

Further Testing

10Log vs 20Log Analysis

While the 10Log vs 20Log analysis shows a more accurate trend correlation when analyzing by means of a 20Log function, testing including higher fuel flow rates would provide a better understanding of the error involved when extrapolating noise values outside of a small range away from the referenced empirical data. A larger range of fuel flow rates would also allow a better understanding of optimal fuel flow rates to use as an empirical reference.

Acoustical Efficiency

With the observance of increasing acoustical efficiencies associated with increasing fuel flow rate for a constant exit area, further testing is required to determine actual causation. Testing of the same format with a multitude of fuel gases would be beneficial. This information would provide more evidence to analyze trends present between fuel gases with different heating values and molecular weights. In addition, fuel blends and inert mixtures would add additional understandings to the phenomena observed. The aforementioned acoustical efficiency testing could potentially yield a more accurate method of predicting multipoint ground flare noise levels.

Combustion vs. Venting

While the combustion versus venting tests provide insight into which frequency ranges are dominated by respective noise mechanisms, it would be beneficial to test a multitude of fuel gases of differing molecular weights and sonic velocities. Adjusting fuel exit areas while maintaining a constant fuel flow rate would yield a better understanding of the driving noise mechanisms of combustion versus venting and would allow a better understanding of the magnitude of impact from combustion noise. This would be facilitated by incrementally decreasing the fuel exit velocity and respective jet noise, while maintaining constant combustion.

References

1. Breysse, P. N. (2006). Noise. John Hopkins School of Public Health.
2. Cunha-Leite, O. (1988, December). Predict flare noise and spectrum. *Hydrocarbon Processing.*
3. Darling, D. (2016). *Noise.* Retrieved from Encyclopedia of Science: http://www.daviddarling.info/encyclopedia/N/noise.html
4. Narasimhan, N. (1986, April). Predict flare noise. *Hydrocarbon Processing.*
5. Reed, R. J. (2001). *North American Combustion Handbook* (Third ed., Vol. I). Cleveland, OH: North American Manufacturing Company.
6. Smith, T. &. (1963, May). Noise Generation by Open Turbulent Flames. *The Journal of the Acoustical Society of America, 35*(5), 715-724.
7. Swithenbank, J. (1972, May). Ecological Aspects of Combustion Devices (with Reference to Hydrocarbon Flaring). *AIChE Journal, 18*(3), 553-560.

479762 Optimization for Flare Operation

Abstract:

Abstract

Flare operation is a challenging task since the flowrate, net heating value, composition of the vent gas and weather conditions vary significantly. The industry is in urgent need of a viable approach to achieve smokeless flaring that can meet the combustion/destruction efficiency standard set by the agencies, while minimize stream and makeup fuel consumption.

In this work, a novel optimization approach was developed based on flare performance models. The controlled variables include flowrate of steam/air assist and flowrate of make-up fuel. For a flaring vent with a specific combustion zone net heating value and particular composition, the most economic solution to achieve smokeless flaring and high combustion/destruction efficiency objective can be identified.

482664 New Manager Prospectives from Johnny Rokisky: Fungicide Formulation Production Superintendent from Syngenta

Abstract:

Johnny Rokisky is the Unit Superintendent of a Fungicide Production Unit for Syngenta in Houston Texas. He received his bachelors of Chemical Engineering from Virginia Tech in 2010 and joined Syngenta after in a Leadership Development Rotational Program. During this 2 years in the rotational program, he worked as a production engineer, a formulation engineer, and a Supply Chain Planner. After graduating from the rotational program, he moved to Omaha to work as a new product engineer. Shoftly after, he was promoted to become a Unit Superintendent at the Houston site.

482665 New Manager Prospectives from Carol Schmidt: A Technical Sales Manager for the Sgl Group

Abstract:

Carol works for SGL Group, as a Technical Sales Manager for the Process Technology/Engineered Solutions department. SGL's product lines focus on non-metal equipment for corrosive services. Mainly, this is heat exchangers with graphite or silicon carbide internals, PTFE bellows and lined piping/columns, and acid synthesis systems. Carol has been doing this for 1.5 years. Carol does not have any direct employee reports, but she manage 12 rep agents in 5 states. Carol is responsible for a sales territory spanning 12 states and she travel about 2-3 weeks of the month. Her responsibilities include, but are not limited to, regularly visiting customers, finding opportunities, forecasting, budgeting, quotations, sales orders, and negotiation.

Carol will speak about managering her group indirectly, prioritizing her workload, and handling all of this dealing with gender/age/ethnic issues as well as children.

482666 New Manager Prospective Carlos Andres Salamanca

Abstract:

Carlos is the current STS-AIChE Chair Elect for 2017, former STS-AIChE Treasurer for 2015 and 2016. He has been an active member of the STS AIChE Young Professionals Board serving as Secretary, Treasurer, Vice-Chair and Chair. He also has represented AIChE with the Offshore Technology Conference events including OTC Houston, Arctic Technology Conference, OTC Brasil, and OTC Asia in different roles. Carlos is also active with Toastmasters, SPE (Society of Petroleum Engineers), and the PMI (Project Management Institute) Houston Chapter.

He has held several positions through different oil and gas projects. Carlos has been working with oil companies in Latin America with focus on Brazil, Colombia and Mexico. His most recent role was with National Oilwell Varco working with challenging projects including oil and gas drilling and production rigs. He also enjoys teaching different subjects. Carlos started his career with Texaco, now Chevron, at the lubricants plant in Bogotá, Colombia.

Carlos earned a BS in Chemical Engineering from the Universidad Nacional de Colombia, and received an MBA from the Bauer College of Business at the University of Houston. Carlos holds a PMP (Project Management Professional) certificate from the Project Management Institute, and an EIT (Engineer-In-Training) certificate from the Texas Board of Professional Engineers. His professional interests focus on Oil and Gas Economics, Project Management, Global Business, Latin American Studies and new Engineering technologies. Carlos likes to read major business publications, leadership, self-improvement, public speaking and management books. He enjoys spending time with family and friends, all sporting events, riding his bicycle, and traveling. Carlos is also fluent in Spanish and is currently learning Portuguese.

Operating Unstable Catalyst – What We've Learned

Jane Liang-Weissgerber
Jonathan Lunn
The Dow Chemical Company

Outline

- Incident Background
- Catalyst Instability
- Summary

DRIVE TO ZERO

Reactor Setup

- Front end de-ethanizer cracker
  - Lead bed with 98% conversion
  - Cleanup bed with 2% conversion
  - Interbed cooler to control conversion into guard bed
- Catalyst was replaced in 2012 after **16** years of service
- No issues with catalyst performance

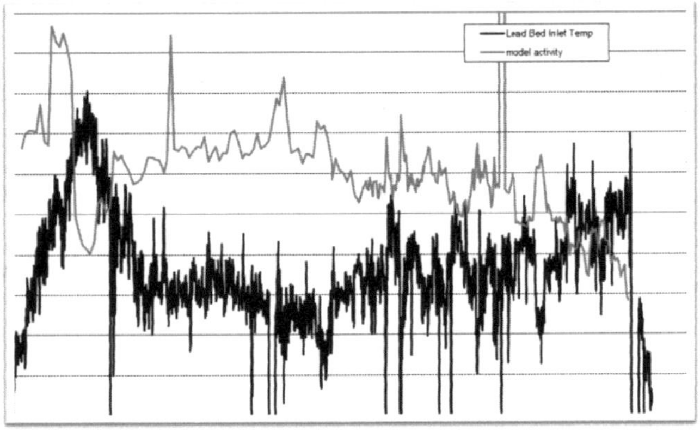

DOW

DRIVE TO ZERO

Incident Background – Reactor Exotherm

- **2014 Major Plant Incident** – *Plant restart* ***6*** *days later*
  - During reactor start-up flow could not be established, leaving pressurized cracked gas in a warm bed
  - **Excessive exotherm** occurred in clean-up bed
  - Signs of ethylene de-composition
- **Problems.....Catalyst Change Out**
  - 4 months post plant incident (return to full rates) – *dP RISING*
  - 1 month later after return to full rates – *EXOTHERM*
  - Started planning for catalyst change out only after **2** years of service

DOW

DRIVE TO ZERO

Incident Background – Reactor Exotherm

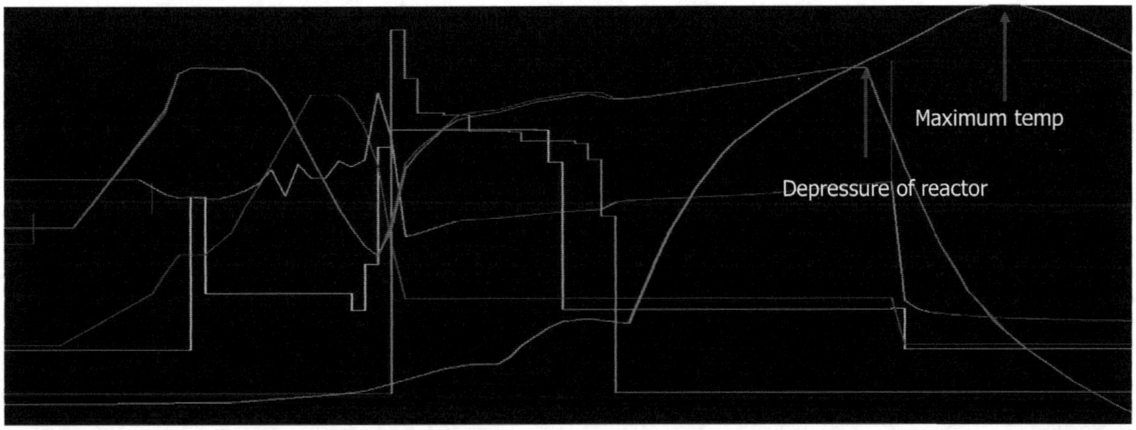

DOW

DRIVE TO ZERO

Incident Background – Root Cause

- ***Process Control***
  - High temperature trip logic did not include certain start-up steps
- ***Operating Discipline***
  - Reactor start-up procedure was not followed
- ***Contributing Causes***
  - The valves were not aborted in the critical start-up step
  - Reactor was started up too warm

DOW

DRIVE TO ZERO

Reactor Exotherm Mechanics

- Adsorption on catalyst governed by H2 > Acetylene > CO >>> C2H4
  - $C_2H_2 + H_2 \rightarrow C_2H_4$
  - $C_2H_4 + H_2 \rightarrow C_2H_6$
  - Other reactions with MAPD and butadiene → not as relevant with front end de-ethanizer configuration
- Operating Window influenced by
  - CO
  - dT (i.e. adiabatic load)
  - Flow (GHSV)
  - Impurities, age of catalyst

Dow

DRIVE TO ZERO

Unstable Catalyst – Multiple Exotherm Incidents

- Additional exotherms (4) initiated by small changes in
  - Feed rate
  - CO fluctuations
  - Radial dT drifts
- Further symptoms
  - dP on the guard bed kept rising
  - Same area as original exotherm would develop hot spot

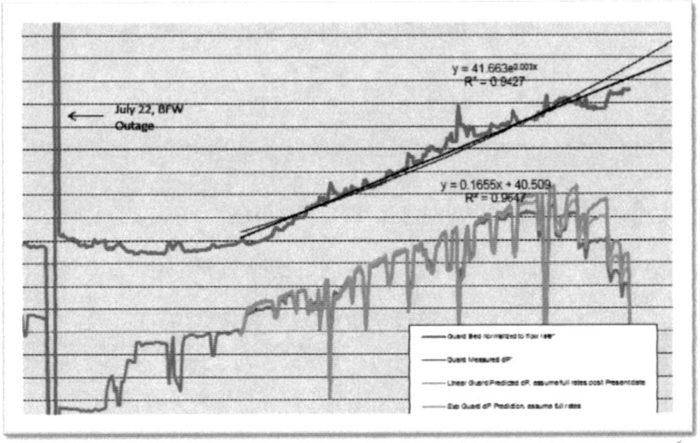

Dow

DRIVE TO ZERO

Comparison of dP Changes

Post Exotherm Incident

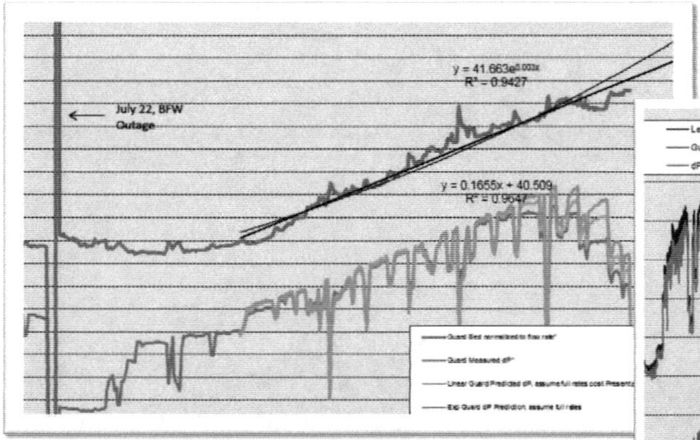

16 Years of Operation

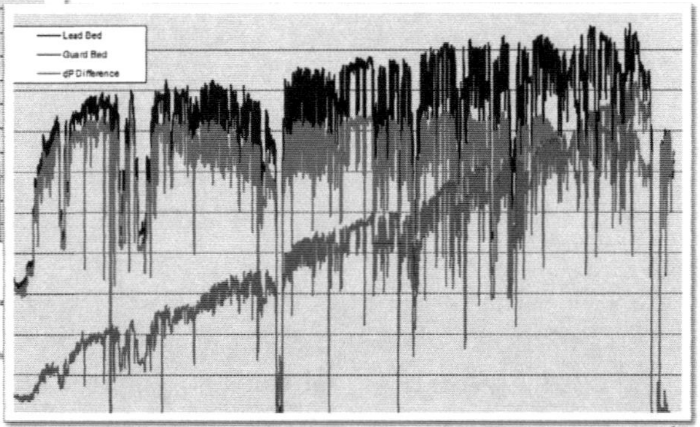

Why was the Catalyst Unstable?

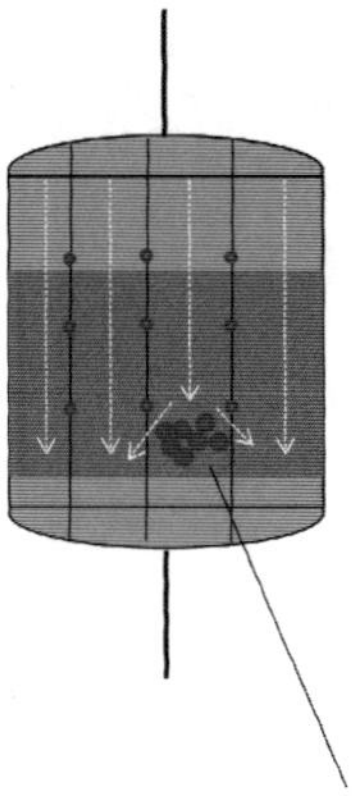

Obstructed Flow

What evidence have we seen that would lead us to believe a hot spot is present?

- Numerous dT "blips" when running lower spec on guard bed outlet
- Runaways/exotherms in the same area as first incident

What could have caused it?

- Ethylene decomposition conditions created during major plant upset event
- Macroscopic coke/carbon likely obstructing flow in that section of the reactor (poor flow ? hot spot formation)
- Climbing dP indicative of loss of void space in reactor that is continuing to worsen

Potential Hot Spot?

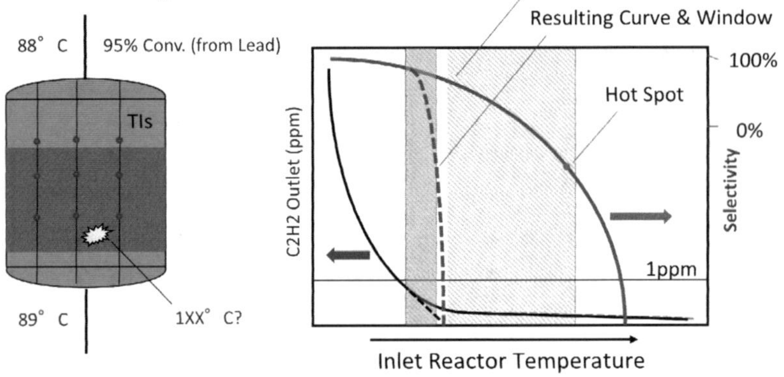

- A small hot spot may not be visible on outlet T (avg. T of all the gas passing through)
- If the hotspot T is near the edge of the operating window, small changes can push it over causing the hot spot to rapidly increase in T and spread across the bed

DRIVE TO ZERO 11

Operating Unstable Catalyst - Overview

- Review Ethylene Specification
- Alarms
- Training and Response Guidelines

DRIVE TO ZERO

Lead Bed vs Guard Bed

- Reactors are setup with inter-bed cooler
- Lead bed 98% conversion, Guard bed 2%
- Moved lead bed closer to runaway conditions to convert more of C_2H_2
- Fine balance of runaway conditions in both lead and guard bed

DOW

DRIVE TO ZERO

Operating Unstable Catalyst – C_2H_2 Spec

- Pipeline specification = 5 ppm
  - Normal operations operated below *1 ppm*
  - *New normal > 3 ppm → derivative plants not as efficient*
  - Mitigate runaway conditions on the reactor
- Developed a new protocol to deal with "off-spec" ethylene
  - Ethylene distribution is integrated across two sites
  - Mitigate big changes on the reactor
- Derivative plants managed to operate with higher/offspec C_2H_2
  - Plants continued to operate
  - Increased understanding of boundaries for the derivative plants

DOW

DRIVE TO ZERO

Operating Unstable Catalyst - Alarm

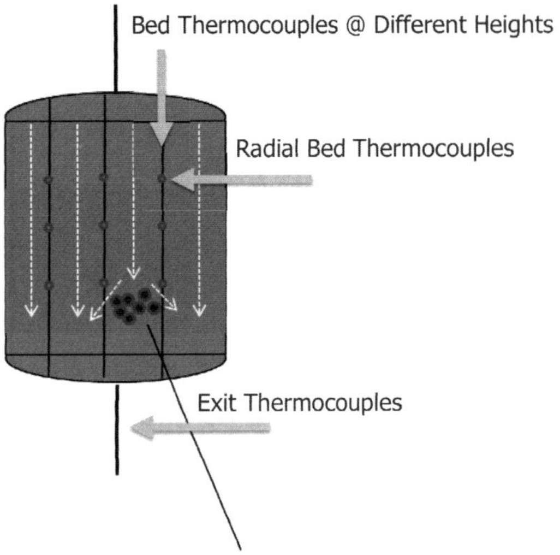

- Exit temperatures – would not detect local hot spot
- Radial temperature alarm to compare drift across thermocouples
- Implemented a monitoring system

Dow

DRIVE TO ZERO

Training Operations – First Line of Defence

- Understand the factors that influence operating window
  - CO
  - dT
  - Furnace load (i.e. feed rate)
- Temperature guidelines on changes
- Reviewed with all operators one on one

Dow

DRIVE TO ZERO

Training Operations – Responding to Temperature Excursions

- Minimizing temperature moves on reactors
  - Allowed C2H2 on the outlet to rise to spec limit before needing to respond
- Monitoring actual dT across entire bed during temperature changes
  - dT would change rapidly because the exit temp had not changed compared to the inlet
- Temperature spread between beds
  - Maintain temperature between beds to allow for response to excursions

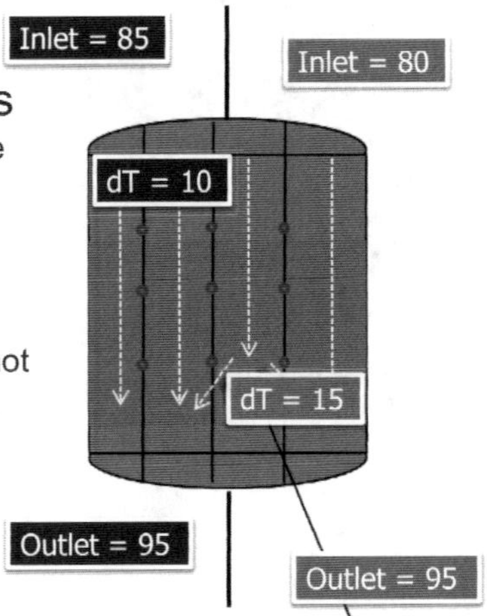

DRIVE TO ZERO

Training Operations – Responding to Temperature Excursions

- Normal response to offspec or change in uncontrollable conditions on reactor – cut rates!
- Not intuitive to **NOT** cut rates as an immediate response
  - Less flow = more residence time in reactor
- Cutting heat only on the guard bed
  - Cutting heat on the lead bed would increase C2H2 to guard bed
  - Hot spot already exist on guard bed, excess C2H2 would increase more temperature
  - Saved compressor trips

DRIVE TO ZERO

What we found when unloading the catalyst....

Summary

- **Everyone** understands the operating conditions and limits
- **Develop** operating discipline for response
- **Collaborate**

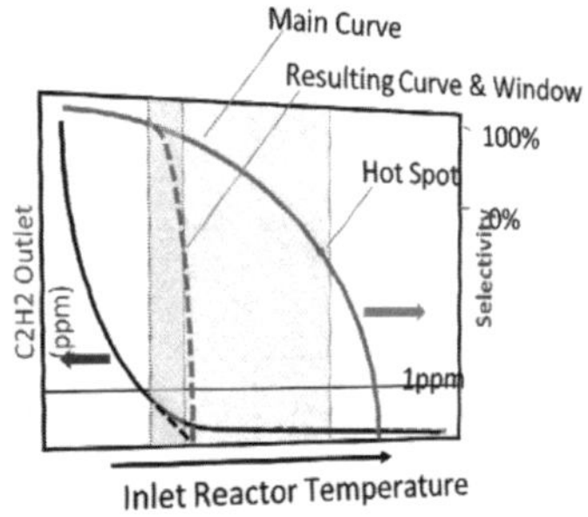

QUESTIONS?

Acknowledgements

- Megan Finstad – Lead LHC Improve Engineer
- Jonathan Lunn – Reactor SME
- Allison Wing - Reactor SME
- Lin Luo - Reactor SME
- Mariam Rasi – LHC Cold End Run Plant Engineer
- Jane Liang-Weissgerber – LHC Senior Run Plant
- Shawn Lefebvre – Cold End Tech Advisor
- Kate Heymans - LHC Improve Engineer

DRIVE TO ZERO

Thank You

485223 Safe Operation in Trickle Phase Reactors

Abstract:

In the Chemical/Refining industry many units operate with high activity nickel (Ni) catalysts. Safe operation of these units is the number one priority. In this presentation we will discuss how to safely operate in trickle phase hydrogenations reactors, share temperature excursion experiences and the lessons learned.

The full and selective hydrogenation units, like C4 butadiene, Pygas 1st stage and aromatics saturation reactors, operate in a range of temperatures and a proper review of the maximum application temperature compared to the onset temperature of methane generation - and potentially run away cracking - is key in determination of a safe operating window for the unit. As with any high activity catalyst defining the desired operating conditions will ensure safe and successful performance.

During the presentation commercial data of multiple units will be shared which will show safe operations and the risks of temperature runaways.

The presentation will highlight the importance of a tailored design, will propose solutions and cover the key hardware requirements for proper flow distribution in the trickle phase flow regime to ensure safe operation.

IMPROVING OVERALL ETHYLENE PLANT PERFORMANCE BY OPTIMISING C2 HYDROGENATION

Authors: Hattachai Aeowjaroenlap[1], Kritsada Chotiwiriyakun[1], Nattawat Tiensai[1], Alejandro Cano[2], Stepan Spatenka[3], Sreekumar Maroor[3], Amit Goda[2]

1. SCG Chemicals, Thailand; 2. Process Systems Enterprise Inc., New Jersey, U.S.A. 3. Process Systems Enterprise Ltd, England, U.K.

Introduction

SCG Chemicals's affiliates produce olefins from naphtha and LPG feedstocks via steam cracking. The cracking process produces a certain amount of acetylene, which is reacted with hydrogen to produce ethylene, helping to maintain ethylene product quality as well as to increase ethylene yield. The reaction is typically carried out in a 2 or 3-stage fixed-bed catalytic reactor with inter-stage cooling.

Side reactions include the production of ethane, which needs to be recycled to the cracker, reducing overall plant throughput and increasing energy use. In tail-end C2 hydrogenation units, acetylene oligomerisation also produces so-called green oil, an oligomer of acetylene that deposits on the catalyst surface and causes catalyst de-activation and loss of selectivity. The challenge for C2 hydrogenation reactor operators is to maximise overall selectivity to ethylene while maintaining product quality.

At ethylene prices of around $1000 per tonne, every additional percent of ethylene production in a 1000 kilotonne per year plant can represent millions of dollars per year of improvement in plant economics. C2 hydrogenation plants are thus prime candidates for optimisation aimed at maximising economic performance.

This paper describes the application of an advanced process modelling and optimisation approach to a typical C2 hydrogenation tail-end process in order to determine optimal inter-bed conditions that maximise the selectivity of the desired reaction. By altering the inlet conditions and thus the conversion in each of the three beds it was possible to realise a 13% increase in ethylene gain from the process and a 10% improvement in process economics.

Approach

In a tail-end C2 Hydrogenation process, C_2H_2 present as an impurity in a C_2H_4 stream is reacted with H_2 to yield C_2H_4 Undesirable parallel reactions include the hydrogenation of C_2H_4 to C_2H_6, and the formation of green oil. The reaction is typically carried out in a 2 or 3-stage fixed-bed catalytic reactor consisting of multiple stages with inter-stage cooling and inter-stage hydrogen injection.

In such a setup, most of the conversion takes place in the first (top) reactor bed, with subsequent beds providing protection to ensure that the acetylene is consumed as far as possible. In an ideal situation the acetylene is all converted to ethylene, with minimum conversion of ethylene to ethane. However, under certain conditions selectivity to production of ethane or green oil is favoured.

The production of green oil in the course of normal operations leads to catalyst deactivation, which decreases the selective hydrogenation of acetylene while increasing the undesired hydrogenation of ethylene. Decrease in activity towards acetylene hydrogenation can be compensated by operating the reactor under more aggressive conditions (higher temperature, higher hydrogen feed rates), but this in turn can cause catalyst deactivation to accelerate and undesired ethane production to increase. Eventually, the degree of catalyst deactivation is such that continued operation is uneconomical and it becomes necessary to take the reactor out of service to regenerate the catalyst. Optimal determination of operating policy throughout a catalyst deactivation cycle can improve process economics by increasing selectivity towards the main reaction and lowering catalyst deactivation rates.

Model-based optimisation

The approach described here involves creating a high-fidelity model of the reactor catalyst beds capable of predicting how reaction rates and catalyst deactivation rates are affected by composition, temperatures and other important attributes to a high degree of accuracy.

Having validated the model against experimental or plant data to ensure its accuracy across the range of conditions to be investigated, the validated model is then used within a mathematical optimisation framework to determine optimal operating conditions. In this case, these are the inter-bed conditions that maximise selectivity of the desired reactions by altering the inlet conditions (temperature, hydrogen feed rates).

The objective of the optimisation is to *maximise economic gain* by *maximising ethylene production and minimizing undesired ethane production.* The decision variables used are the time-varying inlet temperature to each bed, the hydrogen flow to each bed and the run length, with constraints including the maximum bed temperatures and outlet C_2H_2 concentration.

Key aspects to be considered

In order to provide the right level of predictive accuracy, the C2 hydrogenation reactor model needs to include a full representation of bed heat and material balances that take into account the variation in conditions along the length of each bed, as well as the following reaction kinetic aspects:

- The acetylene and ethylene hydrogenation reactions, considering different catalyst active sites for acetylene and for ethylene hydrogenation and reaction mechanism known to depend on reaction conditions (high v. low acetylene concentrations)
- C_4s formation (acetylene dimerisation)
- C_6+ and green oil formation (acetylene hydro-oligomerisation)
- Catalyst deactivation during run, to quantify the loss of activity
- Selectivity change during run, to quantify the loss of selectivity
- Irreversible activity loss due to imperfect catalyst regeneration cycles.

One of the main challenges was that none of the published literature captures sufficiently catalyst deactivation and loss of selectivity, so a semi-empirical model had to be developed and validated against a large amount of historical data.

Determining the kinetic mechanisms

A key challenge in building an effective reactor model is to determine the reaction kinetics (rate laws and kinetic parameters). Two reaction kinetics schemes from literature[1] were used as the basis for the rate laws, and the kinetic parameters in the rate expressions were fitted to plant data. Figure 1 shows a typical reaction scheme formulation for the set of reactions. The main reactions are described here.

The desired reaction is the conversion of acetylene to ethylene:

$C_2H_2 + H_2 \rightarrow C_2H_4$ (Acetylene hydrogenation)

This results in 'ethylene gain' – the production of additional ethylene product.

Undesired reactions are the conversion to ethane, which then needs to be recycled

$C_2H_4 + H_2 \rightarrow C_2H_6$ (Ethane hydrogenation)

and the production of green oil, an oligomer of acetylene

$2C_2H_2 + xH_2 \rightarrow C_4H_{4+2x}$ (Green Oil formation)

The production of ethane effectively leads to 'ethylene loss' as the recycled ethylene limits the ability to process fresh feed. Green oil deposits on the catalyst surface and causes catalyst de-activation and loss of selectivity. The target of all operation is to maximise the ethylene gain, i.e. the selectivity to ethylene.

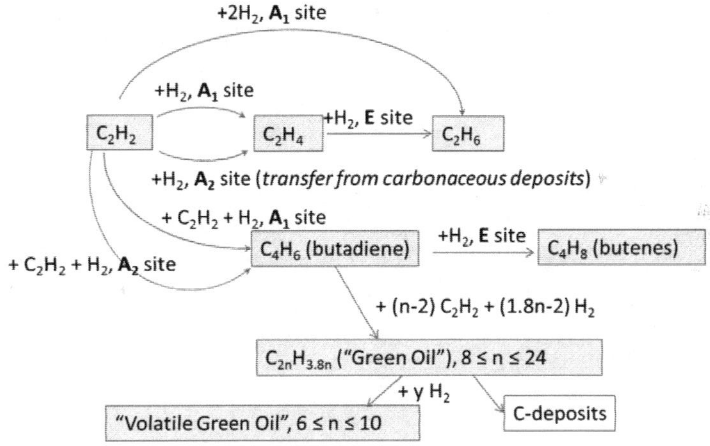

Fig 1. Example reaction schemes from literature

Estimation of kinetic parameters

Once a reaction scheme and rate equations had been selected, the next step was to determine the reaction kinetic parameters. Kinetic parameters cannot be derived from first principles, as they are different from one catalyst to another.

[1] A. Borodzinski and A. Cybulski, Applied Catalysis A: General 198 (2000), 51–66 and A. N. Bos et al., Chem. Eng. Proc. 32 (1993) , 53

They need to be estimated from observed data. This is typically done using a model of the experimental setup, including the proposed reaction set.

Parameter estimation in this context involves the process of fitting empirical parameters, such as the activation energies and pre-exponential constants in reaction rate equations, to observed data. Modern parameter estimation techniques are capable of estimating multiple parameters using data from multiple steady-state and dynamic experiments. The estimation process produces quantitative measures of the degree of confidence associated with each estimated parameter value, providing a quantitative measure of the accuracy of the reaction set model.

The best approach for this is to perform laboratory experiments with a small sample of catalyst pellets, in well-defined conditions, to determine scale-invariant parameters. Plant data typically do not have sufficient variability in operation conditions and there are too many extraneous effects – for example, non-uniform flow and temperature distributions – that mask the true kinetics. It was not possible to perform laboratory experiments in this case, so an additional challenge was how to derive complex reaction kinetics without small-scale experimentation. SCG decided to perform a detailed model-based analysis of historical data, taking account of most relevant guidance from literature.

Parameter estimation was performed on a limited set of kinetic parameters from historical plant data representing six cycles from two reactors using the parameter estimation capabilities in the gPROMS ProcessBuilder modelling software. All in all, over 20 kinetic parameters were estimated for reaction kinetics, where beds 1 and 2 were assumed to have different mechanism than bed 3.

The controlled variables in the plant data included time profiles of the feed rate, feed composition and pressure, as well as the inlet temperature and H2 flow to each bed. The measured variables included time profiles of C_2H_2 and H_2 concentrations at outlet from each bed, temperature measurements at outlet and inside each bed and the C_4s and C_6s product flowrates.

Because parameters were estimated from plant data rather than small-scale experiments, the validated model has limited predictive capability outside current range of operation conditions, but nevertheless is still fit-for-purpose for giving directional advice for determining the optimal operating strategy.

Estimation was performed using a subset of the available data, with other data reserved for comparison of predictions using the fitted parameters. Comparisons with plant data showed that the model reproduced historical cycles sufficiently well to be used in subsequent optimisation. Figure 2 shows the prediction of the validated model (solid line in each case) against plant data for C_2H_2 at the outlet of beds 1 and 2 respectively over a run.

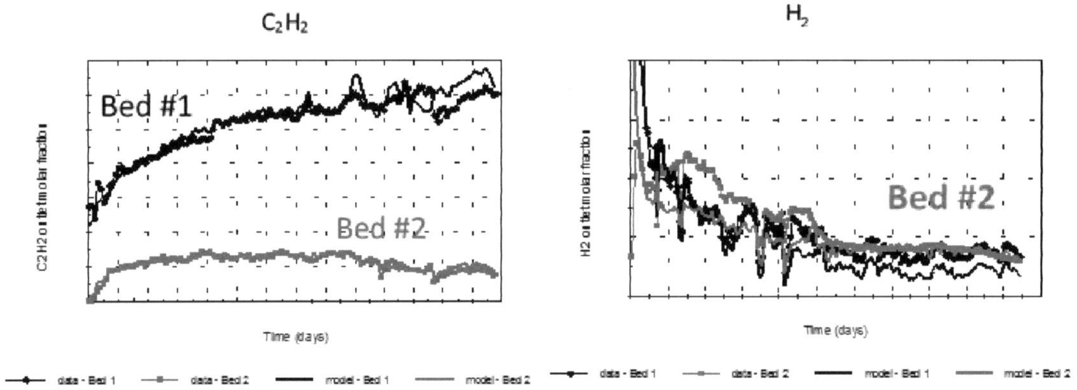

Fig 2. Typical fits to sample data

Constructing the full high-fidelity reactor model

Detailed multi-scale reactor models are increasingly being used to optimise reactor design and operation. Such models typically include high-fidelity sub-models for catalyst particles and the tubes, as well as in some cases shell side hydrodynamics for multitubular reactors. The mass & heat transport phenomena typically taken into account are shown in Figure 3(a); Figure 3(b) shows an example of catalyst geometry and properties definition. Tube and catalyst models are combined with the reaction set models, including the estimated parameters, to provide the full calculation.

Typical applications of such models are:

- Scale-up from laboratory to pilot to commercial plant scale, with a key objective being the maintenance of performance over scales

- Ensuring cost efficiency in investment and operation
- Analysing thermal stability via elimination of hot spots and adjustment of catalytic bed properties (length, activity, shape of particles, etc.)
- Design of cooling system; and catalyst lifetime extensions, through management of catalyst de-activation over the operational cycle.

The detailed SCG catalytic reactor model was developed using the Fixed Bed Catalytic Reactor library within the gPROMS ProcessBuilder advanced process modelling software. A lumped catalyst model was used rather than a full pellet distribution; this was considered to be fit-for purpose given that parameters were being fitted from plant data rather than laboratory experiments. The flowsheet model is shown in Figure 4.

The inputs to the reactor model were feed flow and composition, with manipulated parameters being temperature and H_2/C_2H_2 ratio. The key outputs calculated were product composition, acetylene conversion, ethylene gain, temperature profile in the beds, catalyst activity profiles in the beds, and economic gain.

Simulation analysis and results

Once the model was set up as shown in Figure 4, and the full reaction set incorporated, it was initially used in dynamic simulation mode to explore various operating scenarios, in particular to look at the effect of changing key manipulated variables.

For example, the inlet temperature profile and feed C_2H_2 to H_2 ratio were varied to gauge the effect on product stream compositions, the product outlet conditions, catalyst performance and in particular ethylene gain.

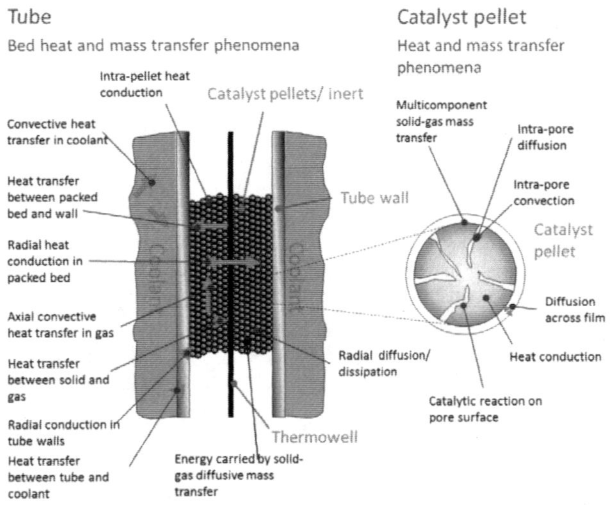

Fig 3(a). High-fidelity reactor models: heat and material flow phenomena taken into account

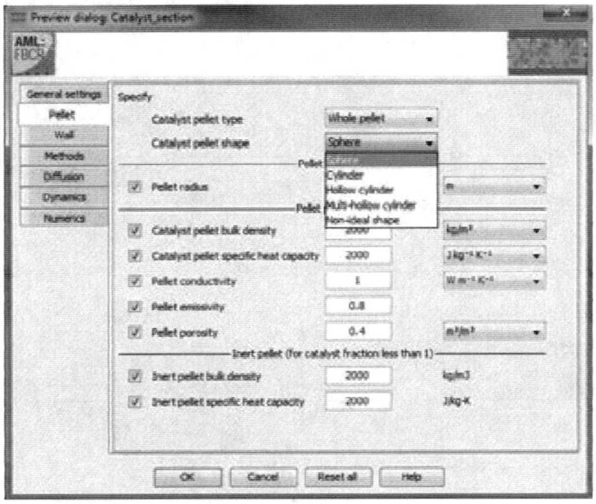

Fig 3(b). Catalyst definition dialog

Figure 5 shows the effects of maintaining the same temperature profile, but decreasing the H_2 to C_2 ratio in the feed. This results in an ethylene gain of 5% in the initial stages of the run.

The results in Figure 5 show that lower H_2/C_2 ratio can initially result in a higher ethylene gain (higher selectivity), but it also enhances catalyst deactivation and therefore affects the selectivity later in the cycle. As such the dynamic optimisation problem needs to consider the whole operation cycle.

Optimisation

Having ascertained the key factors affecting ethylene gain through trial-and-error simulation analysis, a full optimisation run was set up. The objective of the optimisation was to maximise the economic performance over a run, by determining optimal settings for variables such as the inlet temperature, H2 feed rate at each stage (thus the C_2H_2:H_2 ratio entering each bed), the acetylene ppm amount at the bed 2 outlet, and the run length itself.

Once a model is available, it is relatively straightforward to perform an optimisation. This is done by defining an objective function to maximise or minimise, then selecting decision variables and, optionally, defining constraints. The major advantage of dynamic optimisation is that it allows rapid exploration of the full decision space, for many decision variables, in a way that cannot possible be achieved using trial-and-error dynamic simulation.

The objective function can comprise any variables in the model. It is typically an economic objective that sums the values and costs. Here the objective function is defined in terms of the economic gain from C2 reactor:

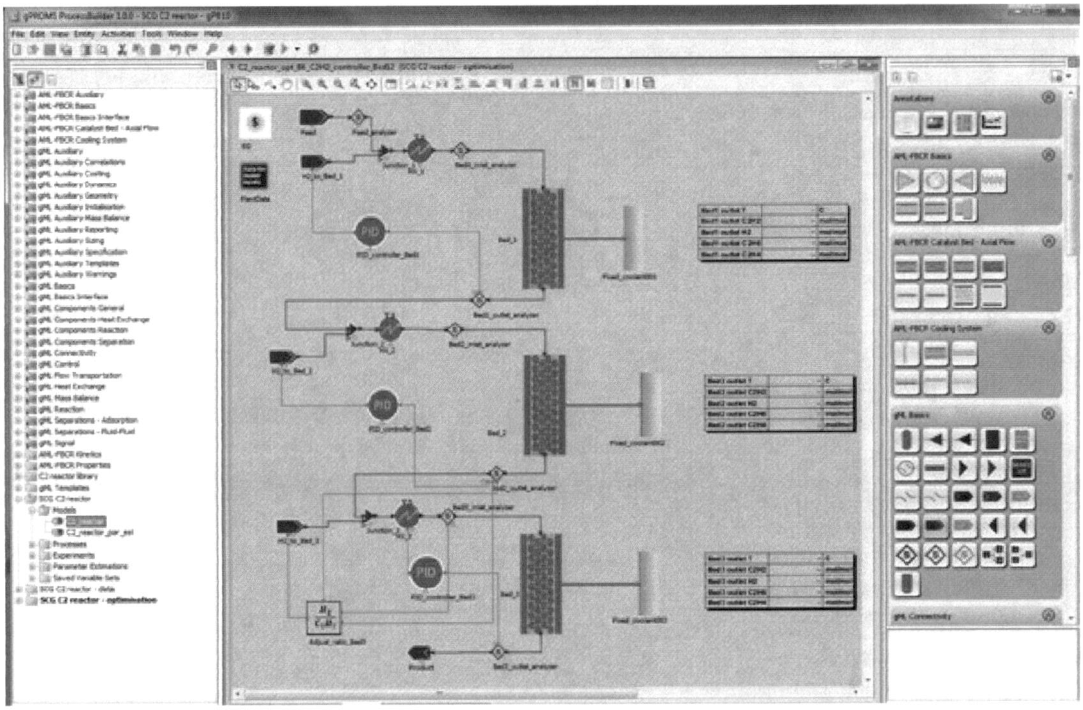

Fig 4. The C2 hydrogenation reactor flowsheet

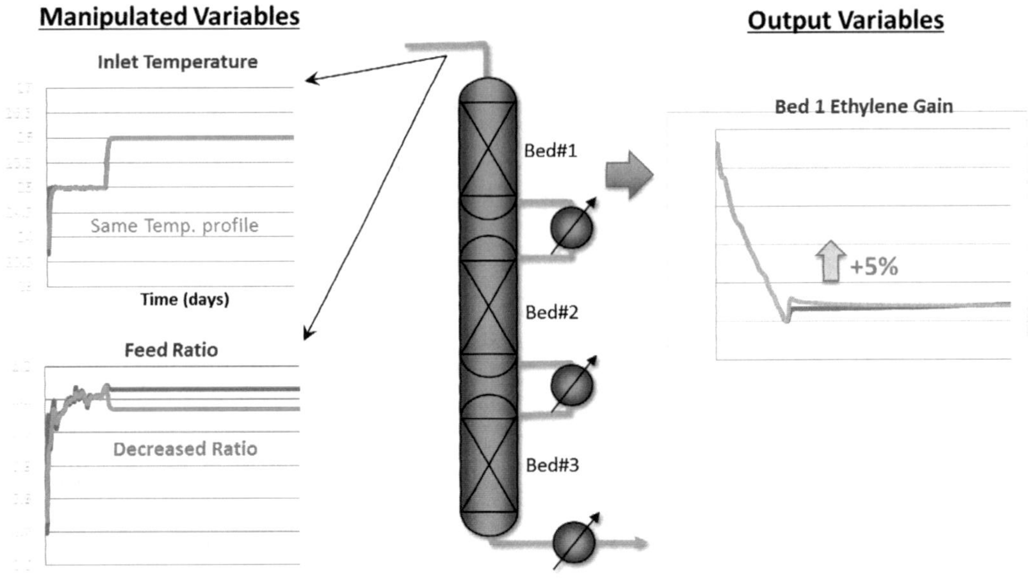

Fig 5. Simulation-based analysis

$$ECG\ [\$/day] = (P_{C2H4} \times Ethylene\ produced) + (Y \times P_{C2H4} \times Ethane\ produced) - (P_{H2} \times Hydrogen\ consumed) - C_{reg} - C_{cat} - C_{sw}$$

Where

P_{C2H4}	Ethylene price
P_{H2}	Hydrogen price
Y	Furnace yield from ethane
C_{reg}	Cost of regeneration per cycle
C_{cat}	Cost of fresh catalyst per cycle
C_{sw}	Switching cost (loss of ethylene) per cycle

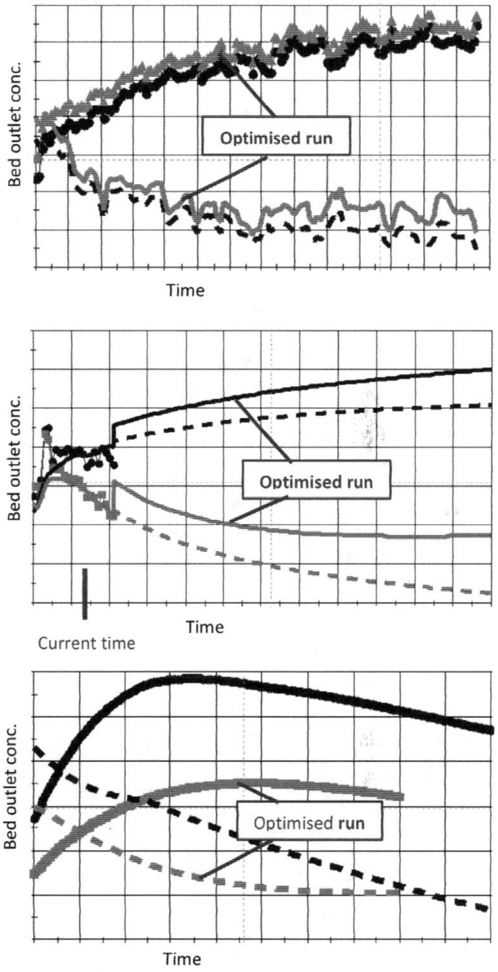

Fig 6. Demonstration on historical cycles

Decision variables can be any specified variable in the model. They can be continuous variables, in which the value can vary in a continuous manner over the course of the optimisation (e.g. a tube length in a design optimisation), or integer variables, in which the decision variable may take only integer or discrete values (e.g. standard pipe diameters, or number of tubes in a reactor).

Decision variables used in this simulation are:

- Run length, discretised into time intervals
- Time-varying controls over run length
- Inlet temperature to each bed
- H2 flow to each bed

The following constraints were implemented:

- Maximum bed temperatures
- Outlet C2H2 concentration

Optimisation approach

Initially the approach was demonstrated on historical cycles. Inputs (flowrate, feed composition, pressure) were read from historical data, then the controls optimised for whole run length (Figure 6, top plot). This approach demonstrated what would be economic gain if selected historical cycles were operated optimally, including the optimal run length choice. Results are shown in Figure 6.

For an offline optimisation of the current (ongoing) cycle, inputs are read from historical data until the current date. From current time onwards, the model inputs are assumed fixed (feed rate, feed composition) and controls are optimised (Figure 6, centre plot). This way an advice can be provided to the operation of the remainder of the run, including optimal run length prediction.

For the optimisation of a new cycle, the time profile of key inputs, such as flowrate and feed composition, need to be assumed and the controls optimised during whole run (Figure 6, bottom plot).

Optimisation results

The optimisation run was executed in order to optimise the temperature and feed ratio profiles. The temperature decision variable was defined as a trajectory over a number of time intervals, and the optimiser allowed to choose the optimum temperature for each interval.

Figure 7 shows the optimised profile. Here it can be seen that the optimal profile is to operate at a lower temperature (brown line) than the usual operating policy (blue line) for a number of days before ramping to the final temperature as conversion and selectivity in the bed decrease. The lower temperature increases selectivity in the first bed, thus reducing production of unwanted ethane and green oil. It also reduces conversion, but this is taken up by spare capacity in the second bed. The lower temperature and corresponding lower rates of reaction in the bed has the effect of reducing catalyst deactivation, hence extending catalyst lifetime. The second plot shows the fixed feed ratio.

The results obtained in the optimisation allowed development of new operating policy that generally reduced loss of ethylene by reducing acetylene conversion in the first bed and increasing the conversion in the second bed.

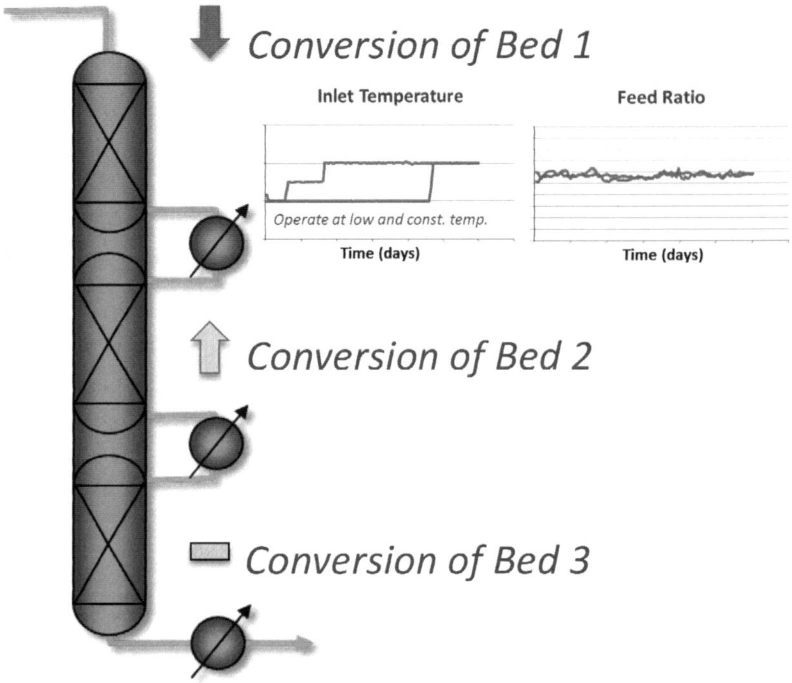

Fig 7. Optimisation-based analysis.

Figure 8 shows the results of application to a real cycle, in which the optimiser was allowed to adjust the feed temperature, feed ratio and run length. Here it can be seen that the ethylene gain increases over the run, to reach achieved a gain of 13% over the standard operation by the end of the run. The corresponding

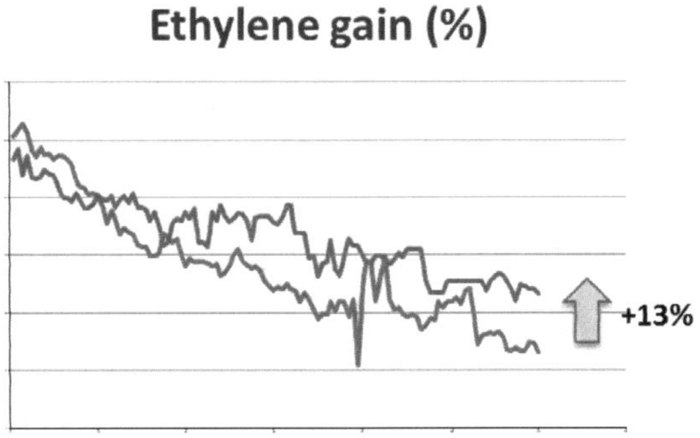

Fig 8(a). the Ethylene gain

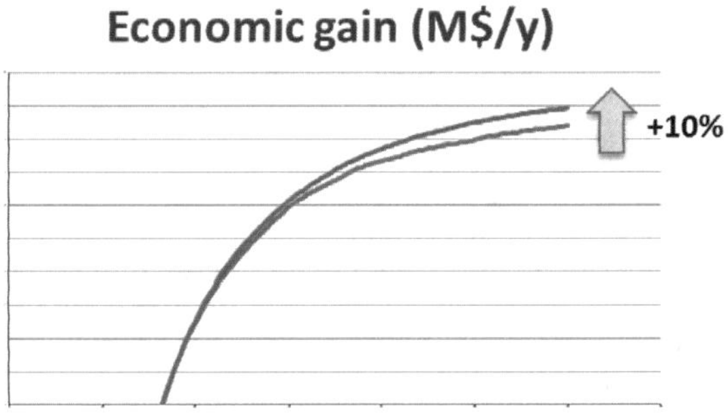

Fig 8(b). Economic gain

economic gain (determined from the objective function calculation) is 10% improvement.

Conclusions

In conclusion, a rigorous model of the acetylene hydrogenation reactor was successfully developed in gPROMS ProcessBuilder. This is capable of explaining the reactor behaviour and helping to create new, optimal operating policy.

The major benefit of applying the model is that it identifies significant improvements in economic gain without the need for capital expenditure. The knowledge gained, as well as the model itself, can be applied to other fixed bed reactors within SCG Chemicals' operations, such as C3 hydrogenation.

There is also potential to connect the model to the process historian for real-time monitoring of catalyst activity and green oil accumulation. The optimisation model can be used for weekly production planning, process condition improvement and process risk assessment.

481101 FHR Air Dryout after 2015 Major Outage

Abstract:

Author: Joseph Bell, Flint Hills Resources Production and Optimization Engineer

Sponsor: Johnny Spell, Flint Hills Resources Technical Manager

The Flint Hills Resources Light Olefins Unit in Port Arthur, TX went through a major outage to execute significant cold train modifications and other TA related repairs / maintenance. Work was executed in the Demethanizer, Cold Train, C2 Recovery, Ethylene Refrigeration, and Propylene Refrigeration systems. After work completion, each of these systems was dried with air in order to achieve a safe, successful startup. This presentation will cover the following:

1. Dry air supply and header plan
2. Dewpoint measurement strategy
3. Process Safety concerns of O2 contaminating the flare system
4. Overall execution results and learnings

AIChE Paper Number: 81e

To Err is Human: Mitigating the Hazards of Dryer Switching Operations

Rob D. Fox
Senior Process Engineer
Equistar Chemicals, LP

Prepared for Presentation at the 2017 Spring Meeting
San Antonio, TX, March 27-30, 2017

To Err is Human: Mitigating the Hazards of Dryer Switching Operations

Author

Rob D. Fox, Equistar Chemicals, LP

Abstract:

Within the chemical process industries regenerable dryer systems have long been used to remove moisture and contaminants from process streams. The design and optimization of these mature technologies is well understood. In these systems, limited run lengths lead to frequent regenerations and it is during the period of transition from one operating mode to another that there is an increased likelihood of a production upset or process safety event to occur if appropriate mitigation steps are not taken. This effect is further amplified by the high level of interaction between operators and the process during switching operations, which may result in higher risks for these activities than previously identified. This paper will explore several of the higher consequence hazards associated with switching process dryers, touch on the frequency of operator error, and provide risk mitigation strategies for dryers with both automated and manual valves.

Overview

In the chemical process industries (CPI) the trend amongst operating companies is to strive for more automation, and less manual operations to improve efficiency. One of the most prevalent manual operations that remain in these processes is the switching of regenerable dryer systems between operation and regeneration modes. These activities typically require multiple steps to be completed in the correct order, and often require the input of both console and outside operators. This high degree of interaction with the process requires considerable communication, and when coupled with the high frequency of dryer regenerations presents an elevated risk for operator error to initiate a process safety event.

All systems are affected differently by human error, so the consistent use of industry validated rates of error within process hazard analyses (PHAs) is critical to understand the risks of dryer systems, and to properly mitigate these risks to acceptable levels. This paper will explore several typical risks of dryer systems in the olefins process, and propose effective mitigations for the hazards faced with both manual and automated valves.

Human Error

One of the most widely used forms of PHA is a layer of protection analysis (LOPA), where the unmitigated frequency and consequence of an event are evaluated to determine the unmitigated risk, and then the independent protection layers (IPLs) are credited to determine the final mitigated risk of a hazard. The Center for Chemical Process Safety (CCPS) has studied the topic of human error in depth; the quotes excerpted below are the basis for the frequency recommendations in their published guidelines [1]:

> "Human error is associated with a significant percentage of incidents in the chemical industry. As such, it is important to consider scenarios initiated by human error in LOPA"
>
> "The frequency at which an error can be expected is a function of both the predicted error rate per task and how many times the task is performed within a given time period."
>
> "The less often a task is performed, the fewer opportunities an individual has to make a mistake. Thus, the annual error rate would tend to decrease as the task frequency decreases. However, when a task is performed less often, the individual is typically less skilled in the task, tending to increase human error rate."

Based on these factors the CCPS recommends using a frequency of human error between 1:1 and 1:100 based on the frequency of the task [1]. In addition to the CCPS recommendations, individual company guidelines should be consulted when evaluating the risks associated with operator error during dryer switching operations.

Another important concept promoted by the CCPS is that of the "hierarchy of controls", which ranks the different types of IPLs based on their effectiveness. Figure 1 on the next page details the hierarchy, where inherently safer designs are considered the most effective process safety mitigations, while administrative controls (e.g., procedures, car seals) are the least. Two categories from the hierarchy that are used widely in the CPI are passive safeguards, which are protective layers that prevent an event from occurring (e.g., certain interlocks), and active safeguards, that react to an event to prevent the consequence (e.g., a pressure relief device (PRD) preventing overpressure). [2]

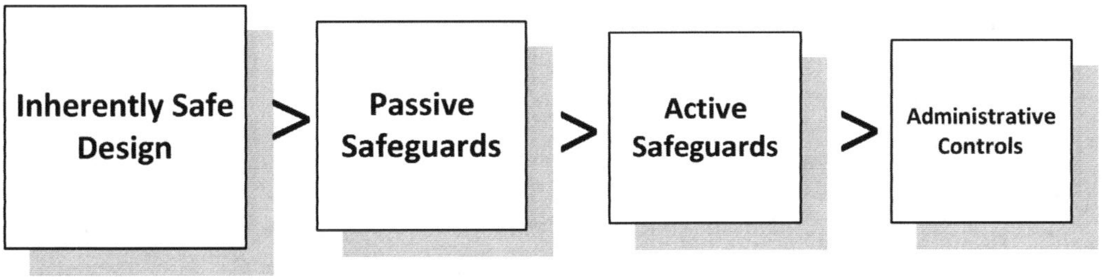

Figure 1: The "hierarchy of controls" as defined by the CCPS. [2]

Hazards of Automated Valve Misalignment

The most prevalent hazards of inadvertent valve operation in an olefins unit are seen with automated valves, where opening the wrong valve can be as simple as hitting an incorrect button. For a better understanding of this hazard, a case will be evaluated for the process gas dryers. A diagram for a typical three dryer system is provided below in Figure 2, where Dryer A is shown in regeneration while Dryers B & C are in service to dry the process gas that is routed forward to the cold box. A typical process downflow/regen upflow arrangement is being utilized for this example. This arrangement presents four distinct valve alignment hazards, which will all result in similar consequences:

- The process inlet valve (A) on the regenerating dryer is opened
- The process outlet valve (B) on the regenerating dryer is opened
- The regen inlet valve (C) on an operating dryer is opened
- The regen outlet valve (D) on an operating dryer is opened

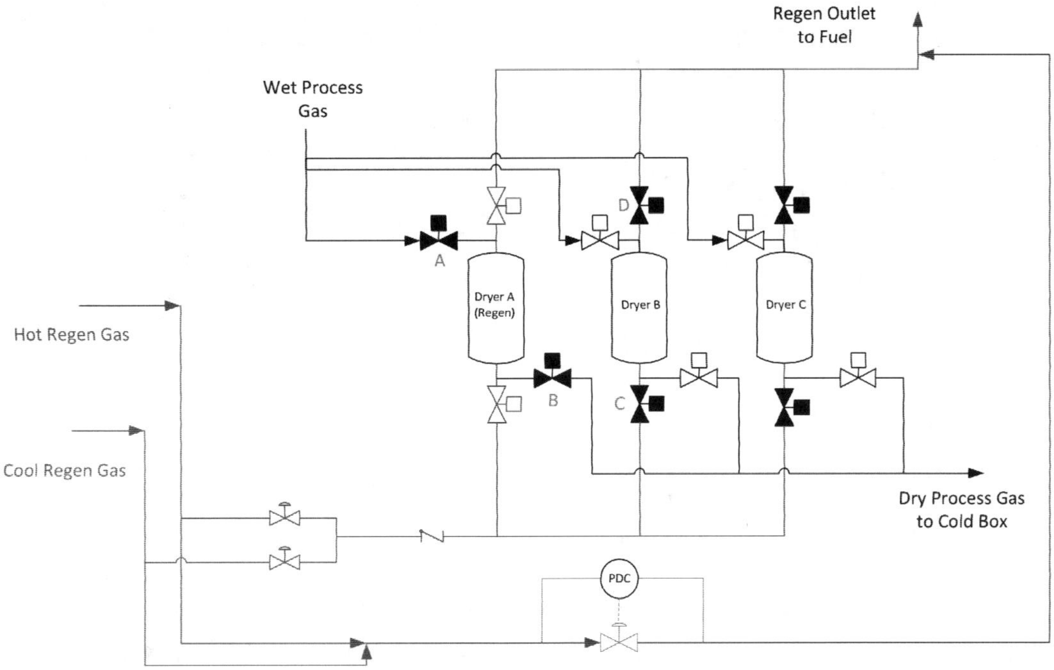

Figure 2: Typical diagram of the process and regeneration connections for process gas dryers.

For all of the misalignment cases above the end result would be routing a significant flow of process gas to the fuel system. Some of the cases above would result in backwards flow through the regen inlet check valve, but unless this valve is specifically designated as safety critical equipment (SCE) and monitored with an inspection and test plan (ITP) caution should be taken before claiming IPLs for this protection. In the unmitigated case of process gas flow to the regen supply system the process gas will still route to fuel, as the regen differential pressure controller will open in an attempt to maintain setpoint.

The unmitigated consequences expected from any of these events are high, with the potential for damage to the cracking furnaces and overpressure of the fuel gas system. The primary driver for the potential damage in the cracking furnaces is the dramatic change in heat input that will be experienced when the fuel gas is replaced with process gas. The higher density and heating value of the process gas will nearly double the heat input to the furnaces and place the furnace tubes at risk of high temperature failure. Typical furnace interlocks will mitigate this risk, but failure of furnace tubes upon interlock initiation is relatively common.

Mitigations for Automated Valves

With the initiating event frequency and potential consequences now understood, the LOPA next determines the required IPLs to reach an acceptable mitigated risk. Individual company risk management standards will vary, but typical industry guidelines would suggest three IPLs are needed for this scenario.

The most straightforward way to mitigate an overpressure hazard is through the use of pressure safety valves, although for the scenario of inadvertently opening the process gas to the fuel system this may not be a feasible solution. The problem lies with the significant flow that must be relieved at low pressure: at a minimum, the fuel system PRDs will need to relieve the full forward flow of process gas from the process gas compressor. If the regen lines are sized large enough, depressurization of the chilling train may be experienced, adding significantly to the required relief rate. For a 1000 kTa ethane cracker (front-end demethanizer arrangement) the minimum relief protection for this case would be approximately 90 in^2 of relief area (API basis) using pressure relief valves (PRVs), requiring four 8T10 PRVs to provide the necessary protection. To evaluate the potential depressurization of the chilling train a dynamic simulation would be needed, and the area required for the expected peak rates may be 2-3 times the steady state minimum.

To minimize the area requirement of these PRVs it may be possible to utilize restriction orifices (RO's) in the regen lines to limit flow, but this comes with consequences for normal regeneration of the dryers. Since either the regen inlet or outlet valve may be inadvertently opened, RO's are required on both lines and their combined effect may limit the flow of regen gas for the dryer.

While it may be possible to mitigate the overpressure hazard with pressure relief devices, the risks to the furnaces and the ensuing process upset will still be experienced during the event. A better solution is to provide a combination of safeguards, so a passive safeguard prevents the event from occurring, and if this fails then the impact is mitigated by an active safeguard.

For a dryer system with automated valves, the most effective passive safeguard is a pressure-based interlock (permissive) that will inhibit movement for both the regen and process valves at the wrong point in the sequence. For a process gas dryer there is a large difference between operating pressure and regeneration pressure, so the pressure of the dryer can be used to determine which valves are appropriate to operate at any point in the sequence; see Table 1 below:

Table 1: Process gas dryer valve permissive logic.

Mode	Regen Valve Movement Allowed?	Process Valve Movement Allowed?
Operating	**N**	**Y**
Regenerating	**Y**	**N**

The type of automated valve used for the dryer can have an impact on the available SIL rating of the permissive installed. For motor operated valves (MOVs) it may only be possible to claim one IPL for the permissive since a single point of failure exists when using relays to break the circuit providing power to the motor (e.g., relays in series can be fused closed by the same event). A higher integrity may be available for air or hydraulically powered valves due to the higher availability of the valves to vent the motive fluid.

If one IPL is available from the permissive, then two more IPLs must be provided to appropriately mitigate the hazard. The use of a SIL-2 trip to isolate the dryer upon detected high pressure in the regen system is an effective active safeguard, as it will contain the high pressure within the dryer and minimize the impact on the regen system. See Figure 3 for a typical installation of both the pressure permissive and the interlock trip.

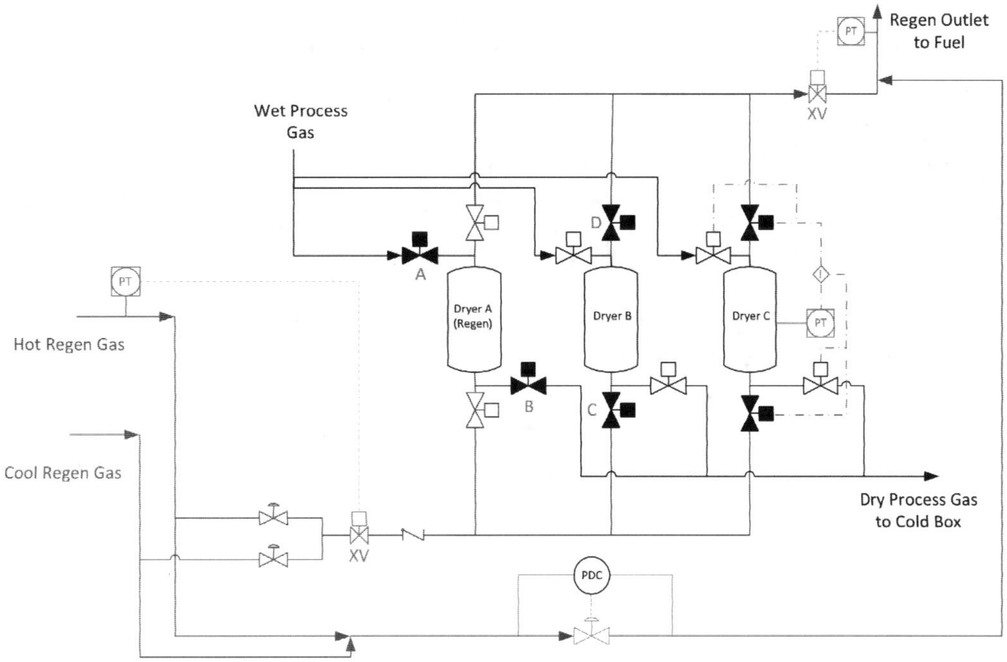

Figure 3: Typical process dryer system with valve permissive and interlock installed.

A positive feature of the system above is that by utilizing common trip valves for multiple dryers the cost of the interlock will be minimized. A simplification made for the sketch above is the required number of pressure transmitters; the final detailed design should incorporate company standards for both availability and reliability to prevent spurious trips. A specific note related to retrofitting this system into an existing unit: the piping specification break from a nominal 300# to a 150# system is typically taken at the outlet flange of the regen out MOV, so the regen outlet line up to the outlet XV will need to be replaced with 300# piping. The regen inlet line should be checked for a similar specification break concern when adding the inlet XV.

The interlock installation diagrammed above is capable of reaching a SIL-2 rating with only single trip valves if an appropriate testing frequency is utilized. The testing will likely be more frequent than once a year, but the system is available for testing any time the dryers are not actively in regeneration so no operational flexibility will be lost.

Hazards of Manual Valve Misalignment

When the valves of a dryer system are manually operated the potential hazards of misalignment are somewhat changed from the cases for automated valves, and the mitigations must change as well. While an operator may begin to open the wrong valve on a process gas dryer during regen, it is highly unlikely that the operator will fully open the valve and present the type of overpressure hazard to the fuel system that was previously discussed.

An interesting example case for the hazards of manual valve misalignment can be found during the regeneration of propylene dryers. To begin the regen sequence the liquid-phase propylene dryers are first drained to an appropriate system and then depressurized to fuel system pressure before starting a cool purge with regen gas (typically tailgas, a mixture of hydrogen and methane). Typically, there is sponge liquid left on the desiccant at the end of the draining/depressurizing process, and the liquid will begin to vaporize when the regen gas purge is started. The light gas atmosphere of the regen gas presents an autorefrigeration hazard to typical propylene dryers, which have historically been built with minimum design metal temperatures (MDMTs) in the range of 30°F to 60°F. At fuel system pressures the saturation temperature of propylene is approximately 20°F, but temperatures as low as -40°F can be experienced when vaporizing propylene in a methane/hydrogen environment.

A necessary tool for evaluating this type of hazard is a minimum allowable temperature (MAT) curve, determined using the criteria of API 579. This curve provides the minimum temperature the vessel can safely withstand at any expected pressure, instead of only at a single value like the minimum design metal temperature (MDMT). An example MAT curve for a typical propylene dryer is given on the next page in Figure 4.

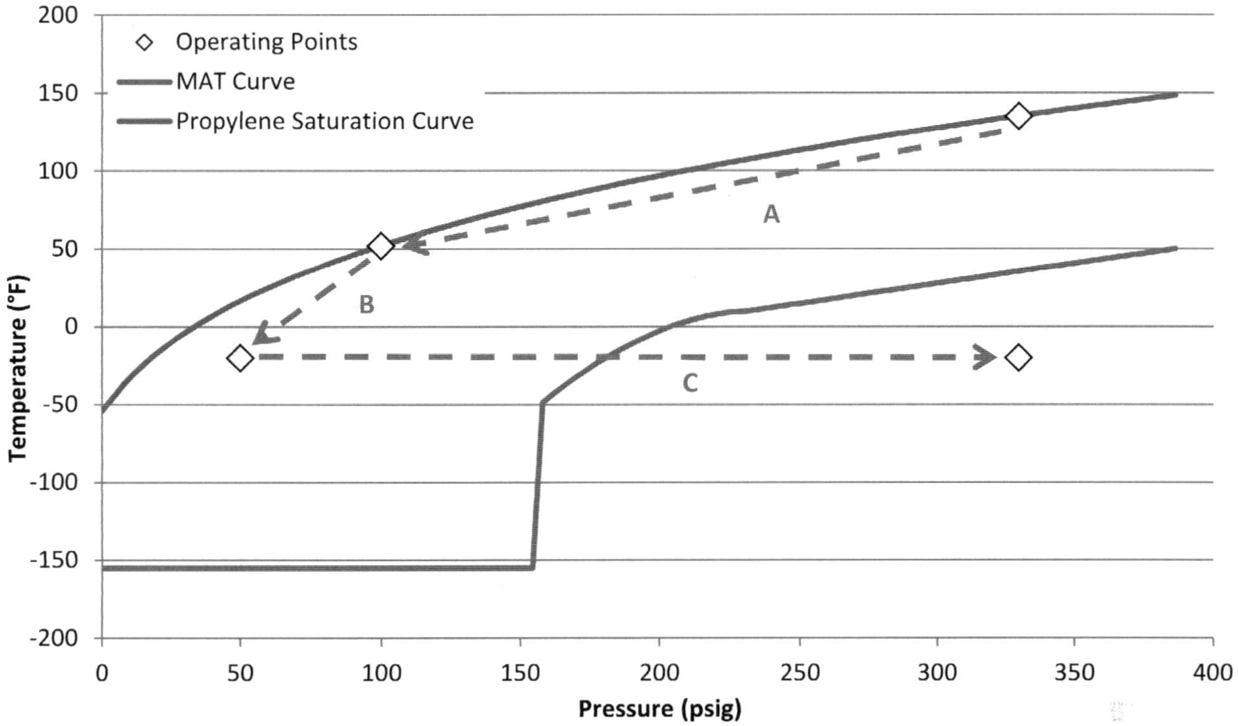

Figure 4: Thermodynamic path of propylene dryer depressurization and start of regen. The temperature follows the saturation curve while the dryer is drained and depressurized (A), becomes colder in the light gas atmosphere once regen gas purging begins (B), and would cross the MAT curve upon repressurization (C) if it were allowed.

The full thermodynamic path of the regen autorefrigeration hazard is shown above; the cold temperatures possible upon the admission of regen gas are acceptable at the low pressure during regeneration, but an operator error during the next step of the regen could open the process back to the low temperature dryer and increase the pressure, crossing the MAT curve. It should be noted that a conscious decision to repressurize the dryer at this point could cause the hazard as well.

Mitigations of Manual Valve Misalignment

For this example it is difficult to prevent the low temperature from occurring, so the only method to ensure a potential brittle fracture event is prevented is to prevent repressurization by inhibiting the movement of the manual process valves. This can be accomplished with a high degree of availability (typically SIL-2 equivalent) by utilizing a captive-key interlock.

With a captive-key interlock, devices are placed on the manual valves that prevent them from being operated unless they have been unlocked with a special key. Multiple keys are utilized in sequence so that the proper valves are operated at all points of the regen. A low temperature permissive can be incorporated into the sequence by utilizing a solenoid box that receives the input signal. A diagram of a typical system for protecting a propylene dryer is on the next page in Figure 5.

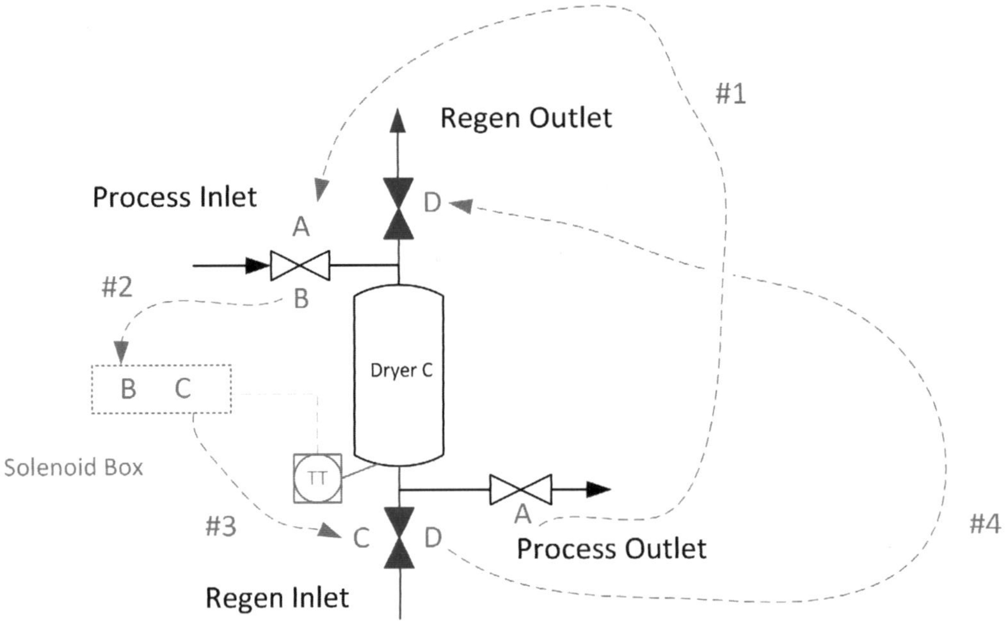

Figure 5: Diagram of a captive-key interlock for protecting a propylene dryer from autorefrigeration hazards during regeneration.

The key sequence proceeds in series through the regeneration, following the basic outline below:

- Close the process outlet valve, which allows Key A to be removed
- Insert Key A in the process inlet valve, close valve and remove Key B
  - Note: Small drain/depressure lines (not depicted) are used to ready dryer for regen gas purge
- Insert Key B in the solenoid box, if temperature is greater than the trip setting, Key C can be removed
- Insert Key C in the regen inlet valve, open valve and remove Key D
- Insert Key D in the regen outlet valve, open valve and proceed with regen

The safety function preventing repressurization is active when attempting to reverse this sequence: if the temperature is too low, Key B will not be released and the process valves cannot be operated, thus preventing the possibility of the repressurization crossing the MAT curve.

Another application where captive-key interlocks can be useful for mitigating manual valve misalignment hazards is for spare reboilers on distillation towers, where the valve sequencing can ensure that the process lines are open back to the tower before allowing the heat supply to be opened.

Summary

The reduction of process safety risk is a common goal within the chemical process industries, and one that requires a drive for continuous improvement. Thorough process hazard analyses play a significant role is driving down process safety risk, and a key factor in accurately assessing hazards is understanding the element of human error. Strong procedures and well-trained operators are vital to the safe and reliable operation of a chemical process, but human error will never be fully eliminated as an initiating event. The hazards presented by valve misalignment must be scrutinized and mitigated. By utilizing consistent frequencies of operator error as initiating events during a LOPA the hazards of operations like dryer switching can be understood. By using interlocks, both automated and captive-key, and permissives the hazards associated with both manual and automated valves can be properly mitigated to an acceptable risk.

References

1. Center for Chemical Process Safety/AIChE. (2014). Guidelines for Initiating Events and Independent Protection Layers in Layer of Protection Analysis. Center for Chemical Process Safety/AIChE.

2. Center for Chemical Process Safety. (2009). Inherently Safer Chemical Processes - A Life Cycle Approach (2nd Edition). Center for Chemical Process Safety/AIChE.

AIChE Paper Number 81f

DECOKING AND CRACKING IN A TWIN CELL FURNACE – DESIGN AND OPERATIONAL CONSIDERATIONS

Harry Wang[1]

Senior Process Engineering Supervisor

TechnipFMC, Claremont, CA

Joel Guillaume

Furnace Technology Manager

TechnipFMC, Houston, TX

Bruce Evans

Director of Pyrolysis Technology

TechnipFMC, Claremont, CA

Prepared for Presentation at the 2017 Spring National Meeting

San Antonio, Texas, March 26-30, 2017

[1] Corresponding author: harry.wang@technipFMC.com

DECOKING AND CRACKING IN A TWIN CELL FURNACE – DESIGN AND OPERATIONAL CONSIDERATIONS

Harry Wang

Senior Process Engineering Supervisor

TechnipFMC, Claremont, CA

Joel Guillaume

Furnace Technology Manager

TechnipFMC, Houston, TX

Bruce Evans

Director of Pyrolysis Technology

TechnipFMC, Claremont, CA

Abstract: Cracking furnaces are the heart of the ethylene plant. As crackers has increased in size, so has the cracking furnace capacity. Ethylene producers purchase various feed stocks ranging from "Gas" feeds to "Liquid" feeds and up to "Heavy Liquid" feeds. Economically, it is important to minimize the number of furnaces in the design of a modern ethylene plant, to have multiple feedstock flexibility and also to have the features to decoking in one cell and cracking in another cell in a twin cell furnace design.

Co-cracking or blended cracking is often used to handle two or more feed stocks in a single furnace. Segregated or "hybrid" cracking (two or more feeds cracked separately in a single cell or in a twin cell furnace) becomes important in modern furnace design for achieving feedstock flexibility while maintaining furnace availability, when it is desired to crack each feed at a specific conversion or severity. In addition, due to the large capacity of modern twin cell furnaces, it is important to consider decoking/cracking in the same twin cell furnace to increase furnace availability.

In 2011, TechnipFMC presented a paper in A.I.Ch.E. Spring Meeting (EPC Conference, Paper Number 96E) titled "Co-cracking vs Hybrid (Segregated) Cracking in Individual Furnace – Design and Operational Considerations". This paper is an extension (Part 2) of the paper presented in 2011 to the even more aggressive scenario of designing, simulating and operating cracking furnaces for the extreme situation of operating one cell in the steam cracking mode and the other cell simultaneously in the steam-air decoking mode.

Simulation of simultaneous decoking and cracking in a twin cell furnace is an important advancement to the simulation of hybrid cracking or co-cracking in an individual furnace. A significant challenge in the design of simultaneous decoking and cracking in a twin cell cracking furnace is how to properly simulate the complete furnace performance due to a large difference in the furnace fired duty and flue gas flow from each cell, as well as a large difference in bridge wall temperature from each cell. In addition to the radiant coils, simulation of furnace firing, flue gas mixing and performance of convection banks also present a significant challenge. Convergence of the simulation of the complete system presents a challenge due to the complexity of the system. How to control temperatures in the decoking cell during this operation is also an important consideration. Material selection of process convection banks which supply the decoking cell is to be carefully evaluated.

For decoking and cracking in the same twin cell furnace, a cracking furnace model has been developed by TechnipFMC called SPLIT-MIXING to simulate decoking and cracking using an integrated EFPS (firebox and radiant section) model and PROVISION (heat exchanger) software. An improved methodology is to utilize TechnipFMC's new generation of furnace simulation software: SPYRO® Suite 7, which is capable to rigorously simulate decoking and cracking in the same twin cell furnace in a single converged file. Additionally, CFD is utilized to verify flue gas flow patterns.

This paper will also discuss key operational aspects of individual cell decoking.

Introduction:

To maintain competitiveness in the current petrochemical market, the ethylene industry has continued to push the economy of scale. As crackers have increased in size, so has the cracking furnace capacity. Current modern furnace designs are routinely 200KTA+ in capacity and will soon be pushing to 300 KTA and beyond. Another large factor in cost competitiveness is the ability to crack a variety of feeds in the ethylene plant. On a recent project there were over 15 individual feeds ranging from ethane, propane, naphtha and up to hydrocracker residue. In addition, recycle streams include ethane and propane, and additionally C3, C4 and C5 streams.

With so many feeds and a limited number of installed furnaces, how can the end users capture both economies of scale and feed stock flexibility with 300KTA+ increments of Mega furnace capacity?

TechnipFMC Process Technology (TechnipFMC) has utilized many methods and techniques in their furnace design over the years to capture economy of scale by maintaining furnace flexibility and maximizing availability. These techniques include Cocracking, Pass/Cell Hybrid Cracking, Independent Cell Operation Decoking (ICD) or called Cracking and Decoking.

This presentation covers the concept of Independent Cell Cracking/Decoke (ICD) from process considerations to SPYRO® Suite 7 modeling, mechanical design considerations and operations.

History of TechnipFMC Process Technology Independent Cell Decoking (ICD) Technology:

TechnipFMC has a long history as a leader in the engineering, design, procurement and construction of twin cell ethylene cracking furnaces. In 1994, the first twin cell furnace with cracking and independent cell decoking (ICD) was in operation. This operation allows the end user to increase both the furnace availability and flexibility. Since then more than 50 twin cell furnaces have been designed and operated with this feature by TechnipFMC.

However, a major issue of this operation is how to accurately simulate this mode in addition to the challenges for the design and operation due to the very complicated system in nature.

In the past, the furnace convection section and radiant firebox had to be decoupled to have the overall furnace simulation with cracking in one cell and decoking in another cell converged. Once the operating conditions are changed, it also takes more time to have the simulation re-converged due to "trial and error" of this complicated system.

Instead of taking this "trial and error" approach to making this complex simulation, TechnipFMC has developed calculation methodology to simulate twin cell furnace with ICD using the latest SPYRO® Suite 7 program which enables simulation of the whole complicated operation in a single converged file. It also becomes much easier- and much faster - to re-converge the file with the change of operating conditions.

Figure 1 is a typical mega twin cell cracking furnace designed by TechnipFMC.

Figure 1. Large Capacity Twin Cell Cracking Furnaces

Types of Independent Cell Decoking (ICD):

The basic types of ICD operation in a twin cell furnace are summarized as follows:

- One cell in cracking mode and another cell in decoking mode
- One cell in cracking mode and another cell in HSSB mode
- One cell in cracking mode and another cell in transition between cracking and HSSB mode and vice versa
- Decoking or HSSB operation can be either Decoke/HSSB with effluent returned to firebox or Decoke/HSSB with effluent to a decoke cyclone and then vented to atmosphere.

Advantages of Independent Cell Decoking (ICD):

ICD mode allows the twin cell furnace design to crack feed at the full capacity and the designed conversion/severity in one cracking cell while the other cell is simultaneously in decoking or HSSB mode.

The advantage of ICD operation is that with such a flexibility of operation, the "very expensive spare" mega furnace may not necessarily be required for the mega plant design thereby minimizing total investment for the mega furnaces. This could be a huge benefit from the investment point of view.

The another advantage of ICD operation is to minimize total cracked gas flow fluctuation downstream of the plant and minimize the fluctuation of feed flow to other furnaces as compared with full furnace decoke/HSSB. This is particularly important for plant expansion such as adding one large capacity twin cell furnace next to many small sizes of existing furnaces. In that case, it is important to have ICD design feature for this large capacity twin cell furnace.

When furnace cracks two different feed stocks from each cell of a twin cell furnace such as cracking ethane in one cell and cracking propane in another cell, the run length from each cell may be quite different (for example, relatively longer run length in propane side and relative shorter run length in ethane side). The advantage of ICD operation is to extend the run length of the cracking cell processing propane feed to increase furnace availability as compared with decoking the full furnace (and bring it out of the production mode) even if the end of run criteria in the propane side has not been reached.

Simulations of Twin Cell Furnaces with ICD operation:

There are significant challenges in the design of ICD twin cell cracking furnace by the proper simulation of the complete furnace performance due to the following reasons:

- Process convection banks and radiant coil banks must process hydrocarbon feed(s), feed(s)+dilution steam in one cell and decoke air, decoke air+steam in another cell, separately, whereas utility banks (economizers, HP steam superheaters) are normally in common. Accurate prediction of heat transfer, process side and flue gas side temperature profiles in the convection and radiant sections is a challenge.

- Fired duty in each cell of a twin cell furnace is quite different during ICD operation. Challenges in the design are the prediction of flue gas mixing in the transition zone from the radiant section of each cell to the common convection section and associated impact on the firebox such as the process crossover temperatures. Prediction of the performance of the coils and passes of the convection section including tube wall and fin temperatures is essential. The same challenges are also applied to the parallel quench exchangers for cracked gas and decoke air/steam.

- Separate CFD modelling in the transition region between firebox and convection can be considered to refine the design.

For ICD operation, a new cracking furnace model has been developed by TechnipFMC called SPLIT-MIXING to simulate this mode using the latest version of TechnipFMC's furnace simulation/optimization software: SPYRO® Suite 7 model, which is capable of simulating ICD twin cell furnace directly. An example of the split-mixing ICD model is illustrated in Figure 2. Alternatively, this system can also be simulated using cracking furnace model such as Firebox/ SPYRO® program linked with other common simulation tools such as ASPEN or PROVISION.

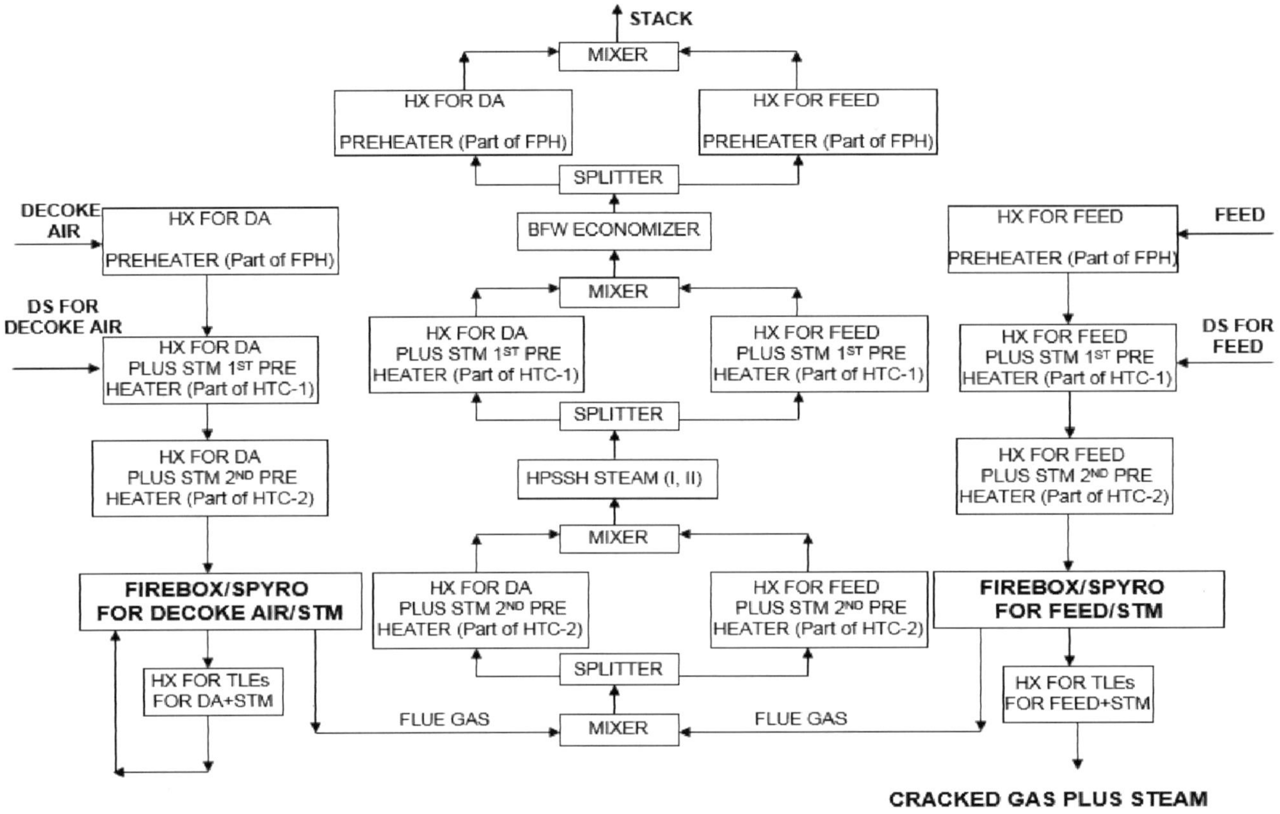

Figure 2. Split-Mixing Cracking/decoking (ICD) Model

SPYRO® Suite 7 is an equation-based simulation-optimization program to simulate almost any type of steam cracking furnace. The program enables flexible simulation of multiple steam cracking furnaces, convection sections, fireboxes and transfer-line exchangers through the flowsheet oriented structure of the program which includes common models like mixers, splitters, adiabatic (cracking) volumes etc.

The salient advantages with SPYRO® Suite 7 simulations for ICD cracking are summarized as follows:

- Rigorous convection section and TLE heat transfer calculations for cracking with ICD can be done automatically and there is no need to link with other external programs such as PROVISION or ASPEN.
- Simulation of cracking and decoking in all the convection section banks and TLE sections is accomplished with SPYRO® Suite 7.
- It can handle simulation of different rows or types of burners and different excess air amount in each cell of a twin cell firebox.
- Program can handle and converge large differences in fired duty between decoke cell and cracking cell.

It has the option to set cracking conversion for gas cracking or severity for liquid cracking in cracking cell and coil outlet temperature in decoke/HSSB cell separately.

- Once a model is built in the SPYRO® Suite 7 environment, calculation due to the changes of operating conditions is much faster. The convergence is relative stable.
- The program can also handle flue gas split and mixing in the convections section banks by the program itself.
- The program can converge all the above in one single file.

A SPYRO® Suite 7 graphic for the sample ICD scheme is given in Figure 3 for main overall flowsheet and Figure 4 for convection section flowsheet.

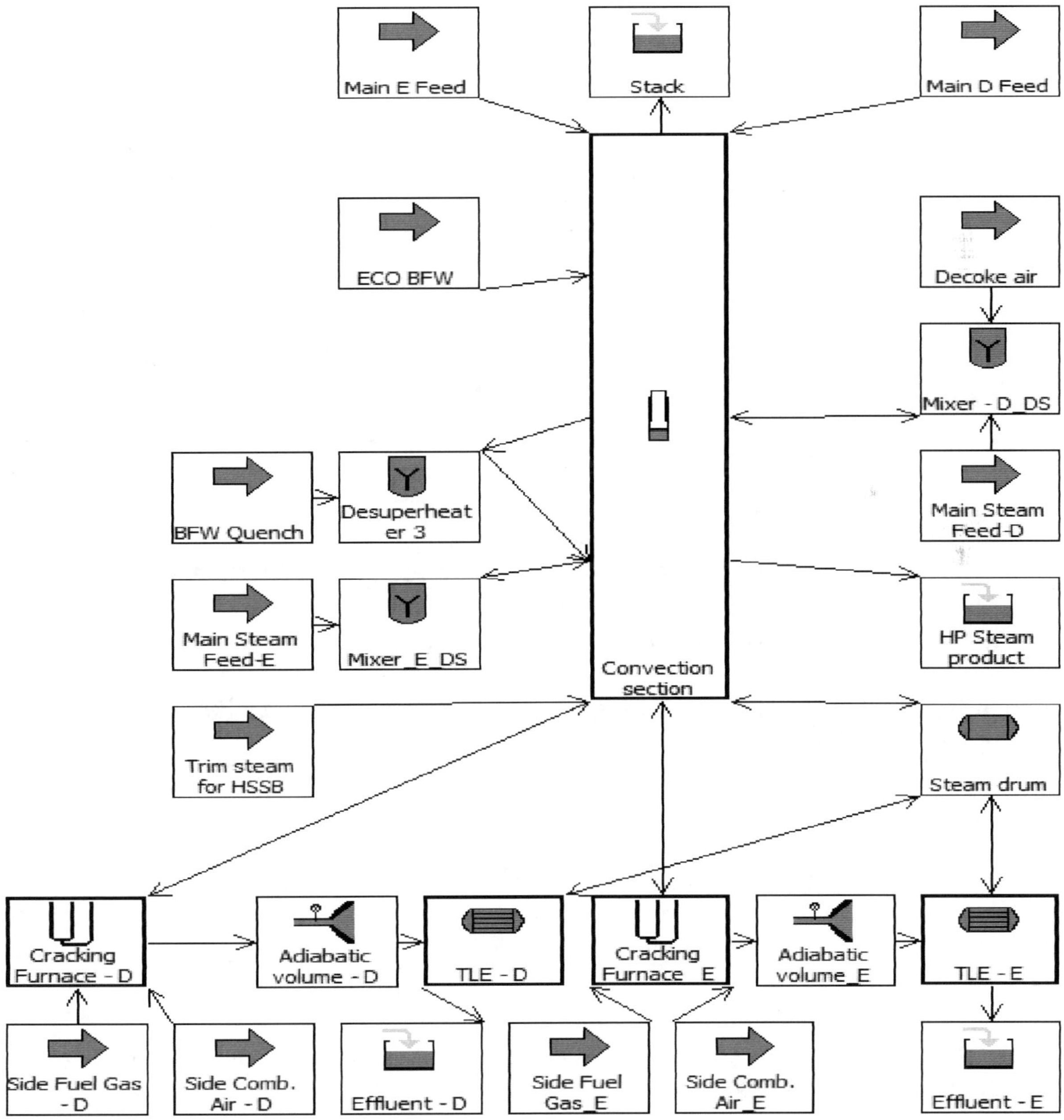

Figure 3. Main Flowsheet of ICD in SPYRO® Suite 7

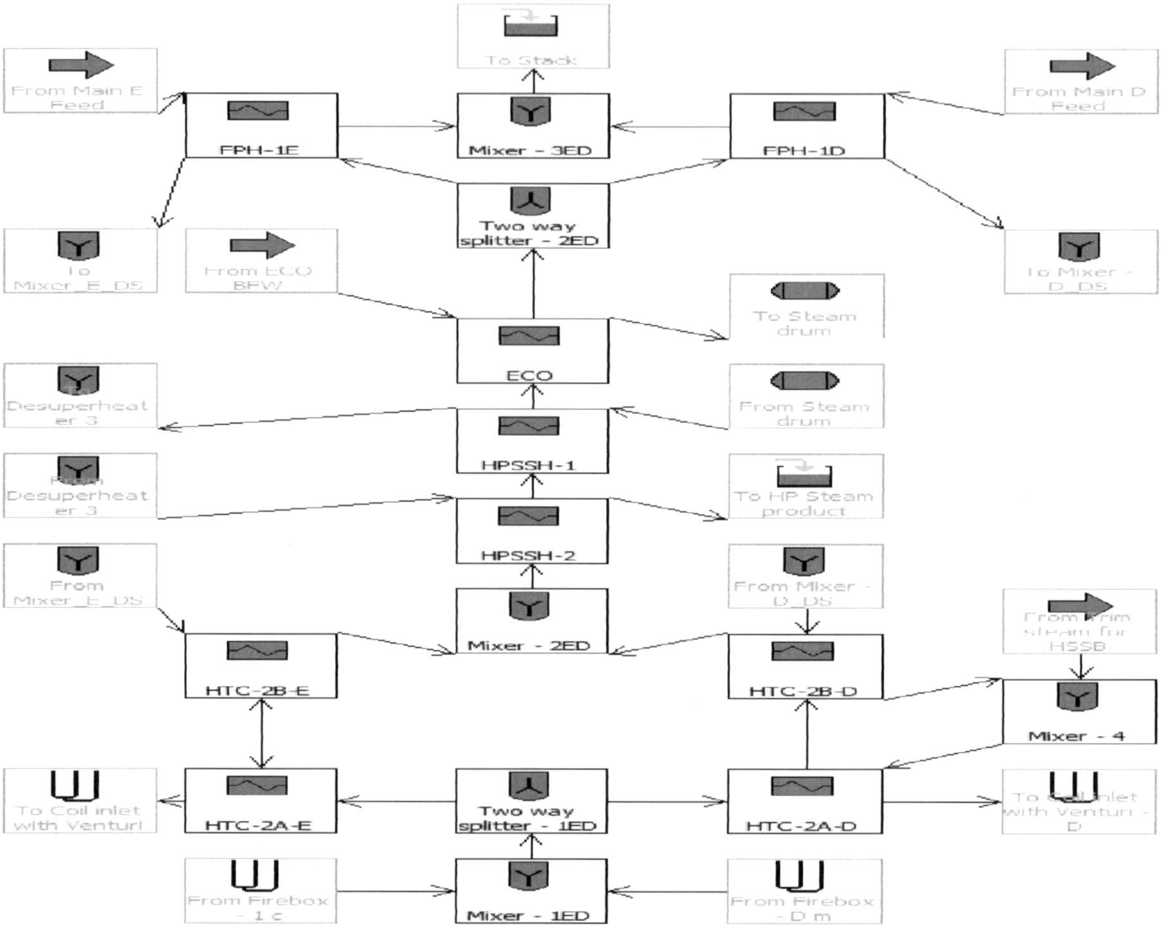

Figure 4. Convection Section Flowsheet of ICD in SPYRO® Suite 7

ICD Design Issues/Considerations:

In actual design of ICD twin cell furnaces, the challenges remain and are to be carefully considered during process design and engineering:

1. Process Design Aspects:

- Process separate convection section passes, radiant coil passes, parallel TLE passes. Ensure all the streams do not combine with each other during this mode of operation.
- How to properly split process banks in the convection section is a consideration. Flue gas interaction at the bridge wall is a consideration. Technip uses CFD and experience from the field data to refine the model.
- Tube metallurgy in the convection section is to be checked for ICD operation. Some material upgrades may be required.

- For both gas and liquid cracking with ICD, consider parallel cracked gas MOVs and decoke MOVs.
- For liquid cracking with ICD, feed vaporization in FPH coil needs to be considered in this operation mode as the flue gas temperature to feed preheater coil is lower for ICD operation than for normal cracking operation.
- For liquid cracking with ICD, it is necessary to consider parallel quench oil fittings.
- For gas cracking with ICD, if BFW preheat in the last transfer line quench exchangers is required, it is important to split BFW flow to the parallel quench exchangers for cracking and decoking/HSSB properly to avoid cracked gas condensation in last TLE. In addition, how to split BFW flow during transition from ICO to ICD and vice versa is also a consideration for smooth transition.

2. Process Control Aspects:

- Provide separate control stations for feed, decoke air, dilution steam (or clean steam) and DMDS to each cell of the furnaces.
- How to control the crossover temperature in the decoking cell from overheating is a consideration since decoke air + steam flow is much lower than the feed + steam flow. The preheating of decoke air + steam is greater than preheating feed+steam in the convection bank due to relatively higher UA.
- During ICD operation, it is also important to adjust the burner air dampers to control excess air to the furnace in the decoking cell with effluent returned to the firebox to prevent from overheating some parts of the convection section.
- It may also be necessary to inject some additional cold steam in the decoking passes of the convection section to cool down the crossover process temperature of the decoking cell.
- Watch firing in the cracking cell to prevent high firing during ICD. Some adjustments may be required for proper control.
- Draft is normally set at the cracking side.
- During ICD operation, separate control of firing to each cell is required as the firing rate from cracking cell and decoking cell is significantly different.

3. Process Safety Aspects:

- Provide positive isolation (EBV or ABV based on design standard) and interlock for feed, decoke air and DMDS during one cell in cracking mode and the other cell in decoking mode to ensure these streams do not combine with each other.
- Provide separate convection section passes, radiant coil passes, parallel TLE passes for decoke air/decoke air+steam and feed/feed+steam/cracked gas. Ensure all the passes do not combine with each other.
- Provide parallel MOV's for decoke air+steam and cracked gas. Provide separate MOV interlocks for each cell for MOV transfer between cracking and decoking within a cell.
- During ICD operation, depending on the design, the effluent can be routed to either a decoke cyclone and eventual venting to atmosphere or directed to the cell's firebox for

combustion. It is necessary to control O2 % at the arch to avoid overheating in the lower process bank feeding decoke air + steam.

- During decoke back to firebox, it is also important to evaluate flame stability while initially introducing steam to prevent from flame blow out, especially utilizing natural gas as fuel.

ICD Operational Issues/Consideration:

Figure 5 is a typical ICD operation in a gas cracking twin cell furnace.

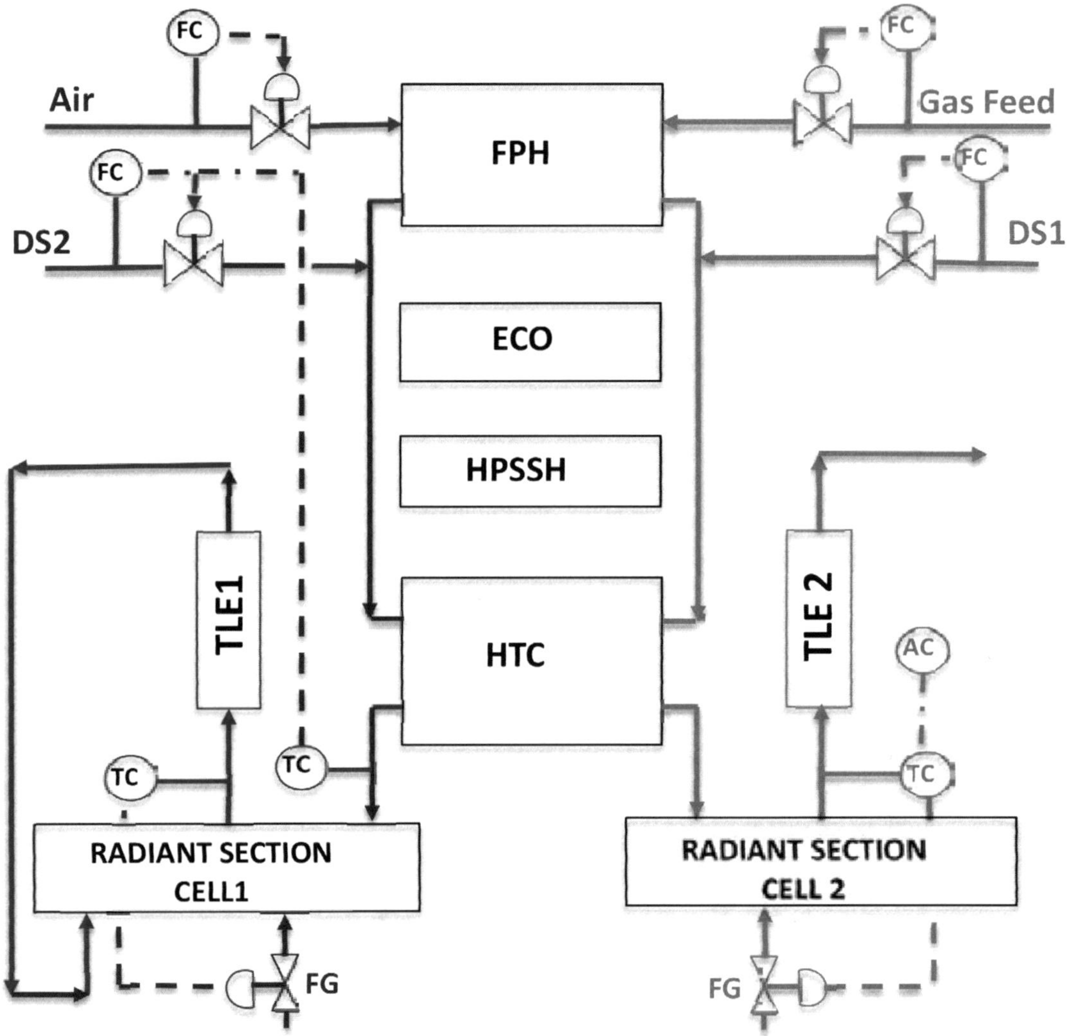

Figure 5. Typical ICD Operation in A Gas Cracking Twin Cell Furnace

Figure 6 is a typical ICD operation in a liquid cracking twin cell furnace.

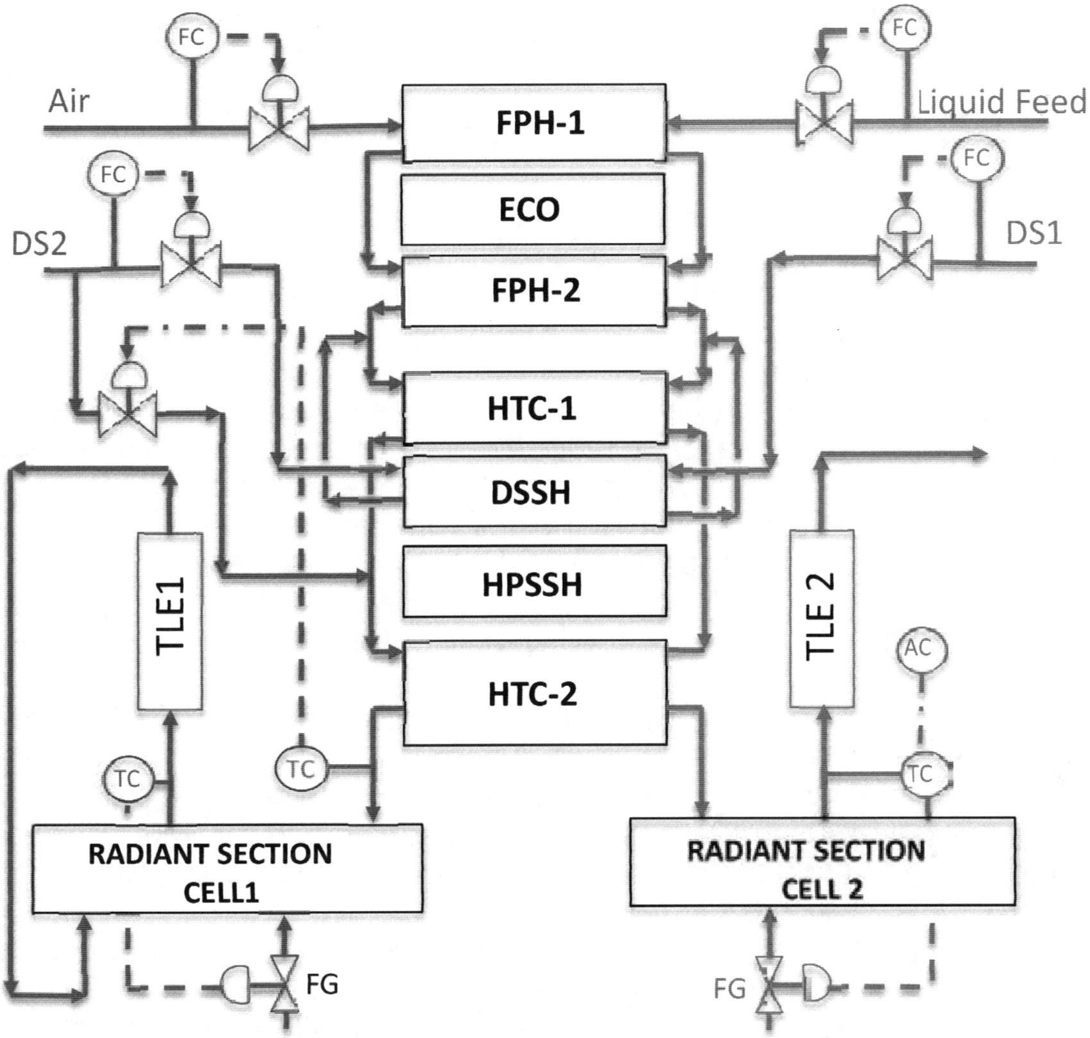

Figure 6. Typical ICD Operation in A Liquid Cracking Twin Cell Furnace

During ICD operation, the following points are recommended for consideration:

- In cracking cell, firing rate is controlled by the once through conversion for gas cracking or severity for liquid cracking resetting the coil outlet temperature which cascade to adjusting accordingly the fuel flow to the burners of that cell.
- In decoking cell, firing rate is controlled by coil outlet temperature.
- The process crossover temperature is limited by the decoke steam flow during decoke steam or trim decoke steam flow injection prior to HTC-2 bank (See Figure 5 and 6).

- Decoke steam/decoke air flow ratio in the decoking cell from start of decoke to end of decoke is a consideration in order to prevent local hot spots along the coils and also higher process crossover temperature in the decoking cell.
- Burner air registers are to be tuned for ICD operation, particularly for decoking back to firebox mode.
- It is important to monitor tube metal temperature in decoking cell. If any local hot spots occur, reduce the air flow in the decoking cell. Some adjustments of the burners nearby may also be required to avoid local hot spots.
- Increase the decoke air slowly. Monitor closely CO2 amount in the decoke effluent. If CO2 amount exceeds the required limit, reduce the decoke air to ensure CO2 is under control.
- Follow decoke procedure properly. The decoke procedure for ICD can be different from the procedure for a full furnace decoke.
- Watch flame stability for the mode of decoking to firebox and in particular with natural gas fuel. Some design improvement for the burners may be required to prevent from flame out when introducing steam to the firebox.
- Firing in the cracking cell will be higher than normal cracking due to lower crossover temperature during ICD operation. A reduction of feed flow or severity may be required.
- During ICD operation with 600 # steam system, BFW vaporization% at the outlet of economizer will be increased. Increased blowdown may be required to avoid too high vaporization within economizer bank.

An example of cracking with independent cell decoking (ICD) is illustrated in Figure 7.

In this case, ethane is cracking in one cell and steam-air decoke is in another cell with decoke effluent to a decoke cyclone and then vented to atmosphere. An ICD scheme was developed. There are parallel passes of the convection section with half of the passes supplied with ethane feed and half of the passes supplied with decoke air. There are also separate firing controls in each cell of the furnace. In this furnace design, ethane feed flow is fixed and ethane conversion is also fixed, which is the same as normal cracking mode. The result is that the firing in the cracking cell increases by about 10% due to lower bridge wall temperature than normal cracking in that cell. The ratio of decoke steam flow to decoke air in the decoking cell is set in order to control the crossover temperature at 1400 ^{o}F. Burner air registers remain unchanged in the cell for decoking. The relative firing rate between cracking and decoking cell is from 5 to 1 to 10 to 1. The BFW flow split between cracking and decoking is about 2 to 1.

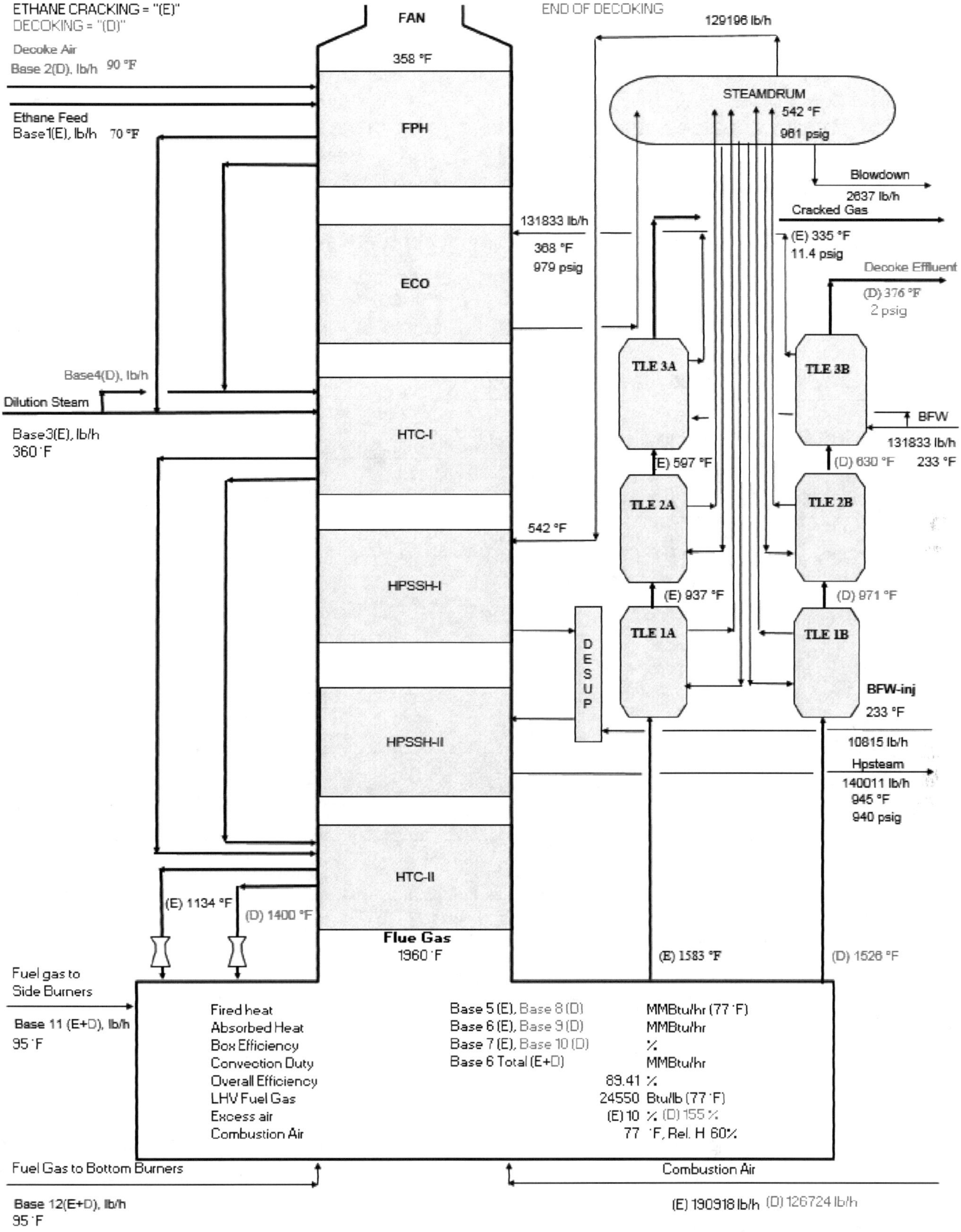

Figure 7. Furnace Performance for Ethane Cracking with ICD

Design Versus Operation:

So far we have discussed design and operating considerations, but how well does this translate to actual field operation? While many applications have been installed, a complete set of operating data that matches the process design parameters is difficult to obtain. Table 1 contains information from a twin cell furnace with ICD. The design data is for an end of decoke scenario, while the operating data is for a start of decoke scenario (HSSB).

Table 1. ICD Furnace Performance Comparison Between Design and Operation

	Design Values (end of Decoke)	Operating Data (start of decoke)
Cracking Cell Feed Flow (t/h)	26.07	26.22
Cracking Cell DS flow (t/h)	13.03	12.2
Decoking Cell DS flow (t/h)	18.75	11.07
Decoking Cell COT (°C)	900	750
Cracking Cell COT (°C)	828	800
Cracking Cell XOT (°C)	566	585
Decoking Cell XOT (°C)	742	740
Trim DS Flow to Decoke Cell (t/h)	0 (6 t/h max)	2.0
Bridgewall Temperature (°C)	1028	950
BFW to Attemperator (t/h)	4.0	2.0

It should be noted that the data is from an operating unit in day to day operation, it is not from a tuned test run with input from the licensor. So what can we glean from this data? First since it is not an exact match to the design process conditions, we have to look at trends but not direct comparison.

The cracking conditions are fairly close to design data, but the decoke conditions are much less severe. Essentially the furnace operation is in a HSSB to firebox mode. The flow of dilution steam to the decoking cell is only 60% of design and the COT is 150 °C less than design. Directionally this will lead to:

- Higher XOT for the decoking cell due to lower flow but same surface area.
- Lower bridgewall temperature due to lower firing in decoking cell
- Less BFW for attemperator due to less flue gas

As can be seen in the operating data the XOT is less than the design, but 2 t/h of cold DS is used to limit the temperature. As expected the bridgewall temperature and BFW low to attemperator are both less than design.

Summary and Conclusions:

For economic reasons, mega furnaces must include features to achieve feed flexibility. Feed flexibility not only includes capability to crack a range of feed stocks, co-cracking or hybrid cracking capability but also ICD features incorporated in a twin cell mega furnace design.

An advantage of applying ICD design feature is that an "expensive" spare furnace may not be required, and furnace availability may be improved. The decision to consider these modes of operation will be influenced by the desired final olefin ratios, individual fresh feedstock consumption, furnace availability and also by the capability of each individual furnace.

Simulation of these cracking modes is challenging due to the geometrical separation of the different fluid paths and the interactivity of these paths. In particular, there is a very large difference in the firing duty from each cell during ICD operation. TechnipFMC Process Technology has developed methodologies to accurately carry out these process designs and simulations to meet client needs.

Safe operation of furnaces with ICD features can be achieved with proven process design, control and operational aspects.

Reference:

[1] Wang, H, Evans, B, Van Goethem, M, "Co-cracking vs Segregated "Hybrid" Cracking in Individual Furnaces – Design and Operational Considerations". Proceedings of 23rd Ethylene Producers Conference, March 13-17, 2011, Chicago, Illinois.

479773 A Better(?) Way to Calculate the Slope of the Equilibrium Line

Abstract:

The calculation of tray efficiencies or the HETP of packed columns requires, among other things, the slope of the equilibrium line. For binary systems the slope is unambiguously defined (although not always rigorously calculated). However, there is no easy way to use a definition of the slope for multicomponent systems because the vapor liquid equilibrium is represented by a surface (in hyperspace if there are more than 3 compounds).

Thus, for multicomponent systems any pseudo-binary method for estimating tray efficiencies and HETP must rely, in part, on an approximate method of estimating the slope. Particularly, information regarding thermodynamic interactions is lost using pseudo-binaries and the slope of the equilibrium line may, therefore, have very limited meaning for a multicomponent system.

In a presentation given at the 2015 National Spring Meeting of the AIChE we looked at four different methods for approximating the slope of the equilibrium line for multicomponent systems. It was shown that these different methods could lead to dramatically different estimates of the equilibrium slope.

Our purpose here is to take further our earlier work and propose a different and, we believe, definitive way to approximate the slope.

The method requires us to define a design component, one per section of the column. The design component can be defined as that which is to be removed from a designated column section. For example: in a distillation to separate the four component mixture ABCD into the AB and CD fractions then the design component for the top section is C and the design component section for the bottom section is B. (In this example these design components are like the heavy and light keys respectively, but that is not always the case, e.g. for column configurations involving side streams.)

The slope then is estimated from two similar flash calculations in which only the design component has its molar amount altered (slightly). This methodology allows proper consideration of thermodynamic interactions and thereby avoids the short comings of adopting a pseudo-binary approach.

We will show that for binary systems this method is exactly equivalent to the rigorous definition of the slope. We will further demonstrate that this method provides a sensible approximation in cases when other published methods have the potential to lead one astray.

478892 Novel Optimization of a Dividing Wall Distillation Column

Abstract:

Novel Optimization of a Dividing Wall Distillation Column

Bailee Roach, Bruce Eldridge

James R. Fair Process Science and Technology Center

University of Texas at Austin

Originally patented in 1949, the dividing wall distillation column has gained renewed attention due to the potential for reduced energy consumption, an estimated thirty to fifty percent over traditional multiple column arrangements, and lower capital investment. Unfortunately, this new interest has not been accompanied by the development of rigorous design approaches that utilize the latest in computational methods. In the current study, optimization and process simulation software was seamlessly integrated to yield minimum energy consuming dividing wall column (DWC) designs over a range of operating conditions and product purity targets. The modeling effort was supported by data collected from the Separations Research Program (SRP) experimental column. Multiple chemical systems were investigated with the study yielding a powerful methodology for future DWC retrofits or new column construction.

2017 AIChE®

SPRING MEETING & 13TH GLOBAL CONGRESS ON PROCESS SAFETY

March 26–30, 2017
Henry B. Gonzalez
Convention Center
San Antonio, TX

17 AIChE Spring Meeting, San Antonio, TX
13TH GLOBAL CONGRESS ON PROCESS SAFETY

17 AIChE Spring Meeting, San Antonio, TX
13TH GLOBAL CONGRESS ON PROCESS SAFETY

Improve Energy Efficiency of DWCs

Manish Bhargava
Licensing Manager
GTC Technology

GTC Technology
- Engineered to Innovate®

Norihito Suzuki
Staff Engineer
TonenGeneral Sekiyu K.K.

TonenGeneral

17AIChE Spring Meeting, San Antonio, TX | 13TH GLOBAL CONGRESS ON PROCESS SAFETY | GTC Technology · Engineered to Innovate® | TonenGeneral

Concept of Dividing Wall Columns

17AIChE Spring Meeting, San Antonio, TX | 13TH GLOBAL CONGRESS ON PROCESS SAFETY | GTC Technology · Engineered to Innovate® | TonenGeneral

Conventional 2-Column Fractionation Sequence For Separating Three Components

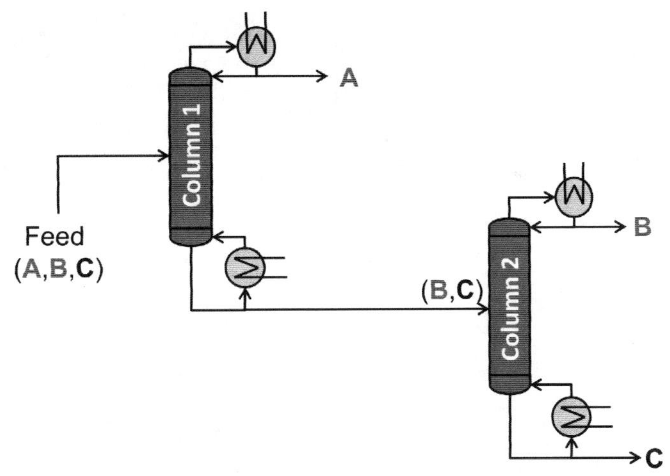

Thermodynamics Inefficiencies in Column 1

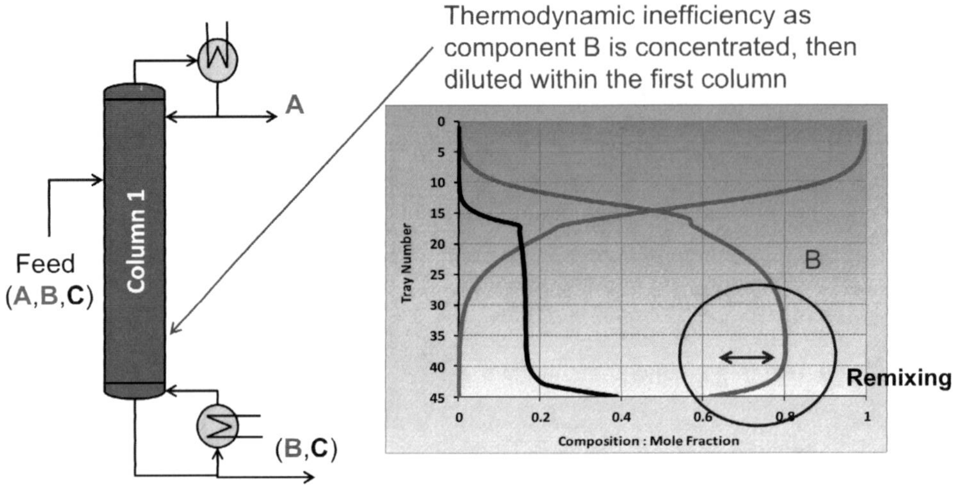

Pre-fractionator Main Column Arrangement

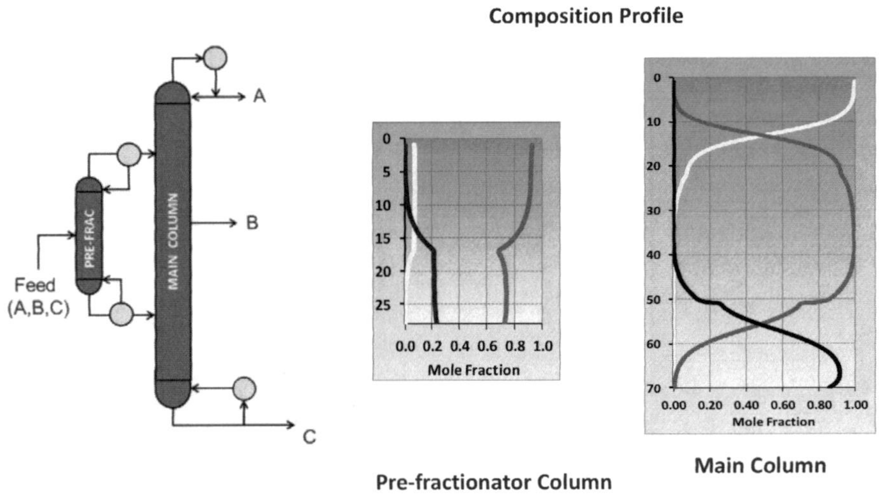

17 AIChE Spring Meeting, San Antonio, TX | 13TH GLOBAL CONGRESS ON PROCESS SAFETY | GT Technology Engineered to Innovate | TonenGeneral

From Prefractionator Arrangement To Dividing Wall Column

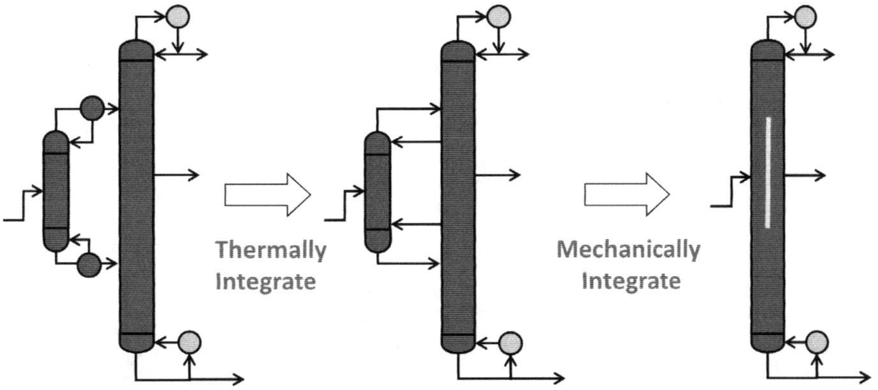

17 AIChE Spring Meeting, San Antonio, TX | 13TH GLOBAL CONGRESS ON PROCESS SAFETY | GT Technology Engineered to Innovate | TonenGeneral

Typical Benefits of GT-DWC℠ (Dividing Wall Columns)

GT-DWC℠ CAPEX is lower by 20-30%

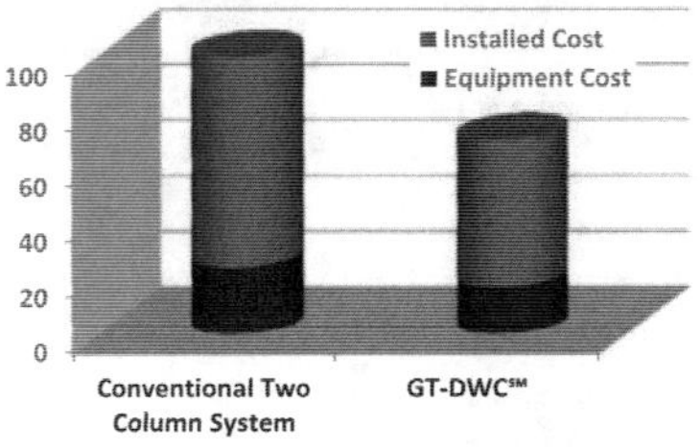

GT-DWC℠ OPEX is lower by 20-30%

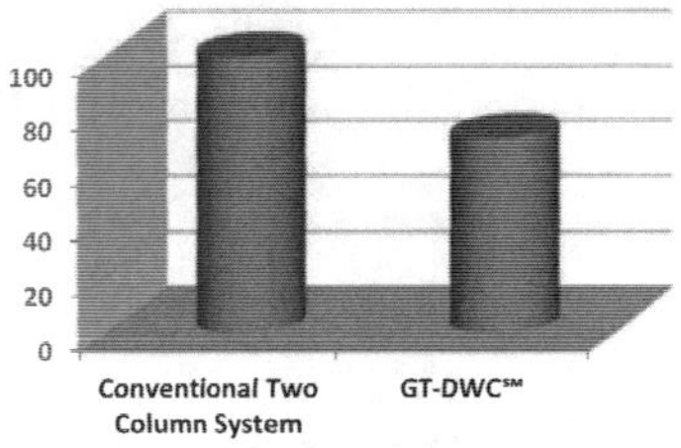

Equipment Count & Plot Space

	Conventional Two Column System	GT-DWC℠
Column	2	1
Reboiler	2	1
Condenser	2	1
Pump	4	3
Plot Space	100%	Approximately 30% less

Typical Dividing Wall Column Applications

- Naphtha Splitters
- Benzene/Toluene/Xylene/C9 Separation
- Debutanizer/Depropanizer
- Depropanizer/Deethanizer
- Reformate Splitters
- Naphtha Splitter/Deisohexanizer
- Dependanizer/Naphtha Splitter

17 AIChE Spring Meeting, San Antonio, TX
13TH GLOBAL CONGRESS ON PROCESS SAFETY
GTC Technology Engineered to Innovate®
TonenGeneral

Case Study 1 : Confidential Client in India
Two-column configuration modified to a single column

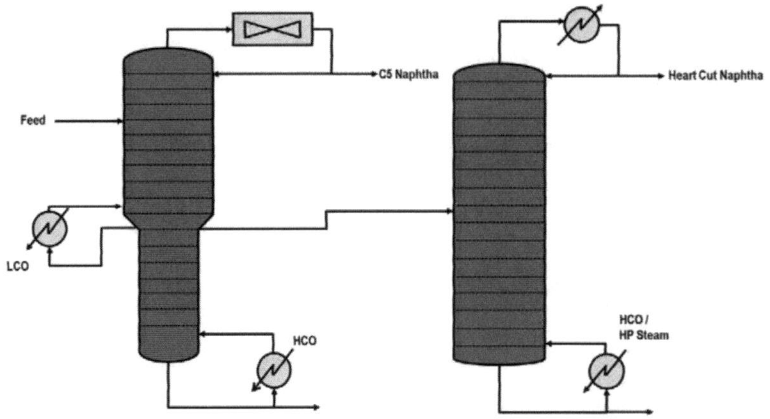

2017 AIChE® SPRING MEETING & 13TH GLOBAL CONGRESS ON PROCESS SAFETY

Objectives for Revamp

- Reduce utility consumption for the separation.
- Produce heart cut naphtha in Naphtha Splitter-1 column, while removing Naphtha Splitter-2 from service.

17 AIChE Spring Meeting, San Antonio, TX
13TH GLOBAL CONGRESS ON PROCESS SAFETY
GTC Technology Engineered to Innovate®
TonenGeneral

Bottlenecks in Column Design

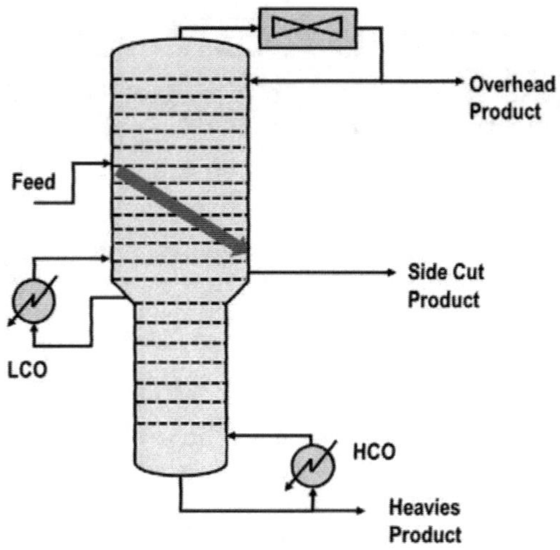

Intermixing of Feed with Side Cut

Lower Diameter Section Puts a restriction on Vapor Liquid Traffic

2017 AIChE® SPRING MEETING & 13TH GLOBAL CONGRESS ON PROCESS SAFETY

Revamp to GT-DWC℠

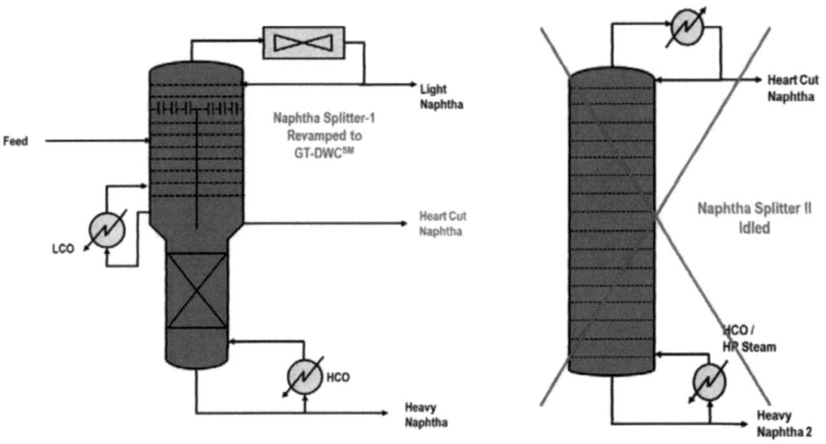

Project Economics

Parameter	Units	Original Design (NS-I and NS-II)	Naphtha Splitter– I after Revamp to GT-DWC℠
Feed	TPH	308.0	308.0
Side Draw Flow Rate	TPH	165.0	165.0
D86 (IBP-FBP)	° C	110.6-170.0	110.0-170.0
Total Heating Duty	MMkcal/hr.	38.0	28.0

Benefits of GT-DWC℠

- Heating duty reduced by 26%.
- HP steam usage eliminated.
- Naphtha Splitter-II is idled.
- Heart Cut Naphtha obtained in Naphtha Splitter-I.

Dividing Wall Column Trays

Top View of Dividing Wall

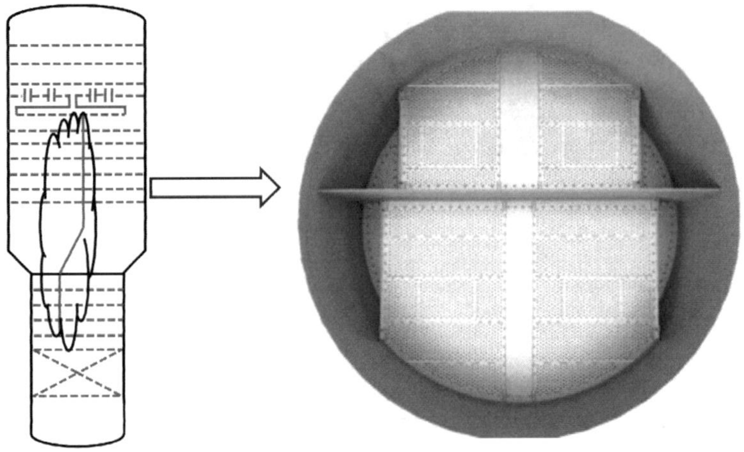

2017 AIChE® SPRING MEETING & 13TH GLOBAL CONGRESS ON PROCESS SAFETY

GTC Internal Liquid Split Distributer

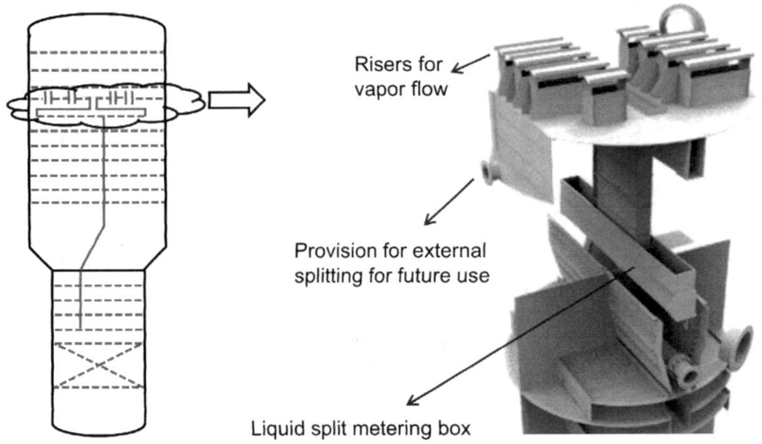

2017 AIChE® SPRING MEETING & 13TH GLOBAL CONGRESS ON PROCESS SAFETY

Case Study 2 : Grassroots Mixed Xylenes Recovery Unit at TonenGeneral (TG), Japan

- TG had an existing unit which produced C7+ product for gasoline blending.
- TG wanted to separate high purity petrochemicals (Toluene, Mixed Xylenes) from the feed.
- TG decided for a DWC solution against a two column configuration because of lower CAPEX and lack of plot space.

17AIChE Spring Meeting, San Antonio, TX
13TH GLOBAL CONGRESS ON PROCESS SAFETY
GTC Technology · Engineered to Innovate®
TonenGeneral

Case Study 2: Mixed Xylenes Recovery

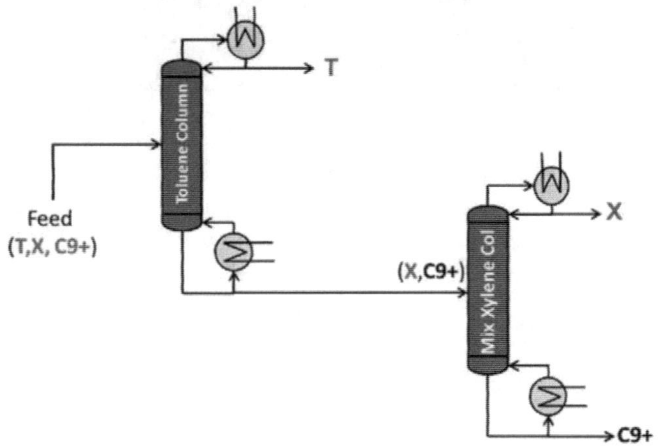

2017 AIChE® SPRING MEETING & 13TH GLOBAL CONGRESS ON PROCESS SAFETY

GT-DWC℠ for Mixed Xylenes Recovery

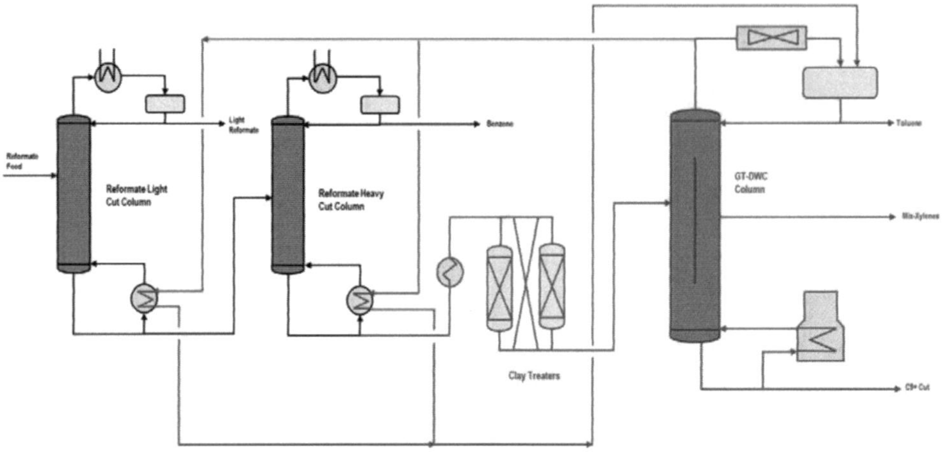

17 AIChE Spring Meeting, San Antonio, TX
13TH GLOBAL CONGRESS ON PROCESS SAFETY
Engineered to Innovate®
TonenGeneral

Comparison: Two Column System vs DWC

Parameter	Units	Two Column Configuration	GT-DWC℠ Configuration
Mix-xylenes product	Kg/hr	29,332	29,334
C8 aromatics	wt%.	99.2	99.3
Reboiler duties	MMkcal/hr.	21.6	17.2
Operating cost savings	%	-	20.0
Capital cost	US $ MM	26.0	21.0
Steam savings due to heat integration	TPH	-	20.0

2017 AIChE® SPRING MEETING & 13TH GLOBAL CONGRESS ON PROCESS SAFETY

17AIChE Spring Meeting, San Antonio, TX
13TH GLOBAL CONGRESS ON PROCESS SAFETY
GTC Technology Engineered to Innovate®
TonenGeneral

Flexibility in Design

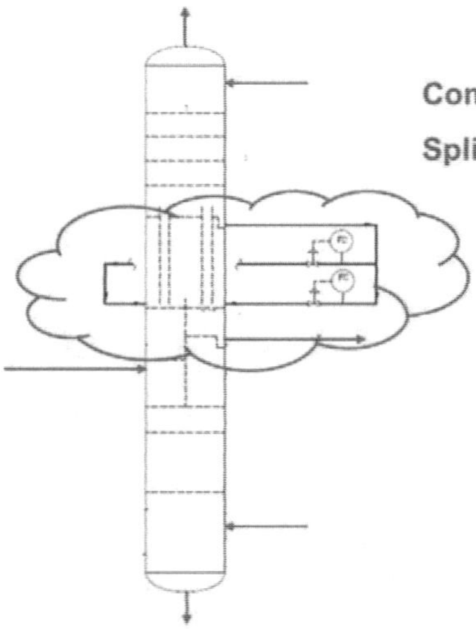

Combination of Internal and External Splitting for Operational Flexibility

2017 AIChE® SPRING MEETING & 13TH GLOBAL CONGRESS ON PROCESS SAFETY

17AIChE Spring Meeting, San Antonio, TX
13TH GLOBAL CONGRESS ON PROCESS SAFETY
GTC Technology Engineered to Innovate®
TonenGeneral

GT-DWC℠ Column Internals at TonenGeneral

Off Center Dividing Wall

2017 AIChE® SPRING MEETING & 13TH GLOBAL CONGRESS ON PROCESS SAFETY

TonenGeneral Mixed Xylenes Recovery Unit

Installation of GT-DWC℠ Column at Chiba Refinery, Japan

Column Operational since March 2016

GT-DWC℠ Internals Manufacturing - Korea

GT-DWC℠ Trays

Testing of GT-DWC℠ Internal Liquid Split Distributer

Heat Integrated GT-TDWC℠

- Conventional side draw column provides limited opportunity for heat integration
- Top dividing wall column gains the thermodynamic efficiency, plus affords meaningful heat integration.

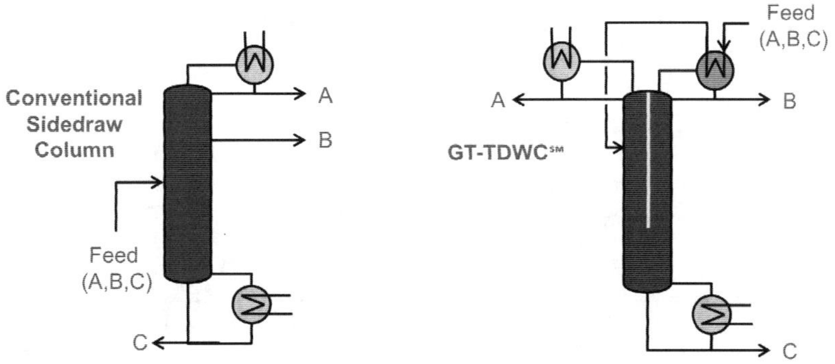

17 AIChE Spring Meeting, San Antonio, TX
13TH GLOBAL CONGRESS ON PROCESS SAFETY
GTC Technology Engineered to Innovate® TonenGeneral

Case Study 3: GT-TDWC℠ for Food Grade Hexane Production (FGH) for BPCL, India

FGH production requires a narrow cut of an intermediate product.

Options for application

1. Conventional 2 Column System : DIH + FGH column
2. Advanced Distillation Option : GT-TDWC

2017 AIChE® SPRING MEETING & 13TH GLOBAL CONGRESS ON PROCESS SAFETY

17 AIChE Spring Meeting, San Antonio, TX — 13TH GLOBAL CONGRESS ON PROCESS SAFETY — GTC Technology · Engineered to Innovate — TonenGeneral

Conventional Option: DIH + New FGH Column

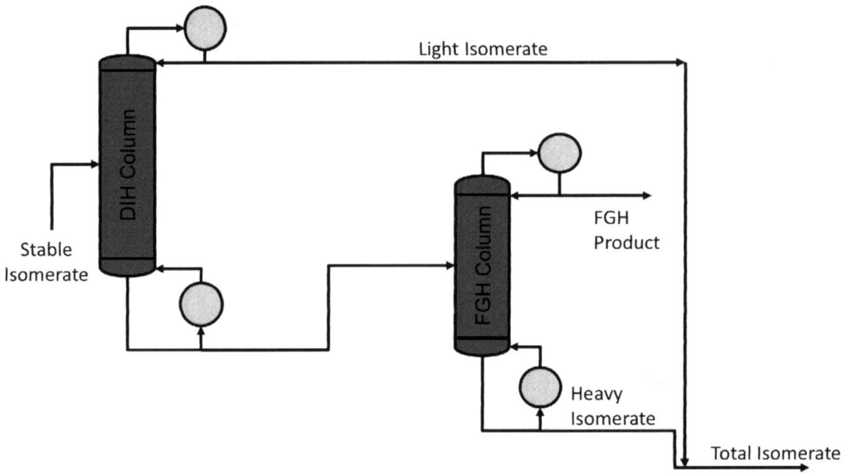

17 AIChE Spring Meeting, San Antonio, TX — 13TH GLOBAL CONGRESS ON PROCESS SAFETY — GTC Technology · Engineered to Innovate — TonenGeneral

GT-TDWC℠ Column for Food Grade Hexane Production

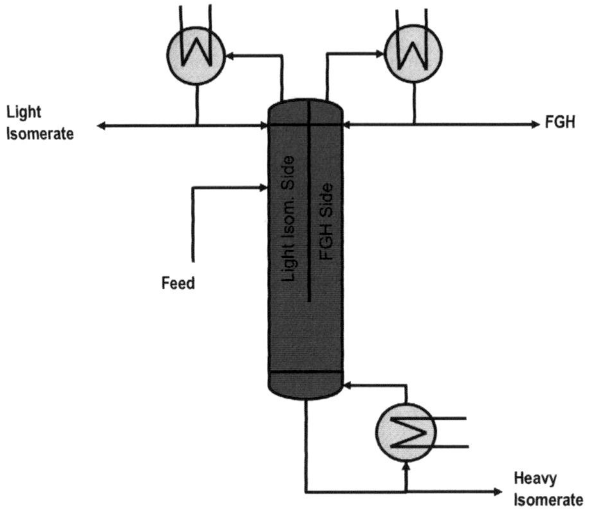

Comparison : GT-TDWC vs Conventional 2 Column System

	Conventional 2 Column System		GT-TDWC
	DIH Column	New FGH Column	
MP Steam Consumption (Tons/Hr)	26.20	8.50	26.46
	Total = 34.7		
Operating Cost (MM USD/year) Steam only	9.69		7.36
Savings (MM USD/year) Steam only	-		2.34

Top Dividing Wall Column Installation at BPCL, India

17AIChE Spring Meeting, San Antonio, TX — 13TH GLOBAL CONGRESS ON PROCESS SAFETY — GTC Technology, Engineered to Innovate® — TonenGeneral

Case Study 4: GT-LPG Max℠ for LPG Recovery

Revamp of Existing LPG Recovery Unit in Texas to GT-LPG Max

Client Objectives:

- Propane recovery in LPG > 97%
- No Refrigeration

17AIChE Spring Meeting, San Antonio, TX — 13TH GLOBAL CONGRESS ON PROCESS SAFETY — GTC Technology, Engineered to Innovate® — TonenGeneral

Case Study 4: GT-LPG Max℠ for LPG Recovery

Product Class	Product Specifications	Feed Composition	
		Component	Wt%
Fuel Gas 3.8%	Minimize Propane	Hydrogen	0.0%
		Methane	0.3%
		Ethylene	0.0%
		Ethane	3.0%
		H_2S	0.5%
		CO_2	0.0%
LPG Product 11.8%	Propane > 90 wt% Ethane < 1 wt% Vapor Pressure < 208 psig @ 100°F	Propene	0.1%
		Propane	11.7%
I-C4 Product 8.3%	Propane < 3.0 wt% N-C4 < 4.0 wt%	i-Butane	8.2%
		i-Butene	0.1%
		1-Butene	0.1%
N-C4 Product 19.6%	Iso-Pentane < 3.0 wt% Iso-Butane < 5.3 wt%	n-Butane	19.6%
		Tr2-Butene	0.0%
Debutanizer Bottoms 56.5%		Heavies	56.6%

Lights are 3.7% of Total Feed

Considerable C5 and Heavy Components

2017 AIChE® SPRING MEETING & 13TH GLOBAL CONGRESS ON PROCESS SAFETY

Conventional Design of LPG Recovery Unit

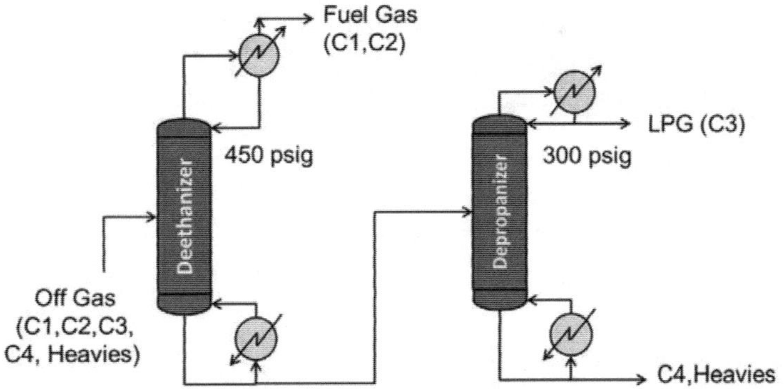

- LPG Recovery – 55%
- Energy – 22.0 MMBtu/hr

GT-LPG MAX℠ - Combined Absorption & Distillation in a Single Column

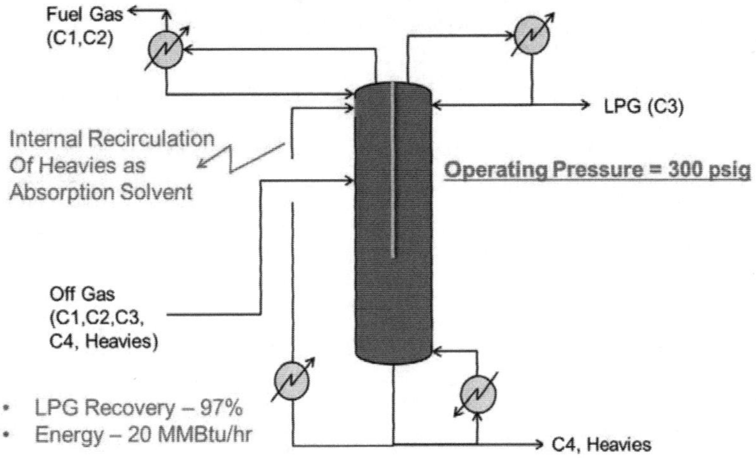

- LPG Recovery – 97%
- Energy – 20 MMBtu/hr

Project Economics of GT-LPG MAX℠

Variables	Existing Configuration	GT-LPG Max℠
Overall Propane Recovery %	55 %	97 %
Total Duty MM kcal/hr	22.0 MMBtu/hr	**20.0 MMBtu/hr**

- Net Benefit/Year $MM 10.1
- Total Installed Cost $MM 15.0
- Simple Pay Back Months 18.0

17 AIChE Spring Meeting, San Antonio, TX
13TH GLOBAL CONGRESS ON PROCESS SAFETY
GT Technology Engineered to Innovate®
TonenGeneral

Summary

- Unconventional distillation applications are an overlooked means to reduce refinery energy consumption
  - GT-DWC reduces 20–30% OPEX through energy savings
  - GT-DWC reduces 20–30% CAPEX by requiring a single column for multi-component separation
- Advanced Distillation schemes can offer CAPEX and OPEX reduction of more than 40% by combining multiple unit operations and heat integration opportunities.

2017 AIChE® SPRING MEETING & 13TH GLOBAL CONGRESS ON PROCESS SAFETY

17 AIChE Spring Meeting, San Antonio, TX
13TH GLOBAL CONGRESS ON PROCESS SAFETY

Improve Energy Efficiency of DWCs With Overhead Heat Integration

Manish Bhargava
Licensing Manager
GTC Technology

Engineered to Innovate®

Norihito Suzuki
Staff Engineer
TonenGeneral Sekiyu K.K.

TonenGeneral

2017 AIChE® SPRING MEETING & 13TH GLOBAL CONGRESS ON PROCESS SAFETY

479915 Distillation of High Purity Products from Feeds Containing Multiple Azeotropes

Abstract:

The basic design principles as well as startup and operation of a distillation systems to be utilized for separating multiple azeotropes for high value materials. We will discuss the determination of critical design factors as well as selection of operating conditions. Details design regarding fabrication, assembly, startup, and operation will be provided.

Cases involving known as well as variable feed compositions will be discussed. The importance of flexible operating conditions will be examined and the relationship to system and operating costs.

Component Trapping with VLE Uncertainty: Principles and Troubleshooting

By

Paul M. Mathias,[1] Henry Z. Kister,[1] Bruce Parker,[2] Lydia Narvaez,[3] Thomas Schafer,[4] Alan Erickson[4]

[1] Fluor Corporation, 3 Polaris Way, Aliso Viejo, CA 92698

[2] Eli Lilly and Company, Lilly Corporate Center, Indianapolis, IN 46221

[3] Fluor Daniel Caribbean, Inc., Suite 500 St. 2 No. 14 Metro Office Park Guaynabo, Puerto Rico 00969

[4] Koch Modular Process Systems, LLC, 45 Eisenhower Drive, Suite 350, Paramus, NJ 07652

Presented at the Kister Distillation Symposium, AIChE Spring Meeting
San Antonio, Texas, March, 2017

Component Trapping with VLE Uncertainty:

Principles and Troubleshooting

Paul M. Mathias,[1] *Henry Z. Kister,*[1] *Bruce Parker,*[2] *Lydia Narvaez,*[3] *Thomas Schafer,*[4] *Alan Erickson*[4]

[1] Fluor Corporation, 3 Polaris Way, Aliso Viejo, CA 92698

[2] Eli Lilly and Company, Lilly Corporate Center, Indianapolis, IN 46221

[3] Fluor Daniel Caribbean, Inc., Suite 500 St. 2 No. 14 Metro Office Park Guaynabo, Puerto Rico 00969

[4] Koch Modular Process Systems, LLC, 45 Eisenhower Drive, Suite 350, Paramus, NJ 07652

Summary

Intermediate components tend to accumulate near the middle of a distillation column in many chemical and petroleum separations. Their accumulation may lead to off-spec products, corrosion, plugging, or periodic cycling. The most common method of removing these intermediate components is by taking them out in one or more a side draws. The addition of such side draws may not go far enough to achieve the desired product specs. If the side-draw approach fails, the expensive addition of a new tower may be required.

The question of whether a side draw is sufficient to remove the accumulating intermediate components, or whether an additional column is needed, depends on the relative volatilities of the intermediate components at tower conditions. Often, there is a high degree of uncertainty in the correlation of the volatilities of these components. The Margules Uncertainty Analysis method is valuable in analyzing these situations, and in guiding engineers to the most effective path.

The case analyzed here is an acetonitrile-water separation tower with t-crotonaldehyde and propionitrile as the intermediate components. There are uncertainties in the correlation of the volatilities of these intermediate components in this system. In the light of these uncertainties, our analysis addresses the question of whether the tight product specs can be achieved in one tower with a side draw or whether more than one tower is needed. The analysis shows that depending on the product specs and the tolerance for a high reflux ratio, there would be situations where one can be certain that one tower will work, while in other cases more than one tower is needed to assure achievement of the required product specs. In the present case, a single distillation tower is judged to be too risky, and hence the more conservative 3-column solution was chosen.

Introduction

This paper studies the options to purify an aqueous solution of ≈ 12 wt% acetonitrile (ACN) by distillation, and to recover about 90% of the ACN at the azeotropic composition (about 83 wt% at 1 atmosphere or 101.3 kPa). The separation is difficult due to intermediate components [propionitrile (PPN) and crotonaldehyde (T-CROTON] that build-up in the distillation column. The solution options are a single column with a side draw, or a more complex flowsheet with multiple columns. An example of a multiple-column solution is to remove all of the water from the ACN, PPN and T-CROTON via a pressure swing distillation, and then separate the dry ACN from the PPN and T-CROTON. The multiple-column solution is the safe solution, but costs more than the single column with a side draw in terms of both capital cost and operating costs. Here we quantify the risk of the single-column solution by introducing uncertainty analysis into the process modeling.

The problem of intermediate components in distillation is a general problem of component trapping that has been discussed extensively by Kister.[1,2] In some cases, the top temperature is too cold and the bottom temperature is too high to allow these components to leave the column at the same rate as they enter. Having nowhere to go, these components accumulate in the column, which causes problems like off-spec products, flooding, cycling slugging, and corrosion. Cycling on a variable time period (typically with frequencies between two hours and two weeks) is a common symptom indicating unsteady state component trapping. Diagnoses of the problem include studying time variation of temperature profiles, taking internal samples, shooting gamma scans, and simulation studies. Solutions comprise raising or lowering the temperature (usually at the expense of recovery), taking side draws, or removing the problem components from the feed. Diagnoses and solutions have been discussed in detail by Kister.[1,2] Here we apply uncertainty analysis to the problem caused by trapping of intermediate components in ACN purification.

Modeling of the distillation separations requires gaining quantitative understanding of the uncertainties in the vapor-liquid equilibrium (VLE) and applying these uncertainties to the problem at hand. While many technologists have highlighted the need to recognize the impact of uncertainties in property models on process design and plant operability[3,4,5,6,7] and several solutions have been suggested and evaluated in the literature,[8,9,10,11,12,13,14] uncertainty analysis is hardly used in chemical-engineering design practice. In fact, Kim et al.[15] observed, "Its practical implementation in a variety of scientific and engineering fields has typically seen less emphasis than it deserves." Mathias[16] recognized that liquid-phase nonideality is the major source of phase-equilibrium uncertainty, and developed an intuitive and easy-to-apply uncertainty-analysis method based upon treating the mixture, for the purpose of activity-coefficient perturbation, as a set of pseudobinaries described by the Margules equation. The Margules perturbation method has been applied to several case studies, including propylene-propane superfractionator, where small changes in relative volatility and product purity have a large effect on design;[16] a dehexanizer column that removes pentane and hexane from a heavier slate of components that are mostly aromatic and the system contains many components (≈ 60); [16] the "textbook problem" of acetone/benzene/chloroform separation, which revealed the relationship between uncertainty analysis and residue-curve analysis;[17] the butanol/water separation, which demonstrated the extension to liquid-liquid equilibrium (LLE) and is a case where the thermodynamic model is inadequate;[18] and ethyl acetate purification, which showed the importance of fitting multicomponent data for complex systems.[19]

The Margules uncertainty methodology used here follows that of the previous studies.[16,17,18,19] The first step is detailed data regression to obtain the best model fit in the region of application (atmospheric pressure, about 101.3 kPa), and quantitative estimation of the correlation uncertainties. This is a necessary first step. The data regression here is mainly applied to the water/ACN binary mixture, with PPN and T-CROTON at low concentration. In the second step, activity coefficients are varied within the process model to relate the design uncertainties to the confidence limits of the correlated phase equilibria.

Margules-Based Activity-Coefficient Perturbation Scheme

The perturbation scheme for the mixture nonideality has been published by Mathias,[16] and hence is only briefly presented here. The approach uses the approximation that the perturbed activity coefficients can be guided by the Margules equation.

$$\ln(\gamma_i) = \ln(\gamma_i^m) + \ln(\gamma_i^p) \tag{1}$$

where γ_i is the activity coefficient of component i after perturbation, γ_i^m is the activity coefficient calculated by the chosen model (the "best" model in the region of interest should be used), and γ_i^p represents the perturbation obtained through the estimated model uncertainty. It should be emphasized that γ_i^m is calculated from the "best" activity-coefficient model, and is not in any way constrained by the pseudo-binary approximation. γ_i^p is given by,

$$\ln(\gamma_i^p) = \delta_i (1-x_i)^2 \left[\frac{|\ln(\gamma_i^m)|}{(1-x_i)^2 + |\ln(\gamma_i^m)|} \right] \tag{2}$$

As previously noted,[16] Equation 2 is phenomenological, but has desired characteristics. The value of $\ln(\gamma_i^p)$ is small when γ_i^m is close to unity (i.e., low departures from ideal behavior), and also is low when the mole fraction of component *i* is close to unity. For components at concentrations close to infinite dilution and with activity coefficients significantly different from unity, the fractional change in γ_i resulting from the perturbation asymptotically approaches $\{exp(\delta_i) - 1\}$. In order to apply the perturbation method, the value of δ_i for each component must be related to the estimated uncertainty in its activity coefficient. This thorough data analysis is a necessary element of the perturbation scheme and is discussed in detail later in the paper.

Correlation of VLE in the Water/ACN/PPN/T-CROTON System

As in the previous uncertainty-analysis case studies,[16,17,18,19] the perturbation scheme has been implemented as a user activity-coefficient model in the NRTL-RK property option in Aspen Plus V8.8. The NRTL-RK property option uses the NRTL activity-coefficient model[20] and the Redlich-Kwong equation of state[21] for the vapor phase. The pure-component properties come from the Aspen Plus V8.8 pure-component database. This study is at a pressure of 1 atm (101.3 kPa), the temperatures of interest are approximately 70-100 °C, and at these temperatures the correlations for the vapor pressures of these four compounds from the Aspen Plus database are expected to be accurate to better than 1 %. The reason for this confidence is that the Aspen Plus pure-component properties come from the DIPPR 801 database,[22] and the correlated vapor pressures have been reported to be accurate to <1%. The Redlich-Kwong equation estimates the vapor-phase fugacity coefficient, but this thermodynamic quantity is expected to be close to unity and accurately predicted since the pressure of interest is low, ≈1 atm. For this application, the phase-equilibrium uncertainty is largely determined by the uncertainties in the component activity coefficients.

The NRTL model has two temperature-dependent binary parameters (α_{ij}^{N} and τ_{ij}). α_{ij}^{N} is a symmetric binary parameter and τ_{ij} is an asymmetric binary parameter,

$$\tau_{ij} = a_{ij} + b_{ij}/T + e_{ij}\ln(T) + f_{ij}T \quad (3)$$

$$\alpha_{ij}^{N} = c_{ij} + d_{ij}(T - 273.15) \quad (4)$$

where, T is the temperature in Kelvin.

The K-value of component i in a mixture (K_i) at vapor-liquid equilibrium (VLE) is defined to be the ratio of the vapor and liquid mole fractions at equilibrium, and the relative volatility between components i and j (α_{ij}) is the ratio of the K-values.

$$K_i \equiv \frac{x_i}{y_i} \quad (5)$$

$$\alpha_{ij} \equiv \frac{K_i}{K_j} \quad (6)$$

Note that the superscript "N" has been added to the NRTL parameter α_{ij}^{N} in order to distinguish it from the relative volatility, α_{ij}.

The Data Regression System (DRS) in Aspen Plus V8.8 has been used to obtain optimum values of the NRTL binary parameters at pressures at about 1 atm (101.3 kPa), which corresponds to the temperature range of ≈ 70-100°C. Phase-equilibrium data were obtained from the NIST-TDE[23] database, and key VLE data have been measured as part of this effort. The water/ACN binary is well-studied systems since a large number of datasets are available, however, the four binaries consisting of PPN and T-CROTON with water and ACN have only a few data. In fact, data for systems with PPN are very poorly covered, and hence this study has mainly focused on the effect of T-CROTON as the representative intermediate component.

Table 1 presents the 14 components considered in the Aspen Plus simulation together with their boiling points. The table also shows the estimated infinite-dilution activity coefficients in water and ACN at a representative temperature of 75°C; these correlations/estimations are discussed later in this paper. The last column of Table 1 designates components other than water and ACN as "light" or "heavy." The "light" components (1-4) have normal boiling points lower than that of ACN. These components will always have K-values greater than unity, and hence will split predominately to the distillate in the distillation column. The "heavy" components (10-14) have normal boiling points greater than that of water, and also are fairly ideal with water (infinite-dilution activity coefficients close to unity). These "heavy" components will always have K-values less than unity, and therefore will largely split to the bottoms stream of the distillation column. The "intermediate" components have normal boiling points higher than that of ACN, and form fairly ideal mixtures with ACN, consequently their K-values are likely to be below unity at the top of the column (in the vicinity of the water/ACN azeotrope). Further, the "intermediate" components have T_b values close to that of water and also water infinite-dilution activity coefficients significantly greater than unity (see Table 1), and therefore their K-values will be greater than unity at the bottom of the column. As a result, the intermediate components have K-values that cross unity and tend to build-up in the column.

Table 1. Components in the acetonitrile purification Aspen Plus model.

Component Name	T_b (°C)	Component ID	γ^{∞} at 75°C in		Impurity Category
			Water	ACN	
1. Acetaldehyde	21.0	ACETALD	9.0	1.1	Light
2. Dichloromethane	39.8	DCM	235.9	1.2	Light
3. n-Propionaldehyde	49.0	PROPALD	32.2	1.3	Light
4. Acrolein	52.7	ACROLEIN	17.1	1.1	Light
5. Acetonitrile	81.6	ACN	11.7	1.0	-
6. cis-Crotonaldehyde	88.7	C-CROTON	32.5	1.4	Intermediate
7. Propionitrile	97.4	PPN	41.6	1.0	Intermediate
8. Water	100.0	WATER	1.0	6.1	-
9. trans-Crotonaldehyde	102.5	T-CROTON	32.5	1.4	Intermediate
10. Acetic Acid	117.9	ACETACID	1.3	1.9	Heavy
11. 2-Methoxyethanol	124.4	MEOXYETH	2.7	2.0	Heavy
12. Morpholine	128.0	MORPHOL	1.1	1.0	Heavy
13. Acetamide	221.2	ACETAMID	1.0	1.0	Heavy
14. 4-Ethylmorpholine	239.4	ACETMOR	1.1	1.0	Heavy

Crotonaldehyde has two isomers, *cis* and *trans*. The study of McGreer and Page[24] found that cis-crotonaldehyde is far less chemically favored than trans-crotonaldehyde, by about 1:50. Hence, although cis-crotonaldehyde is available in the simulation file, its properties are not considered reliable, and crotonaldehyde has been treated as entirely the trans isomer (T-CROTON)

It is clear that the system of interest to this study consists of 4 components: water, ACN, PPN and T-CROTON. For the other binary mixtures, the NRTL parameters were obtained by regressing data from NIST-TDE,[23] or were estimated using the UNIFAC method.[25] It is noted that the K-values of the other components (1-4 and 10-14 in Table 1) have negligible effect on this study.

Table 2. NRTL parameters for the water/ACN/PPN/T-CROTON system. See Eqs. 3 and 4 for a description of the parameters. Parameters e_{12}, e_{21}, f_{12}, f_{21} and d_{12} are always equal to zero.

Component 1	ACN	PPN	WATER	ACN	ACN	PPN
Component 2	WATER	WATER	T-CROTON	PPN	T-CROTON	T-CROTON
a_{12}	-2.34	-4.53	5.38	0	0	0
a_{21}	2.39	2.99	-2.34	0	0	0
b_{12}	1165.5	2041.1	-751.4	27.1	55.1	0
b_{21}	-199.2	-9.9	914.6	-30.0	55.1	0
c_{12}	0.45	0.41	0.3	0.3	0.3	0.3

Table 2 presents the NRTL parameters for the water/ACN/PPN/T-CROTON system. The water/ACN binary is well covered by experimental VLE data. The binaries of water with PPN and T-CROTON are also covered by VLE and LLE data; it is noted that these two binaries have regions of liquid-liquid immiscibility, but the concentrations of PPN and T-CROTON in the process are too low to exceed their solubility limit. No data were found in the literature for the other three binaries. Data for the water/ACN/T-CROTON ternary were measured as part of this project.[26] The NRTL parameters for the ACN/PPN are estimated based upon UNIFAC predictions.[25] Generally, UNIFAC predictions are not accurate, so future measurement was considered pending outcome of this study. The NRTL parameters for the PPN/T-CROTON binary have been defaulted to zero. UNIFAC estimations[25] indicate that this is a fairly ideal system (infinite-dilution activity coefficients ≤ 1.2). Minor components have a negligible effect on overall mixture nonideality, and it was decided that the assumption of ideality is the preferred choice.

Key comparisons of model results to experimental data are provided next in order to estimate the VLE uncertainties. Figure 1 compares the calculated relative volatility in the water/ACN binary to data from 9 sources available in NIST-TDE.[23] The model describes the data, including the azeotropic region, accurately, and there is no indication of model bias. The dashed lines show $\pm$ 20% from the model calculation. It is noted that that the water/ACN binary is not a significant cause of design uncertainty, and no uncertainty perturbation has been applied to the water/ACN binary. Figure 2 presents comparisons between model and data[26] for the K-value of T-CROTON at low concentration (effectively infinite dilution) in the water/ACN binary mixture at 1 atm (101.3 kPa). The dashed lines indicate $\pm$ 30% from the model results. Most of the data points fall within the dashed lines, and this may be considered to be the combined uncertainty (about 3 standard deviations) of the model K-values for T-CROTON. The vertical line in Figure 2 at 83 wt% shows the ACN/water azeotrope, which corresponds to the distillate composition from the column, and it is clear that both the model and experimental data indicate that the K-value of T-CROTON drops below unity before reaching the top of the column, which, of course, is the reason that T-CROTON is an intermediate component. Figure 3, which shows the K-value of PPN, is analogous to Figure 2. While the calculated results are equivalent for PPN and T-CROTON, there are no experimental data available for PPN, and hence the perturbation studies here are limited to T-CROTON, with the expectation that they will be approximately applicable to PPN.

Figure 4 provides a visual indication of the effect of $\delta_{T\text{-}CROTON}$ (hereafter simply referred to as δ, since the Margules perturbation in this work is only applied to T-CROTON). The bottoms concentration is nearly-pure water and the distillate concentration corresponds to the water/ACN azeotrope. Referring to Eq. 2 and Figure 2 it is clear that for a given value of δ, the perturbation at the bottom of the column (higher T-CROTON activity coefficient) will be higher than at the top of the column, and that a value of $\delta = 0.4$ is a "full" perturbation, corresponding effectively to the 30% uncertainty in Figure 2. Note that only positive values of δ are relevant to the study because increased values of δ will tend to increase the concentration of T-CROTON in the distillate, which makes it more difficult to meet the design goal of low T-CROTON concentration in the distillate.

Two design studies have been conducted to evaluate the effect of T-CROTON VLE uncertainty on the performance of the single column, and these are described next.

Design Study 1 – Single Column with Side Draw

Figure 5 presents the single column with side draw used for the first set of design studies. The feed flow is 770 kg/hr and is an aqueous mixture containing 12 wt% ACN, and with minor impurities. The impurity levels of PPN and T-CROTON in the feed are 100 ppmw and 20 ppmw, respectively. The column has 35 theoretical stages including a reboiler and total condenser. The distillate-to-feed ratio is set at 0.13, which corresponds to $\approx$ 91% ACN recovery at the azeotropic composition. The reflux ratio is varied to obtain a T-CROTON concentration in the distillate of 1 ppmw. The feed stage is varied from 12 to 24. The base case is the feed to stage 20, and for this case the calculated reflux ratio is 7.0. Figure 6 shows the vapor and liquid concentration profiles (weight fraction) for ACN, PPN and T-CROTON in the distillation column for the base case. The concentration of ACN rises with stages toward the condenser (here in decreasing stage number) and quickly reaches the water/ACN azeotropic concentration just a few stages above the feed stage. The concentration profiles of PPN and T-CROTON show the characteristics of intermediate components since they both have concentration peaks below the feed stage. Note that the concentration peak of T-CROTON is about three orders of magnitude above the distillate and bottoms concentrations. The distillate and bottoms concentrations of both PPN and T-CROTON are both small since most of these two components are removed in the side draw. Figure 7 emphasizes these points by showing the K-value profile of T-CROTON alongside its concentration profile. Note that the relative values of the vapor and liquid concentrations switch as the K-value moves from less than one at the top of the column (lower stage numbers) to greater than one at the bottom of the column.

Figure 8 studies how the reflux ratio needs to be adjusted to attain the spec of 1 ppmw T-CROTON in the distillate as the perturbation parameter (δ) and the number of rectification stages are varied. Increasing the number of rectification stages is helpful in reducing the required reflux ratio, and this can inferred from Figure 6 (and seen in Figure 8) because the concentrations of PPN and T-CROTON decrease monotonically in the rectification section of the column, but the effect is relatively weak. On the other hand, increasing δ (increased activity coefficient and therefore K-value of T-CROTON) results in a strong increase in the required reflux ratio. Very large reflux ratios cause problems, for example requiring large-diameter columns and significant increases in reboiler duty, and are usually not accepted in industrial process-design practice

Figure 9 studies the effect of the draw rate for the base case (Figure 5 with reflux ratio of 7.0 and feed to stage 20). It is clear that varying the draw rate has a strong effect on the concentration of T-CROTON in the bottoms (because it changes the total amount of T-CROTON removed in the side draw), but it has only a weak effect on the concentration of T-CROTON in the distillate. Since the main goal here is to reduce the concentration of PPN and T-CROTON in the distillate, the next study is done without the side draw.

Design Study 2 – Single Column with no Side Draw

Figure 10 presents the single column with no side draw used for the second set of design studies, and with the total number of stages increased from 35 to 40. Other design conditions are the same as in Figure 5. Figure 11 is qualitatively the same as Figure 8, which supports the conclusion that the side draw affects the bottoms concentration of T-CROTON, but has only a weak effect on the key result we desire, which is minimizing the concentration of T-CROTON in the distillate.

Figure 12 presents the results of a study where the reflux ratio has been fixed at 10.0, and the perturbation parameter (δ) and the number of rectification stages are varied. Similar to Figure 8, Figure 12 shows that it is beneficial to increase the number of rectification stages as this reduces the concentration of T-CROTON in the distillate. However, the key result is that the concentration of T-CROTON in the distillate is a very strong function of the uncertainty in the VLE. As δ increases not only does the concentration of T-CROTON in the distillate increase beyond the target value (1 ppmw), but it can also exceed the feed concentration (20 ppmw), in effect concentrating undesired impurities in the product!

The single-column solution was rejected as being too risky.

Conclusions

The purpose of this study was to apply the Margules uncertainty method to an industrially important component-trapping case, i.e., purification of acetonitrile used as a solvent in the pharmaceutical industry. Small amounts of T-CROTON and PPN turn a simple distillation into a very challenging problem. The study first focused on the VLE correlation, and determined that the key uncertainty is the K-values of the intermediate components, PPN and T-CROTON. Since K-value data are only available for T-CROTON, this component was chosen as the representative intermediate key component. The first set of studies indicated that the target of 1 ppmw T-CROTON in the distillate can be met, at least in the simulation model, but at a significant increase in reflux ratio if the T-Croton K-value is increased within the established uncertainty. The second set of studies concluded that increasing the K-value of T-CROTON within the established uncertainty caused a significant increase in the T-CROTON concentration in the distillate, and the product concentration could even exceed the feed concentration. The three-column solution was the eventual low-risk process solution selected.

List of Tables

List of Figures

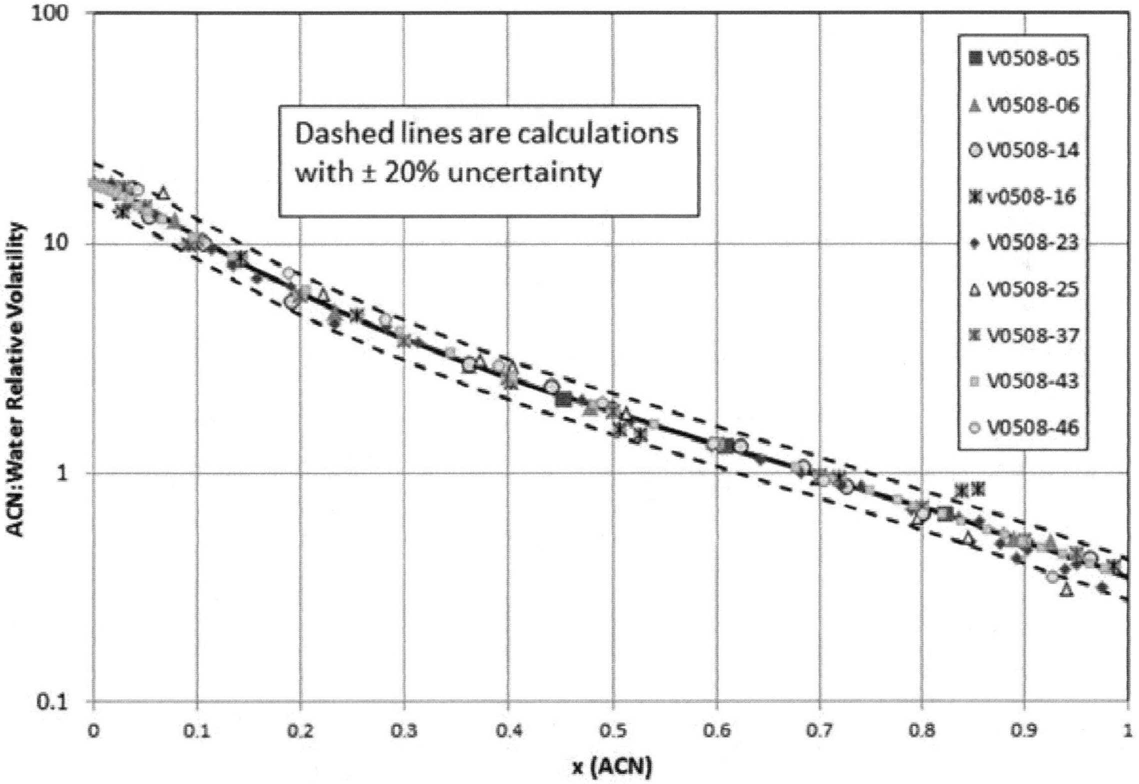

Figure 1. Water/ACN relative volatilities at 1 atm (101.3 kPa). Comparison of model calculations to 9 data sources from NIST-TDE.[23]

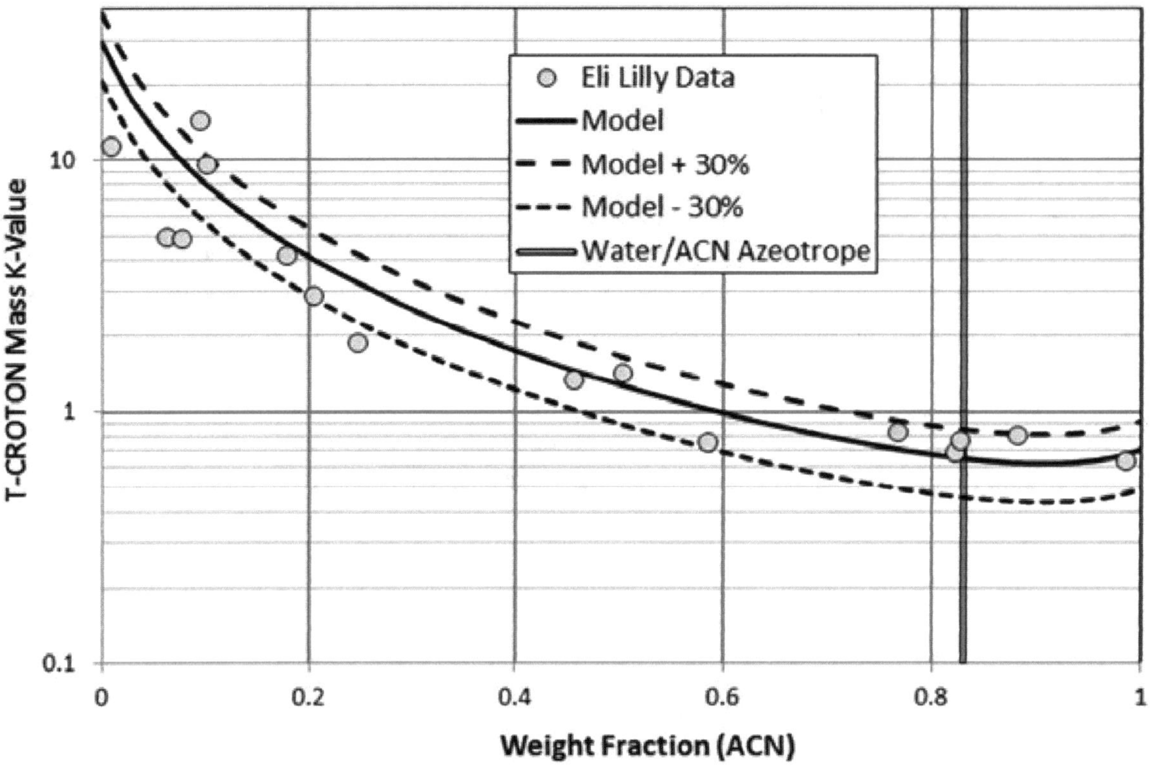

Figure 2. K-Value of T-CROTON at low concentration in water/ACN at 1 atm (101.3 kPa). Comparison of model calculations to Eli Lilly data.[26] The dashed lines show ± 30% from the model results. The vertical line indicates the water/ACN azeotrope.

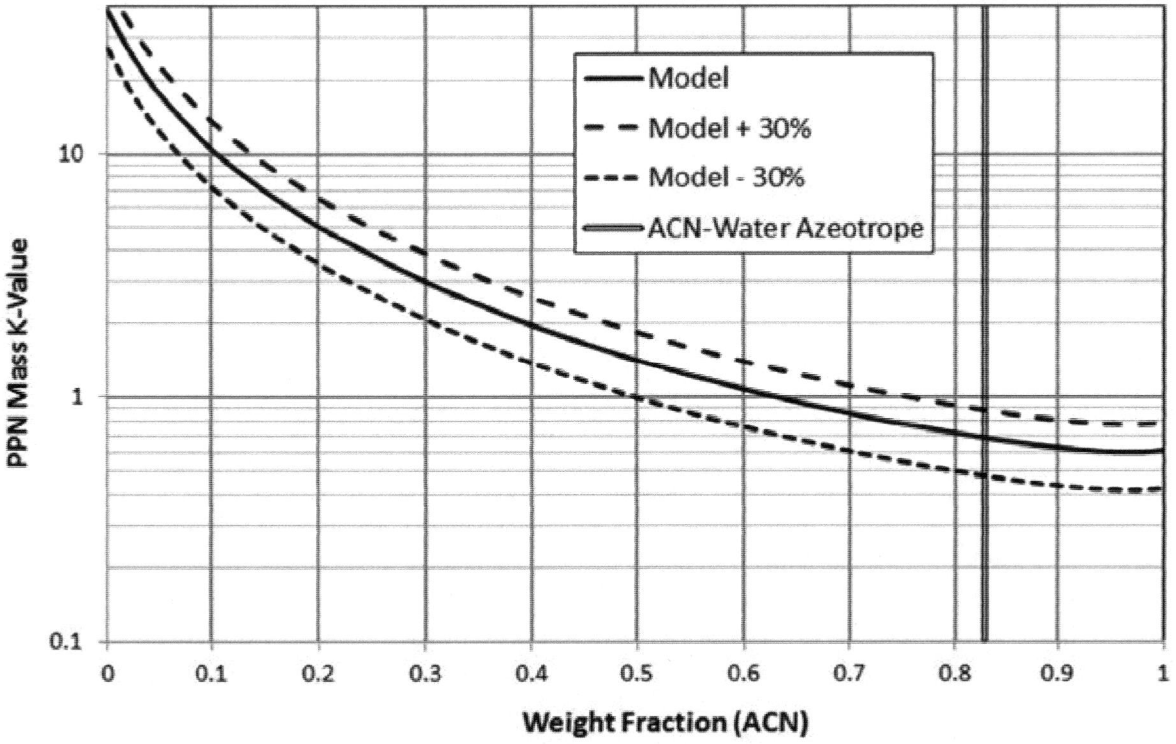

Figure 3. K-Value of PPN at low concentration in water/ACN at 1 atm (101.3 kPa). No experimental data are available. The dashed lines show ± 30% from the model results. The vertical line indicates the water/ACN azeotrope.

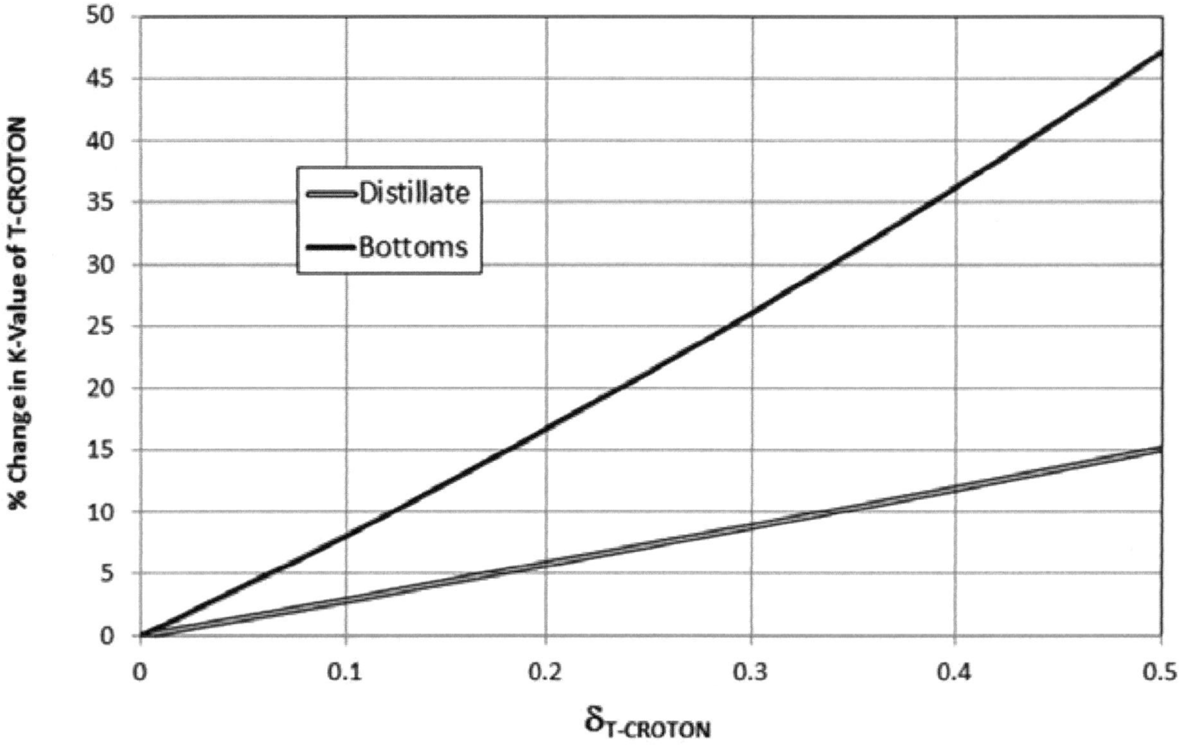

Figure 4. Effect of T-CROTON perturbation parameter (δ) on the percentage change in the K-value of T-CROTON. The bottoms concentration is nearly-pure water and the distillate concentration corresponds to the water/ACN azeotrope.

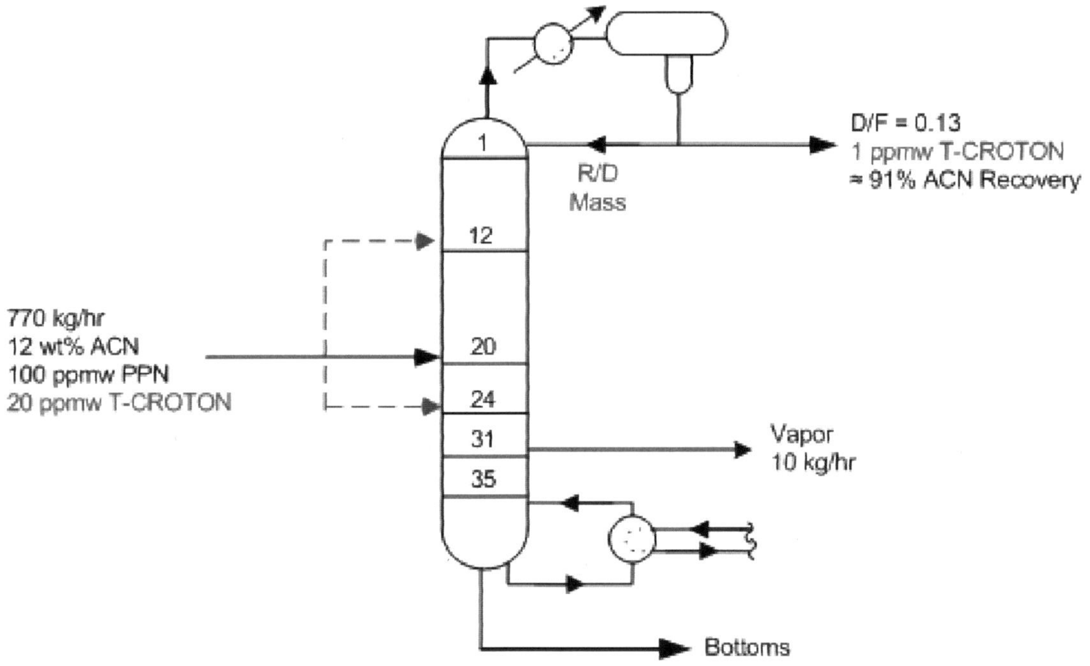

Figure 5. Single column with side draw used for perturbation studies-1. The column operates at 1 atm (101.3 kPa).

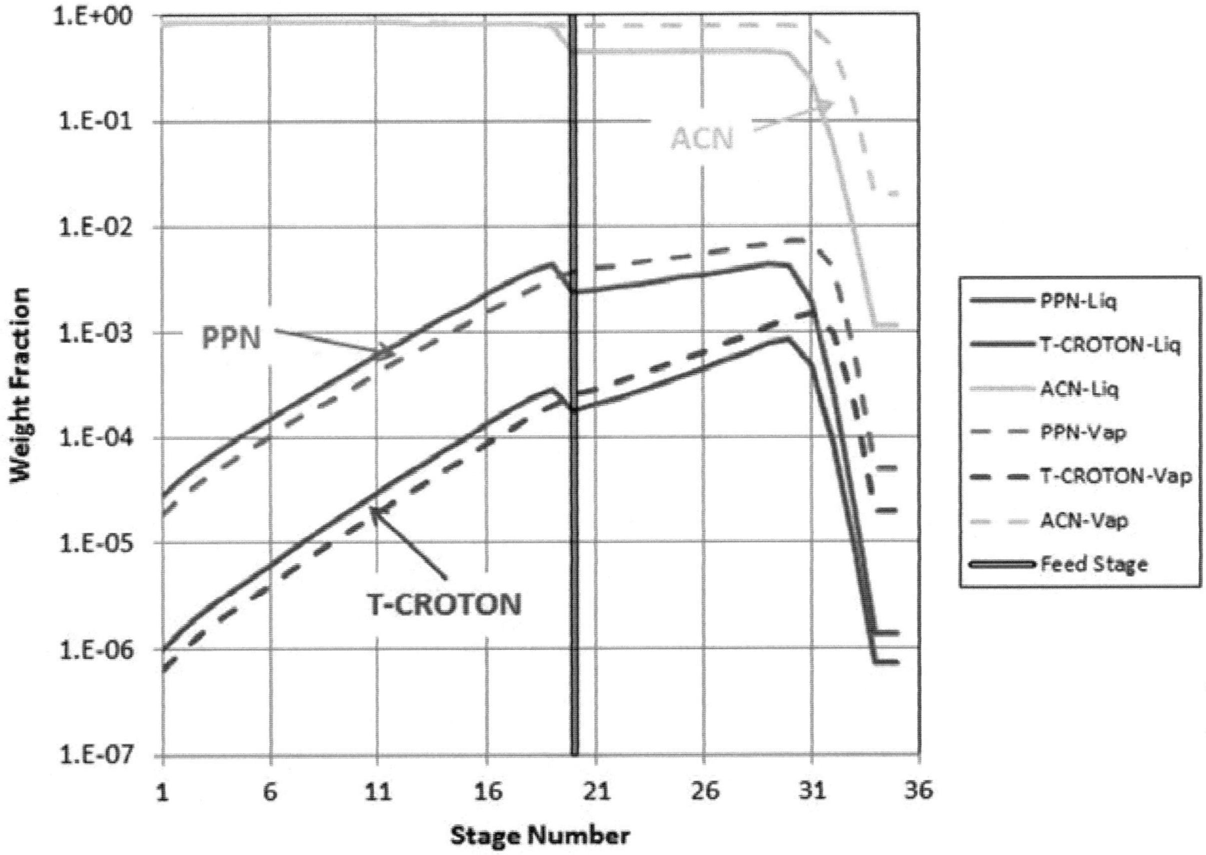

Figure 6. Vapor and liquid concentration profiles for ACN, PPN and T-CROTON in distillation column corresponding to Figure 5, and with feed to stage 20.

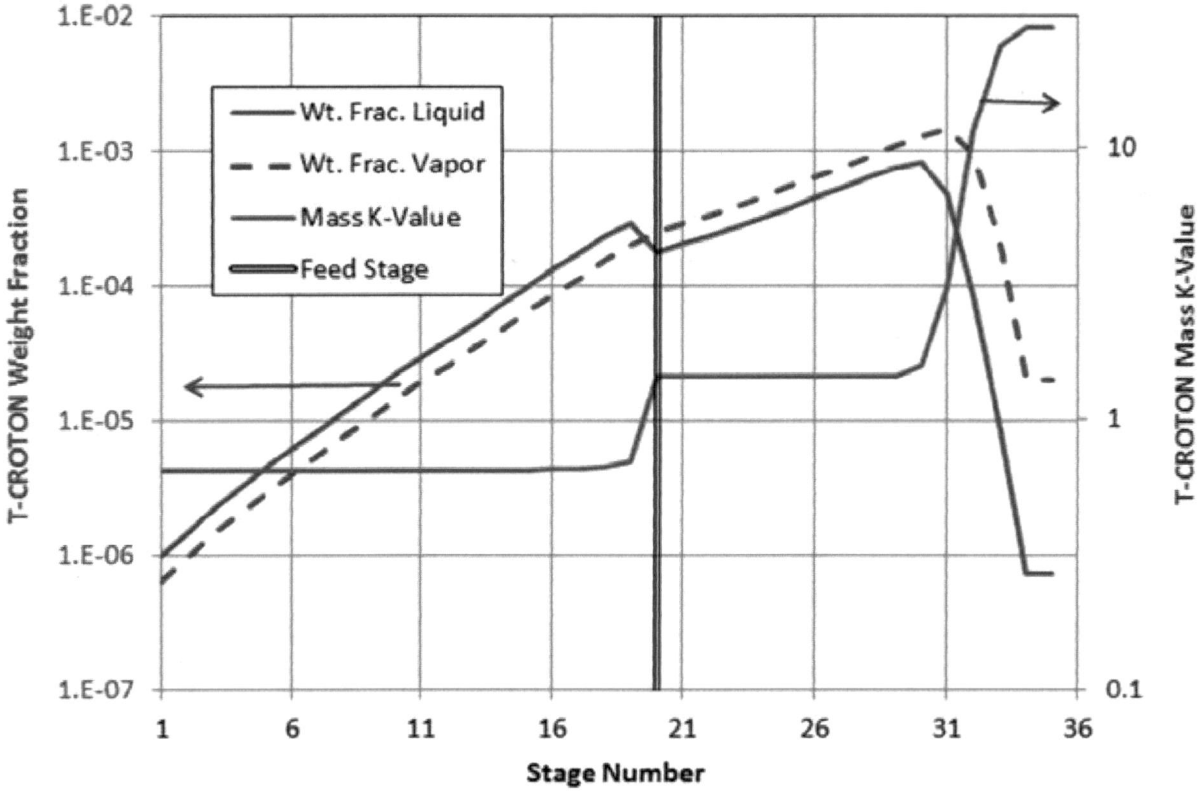

Figure 7. Mass-fraction and K-value profiles of T-CROTON in the base case. The specifications are the same as those in Figure 6.

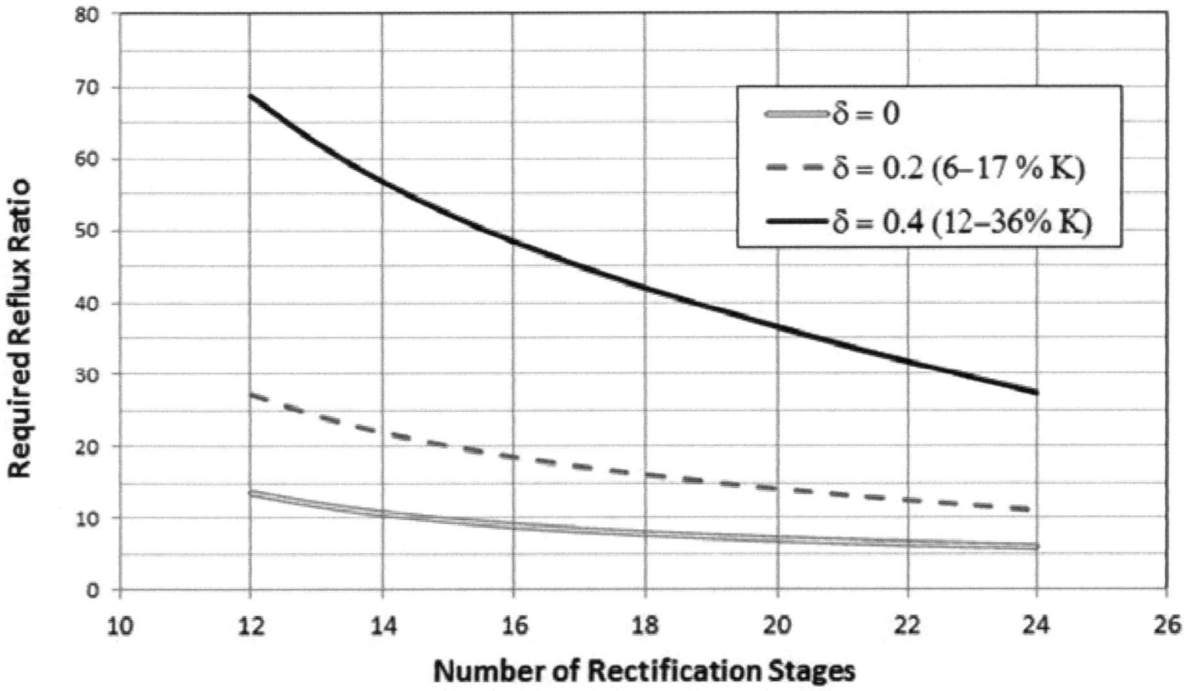

Figure 8. Effect of perturbation parameter (δ) and number of rectification stages on the reflux ratio required to meet the spec of 1 ppmw T-CROTON in the distillate. See Figure 5 for details.

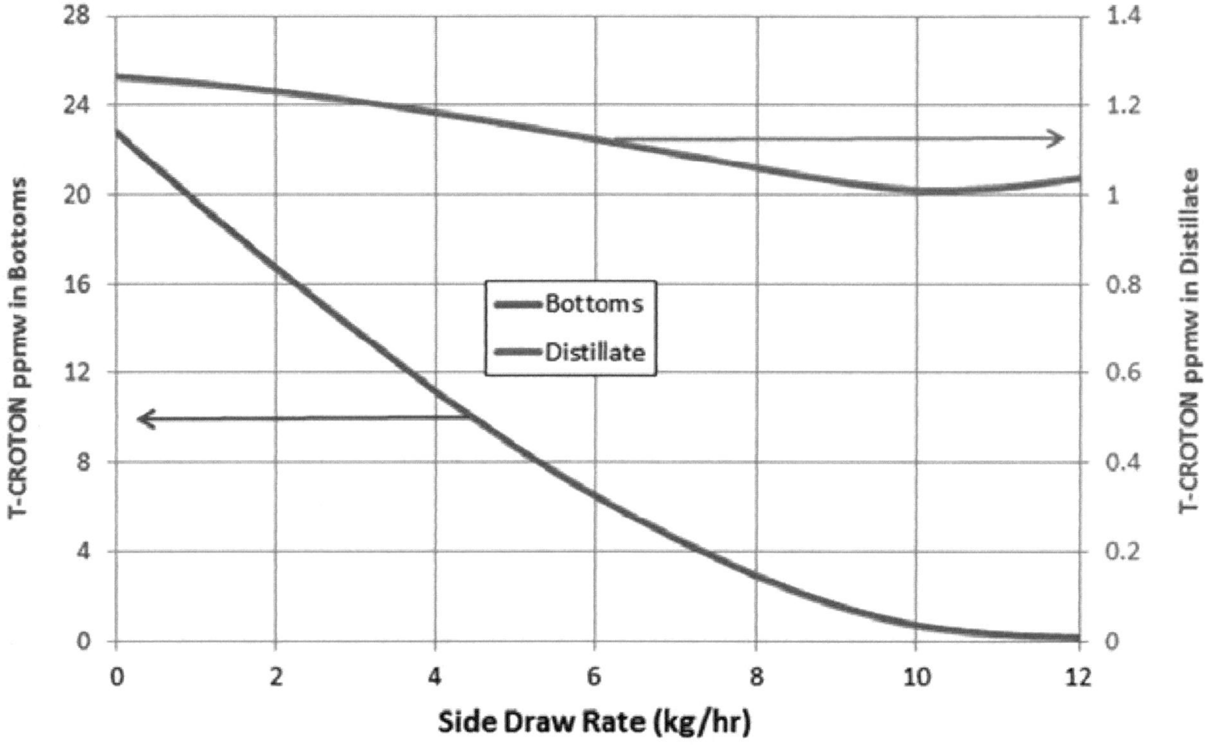

Figure 9. Effect of modifying side-draw rate. The feed is at stage 20 and the reflux ratio is fixed at 7.0.

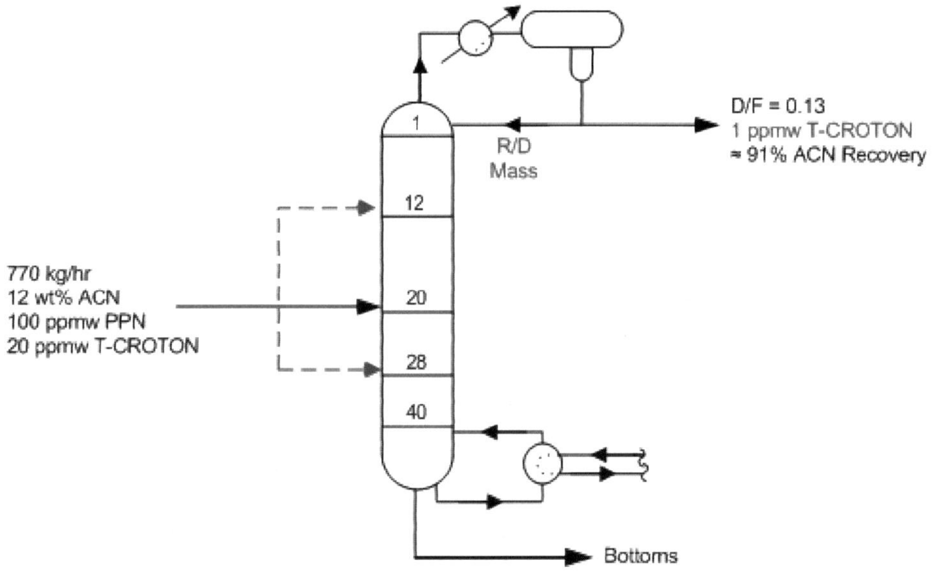

Figure 10. Single column with perturbation studies-2. The difference from Figure 5 is that the side draw has been eliminated.

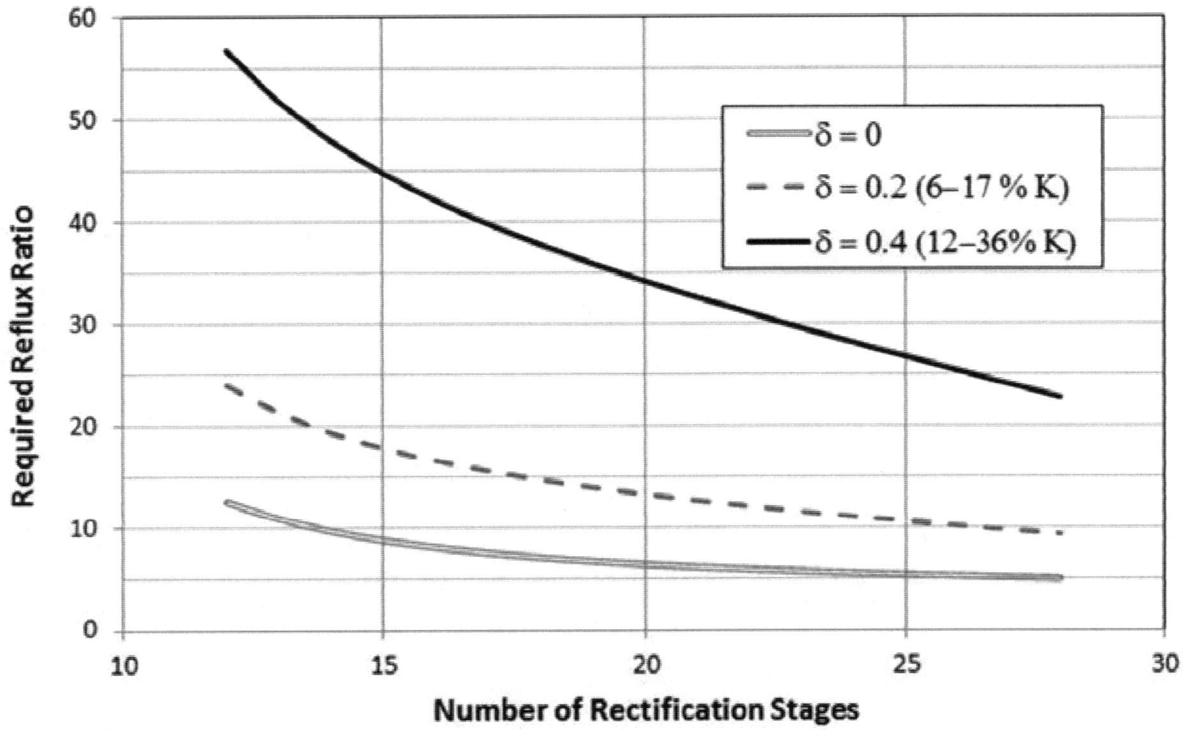

Figure 11. Repeat of study shown in Figure 8 with no side draw and total number of stages increased from 35 to 40 (See Figure 10).

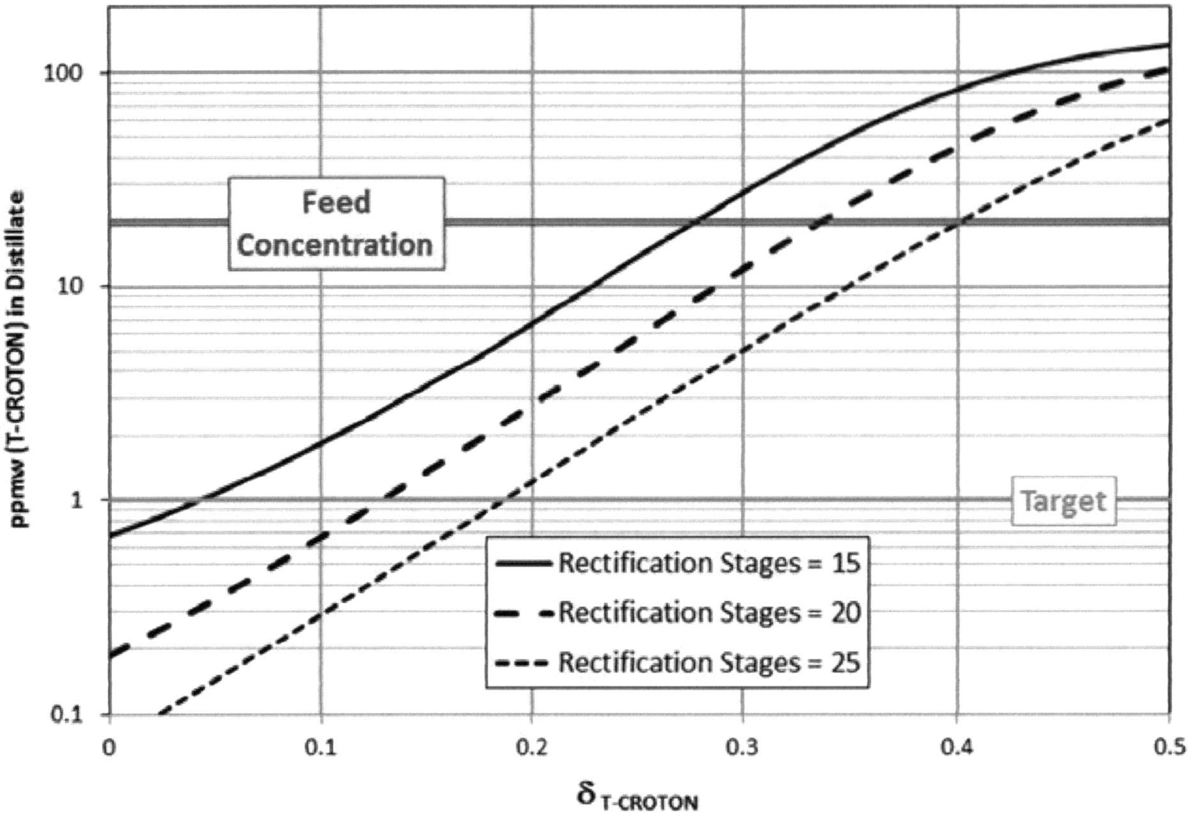

Figure 12. Effect of perturbation parameter (δ) and number of rectification stages on concentration of T-CROTON in distillate. Referring to Figure 10, the reflux ratio is fixed at 10.

Literature Cited

[1] Kister, H. Z. Components Trapping in Distillation Towers: Causes, Symptoms and Cures," *Chem. Eng. Prog.* **2004**, 100, 22-33.

[2] Kister, H. Z. *Distillation Troubleshooting*. John Wiley and Sons, Inc., Hoboken, NJ, **2006**.

[3] Streich, M.; Kistenmacher, H. Property Inaccuracies Influence Low Temperature Designs, *Hydrocarb. Proc.*, **1979**, 58, 237-241.

[4] Mah, R. S. H. Effects of Thermophysical Property Estimation on Process Design, *Computers Chem. Eng.*, **1977**, 1, 183-189.

[5] Zudkevitch, D. Imprecise Data Impacts Plant Design and Operation, *Hydrocarbon Process*. **1975**, 54, 97–103.

[6] Zudkevitch, D.; Gray, R. D. Impact of Fluid Properties on the Design of Equipment for Handling LNG, *Adv. Cryog. Eng.* **1975**, 20, 103–123.

[7] Larsen, A. H. Data Quality for Process Design, *Fluid Phase Equilib.*, **1986**, 29, 47–58.

[8] Zeck, S. Thermodynamics in Process Development in the Chemical Industry - Importance, Benefits, Current State and Future Development, *Fluid Phase Equilib.*, **1991**, 70, 125–140

[9] Macchietto, S.; Maduabeuke, G,; Szcepanski, R. Exact Determination of Process Sensitivity to Physical Properties, Fluid Phase Equilib., **1986**, 29, 59-67.

[10] Reed, M. E.; Whiting, W. B., Sensitivity and Uncertainty of Process Designs to Thermodynamic Model Parameters: A Monte Carlo Approach, *Chem. Eng. Commun*. **1993**, 124, 39–48.

[11] Whiting, W. B. Effects of Uncertainties in Thermodynamic Data and Models on Process Calculations, J. *Chem. Eng. Data*, **1996**, 41, 935–941.

[12] Whiting, W. B.; Vasquez, V. R.; Meerschaert, M. M. Techniques for Assessing the Effects of Uncertainties in Thermodynamic Models and Data, *Fluid Phase Equilib.* **1999**, 158–160, 627–641.

[13] Xin, Y.; Whiting, W. B. Case Studies of Computer-Aided Design Sensitivity to Thermodynamic Data and Models, *Ind. Eng. Chem. Res*. **2000**, 39, 2998–3006.

[14] Whiting, W. B.; Vasquez, V. R.; Meerschaert, M. M. Techniques for Assessing the Effects of Uncertainties in Thermodynamic Model and Data, *AIChE J.*, **1997**, 43, 440–447.

[15] Kim, S. H.; Kang, J. W.; Kroenlein, K.; Magee, J. W.; Diky, V.; Frenkel, M. Online Resources in Chemical Engineering Education: Impact of the Uncertainty Concept from Thermophysical Properties, *Chem. Eng. Educat.*, **2013**, 47, 48-57.

[16] Mathias, P. M., Sensitivity of Process Design to Phase Equilibrium – A New Perturbation Method Based Upon the Margules Equation. *J. Chem. Eng. Data*, **2014**, 59, 1006-1015.

[17] Mathias, P. M. Effect of VLE Uncertainties on the Design of Separation Sequences by Distillation – Study of the Benzene-Chloroform-Acetone System. *Fluid Phase Equilib.*, **2016**, 408, 265-272.

[18] Mathias, P. M. Effect of Phase-Equilibrium Uncertainties on the Separation of Heterogeneous Azeotropes – Application to the Water/1-Butanol System. *J. Chem. Eng. Data*, **2016**, 61, 4077-4084.

[19] Mathias, P. M.; Kister, H. Z. Effect of Phase-Equilibrium Uncertainties on Ethyl Acetate Purification, submitted for publication to *J. Chem. Eng. Data*, February **2017**.

[20] Renon, H.; Prausnitz, J. M. Local Compositions in Thermodynamic Excess Functions for Liquid Mixtures. *AIChE J.*, **1968**, 14, 135-144.

[21] Redlich, O.; Kwong, J. N. S. On the Thermodynamics of Solutions. V An Equation of State. Fugacities of Gaseous Solutions. *Chem. Rev.*, **1949**, 44, 233-244.

[22] Rowley, R. L.; Wilding, W. V.; Oscarson, J. L.; Yang, Y.; Zundel, N. A. DIPPR® *Data Compilation of Pure Chemical Properties, Design Institute for Physical Properties*, AIChE, New York, NY, **2010**.

[23] Frenkel, M.; Chirico, R. D.; Diky, V.; Yan, X.; Dong, Q.; Muzny, C. ThermoData Engine (TDE): Software Implementation of the Dynamic Data Evaluation Concept. *J. Chem. Inf. Model.*, **2005**, 45, 816-838.

[24] McGreer, D.E.; Page, B. D. cis-Crotonaldehyde and Related Compounds. *Can. J. Chem.*, **1969**, 47, 866-867.

[25] UNIFAC is an estimation method based upon group contributions. Several UNIFAC models are available in Aspen Plus V8.8. In the present study, since mainly approximate estimations are required, the base UNIFAC model has been used. Fredenslund, Aa.; Gmehling, J.; and Rasmussen, P. *Vapor-Liquid Equilibria Using UNIFAC, A group-Contribution Method*, 3rd Impression, Elsevier, Amsterdam, **1977**.

[26] Parker, B. Data measured in Eli Lilly Corporate Center Laboratory, August, **2016**.

The Successful Application of a Partition Flow Spray Tray (PFST®) in C4 Extractive Distillation

Zhuxiang Wang, Enxia Shang, Wei Hong, Tianjin ChuangJu Technology Co. Ltd, Tianjin, P. R. China

Yan Zhuang, Advanced Column Technology, Perth, Australia

The crude C4 hydrocarbons production, as a by-product of the ethylene industry, exceeds 6 million tonnes per annum in China. As the first step in crude C4 upgrade, Butadiene is recovered from the crude C4 using extractive distillation. The extractive distillation columns used to extract butadiene from the crude C4 presently have issues of high pressure drop, lower than design throughput, poor separation and high steam consumption due to its high foaming tendency of the solvent and a high liquid to vapour load ratio[1][2].

This paper presents three industrial case studies using ChuangJu's proprietary partition flow spray tray (PFST®) design to solve the above mentioned problems. The PFST® is designed with a partition plate to catch the liquid that has been sprayed out of the vertical sieve caps after mixing with the vapor. The liquid is then channeled down to the next tray by a multiple down comer and suspended settling trough design to handle the high liquid load, thus eliminates the problem of liquid back mixing on the tray. The settling troughs provide extra residence time for liquid and vapour disengagement. The spray action of the vertical sieve cap suppress foaming, thus reducing the tendency of vapour underflow back mixing and flooding caused by foaming. With this design, the vapor and liquid phases come to contact counter-currently with the maximum mass transfer driving force, everywhere on the tray with little liquid flow gradient or concentration gradient, so the tray efficiency is increased.

In the first case study, PFST® trays were used to revamp two butadiene extractive distillation columns and one de-gassing column for the Ningbo Haode Chemicals. Before the revamp, float valve trays were used and the maximum throughput achieved was only 70% of the design value. After the revamp, the feed rate is running at 140% of the design value with stable operation.

The other two C4 separation plants, which were designed, fabricated, installed and commissioned by ChuangJu, are running at above design capacities with good separation and recovery performance.

Keywords: partition flow spray tray, C4 extractive distillation, butadiene, foaming system, PFST®

INTRODUCTION

ChuangJu's R&D on tray column design focuses on solving the problems of high pressure drop, vapour phase uneven distribution caused by the hydraulic gradient on the tray, liquid back mixing, as well as flooding caused by foaming etc. Within the series ChuangJu spray tray designs, the PFST® tray is especially suitable for the C4 extractive distillation in that it can minimise liquid phase back-mixing under a high liquid load and it can also supress foam formation[3]

Figure 1: A picture of a PFST® tray

THE DEVELOPMENT OF THE PARTITION FLOW SPRAY TRAY

Conventional trays, such as a F1 float valve tray, can be struggling to handle a system with a high liquid load, susceptible to foaming and components with very close relative volatilities. The reasons are:-

- The liquid and vapour are contacted in a cross-flow pattern. Due to the concentration gradient of the liquid flow, the mass transfer driven force is reduced along the liquid flow path, which leads to a lower tray efficiency.
- High liquid load leads to higher hydraulic gradient which can cause uneven vapour flow distribution.
- For a foaming susceptible system, the conventional tray design does not have a foaming suppression mechanism, other than adding foaming suppression agents, so it is easy to run into a flooding.
- Some of the conventional tray designs have "mass transfer dead zones" near the column wall where there exist stagnant flow, side flow and circular flow, which can lower the tray efficiency.

The PFST® tray design uses multiple down comers and spray contactors to handle a high liquid load and foaming issue. By using a partition plate to separate the portion of liquid that has been contacted with the vapor, from the liquid coming directly from the tray above, thus reduces liquid phase back mixing, which increases mass and energy transfer efficiency.

THE MAIN CHARACTERISTICS OF THE HIGH PERFORMANCE PFST® TRAY

Multiple Down Comer Design

For a high liquid load situation, multiple down comer design can shorten the length of liquid flow path, thus reduce the hydraulic gradient on the tray, which make the vapor rise more evenly across the tray. It also can effectively control the liquid layer height on the tray within an optimized range.

Spray Type Contactors

The spray type contactor offers higher mass and energy transfer efficiency by dispersing the liquid phase into a large amount of small droplets, which creates more interfacial contact areas.

Foaming Suppression

The spray type tray has a built-in foaming suppression mechanism which makes it a preferable choice for a foaming susceptible system. The droplets sprayed out of the spray contactors falling back to the tray floor which can break some of the bubbles.

Reduced Liquid Back Mixing

By using a partition plate to catch the liquid that spray out of the spray contactors, ie to separate the portion of liquid that has been in contact with the vapor, from the portion coming directly from the tray above, there is much less chance for liquid phase back mixing. There is also much reduced hydraulic gradient, concentration gradient and temperature gradient on the tray. As a result, each of the vertical spray contactors has the same point efficiency achieved by a uniform and high mass transfer driving force across the tray.

Lower Pressure Drop

In a spray type tray, the vapor flow does not have to pass through the liquid layer on top of the tray. The vapor phase pressure drop only need to counteract the kinetic energy used to lift the portion of liquid into the spray contactors, which leads to lower pressure drop compared with a conventional tray operating in the froth operating regime. In a low vapor load situation, the pressure drop of the spray tray is comparable to that of a F1 float valve tray. In a high vapor load situation the pressure drop of the spray tray is 20 to 30% lower than that in a F1 float valve tray. The higher the vapor load, the higher the pressure drop reduction by using a spray tray.

Higher Capacity

In a spray type tray, because of the structure of the spray contactor, the liquid-vapor mixture sprays out of the cap either horizontally or downwardly, as per the figure depict. This increases the traveling path and time for the disengagement of the phases, thus reducing liquid entrainment. With the same tray open areas, the allowable vapor velocity can be 1.5 to 2.0 time higher than that in a conventional tray, so the capacity can be much higher.

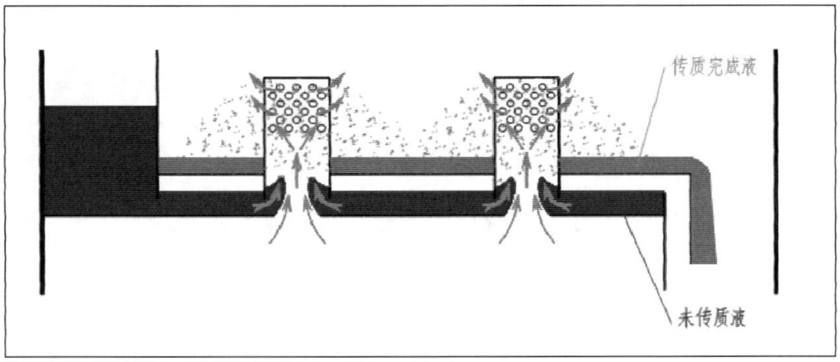

Figure 2: How does a Partition Flow Spray Tray Work

CASE STUDY 1: NINGBO HAODE CHEMICAL INDUSTRY LTD.

附图六 宁波昊德 C4 分离单元各塔

The Ningbo Haode owns and operates a facility that produce 47,000t/a of isobutene. The facility consists of a C4 separation, n-butene isomerisation unit, a MTBE synthesizing unit and a MTBE cracking to isobutene unit. The capacity of the C4 separation unit is 120,000t/a, using an extraction solvent of MEK and NFM[4]. This system uses a high liquid to vapour load; is susceptible to foaming; and is difficult to fractionate due to similar relative volatilities. The extractive distillation towers A&B and the degassing tower are the key pieces of equipment, as seen in the Figure above.

The original design of the columns used a fixed valve tray. After the start up in March 2016, the unit could only run at 70% of the design capacity. ChuangJu was invited to investigate and solve the problem. After site visit with ChuangJu's technical personnel, they identified the key the issues causing the low throughput, including: the liquid flow cross section area is too small; the tray separation efficiency is low, and there is severe foaming tendency which make columns easily flooded.

Based on these findings, ChuangJu selected their PFST® tray to revamp the two extractive distillation columns and the solvent degassing column. The revamped columns were started-up successfully in June 2016. The achieved throughput is increased by 70% compared with that before the revamp, and has been running at 30% above its design capacity since then and there is still upward potential for higher throughput.

It can be seen from Table 1, both the steam consumption and the solvent loss due to entrainment are reduced. The sheducled maintenance time was alse reduced by three times a year. The facility can consistently produce on-spec product throughout since the completion of the revamp. The combined economic savings equals RMB13.7 million, while the revamp cost was only RMB 2 million, which means the pay-back time is less than two month.

Table 1: Ningbo Haode C4 Processing Column Revamping Results

Parameters		**Before Revamp**	**After Revamp**
Type of tray		Fixed Valves	PFST®
Feed rate, kg/h	Distillation Columns A/B	12,560	21,600
	De-gassing Column	124,660	226,930
Temperature℃	Distillation Columns A/B	142.8	140.3
	De-gassing Column	156.8	154.4
Steam Consumption kg/t feed	Distillation Columns A/B	520	295
	De-gassing Column	53.8	37.5
Solvent entrainment, ppm	Distillation Columns A/B	1,000	0
	De-gassing Column	1,000	0
Product Butane Content %		96%	96%
Total olefin content%		90%	90%
Scheduled maintenance times per year		4	1

CASE STUDY TWO:- 100,000 T/A RAW C4 DOWNSTREAM PROCESSING , LUOYANG PETROCHEMICAL HONGDA INDUSTRY LTD.

Figure 4: Raw C4 Process at Louyang Petrochemical Hongda Industry Ltd

In September 2013, the Louyang Petrochemical Hongda Industry Ltd started its 100,000t/a Raw C4 Downstream Processing Facility, which includes an extractive distillation, a de-gassing and a solvent regeneration column. ChuangJu designed and manufactured all the three columns. The extractive distillation column and the de-gassing column used ChuanhJu's PFST® tray. The unit reached the design capacity and met or exceeded all products specification within 65 hours of start-up with steady operation.

Table 2: Louyang Petrochemical Hongda C4 Facility Design vs Achieved Results

	Design	Achieved Results
Capacity, t/a	100,000	100,000
Extractive Distillation Column Temperature 0C	132	118.8
De-gassing Column Temperature 0C	172.9	165
Steam saving, t/d		48
Iso-butane product Content, %	≥96	98
Olefin Content, %	≥60%	66

CASE STUDY THREE:- 300,000 T/ARAW C4 DOWNSTREAM PROCESSING UNIT, HEBEI HAITE WEIYE PETROCHEMICAL LTD.

Figure 5: 300,000t/a Raw C4 Downstream Processing Facility at Hebei Haite Weiye

In October 2014, the Hebei Haite Weiye Petrochemical Ltd started its 300,000t/a Raw C4 Downstream Processing Facility, which includes an extractive distillation, a de-gassing and a solvent regeneration columns. ChuangJu supplied PFST® trays for the extractive distillation column and the de-gassing column. The unit reached 150% of the design capacity and the tray pressure drop is averaged at 550Pa. After a successful start-up, the product quality and energy consumption all met or exceeded design specification.

CONCLUSION

The PFST® high efficiency spray tray technology has been successfully commercialised in C4 separation facilities such as the design and supply of the three distillation columns for Luoyang Petrochemical Hongda, the revamping of the extractive distillation and de-gassing columns for Ningbo Haode, as well as several other C4 downstream treatment applications.

The successful industrial application proved that the PFST® tray can handle large diameter columns with difficult process condition such as high liquid to vapour ratio and foaming, which opens further possibility for similar industrial applications.

The following points can be drawn from the two case studies:

1. By using a spray contactor the mass transfer is intensified by operating in a stable and controlled spray regime, without the risk of running into jet-flooding or high liquid entrainment.
2. By separating the fresh liquid from the tray above and the liquid coming out of the spray contactors, using a partition plate, the liquid phase back mixing is eliminated, which ensures the maximum driving forces from counter-current contact of liquid and vapour is realised.
3. The multiple down comer trough design gives more time for liquid/vapour disengagement thus reduces vapour recycle. This is especially important for a foaming susceptible system.
4. The multiple down comer design also shortens the liquid flow path, thus reduces the hydraulic gradient, which leads to an even vapour flow distribution. This design also eliminates stagnant flow, short circuiting flow as well as circular flow on the tray.

References:

[1] Zhang Long. “The Principles and Technologies for Downstream Processing of the C4 and C5 Fractionates” , The Chemical Industry Press, Beijing, China, 2011

[2] Diego B，Candido D A et al. Process for separation of butadiene by pluralatage extractive distillation: US，412845. 1978.

[3] Wang Zhuxiang, et al, “A Partition Flow No Back Mixing Spray Tray”, Chinese Patent：CN 102961888 A，2013-03-13.

[4] Wang Wenhua, Ren Wanzhong, The Research and Development of Extractive Distillation on the Separation of alkanes and alkenes in Raw C4 Mixture, “The Journal of Technology and Development of Chemical Industry”, 1999(2):20-23.

480648 Research and Development about a New Catalytic Distillation Structured Packing

Abstract:

Research and development about a new catalytic distillation structured packing

Our team developed a new type of modular catalytic structured packing Winpak-C. Systematic studies like theoretical simulations and experimental validations were carried out to guide the industry production, which has successfully realized the industrial application.

In the process of development, packing property, pressure drop, liquid holdups, gas loading velocity were tested. Based on which, a revised irrigated pressure drop model were proposed and fitted well with the experimental results.

In order to meet the production requirement, we optimized the dry pressure drop of Winpak-C with the help of Solidworks, Gambit, Fluent and Tecplot 360. By analyzing the flow pattern near the packing junction, we proposed that reducing the height of the reactive elements to allow the gas streamlines become more gradual. The model and the optimization method were validated by the experiments. The irrigated pressure drop and other system also proved that the model is credible.

We also compared axial and radial dispersion coefficient of the reactive elements. Based on the RTD curves which obtained by the conductometer, residence time, dimensionless variance, skewness, axial and radial dispersion coefficient can be calculated which highlight the advantages of Winpak-C compare with traditional catalytic distillation structured packing.

We will introduce some industrial application cases in different process base on our new catalytic structured packing in terms of successfully applications in MTBE production, light gaso-line etherification, isomerization reactions.

480357 Chemical Kinetics Mechanism Modeling of Sour Gas Combustion

Abstract:

Sour gas combustion is very important for gas processing plants and refineries. Combustion behavior of sour gas is different from other hydrocarbon gases due to the lower heat capacity of sulfur containing components and different radical pool, etc. However, researches on how to combust sour gas stably, effectively, while producing minimal emission is still a big problem. Chemical kinetic modeling provides us an efficient way to assess to this approach. The reaction mechanism is a combination of exciting mechanism of oxidation of C1-C3 hydrocarbons and H_2S. Optimization work is done to compensate the radical pool changes and interaction between C containing and S containing component. The optimized mechanism was validated against real experiment data to achieve a reliable overall mechanism.

480407 Performance Modeling of a Reverse Osmosis Element for Water Purification

Abstract:

Reverse osmosis (RO) membranes provide reliable and efficient separation of dissolved ions from aqueous feeds and are utilized heavily in large-scale water purification operations. Often, these systems will have strict production requirements on pure water as well as complex feed streams containing multiple dissolved ions. To assist with end users' selection process on which RO products best fit their application, FILMTEC™ produces and maintains a software package that can be used to predict the performance of an RO system with regard to operating temperatures, feed pH, ion composition, and production output.* These performance predictions are functions of the various element construction parameters as well as overall system design. In this work, a newly developed membrane element designed for superior antifouling performance is tested over a large concentration range of total dissolved solids. The experimental results are modeled using both empirical and fundamental relationships, and additional performance data is also discussed.

*FILMTEC is a wholly owned subsidiary of The Dow Chemical Company.

SOLID-LIQUID SEPARATION TECHNOLOGY FOR REMOVING CONTAMINANT FINES FROM WATER SCRUBBING, CLARIFIER EFFLUENTS AND GREY WATER

PAPER NUMBER 480801

Barry A. Perlmutter, President & Managing Director
BHS-Sonthofen Inc.
14300 South Lakes Drive
Charlotte, North Carolina 28273-6794
Phone: 704.845.1190
Fax: 704.845.1902
E-mail: barry.perlmutter@bhs-filtration.com
Blog: perlmutterunfiltered.com

ABSTRACT

Treating water for environmental and process benefits is becoming more important in the oil and gas, energy, bioenergy, biochemical and chemical industries. As these operations are looking to become more cost efficient, water usage is being investigated. One critical aspect of these processes is the solid-liquid separation technology that is used to keep the water free of solid contaminant fines.

This paper discusses thin-cake candle filter technology as an alternative to filter presses and bag and filter cartridges for removing solid contaminant fines from the water streams. These contaminants originate from various sources and are generally less than 1 micron (um) in size, which makes their removal very difficult. Candle filter technology and the process of thin-cake building, is a new approach for high-efficiency and cost-effective fines removal.

The paper begins with the problem definition and a discussion of the bench-top laboratory tests that are conducted for problem analysis, process development, technology selection and scale-up. The tests include pressure, filter media, filter aids, and similar process parameters. The candle filter technology is examined for its ability to filter and dry the contaminants to meet the standards of non-hazardous and "no free liquid" for solids disposal. Cake washing is also discussed, as, in some cases, soluble or insoluble components or hazardous solvents must be removed from the solid cake prior to disposal. Three different case histories are presented for purge water treatment units using filter aids, concentration without filter aid, and grey water concentration. Finally, there is a discussion of the impacts of chemical usage in clarifiers on filtration efficiency and troubleshooting.

INTRODUCTION

Water usage has many applications at refineries, gas plants, chemical plants and bioenergy /biochemical facilities. Wet scrubbing at refineries, for example, reduce sulfur dioxide (SO_2/ SO_3) and are the primary and fine particle removal mechanisms from the flue gas. In gasification, water carries the coal fines and catalyst fines during the process. Clarifiers are used in many applications to settle/concentrate to sludge and remove the solids from the process. While the plant location and type of removal process may vary, there is a need to remove the fine particulate matter from the water streams for either reuse or disposal. The particulate matter can be catalyst fines and other contaminants in the feedstock. It is difficult or impossible to remove particles of this size in settling tanks, hydrocyclones or centrifuges, so the particles must be removed by pressure filtration. The use of thin-cake candle filter technology has been proven to be a cost-effective and reliable approach for removing the contaminants, recovering the scrubbing liquids and drying the cake for non-hazardous and "no free liquid" landfill disposal.

PROBLEM DEFINITION

Various catalyst, corrosion products and carbonized particles are carried into the gas and are captured by the water with flow rates as high as 200 m^3/hr. The fine particles are less than 1 micron and can be up to 1% solids in the slurry. Bag and filter cartridges at these solids concentrations are uneconomical and could result in costs of over $500,000 per year. Similarly, the use of filter presses has high operating and maintenance costs. This paper discusses specifically the removal of these fines and provides three illustrative case histories.

TECHNOLOGY OF CLARIFICATION & RECOVERY OF SOLIDS FROM SLURRIES

Candle Filters are installed for clarification and recovery applications from liquids with low solids content. They provide clean fluids to 0.5 micron with either a dried cake (no free liquids) or a concentrated slurry.

Candle Filtration Technology

A candle filter is a pressure vessel filled with tubular filters called filter candles (Figure 1). A typical filter candle is comprised of a dip pipe for the filtrate and pressurized gas, a perforated core with supporting tie rods, and a filter sock (Figure 2). The filtrate dip pipe runs the length of the candle and ensures high liquid flow, as well as maximum distribution of the gas during cake discharge. The tie rods create an annular space between the filter sock and the perforated core, which helps to maintain a low pressure drop during operation and promotes efficient expansion of the filter sock during cake discharge. The filter sock is installed over the candle, and can be made of various synthetic materials, capable of removing particles smaller than 1 μm. As the cake builds during operation, the candle filter's removal efficiency increases, enabling removal of particles as small as approximately 0.5 μm.

Figure 1: Candle Filter Complete

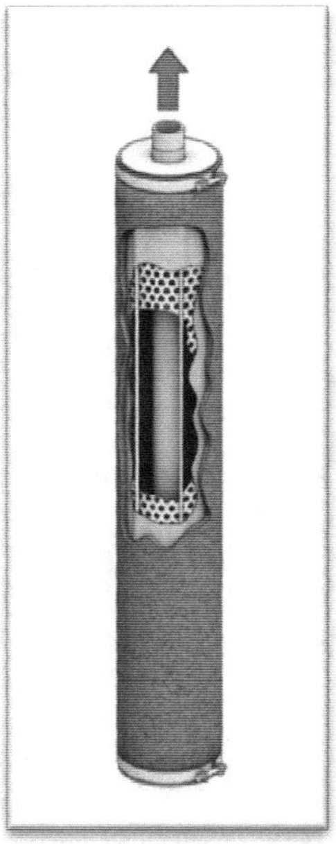

Figure 2: Candle Filter Element

The candles are installed in a pressure vessel constructed of stainless steel or another alloy. Within the vessel are horizontal manifolds called candle registers. Each candle is connected to a register with a positive seal to prevent bypass. Depending on the filter size, each register may contain 1–20 candles and the total candles can be over 200 candles per vessel. The liquid filtrate and pressurized gas flow through the register; automated valves ensure optimum flow in both directions.

During operation, a feed pump forces the slurry into the bottom of the pressure vessel. The solids build up on the outside of the filter sock, while the liquid filtrate flows into the candle, through the registers, and out of the vessel. This process continues until the maximum pressure drop, design cake thickness, minimum flow, or maximum filtration time is reached.

The cake is washed to remove impurities and residual mother liquor, and then dried. Next, low-pressure gas enters the individual candles and expands the filter socks. This process breaks apart the dry cake, which detaches from the filter sock (Figure 3) and falls into the vessel cone. The cake can also be discharged as a concentrated slurry without drying.

Candle filters are used for thin-cake (5–20 mm) pressure filtration applications. They are best suited for filter cakes that are vertically stable because of the orientation of the candles.

Figure 3: Candle Filter Cake Discharge

Figure 4: Pocket Leaf Filter Testing

DETERMINING CANDLE FILTER SYSTEM DESIGN

Laboratory testing at a constant slurry flow is used to determine the size and design of a clarification system for processes with extremely low concentrations of solids. The test evaluates the filter media, operating pressure, and cake thickness to determine the optimum clarification system design and size.

A lab-scale leaf filter (Figure 4) is used as the test filter. Various filter media can be installed, depending on the desired filtrate clarity, filtration flux rate (time), cake thickness, and cake discharge rate, as well as compatibility with the process. A peristaltic pump supplies a constant flow of slurry to the filter, and a pressure gage measures the change in pressure (delta P) across the filter media.

The washing analysis consists of running a liquid at the same or different flowrate as the test flowrate through the peristaltic pump, or adding wash liquid to the leaf filter directly and applying air or gas (nitrogen, argon, etc.) pressure until the target pressure drop or quality (conductivity, acid level, etc.) is reached

The cake is then dried with pressurized gas or air. The amount of compressed gas is regulated to prevent the use of excess gas which could lead to overly optimistic moisture targets and uneconomical compressed gas volumes when scaled up.

WATER SCRUBBING AND CLARIFIER EFFLUENT USING FILTER AIDS: ALTERNATIVE TO FILTER CARTRIDGES

In this application, a wet scrubber is used to remove catalyst fines and sulfur dioxide from a fluid catalytic cracking unit flue gas. A purge stream is sent from the wet scrubbing system to a purge water treatment unit and then to disposal. The candle filter system removes catalyst particles fines from the clarifier effluent stream prior to discharge. The flow rate is 70 m^3/hour with a solids concentration of 190 ppm. The particle size is below 0.5 micron down to 0.06 micron. The filtrate specification for discharge is less than 40 ppm. Initially, filter cartridges were planned but the cost of change out alone, based upon the above parameters was estimated to be over several million dollars not including the cost of cartridge inventory, disposal and manpower for cartridge change out, assumed to be every shift.

BHS conducted laboratory tests, as previously described including the use of filter aids due to the size of the particles being less than 0.5 microns. The installation is shown in Figure 5 and consists of two BHS candle filters, each with 100 m^2 of filter area and a filter aid precoat package. The cake discharged is shown in Figure 6 while the filtrate is much below the specification at 5 – 8 ppm.

Figure 5: BHS Candle Filters, 2 x 100 m^2

Figure 6: BHS Candle Filters, 2 x 100 m^2, cake discharge

WATER SCRUBBING-CLARIFIER EFFLUENT CONCENTRATING WITHOUT FILTER AIDS: ALTERNATIVE TO BACKWASHABLE METAL CARTRIDGES

In this application, the wet scrubber water purge is pumped to a clarifier for primary removal of large catalyst solids. The clarifier overflow discharges to an oxidation tower and the effluent from the tower is pumped through the effluent filter and heat exchanger prior to discharge.

The flow rate is 46 m^3/hour with a solids concentration of 200 ppm. The particle size is below 1.0 micron. The filtrate specification for discharge is less than 15 ppm. Initially, backwashable metal cartridges were planned but the low dirt holding capacity of the metal cartridges and the frequent backwash, which is sent back to the clarifier, was estimated to be uneconomical. Therefore, the decision was to look at the BHS candle filter technology solution.

As this process required a concentrated slurry rather than a dried cake, the BHS process solution was concentrating candle filters without filter aid. The installation is shown in Figure 7 and consists of two BHS candle filters, each with 25 m^2 of filter area. The benefits of the candle filter solution included clear filtrates to less than 1 ppm as well as less backwash to the clarifier as the candles are “cleaned” with compressed gas and not liquid.

Figure 7: BHS Candle Filters Downstream of Clarifiers

GREY WATER CONCENTRATING

In this application, grey water containing catalyst fines, coal fines, char and other non-reactive solids required clarification to 0.5 microns. The filtered water is sent back to the gasifier process while the solids are sent to the wastewater plant. The grey water flow rate is 180 m^3/hour with a solids concentration of 200 ppm.

BHS conducted lab testing to determine the filtration flux rate. It was also determined that due to the flow rate and low solids it was a more cost-efficient approach to do the final solids drying in a filter press rather than on the candle filters. Figure 8 shows the final sizing and Figure 9 shows the completed turnkey skid package. The concentrating candle filters produced a clear filtrate of 0.5 microns as well as reduced the feed to the wastewater plant to 10% solids at 300 gallons every 2 hours compared with 800 gpm at 200 ppm.

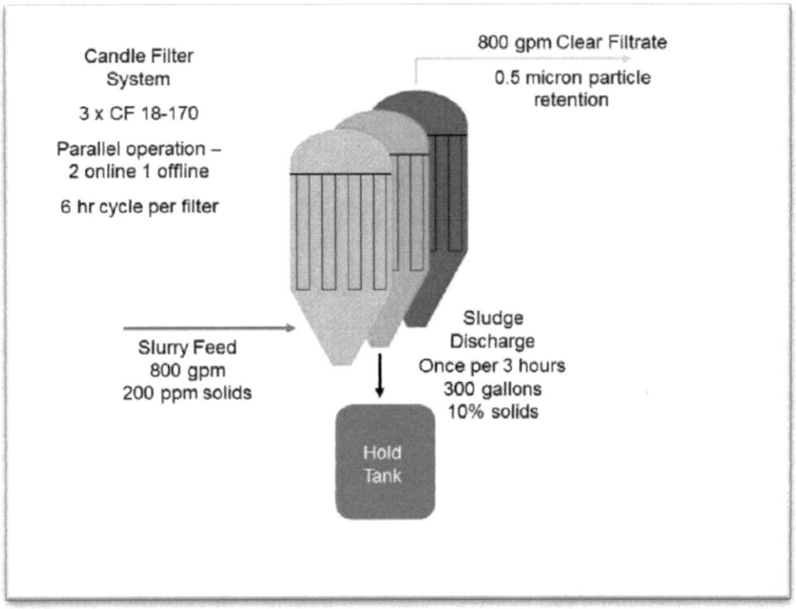

Figure 8: BHS Candle Filter Sizing for Grey Water

Figure 9: BHS Candle Filter Grey Water Installation

FILTRATION SYSTEM TROUBLESHOOTING

There are normally three main areas that must be examined when the filtration system appears not to be functioning properly. These areas include: (1) the filter itself for mechanical reasons, (2) the equipment around the filter is not working and (3) the filter operational procedures are not correct. Let's examine these groups but more importantly, it is necessary to separate the symptoms from the causes. As will be seen, excessive solids in the filtrate (the symptom) has many causes (damaged filter cloth, PSD changes from the reactor, post-precipitation due to temperature changes, etc.).

First, there could be a failure of the equipment itself such an internal components, seals, etc., all as is normally described in the O & M preventative maintenance section from the supplier. This should be the first item to be checked.

Secondly, the filtration system is part of the entire process including the upstream and downstream equipment. For example, are the reactors performing correctly in terms of agitation, temperature control, etc. in order to produce the specified crystals? Are the filter aid systems in tune for mixing, feeding, flow rates, solids loading, etc.? Are the valves and instruments operating correctly and reading the correct variables (calibrations), etc.? Next, what about the pumps that feed the slurry and washing liquids as well as the compressors that feed the gas streams for drying and cake discharge. Finally, are their interlocks in the control system or a control communication problem that are not being recognized that are causing the filter problem.

Finally, there may be process or operational procedures that are resulting in filtration problems. For example, the particle sizes may have changed, the amount of solids in the slurry may have changed, the cake compressibility may have changed, etc. In terms of the operation, has the filtration pressure changed timers changes, speed changed, etc.?

Troubleshooting The Water Scrubbing-Clarifier Effluent Concentrating System
In a clarification application previously discussed, the system was installed and started-up and successfully ran for over one-year. Inexplicitly, the performance changed drastically and the filter media began plugging very quickly during the cycles. The process engineers explained that there had been no change in the process and began to investigate the upstream conditions using the list below.

A. Clarifier overflow with no coagulant / no flocculants
B. Clarifier overflow with only coagulant / no flocculants
C. Clarifier overflow with both coagulant and flocculants
D. Clarifier overflow with only flocculants / no coagulant

After months of work, the process engineers narrowed down the changes to the chemical supplier of the flocculants and coagulants. BHS and the client began lab testing again using the initial slurry along with the chemical supplier and their flocculants and coagulants. The resulting testing showed that the chemical change caused the larger particles to settle out quickly and therefore only the smaller particles reached the filtration system which was blinding the filter media. The client and the chemical company eliminated the flocculants and reduced the amount of coagulant used (better for the client and not so good for the chemical supplier) which produced a smooth and consistent particle size distribution and better filtration rates. Once again, we see the systems and holistic approach to process filtration.

SUMMARY & TAKE-AWAYS

Sherlock Holmes and Dr. John Watson are fictional characters of Sir Arthur Conan Doyle. Process engineers who live in the real-world can learn many things from the two of them for solving process filtration problems. One example that Holmes proves time and again is that there is no benefit to "jumping to conclusions." As discussed in the illustrations, there are many choices of filtration technologies to achieve the necessary level of quality with a cost-effective and a reliable process. Also, in terms of troubleshooting, there are many causes of filtration symptoms. With careful analysis and taking a creative and "non-conventional approach" by looking at what is behind the process data and the operating parameters, the optimum solution will be realized.

480336 Utilization of Flare Gas for Direct Non-Oxidative Production of Liquid Fuels and Chemicals

Abstract:

Natural gas flaring produces 400 million tons of greenhouse gas emissions annually [1]. The current gas-to-liquids technology such as Fischer-Tropsch process requires two stages with synthesis gas ($CO + H_2$) as intermediate, which is extremely capital-intensive and not economically viable for stranded gas utilization. Other technology, e.g., the oxidative coupling process is difficult to scale up. This project will develop a process that enables direct and continuous conversion of flare gas into liquid hydrocarbon fuels. In this process, an initiator was used to facilitate the dissociation of alkanes, which can be converted into liquid hydrocarbons (benzene, toluene, xylenes, mesitylenes, etc) using highly selective zeolite catalysts. This direct conversion will considerably simplify the process, decrease the capital cost and avoid the use of expensive precious metal catalysts and corrosive activators. Therefore, the novel one-step gas-to-liquids process is obviously superior to the current state-of-the-art processes. Moreover, it is expected that the techno-economics of the one-step process will be attractive. This technology has the potential to eliminate flaring for the shale gas industry.

References

[1] http://www.theenergyexchange.co.uk/partner/global-gas-flaring-reduction-ggfr.

480113 Emission Conscious Scheduling of Crude Unloading, Transferring, and Processing for Petroleum Refineries

Abstract:

Emission reduction is becoming increasingly important for petroleum refineries as the increasing concerns of environmental concerns and regulations. Petroleum refinery uses crude oils to produce various fuels such as gasoline, aviation kerosene, diesel, heavy fuel oil, and chemical raw materials such as naphtha and benzene. Crude planning and scheduling is an important part of petroleum refinery due to the potential realization of large cost savings. Cost-effective solution strategies require emission reductions to be addressed from the entire plant point of view, where emission source generations and possible utilizations should be well balanced. But it is a very challenging task due to the large-scale complex manufacturing system, which consists of over 17 major facilities with more than 85 main operation units, and thousands of process streams. The cost-effective emission-reduction strategies should focus on the comprehensively study of material, energy, and information exchanges within the entire manufacturing system, so as to identify the best solutions to the entire plant.

In this paper, a new methodology framework together with a new general scheduling model have been developed for emission conscious crude unloading, transferring, and processing (ECUTP) operations to achieve profitable emission-reduction scheduling (PERS) from the entire system point of view. To identify PERS strategies for refinery plants, a production scheduling model covers both front-end and refinery has to be involved. The key is to optimize the overall material and energy flows to make the plant net profit maximum, meanwhile to ensure emission source generations and utilizations to be smartly balanced. Other manufacturing constraints, such as operation specifications and inventory limits, should also be satisfied as well. Major air emissions from refineries: such as CO_2, volatile organic compounds (VOC), nitrogen oxides (NO_X), and particulate matters (PM) are characterized and quantified.

The scheduling model is a large-scale mixed integer nonlinear programming problem (MINLP). The efficacy of the proposed emission-considered scheduling model has been demonstrated by different case studies. PERS thrusts target on emission source reductions in a profitable and systematic way, which are economically attractive, environmentally benign, and technologically viable for petroleum refineries. It may also help emission generators (not just limited to petroleum refineries) to proactively and systematically reduce their emissions to meet increasingly strict economic and environmental challenges.

480023 Conceptual Design and Cost Estimation of an Offshore Autonomous Once-through Gas-to-Liquid Process Combined with Ammonia Synthesis

Abstract:

Conceptual design and cost estimation of an offshore autonomous once-through gas-to-liquid process combined with ammonia synthesis

A novel process concept is proposed for converting natural gas to liquid Fischer-Tropsch products and ammonia (Figure1). An autothermal reformer with enriched air as oxidant is applied for synthesis gas (syngas) production, and because of the inert nitrogen a once-through Fischer-Tropsch synthesis is the preferred option. In order to maximize the syngas conversion and the production of heavy hydrocarbons, a staged reactor path with distributed hydrogen feed and product withdraw is proposed. The hydrogen is produced by steam methane reforming in a heat exchange reformer (gas heated reformer), heat integrated with the hot effluent stream from the autothermal reformer. The nitrogen needed for ammonia synthesis is produced from air membrane. Tail gas from the last Fischer-Tropsch stage is sent to a gas turbine for power production. The hot exhaust gas from the gas turbine is used for natural gas preheating. The process is autonomous in the sense that it is self-sufficient with power and water, and therefore well suited for production in remote locations such as a floating production unit. The process concept is simple and inexpensive since cryogenic air separation and fired heaters are not required. For the Fisher-Tropsch synthesis microchannel reactors are considered. Cost evaluation is carried out to find out the economic feasibility of the process.

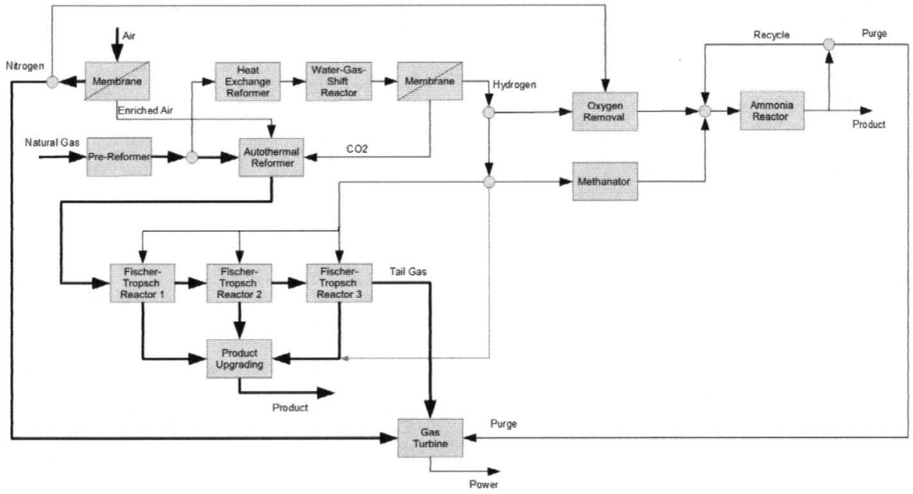

Figure 1: Block flow diagram of the proposed process concept; water and steam are not shown

480897 Study of Liquid Dispersion in Winpak Modular Catalytic Structured Packing: Experiments Vs. CFD Simulations

Abstract:

The response technique was employed to investigate the axial and radial distribution of the liquid phase in Winpak modular catalytic structured packing which was named Winpak-C. Residence time distribution (RTD) experiments were performed at ambient temperature and atmospheric pressure. The axial and radial dispersion coefficients were obtained by solving a two-dimensional dispersion model analytically from the RTD curves. Based on proper assumption, a three-dimensional model of reaction elements in MCSPs was established to determine the dispersion characteristics with the use of computational fluid dynamics (CFD). Porous flow was introduced by using a specification of an extra body force to account for liquid flow through the reaction element. These measured or simulated dispersion coefficients enable the reliable and predictive design and construction of scale-up models.

481499 Life Cycle Analysis of Hydrogen Gas Used for Biogas Upgrading Process in Expanded Granular Sludge Bed Reactor

Abstract:

<align="center">Life Cycle Analysis of Hydrogen Gas Used For Biogas Upgrading Process in Expanded Granular Sludge Bed Reactor

Shruti S. Karambelkar, Haider Al-Rubaye, Manohar M. Shivashankaraiah, Joseph D. Smith, PhD.

*Chemical & Biochemical Engineering Department, Missouri University of Science and Technology, Rolla, USA*Abstract

The high demand for energy around the world and the environmental effects resulting from the use of the fossil fuels, makes many industrial sectors focus on the alternative energy. Biogas Production from anaerobic digestion forms an essential part of alternative and renewable energy development.

The biogas mainly consists of methane, carbon dioxide and trace amounts of hydrogen sulfide and ammonia. Hydrogen gas produced in the first two steps of the anaerobic digestion - Hydrolysis and Acidogensis is released to the atmosphere during the process. Injecting the same amount of released Hydrogen into the last two steps of the Anaerobic Digestion - Acetogensis and methanogenesis, improves Methane composition in the biogas, up to 88%. On injecting Hydrogen, it reacts with the produced carbon dioxide to produce more methane, thus, enhancing methane in the biogas. This increase in the methane composition will be accompanied by a high production rate since the specific volume of the methane is four times more than hydrogen and the volumetric energy content for methane is higher (40 KJ/L) than that of hydrogen (12.78 KJ/L).

Life Cycle analysis has been conducted to check the feasibility of using Hydrogen gas to upgrade the biogas so it could be accounted as a natural gas and used in all natural gas equipment.

478350 Permylene™ Membrane Technology By Imtex Membranes Corp. - the Cost Effective Alternative to Distillation for Olefin Purification and Extraction

Abstract:

Imtex Membranes Corp. is a leading innovator of advanced membrane separation technology currently focusing on olefin/paraffin separation for application in the petrochemical and refining industries. Following years of research and development, Imtex currently possesses the only high performance, commercially viable alternative to traditional distillation techniques for olefin purification and extraction - Permylene™. The technology's potential for substantial economic and environmental benefits is expected to set new standards across the industry.

For over 30 years the petrochemical industry has been searching for commercially viable alternatives for olefin extraction and purification. The capital and operating costs associated with conventional separation via distillation have hindered project returns and ongoing operational profitability.

Imtex has now developed such an alternative that significantly reduces upfront capital investment and ongoing operational costs. Using proprietary membrane technology, this system is also able to perform separations that have to date been deemed impractical or too costly to pursue. This revolutionary breakthrough is seen as the next advancement in olefins processing technology, offering a high performance alternative to the status quo. Imtex is currently working with several industry leaders to deploy demonstration systems in their petrochemical facilities.

Imtex continues to develop membrane separation technology for a range of applications involving C_2, C_3 and C_4 olefins. Imtex is actively searching for additional demonstration opportunities within the industry, which is the first step towards users realizing the significant benefits the technology will deliver.

Imtex will present the characteristics, attributes and benefits of the Permylene technology, and will illustrate several comparative cases that demonstrate the commercial investment and operational cost advantages of the technology versus conventional olefin/paraffin separation technologies.

Hydrogen Storage and Transportation using SPERA Hydrogen Process for realizing Low-Carbon Society

Yusuke Nakajima, Hironori Kawai and Tomoyuki Mikuriya
Chiyoda Corporation, Yokohama, Japan

Introduction

Hydrogen is a clean energy produced from any form of primary energies. It is expected that hydrogen will promote the utilization of renewable energies and solve the problems related to CO_2 emissions.

The establishment of storage and transportation technologies, similar to those used in the petroleum or gas industries, is required for the global utilization of hydrogen.

Chiyoda Corporation has developed the organic chemical hydride (OCH) method using a toluene (C_7H_8) / methylcyclohexane (MCH: C_7H_{14}) system as the basis for the large-scale storage and transportation of hydrogen energy, and demonstration tests were successfully carried out to verify its expected performance. We call the MCH in this system "SPERA (Latin word for hope) Hydrogen", which reflects our ambition to establish a hydrogen supply chain.

At the AIChE Spring Meeting in 2014, we already reported the OCH method and its development status[1). This paper introduces the hydrogen supply chain concept, results of the demonstration tests, current activities and future plans of the SPERA Hydrogen process towards a low carbon society.

1. Hydrogen Supply Chain

Figure 1 shows the hydrogen supply chain concept which was proposed in Japan in 2004, by using a hydrogen storage and transportation system based on the OCH method[2-3)].

The hydrogen supply chain concept will be developed in two steps. In the first stage, hydrogen from fossil resources, called clean hydrogen, is utilized with the combination of carbon dioxide capture and storage (CCS), enhanced oil recovery (EOR) or enhanced coal bed methane (ECBM).

In the second stage, hydrogen from renewable energy (wind power, hydraulic power, solar power etc.) will gradually be introduced in order to realize a low-carbon society.

2. Organic Chemical Hydride Method

The OCH method is a way to store and transport hydrogen under ambient temperature and pressure by using a Liquid Organic Hydrogen Carrier. Hydrogen is fixed to aromatic compounds such as toluene by the hydrogenation reaction, whereby toluene is converted to MCH. Hydrogen is generated by a dehydrogenation reaction from MCH at the hydrogen user site, whereby recovered toluene is recycled back to hydrogen production site. The reaction formula is shown below.

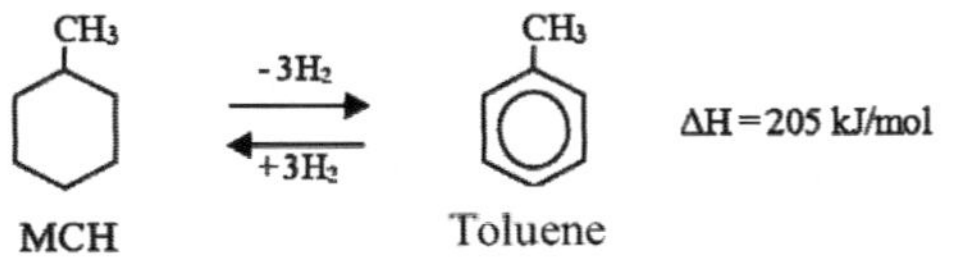

Since both toluene and MCH are components of gasoline and are in liquid state under ambient conditions, gaseous hydrogen can be stored in MCH on a large scale and transported in the same way as gasoline.

Table 1 shows the comparison of physical properties of typical organic hydride systems. Chiyoda Corporation has employed the toluene and MCH system, because these components can be kept in a liquid state in the temperature range of -95 to 101℃. Therefore, the system does not require a solvent, unlike other systems which need a solvent to maintain a liquid state.
The OCH method has the following features:

(1) Hydrogen gas volume is reduced to <1/500 of that under ambient conditions

Hydrogen has the lightest molecule and has the largest volume by weight. It is explosive and therefore has potential risk when stored in a gaseous state. In order to reduce the volume to under 1/500 by compression, over 500 bar is required. However, since the OCH method uses chemical reactions, the hydrogen is stable in the MCH liquid, and the gas volume is reduced to < 1/500 under ambient conditions.

(2) Negligible hydrogen loss under long term, large-scale storage

This method does not require high pressures or extremely low temperatures, as with LNG (-163℃) or liquefied hydrogen (-253℃). Toluene and MCH exist as liquid under ambient conditions, therefore, there is no need for boil-off gas treatment.

(3) Utilization of existing infrastructure

Toluene and MCH are components of gasoline and are not corrosive, therefore, conventional gasoline or chemical storage tanks or tankers can be utilized for storage and transportation. This feature decreases infrastructure preparation costs and reduces the risk of large-scale hydrogen use to those associated with gasoline.

3. Dehydrogenation Catalyst

Since the 1980s, it has not been possible to commercialize the MCH dehydrogenation process due to severe catalytic deactivation. Chiyoda Corporation has developed a new dehydrogenation catalyst (SPERA Hydrogen catalyst) and a new dehydrogenation process by scale-up engineering technology.

The SPERA Hydrogen catalyst (S-Pt/Al_2O_3) has uniformly well-dispersed nano-size Pt clusters (< 1nm) on an alumina support . This well-dispersed Pt catalyst enables high reactivity and prevents catalytic deactivation by coking.
Figure 2 shows the result of the catalyst performance test. Over a period exceeding 10,000 hours, the catalyst achieved over 95% MCH conversion, over 99.9% toluene selectivity and over 1,000 $Nm^3/h/m^3$-cat of hydrogen production at a bench facility.

4. SPERA Hydrogen Process

Chiyoda Corporation has developed the new dehydrogenation process using this developed catalyst, and also improved the hydrogenation process based on commercialized technology. Figure 3 and Figure 4 shows simplified process flows of the hydrogenation and dehydrogenation process respectively. The remarkable feature of the SPERA Hydrogen process is its simple process configuration. Each process feedstock reacts in the fixed bed tubular reactors in vapor phase, and high product yield is achieved with simple vapor and liquid separation drums.

Hydrogenation process

The toluene feed is vaporized (Eq. 1) and mixed with hydrogen including recycled gas. The feed mixture is superheated to reaction temperature (Eq. 2) and enters the top of the hydrogenation reactor (Eq. 3), which is a fixed-bed tubular type reactor charged with the hydrogenation catalyst. In the tubes, toluene reacts with hydrogen to produce MCH. Hydrogenation is an exothermic reaction, so in order to control the reaction temperature, the generated heat is removed by a cooling medium, such as a steam generation system.

The effluent gas from the reactor is cooled and the condensed MCH is separated (Eq. 6), with the hydrogen gas being recycled to the reactor. The liquid MCH product is sent to a storage tank.

Dehydrogenation process

The MCH feed is vaporized (Eq. 11), and superheated in a charge heater (Eq. 12) before entering the dehydrogenation reactor (Eq. 13). The reactor is of a fixed bed tubular type charged with the dehydrogenation catalyst in the tubes. Dehydrogenation is an endothermic reaction, so a heating source is required. A hot oil system is typically applied (Eq. 14) to supply this reaction heat.

The effluent gas from the reactor is cooled to separate the hydrogen and toluene (Eq. 16). The separated hydrogen is compressed and sent to users (Eq. 17). Product H_2 treatment (Eq.18) is the optional facility to meet any product hydrogen specifications.

5. Demonstration test of the SPERA Hydrogen Process

Since April 2013, demonstration tests have been conducted in a 50 Nm^3-H_2/h capacity demonstration plant located in Yokohama, Japan. The plant has both hydrogenation and dehydrogenation units, and the conversion between toluene and MCH has been conducted repeatedly. Figure 5 shows this demonstration plant. Although the plant capacity is only 50Nm^3/h-H_2, the catalyst and reactor tubes are the same as those of a commercial plant. Therefore, the process is able to be scaled-up by increasing the number of catalyst tubes.

Figure 6 shows the performance of the hydrogenation and dehydrogenation processes. The upper graph shows operation load and the lower graph shows the performance of each process (hydrogenation and dehydrogenation) performance. The ratio to equilibrium is a ratio of real MCH or toluene conversion to equilibrium conversion. Excluding a period to change operating conditions, we have achieved almost 10,000 hours of successful and trouble-free operation with excellent performance. In addition, the stable recycling of toluene and MCH was also achieved, which has proved that the SPERA Hydrogen process is a feasible technology.

6. Current activities and future plans

6.1 Global hydrogen supply chain demonstration

The next target is to conduct a demonstration of the global hydrogen supply chain, between the hydrogen production site (abroad) and the end user (Japan). In the operations, the produced toluene/MCH are repeatedly transported by shipping, and the transported hydrogen is used in thermal power plants. Through these demonstrations, the actual chain operations are confirmed and then, the global hydrogen supply chain can become real.

6.2 Utilization of renewable energy

To realize a low-carbon society, hydrogen must be produced from renewable energies. There are several technical challenges for harnessing renewable energies. The first is utilizing the surplus energy from wind or solar power, which is generated by the gap between power output and power demand. The second is utilizing unstable power, especially as observed in wind power, which changes from day to day and by the seasons. Furthermore, the renewable energies are unbalanced in global supply and demand. Therefore, the SPERA Hydrogen will be an energy carrier that absorbs this surplus and unstableness of renewable energy and make up any energy imbalances.

In order to introduce renewable energies into in the SPERA Hydrogen process, we have tested and planned the integration of water electrolysis using wind power and SOFC (Solid Oxide Fuel Cell) power generation. The overall scheme of the plan is shown in Fig. 7. A wind power simulation system, alkaline-water electrolyzer and hydrogen purification unit are combined with the existing demonstration plant (described in chapter 5).

The wind power simulation system is applied to an alkaline-water electrolyzer to test the hydrogen generation using unstable renewable energies. This unstable power generates "unstable flow of hydrogen", therefore, a control system that follows the unstable flow will need to be developed. The obtained data will be used to commercial scale-up. The produced hydrogen in the electrolyzer contains impurities (alkaline mist, water, oxygen), so the hydrogen purification unit is applied for the stable operation of the toluene hydrogenation.

The SOFC generates power and surplus heat, the surplus heat can be utilized as heating source of dehydrogenation reaction. The required amount of impurities' removal for the SOFC and the recovery of the surplus heat from the SOFC are studied in laboratory tests. After establishment of the overall integrated system (Fig. 7), variable wind power can be stably supplied using SPERA Hydrogen as a buffer.

6.3 Dehydrogenation facility for hydrogen refueling stations

Considering its ease of handling, the SPERA Hydrogen can be supplied for many purposes. In the future, the demand of hydrogen will increase, not only for industrial purpose, but also for use in logistics, residential and commercial buildings, and so on.

In order to meet this demand, we plan to use the SPERA Hydrogen to supply hydrogen to FCVs (Fuel-Cell Vehicles) by establishing dehydrogenation facilities for hydrogen refueling stations.

Japanese government plans to introduce more than 800,000 FCVs by 2030, and hydrogen refueling stations have started to open and operate in major cities.

As a hydrogen-supply facility for the refueling stations, Chiyoda Corporation has planned an on-site type dehydrogenation facility. As the refueling stations are located in towns, the hydrogen-supply facilities need to be compact, and the produced hydrogen must be purified to meet the specifications for FCVs. As a purification unit, PSA (Pressure Swing Adsorption) is applied and the off-gas from PSA is used for fuel gas as heating source of dehydrogenation.

We plan to start operation of the demonstration plant in Yokohama, Japan in 2017.

7. Conclusion

Chiyoda Corporation proposes a hydrogen supply chain using the SPERA Hydrogen process. The basic technologies were established by achieving around 10,000 hours of successful demonstration operation in Yokohama, Japan. To realize a low-carbon society, we are planning to conduct demonstrations for the global hydrogen supply chain, the utilization of renewable energy and the dehydrogenation facilities for hydrogen refueling stations. These current activities and future plans are subsidized by the New Energy and Industrial Technology Development Organization (NEDO).

References

1) Shiraga M. et al. (2014) AIChE Spring Meeting and Global Congress on Process Safety, 89-C
2) Okada Y. et al. (2004) 15th World Hydrogen Energy Conference, 30B-5
3) Okada Y. et al. (2006) Int. J. Hydrogen Energy, 31, 1348

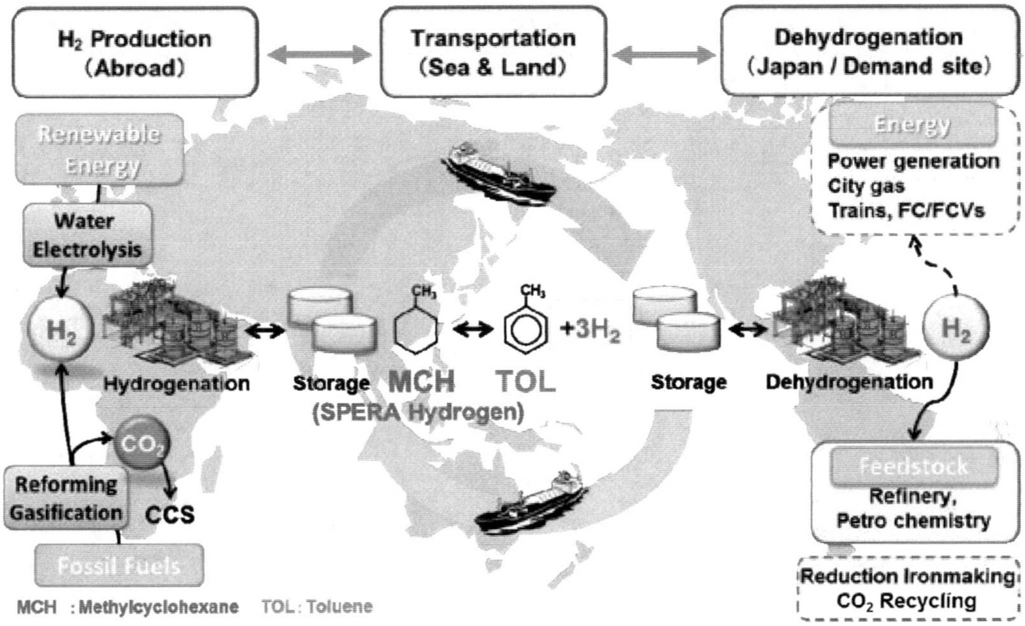

Fig. 1: Hydrogen supply chain concept

Table1: Physical properties of typical organic hydride system

	MCH−Toluene		Cyclohexane −Benzene		Decalin−Naphthalene	
	MCH	Toluene	Cyclohexane	Benzene	Decalin	Naphthalene
Molecular formula	C_7H_{14}	C_7H_8	C_6H_{12}	C_6H_6	$C_{10}H_{18}$	$C_{10}H_8$
Chemical equation	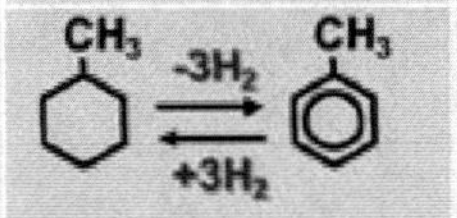 ΔH = 205 kJ/mol		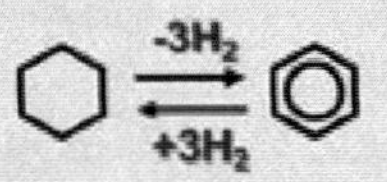 ΔH = 206 kJ/mol		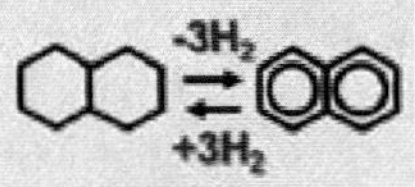 ΔH = 332 kJ/mol	
Molar mass(g/mol)	98.2	92.1	84.2	78.1	138.3	128.2
Phase @room temp.	Liquid	Liquid	Liquid	Liquid	Liquid	**Solid**
Density(g/cm^3)	0.77	0.87	0.78	0.87	0.90	0.98
Melting point (Degrees C)	-127	-95	7	6	Cis:-43 Trans:-30	80
Boiling point (Degrees C)	101	111	81	80	Cis:195 Trans:186	218
H_2 store density (wt%)	6.2	–	7.2	–	7.3	–
H_2 store density (kg-H_2/m^3)	47	–	56	–	65	–

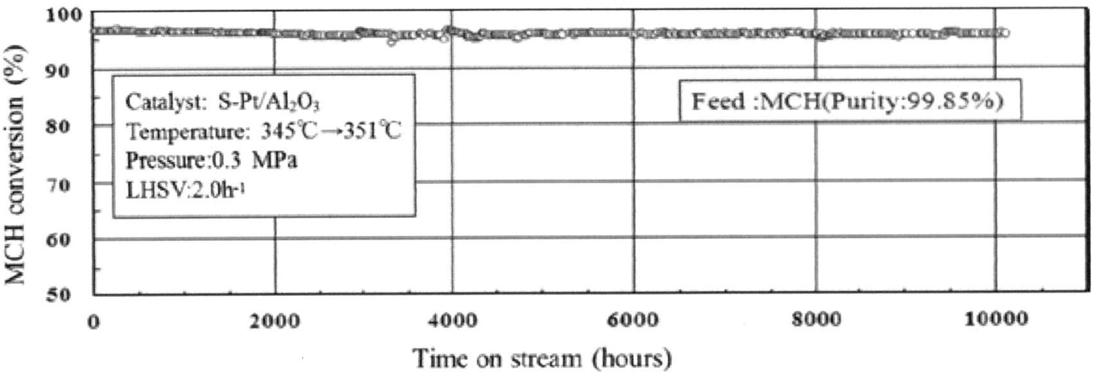

Fig. 2: Result of dehydrogenation catalyst performance test

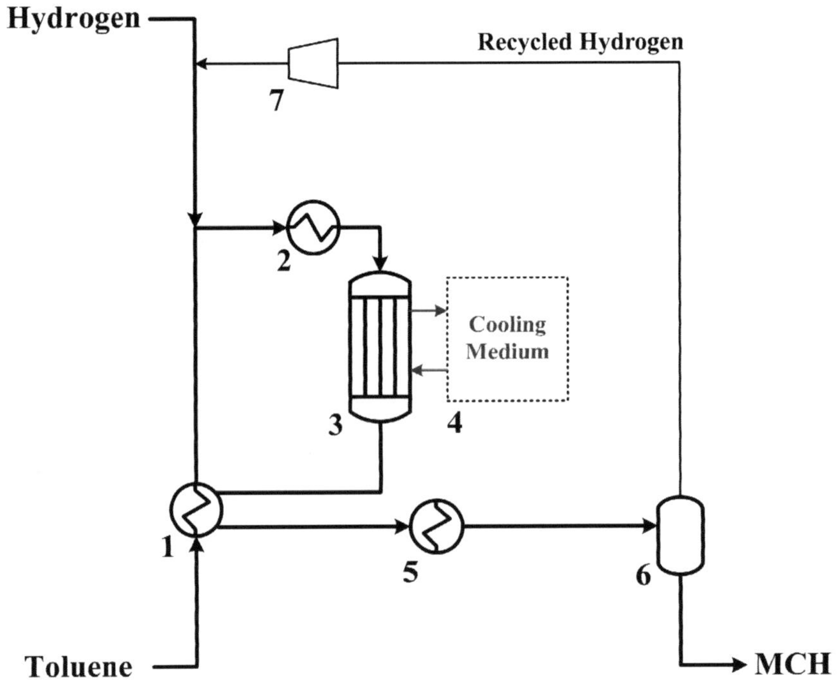

Fig. 3: Hydrogenation process

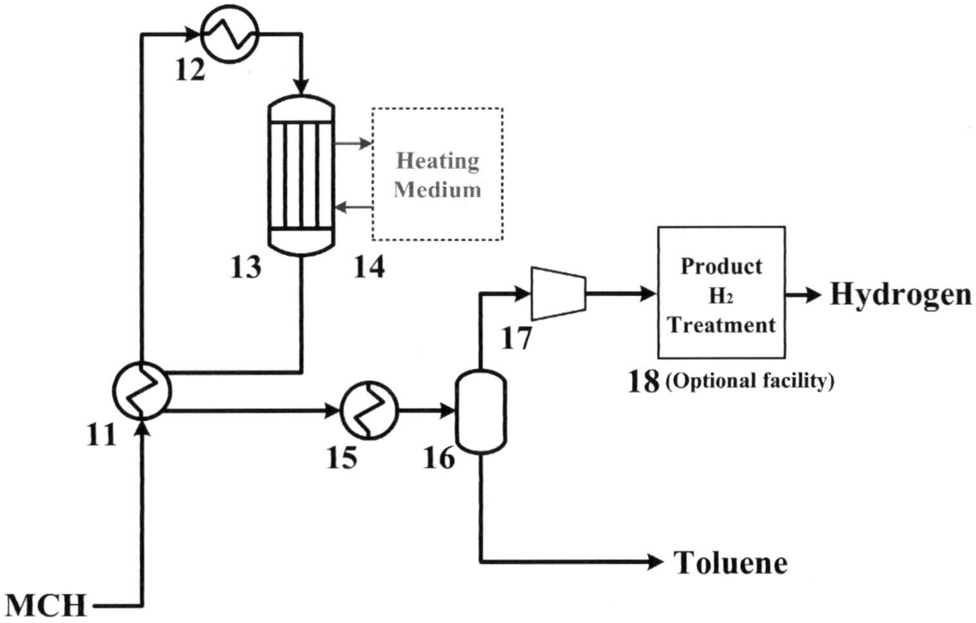

Fig. 4: Dehydrogenation process

Fig. 5: Demonstration plant

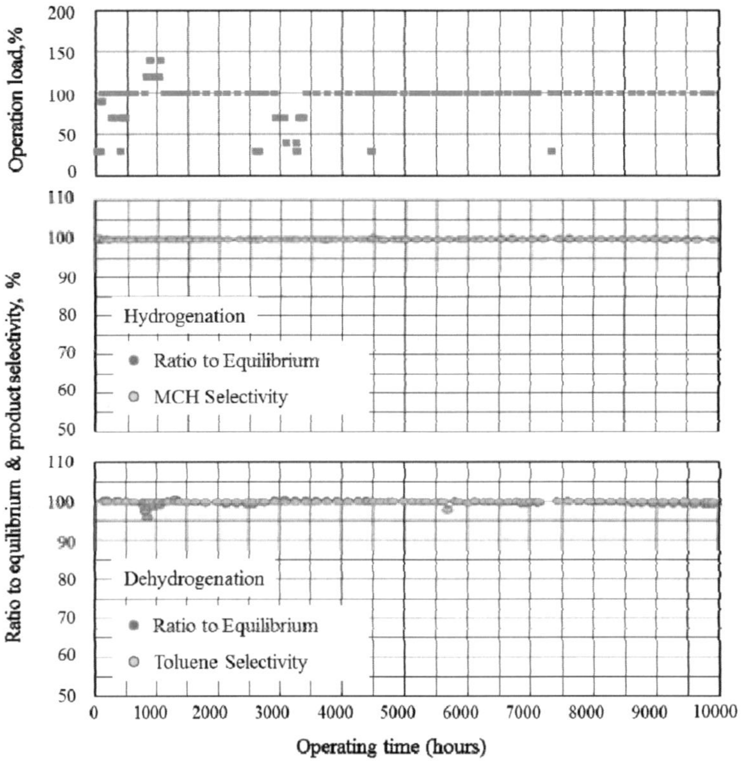

Fig. 6: Performance of hydrogenation & dehydrogenation process

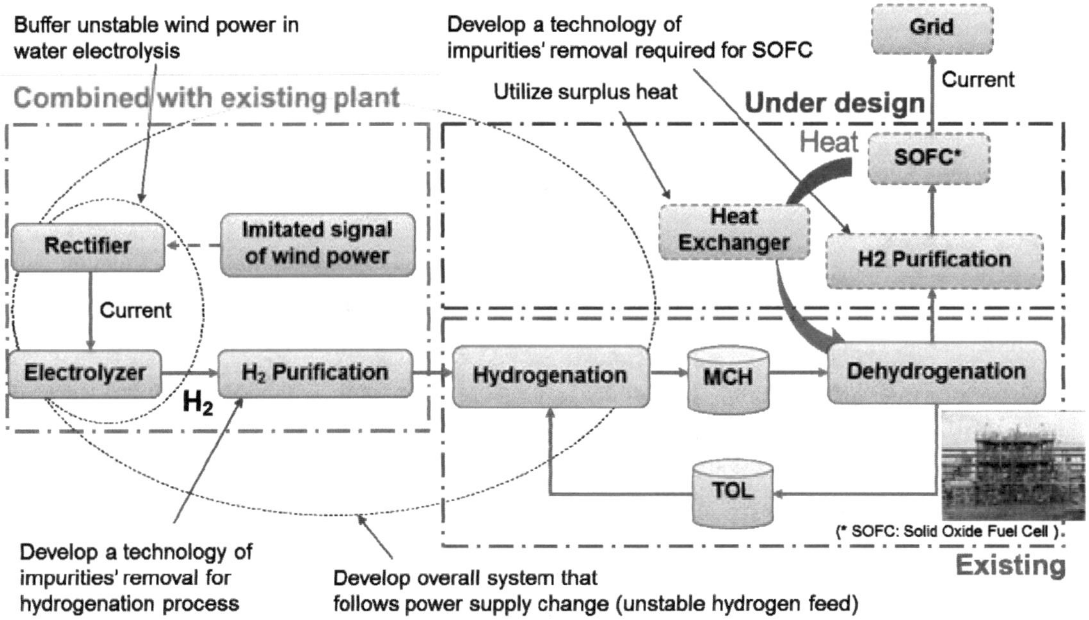

Fig. 7: Demonstration plans for utilizing renewable energies

482575 Personal Branding

Abstract:

Do you have a personal brand? Are you taking a proactive approach to managing your career? Your personal brand is the unique promise of the value and impact you create. It is what shows your authenticity, differentiates you from your peers and creates the picture for those who are making decisions about you. Are you the world's best kept secret? Or are you known as someone who is the "go to" person, who is invaluable for "fill in the blank"? We have control over what people think when they hear or see our name online (on LinkedIn, etc.), in our communities and within our careers. In this session we will discuss how to stand out, highlight your strengths and create the vision you want people to have of the brand you call you (extroverts and introverts welcome).

480396 Exploring the Catalytic Options in Low-Mid Pressure Diesel Hydrotreating

Abstract:

Regularly revised, worldwide diesel fuel specifications continue to press impurity levels lower and product quality higher. Such fuel specification revisions naturally encourage the semi-continuous development of superior hydrotreating catalysts, to satisfy the new requirements and improve the economics of hydrotreacting where applicable. The goal of this talk is to draw rational boundaries where the most active examples of supported NiMo and supported CoMo catalysts ought to be applied, given the inhibiting influences of organic nitrogen species and aromatics species, for example. Included in the scope shall also be application of both types of catalysts simultaneously, where a bed of one of the catalysts is loaded above the other. In consideration of the pressing need for improved hydrotreating catalysts, Honeywell UOP has recently commercialized an entirely new portfolio of supported hydrotreating catalysts under the Unity(TM) trademark. Several of these catalysts shall be featured in this presentation, since they were developed for diesel hydrotreating in the low-to-mid pressure range.

480176 Next-Generation Utilities Optimisation for Refineries and Large-Scale Chemical Production Sites

Abstract:

Refineries and chemical production sites are major consumers of energy in the form of electricity, steam and hydrocarbon feedstocks. Given that tariffs, costs and demands are constantly changing, there is much scope for optimising on-site production, conversion and distribution of energy to minimise cost and emissions, by managing the options available in the most cost effective way while meeting all constraints of the system. However many current 'optimisation' applications do not provide full capabilities for maximising economic value or minimising emissions, particularly in situations where demands from major consumers and the prices of the various components can change on an hourly basis.

This paper describes an advanced optimisation platform for managing and optimising utility operation that not only helps planners rapidly optimise equipment selection and load allocation to improve overall efficiency and reduce emissions and operating costs, but also presents operators with a ranked list of possible actions from which they can choose the best course of action given the situation 'on the ground'.

The approach uses medium-fidelity models of the utilities system and major devices that are coupled with plant operating data via data validation and reconciliation facilities. An advanced optimisation system capable of both continuous and integer decisions determines the economically optimal operating point taking into account equipment and operational constraints, including availability. The resulting mathematical problem is solved within an equation-oriented framework, providing robustness and speed of execution well in advance of most current systems.

Because of the speed of solution, a key feature is the ability to run multiple optimisations and provide operators with a ranked list of potential combinations of changes and their corresponding benefits, within a dashboard tailored for the site. This allows operators clearly to evaluate and discuss which changes are the best to apply when, resulting in advice that is practical and easy to implement and verify. It addresses one of the biggest obstacles to practical realisation of the benefit of optimisation systems, which is gaining operator buy-in to proposed changes.

The paper is illustrated with an examples of implementation on a major European chemical manufacturing site.

478714 Study on Regional Air Quality Impact from Olefin Plant Shutdown Operations

Abstract:

Ground-level ozone is a pervasive air pollutant, which can be potentially aggravated by flaring emissions from olefin plant shutdown operations. Although, flaring is crucial to chemical plant safety, the excessive flaring, for instance the intensive flaring from plant shutdown operations, emits huge amounts of ozone pollution precursors (VOCs and NOx). This will cause tremendous industrial material loss and environmental damage; meanwhile result in potential adverse impact to the regional air quality. Thus, the plant shutdown flare emissions should be minimized at any possibility. In this paper, plant-wide dynamic simulations for flare emissions with regional air-quality modeling are coupled together to quantify the air-quality impact and investigate potential opportunities to mitigate the adverse air-quality impact during plant shutdown operations. It proposed a systematic methodology on air-quality conscious study for shutdown operations of an olefin plant. Through studies of original and optimal cases, it shows that the original plant shutdown operation procedure has a serious and negative effect on the regional air quality; however, the optimal shutdown operation procedure would significantly reduce such impact together with the saving of tremendous emission sources. This study can provide valuable and quantitative supports for decision makings involving multiple stakeholders, including environmental agencies, regional plants, and local communities.

2017 AIChE Spring Meeting & 13th GCPS Global Conference on Process Safety

Up-to-date Engineering Solutions to Estimate and Minimize Thermal Breathing Losses from Storage Tanks

Dr. D. Moncalvo & Dr. M. Davies & Dr. T. Heidermann

Some examples of Tank Vacuum Collapse

Maybe you won't hear about it in the news but it surely costs lots of money!

PROTEGO
for safety and environment

Tank Breathing in API 2000 7th Ed. and ISO 28300

Tank breathing in API 2000 7th Ed. and ISO 28300
is distinguished into **Outbreathing** and Inbreathing

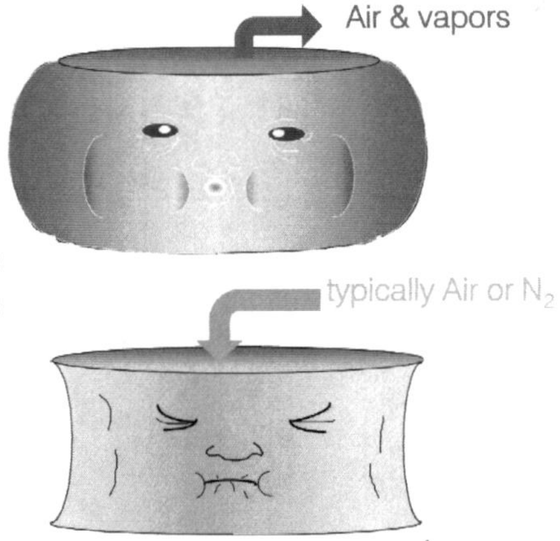

These standards distinguish between two scenarios:

- Normal in- and out-breathing
  - Liquid pumping
  - Weather changes (sun exposure, rain)
  - Internal failure of heat transfer devices
  - Control valve failures
  - Inert purge and utility failures
  - Vaporization or condensation
  - Steam collapse
  - Other causes
- Emergency out-breathing (fire exposure)

PROTEGO
for safety and environment

Tank Thermal Breathing in API 2000 7th Ed. and ISO 28300

$$\dot{V}_{thermal\ out} = Y \cdot V_{tk}^{\ 0.9} \cdot R_i$$

$$\dot{V}_{thermal\ in} = C \cdot V_{tk}^{\ 0.7} \cdot R_i$$

Insulation factor R_i

$$R_i = 1/[1 + h \cdot l_{in}/\lambda_{in}]$$

Not considering condensation
or evaporation of product

Latitude		Y Factor	C Factor			
			Vapor pressure			
			Hexane or similar		Higher than hexane or unknown	
			Average storage temperature			
			< 77°F	≥ 77°F	< 77°F	≥ 77°F
0 - 42°	Houston	0.32	4	6.5	6.5	6.5
42° - 58°	Rotterdam	0.25	3	5	5	5
58° - 90°	Norway	0.20	2.5	4	4	4

PROTEGO
for safety and environment

Tank Thermal Breathing in API 2000 7th Ed. Annex A (*API 2000 5th Edition*)

Tank capacity	Thermal breathing (*SCFH*)	Max. heat transfer rate
Between 20 000 und 180 000 bbl.	2 SCFH each ft^2 of roof and shall area	Between 50 and 100°F/h
Below 20000 bbl.	1 SCFH each bbl. of tank volume	100°F/h

Annex A method is applicable, if following ***Limitations*** are satisfied:

Tank volume is less than 180 000 bbl.
Max. operating temperature in vapor space is 120°F
Tank is uninsulated
Temperature of stored products and feed is below boiling temperature at max. oper. pressure

Not considering condensation or evaporation of product

PROTEGO
for safety and environment

Tank thermal breathing in the scientific literature

Model	Year	Thermal modeling	Evaporation modeling	Condensation modeling	Sensitivity analysis	Measurements
Sigel et al.	1983	X		X		X
Förster et al. (*PTB model*)	1984	X				
Fullarton et al.	1987	X	X	X		X
Holtkötter et al.	1997	X		X		X
Salatino et al.	1999	X				
Brooks (DIERS pres)	2012				X	
Davies and Moncalvo (1)	2016	X		X	X	
Davies and Moncalvo (2)	2016	X	X	X		X

(1)	Journal of Loss Prevention in the Process Industry , volume 43, doi: 10.1016/j.jlp.2016.06.006
(2)	12th Global Congress on Process Safety (GCPS), Houston, April 2016, ISBN: 978-0-8169-1096-0

PROTEGO
for safety and environment

Davies and Moncalvo (2016)

- Balance of energy within the tank

$$\left(M_{vap}c_{vap} + M_{liq}c_{liq} + M_{air}c_{air}\right)\frac{dT_B}{dt} = -h_{int}A(T_B - T_E) - \dot{M}_{air}c_{air}(T_B - T_{amb})$$

- Balance of energy at the tank walls

$$M_E c_E \frac{dT_E}{dt} = h_{int}A(T_B - T_E) - h_{rain}A(T_E - T_{rain}) + \rho_B \dot{V}_{cond}\Delta h_{cond}$$

- Balance of mass within the tank

$$-\left(V_{air} + V_{vap}\right)/T_B \cdot dT_B/dt + \dot{V}_{cond} = \dot{V}_{air}$$

- Condensation model

$$\dot{V}_{cond} = \beta \cdot 1/p_B \cdot dp_{sat}/dT \cdot [T_B - T_E]A$$

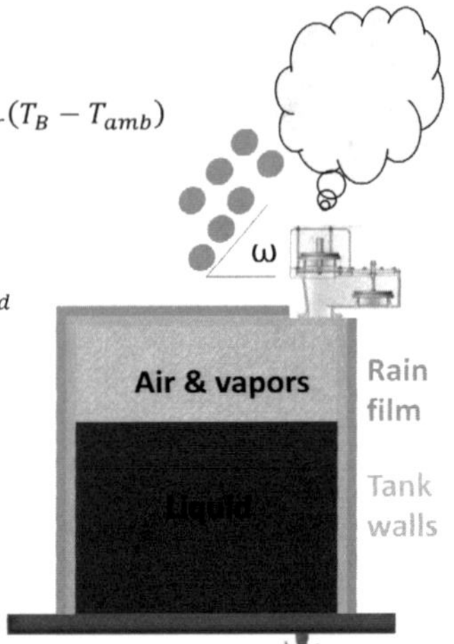

Source	Journal of Loss Prevention in the Process Industry, volume 43, doi: 10.1016/j.jlp.2016.06.006

PROTEGO
for safety and environment

Some realistic examples of thermal inbreathing

	Tank 1	Tank 2
Tank height	49.2 ft.	
Tank diameter	49.2 ft.	82.0 ft.
Tank volume (V_{tank})	16,650 bbl.	46,250 bbl.
Initial temperature (T_0)	120°F	
Ambient temperature	60°F	
Average storage temperature	68°F	
Rain intensity	46.1 lb/h ft²	
Rain inclination	68°	
Wall material and thickness	Stainless Steel; $^1/_5$ in	
Insulation	Mineral wool	
Stored product	Dry Air Saturated Humid Air Saturated Heptane / Air Saturated Methanol / Air	

PROTEGO
for safety and environment

Results for ***uninsulated*** *tank 1 (D = 49.2 ft.)* and *tank 2 (D = 82.0 ft.)*

Required Inbreathing [SCFH]	API 2000 Main body (max)	API 2000 Main body (min)	API 2000 Annex A	PTB (numerical)	Davies Moncalvo *without* condensation	Davies Moncalvo *with* condensation
Dry air	37,179	23,237	16,659	30,931	31,258	31,258
Humid air				30,730	30,682	48,138
Sat. Heptane / Air				19,576	19,608	31,640
Sat. Ethanol / Air				24,761	24,794	52,825

Required Inbreathing [SCFH]	API 2000 Main body (max)	API 2000 Main body (min)	API 2000 Annex A	PTB (numerical)	Davies Moncalvo *without* condensation	Davies Moncalvo *with* condensation
Dry air	76,013	47,508	35,910	68,807	69,576	69,576
Humid air				67,978	68,149	102,569
Sat. Heptane / Air				41,455	41,680	65,512
Sat. Ethanol / Air				53,485	53,807	108,608

PROTEGO
for safety and environment

Worst realistic cases for rain intensity on *Tank 1*

Required Inbreathing [SCFH]		Rain intensity [lb/h ft²]		API 2000 Main body (max)	Davies Moncalvo *without* condensation
Dry air Tank 1	Scenario 1	46.1	API 2000	37,179	31,258
	Scenario 2	62.4	US historical highest (Holt, MO)		34,020
	Scenario 3	85.9	Worldwide hist. highest (Cherrapunji, IND)		36,910

Lesson learned:

API 2000 7th Ed. Main Body is safe for non-condensing products even under extreme weather

PROTEGO
for safety and environment

Reducing Tank breathing: Insulation Layer and Tank Wall Thickness on *Tank 1*

Required Inbreathing [SCFH]	Insulation Thickness [in]	API 2000 Main body (max)	Davies Moncalvo without condensation
Dry air Tank 1	uninsulated	37,179	31,258
	1	8,905	1,364
	2	5,058	949
	4	2,714	551

Required Inbreathing [SCFH]	Wall Thickness [in]	API 2000 Main body (max)	Davies Moncalvo without condensation
Dry air Tank 1	0.2	37,179	31,258
	0.25		28,989
	0.3		27,221
	0.4		24,404

Lessons learned:

Increasing tank wall thickness may not suffice to reduce inbreathing requirements

Perform an engineering and economical study to find out the best insulation strategy for your tank

PROTEGO
for safety and environment

Tests on the influence of overpressure on tank leakage losses

- Leakage, tested with PROTEGO® portable test rig, may begin at 75% of set pressure as per API 2000 5th Edition Appendix C
- The valves were supposed to be fully open at 1 oz./in² (4.3 mbar) gauge.
- The first test was performed at 100 % overpressure
- The second test was performed at 10% overpressure

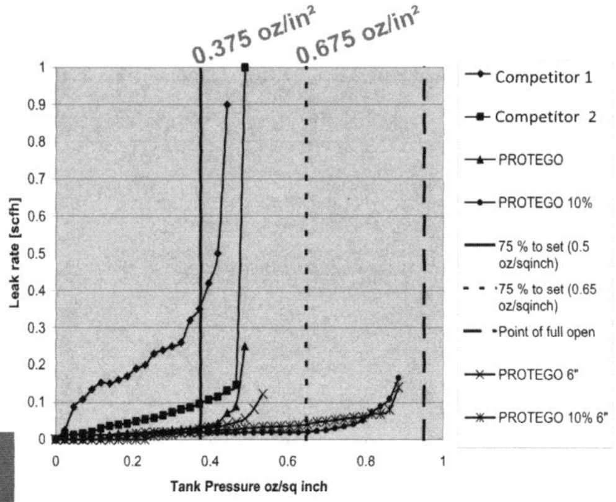

Relieving pressure [oz/in²g]	Overpressure ($\%_{overpr}$)	Set pressure [oz./in² a]	Leakage Pressure [oz./in² g]
1.0	100% (1)	0.5	0.375
1.0	10% (0.1)	0.9	0.675

$$p_{rel\,g} = p_{set\,g}(1 + \%_{overpr})$$

$$p_{leak\,g} = p_{set\,g} \cdot \%_{leak}$$

PROTEGO
for safety and environment

Impact of test results in improvement of API 2000 standard

- Two main results from this research test:
  - Avoid choice of valves requiring large overpressures
  - Leakage pressure at 75% set pressure underestimates leakage losses

- Improvement of API 2000 Standard between 5th (1998) and 7th Edition (2014)

Requirements	API 2000 5th Edition	API 2000 7th Edition
Leakage pressure	75 %	90 %
Blowdown	0 %	Dependent on design

- Additional drawbacks to be considered when selecting a PRV with large overpressures:
  - Leakage pressure of PRV may be higher than set pressure of blanketing devices
  - PRV with low set pressures tend to cycle more (additional cost of spare parts)

PROTEGO
for safety and environment

Conclusions

API 2000 7th Ed. Main Body is safe for tanks with non-condensing gases even under extreme weather

Do not underestimate condensation in atmospheric tanks and perform an engineering study, especially if your tank contains volatile products, like (m)ethanol or gasoline

Increasing tank wall thickness may not suffice to reduce inbreathing requirements

Perform an engineering and economical study to find out the best insulation strategy for your tank (trade-off between costs of insulation purchase and maintenance against savings on breathing losses)

An effective minimization of product losses suggests the selection of PRV with minimum overpressure: the authors support the 10% overpressure as standard also for PVRV

PVRV with low overpressures usually guarantee also a chatter-free opening and avoid conflict with blanketing valves and ERV.

PROTEGO
for safety and environment

Thank you for your kind attention!
Reproduction, in part or in full, subject to prior written approval of Braunschweiger Flammenfilter GmbH.
Braunschweiger
Flammenfilter

480314 Carbon Capture and Reuse in Industrial Scale Continous Biological Processes

Abstract:

Biological processes offer of the lowest energy consuming options for capturing and recycling carbon dioxide into useful sellable or stable products. As tools and technologies continue to advance in transgenics, the possibilities for industrial scale continuous biological processing is becoming a reality. Understanding the state of the art and pathways to address the challenges of scaling-up biological processes for carbon capture and sequestration are critical for advancement of these technologies. This presentation will review the state of the art and evaluate how tools in biochemical engineering (CRISPR) and process engineering will continue to advance biological processes that ultimately rival conventional chemical, petrochemical, and petroleum processes in capability.

479999 Multiple Steady States in Oxidative Steam Reforming of Methanol (OSRM) over CuO-ZnO-Al2O3 Catalyst

Abstract:

Multiple steady states in oxidative steam reforming of methanol (OSRM) over CuO-ZnO-Al2O3 catalyst

Young Shin Jang, Jung Hyun Kim, Jitae Lee, Dong Hyun Kim*

Department of Chemical Engineering, Kyungpook National University

Daegu, 702-701, Korea

1. Introduction

Fuel cell cars are operated on compressed hydrogen. The pressure of the stored hydrogen in the cars is extremely high, around 700 bar, to extend the range between refueling. Nevertheless the range is still relatively short (~ 400 km, Hyundai fuel cell car) compared to ordinary cars using gasoline or diesel. Moreover the fuel cell cars need a new social infrastructure of hydrogen stations. These problems as well as safety issues of handling highly compressed flammable gas act as barriers for widespread use of fuel cell cars. An alternative to storing hydrogen onboard is to produce it onboard from readily available liquid fuel.

Methanol can be regarded as a hydrogen carrier as it can be easily reformed with steam (SRM) to produce hydrogen by

$CH_3OH + H_2O \rightarrow CO_2 + 3H_2$, $\Delta H = 49.5$ kJ/mol (1)

The reaction is endothermic and the SRM reactor has a heat exchange type structure to supply the reaction heat efficiently. On the other hand the reaction heat can be supplied in situ by oxidizing a part of methanol by adding O_2 into the SRM feed. This is called the oxidative steam reforming of methanol (OSRM). As the amount of the heat generation depends on the O_2 content in the feed, an autothermal system (ATR), in which the net enthalpy change is zero, can be made

$CH_3OH + 0.8\ H_2O + 0.10\ O_2 \rightarrow CO_2 + 2.8\ H_2$, $\Delta H = 0$ kJ/mol (2)

In practice, the O_2 content relative to CH_3OH is higher than 0.1 to account for the heat loss of the OSRM reactor. There is still disagreement in the literature on how the O_2 reacts with CH_3OH. Two reactions have been suggested.

$CH_3OH + 0.5\ O_2 \rightarrow CO_2 + 2\ H_2$, $\Delta H = -192.5$ kJ/mol (3)

$CH_3OH + 1.5\ O_2 \rightarrow CO_2 + 2\ H_2O$, $\Delta H = -676$ kJ/mol (4)

Majority of studies on OSRM assumed Eq. (3), partial oxidation of methanol (POM), as the heat generating reaction [1-5], while a few studies have shown that Eq. (4), combustion of methanol (COM), as the oxidation reaction [6,7]. It is one of the objective of this study to experimentally determine which of the two reactions takes place during OSRM.

The exothermic reaction, Eq. (3) or Eq. (4), can give rise to a multiplicity in steady states of OSRM. In the present study, we first report such multiple steady states of OSRM observed experimentally and analyze the states to unravel the heat generation and removal in the OSRM reactor.

2. Experimental

The reactor was made of 10 cm copper tube (3.175 mm OD, 1.55 mm ID). The catalyst in the reactor was a commercial $CuO\text{-}ZnO\text{-}Al_2O_3$ (Synetics 33-5) of a size of 0.2 - 0.3 mm (50 mg) diluted with the same volume of glass beads of the same size. The feed rate was 100 ml/min (CH_3OH:H_2O: O_2:He= 15:20:3:62). The reaction temperature was measured at the outer surface of the reactor. The temperature variation along the reactor was negligible because of the high thermal conductivity of the reactor and the short catalyst bed (~ 4 cm). The reaction product was analyzed online with a gas chromatography. The reaction products were mainly H_2 and CO_2 with less than 0.5 % CO.

3. Results and discussions

Fig. 1. shows the methanol conversion with respect to the reaction temperature. With increasing temperature from 160 °C, the conversion followed branch I until 250 °C where the conversion (X) was 0.13 and the products were H_2O and CO_2, showing that COM took place in the branch I. Above 250 °C, the conversion jumped to B at 275 °C where the conversion was 0.77 and with further increase in the temperature, the conversion followed the branch II from B to C. On decreasing the reaction temperature from C, the conversion followed first branch II and then to branch III until D where the reaction temperature was 220 °C and the conversion was 0.19. A further decrease in the temperature from D quenched the reaction to E where T=209 °C and X=0.02.

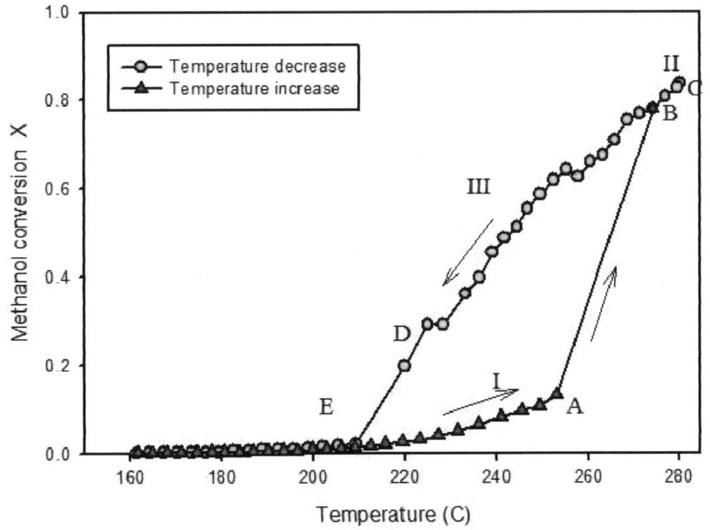

Figure 1. Multiple steady states in oxidative steam reforming of methanol

The O_2 conversion (not shown) was almost 1.0 at A. If O_2 were consumed by POM, Eq. (3), the methanol conversion would have been at least 0.4. The observed X=0.13 at A only corresponds to the combustion of methanol, disproving that POM is an independent single reaction. On branches II and III, the O_2 conversion was always complete and H_2 was produced in proportional to the methanol conversion, showing that SRM followed COM in the catalyst bed.

The multiple steady states were analyzed in terms of heat generation and removal of the reactor [8]. For this purpose, we obtained experimentally the kinetics of COM and the overall heat transfer coefficient between the catalyst bed and its surrounding. The details of the analysis and comparison between the simulation and experiments will be presented at the conference.

4. Conclusions

In oxidative steam reforming of methanol (OSRM), the reaction between oxygen and methanol is experimentally found to be combustion of methanol (COM), Eq. (4), and hydrogen is formed only by the steam reforming of methanol (SRM), Eq. (1). No experimental evidence for the occurrence of the partial oxidation reaction (POM), Eq. (3), has been observed. Multiple steady states of OSRM were experimentally observed and analyzed to be the results of balancing the heat generated by combustion and the heat consumed by steam reforming as well as the heat loss to the reactor surrounding.

References

1. S.Velu , K. Suzuki, M.P. Kapoor, F. Ohashi, T. Osaki, Applied Catal. A: General 213 (2001) 47.
2. J. B. Wang, Chia-Hao Li, Ta-Jen Huang, Catalysis Letters 103(2005).
3. M. Turco G. Bagnasco, U. Costantino, F. Marmottini, T. Montanari, G. Ramis, G. Busca, J. of Catal. 228 (2004) 56.
4. M. Turco, G. Bagnasco, C. Cammarano, P. Senese, U. Costantino, M. Sisani, Applied Catal. B: Env. 77 (2007) 46.
5. S. Patel, K.K. Pant, Applied Catal. A: General 356 (2009) 189.
6. J. Agrell, M. Boutonnet, J. L.G. Fierro, Appl. Catal. A: General, 253 (2003) 213.
7. M. Lyubovsky, S. Roychoudhury, Appl. Catal. B: Environmental, 54 (2004) 203.
8. H. Fogler, Elements of Chemcal Reaction Engineering, 4th ed. Prentice Hall (2006) 533.

480965 Technology Advancement in the Field of Ethylene Oxide and Ethylene Glycol

Abstract:

SABIC is the world largest MEG producer in petrochemical sector and most of its MEG is being produced through the hydrolysis of EO. Ethylene oxide (EO) is a simplest form of epoxides and due to its molecular structure, it acts as a versatile intermediate for a variety of chemicals such as; monoethylene glycol (MEG), di- and triethylene glycols, ethoxylates, ethanol amines, and glycol ethers. The growing demand of MEG owing to its downstream utilization in polyester industry has prompted the research of an efficient, cost effective, and environmentally benign EO production process.

Currently, most of the World's MEG is being produced through conventional route; however, there is an increasing trend to exploit new routes, and feedstocks such as coal, bio-based, and Shale gas. The production capacity of MEG is expected to increase to 36 million tons by 2024. The heavy dependence of MEG producers on ethylene and its scarcity in some parts of the world promoted the need of alternate feedstock.The technological as well as commercial aspects will be discussed along with the future growth of EO/EG business.

483156 Effective Networking

Abstract:

Networking - we have all heard the word, all been told it is important, but do we really know what it means or how to do it effectively? How many times have you been to an event, handed out 20-30 business cards and come away feeling like it was a waste of time. Networking is important and it can help you both professionally and personally, but it needs to be Effective Networking. The best networkers know that networking is not about what you can gain, it's about what you can give, building relationships and developing trust. Every contact is a good contact, the timing just might not be right. This workshop will show you the basic do's and don'ts of networking, explain what Effective Networking is, show you how to perfect your skills and give you time to practice. You will also leave with some tips for awkward networking situations. After the workshop, join us in a special section of the Conference Happy hour and practice the skills you learned in this workshop and the earlier leadership workshops on Mentoring, Unconscious Biases and Personal Branding.

Lori McDowell, workshop leader has been practice effective networking for years. As a sales and business development professional, she has used it to help her business as well as gain lifelong friends. Former President of the Women's Energy Network, Board Member of the National Business Development Association, and active AIChE volunteer and this meetings MPC, Lori will share the skills she has developed over the course of her career.

2017 AIChE®

SPRING MEETING&

13TH

GLOBAL CONGRESS

ON PROCESS SAFETY

March 26–30, 2017
Henry B. Gonzalez
Convention Center
San Antonio, TX

17AIChE
Spring Meeting, San Antonio, TX

13TH GLOBAL CONGRESS ON PROCESS SAFETY

17AIChE
Spring Meeting, San Antonio, TX

13TH GLOBAL CONGRESS ON PROCESS SAFETY

Advanced Analytics to Discover the Root Cause of Multi-variate Control Issues

Steven Obermann
Metso ExperTune
March 28, 2017

17AIChE Spring Meeting, San Antonio, TX | 13TH GLOBAL CONGRESS ON PROCESS SAFETY

What is a Control Issue? (Upset)

- Something that disturbs the steady state behavior of an operating process.
- Something that causes a cycle in an otherwise constant system.

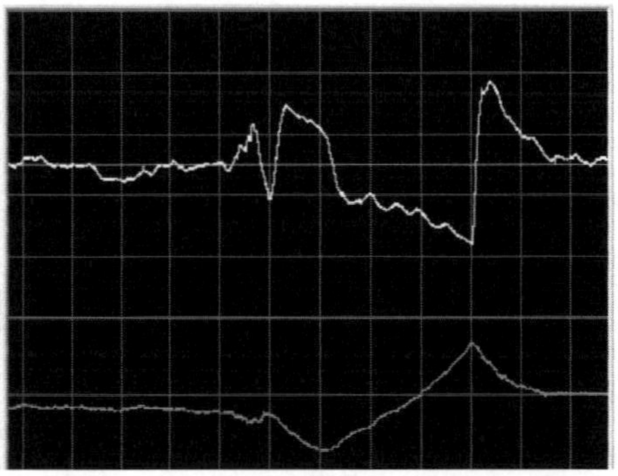

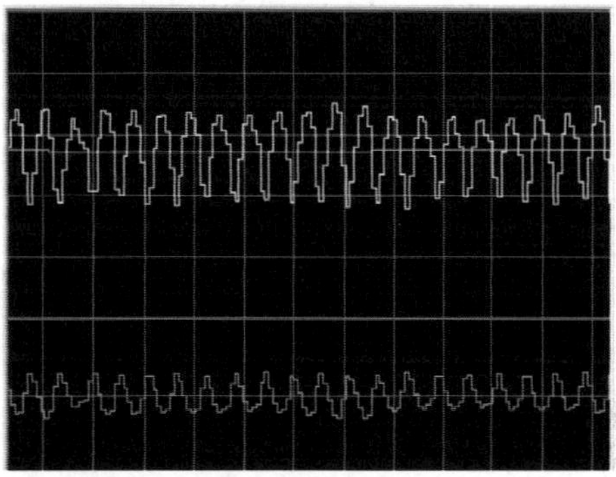

2017 AIChE® SPRING MEETING & 13TH GLOBAL CONGRESS ON PROCESS SAFETY

17AIChE Spring Meeting, San Antonio, TX | 13TH GLOBAL CONGRESS ON PROCESS SAFETY

Where do these upsets come from?

- Poor controller behavior
- Equipment Malfunction
- Process Conditions
- Environmental changes
- Interactions
  - Process Design
    - Recycle Loops
    - Energy Conservation
    - Common Headers
  - Control Design
    - Ratio Control
    - Calculated Setpoints

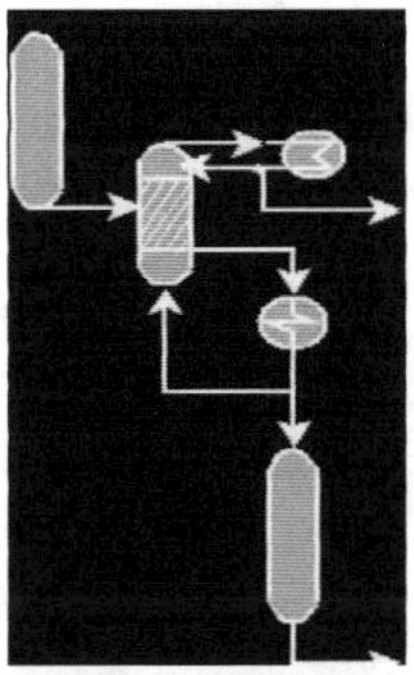

2017 AIChE® SPRING MEETING & 13TH GLOBAL CONGRESS ON PROCESS SAFETY

17 AIChE Spring Meeting, San Antonio, TX

13TH GLOBAL CONGRESS ON PROCESS SAFETY

Why do we avoid getting to the root of these problems?

Lots of process measurements involved
Can be complex
Time consuming
Reduced staff
Needs technical depth & wide process understanding

Often the resources needed are not available

Modern tools are available that remove the barriers

2017 AIChE® SPRING MEETING & 13TH GLOBAL CONGRESS ON PROCESS SAFETY

17 AIChE Spring Meeting, San Antonio, TX

13TH GLOBAL CONGRESS ON PROCESS SAFETY

What do we need to get to the source of these upsets quickly and accurately within the current resource limits?

- "Big data"
- Sophisticated numerical methods
- Intuitive visual rendering of number crunching results

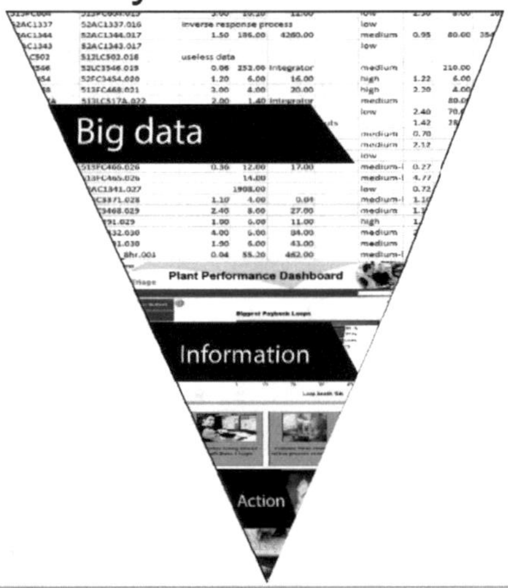

2017 AIChE® SPRING MEETING & 13TH GLOBAL CONGRESS ON PROCESS SAFETY

17 AIChE Spring Meeting, San Antonio, TX — 13TH GLOBAL CONGRESS ON PROCESS SAFETY

You already have the data!

- All important process variables are being recorded.
- Huge amounts of data are available.

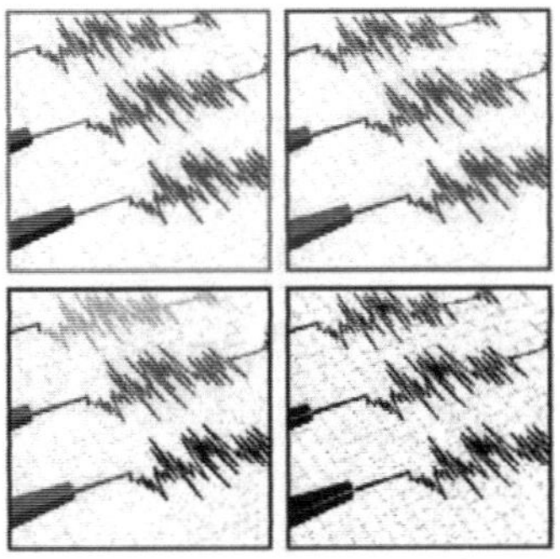

17 AIChE Spring Meeting, San Antonio, TX — 13TH GLOBAL CONGRESS ON PROCESS SAFETY

Apply Some Numerical methods

- Correlate process variables (measurements, control outputs).
- Use frequency information.
- One against many.
- Many against many.

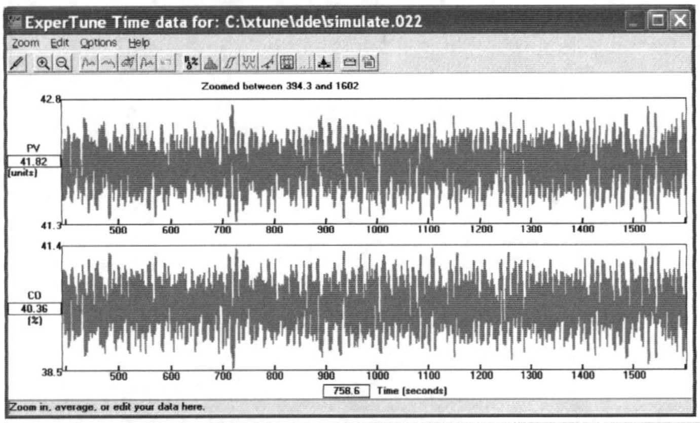

17 AIChE Spring Meeting, San Antonio, TX

13TH GLOBAL CONGRESS ON PROCESS SAFETY

Statistically proven – No interaction

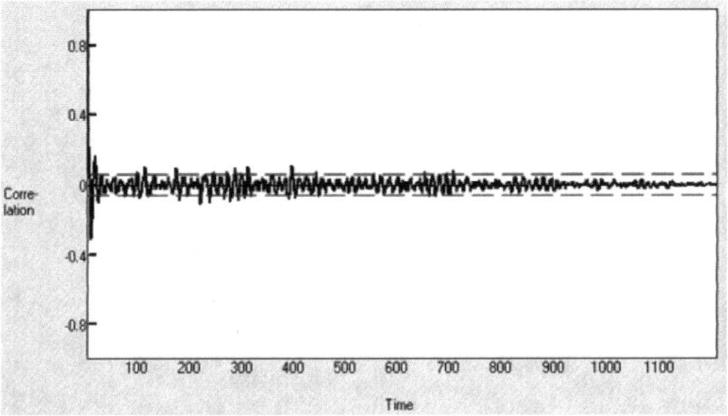

2017 AIChE® SPRING MEETING & 13TH GLOBAL CONGRESS ON PROCESS SAFETY

17 AIChE Spring Meeting, San Antonio, TX

13TH GLOBAL CONGRESS ON PROCESS SAFETY

Interaction exists between these variables

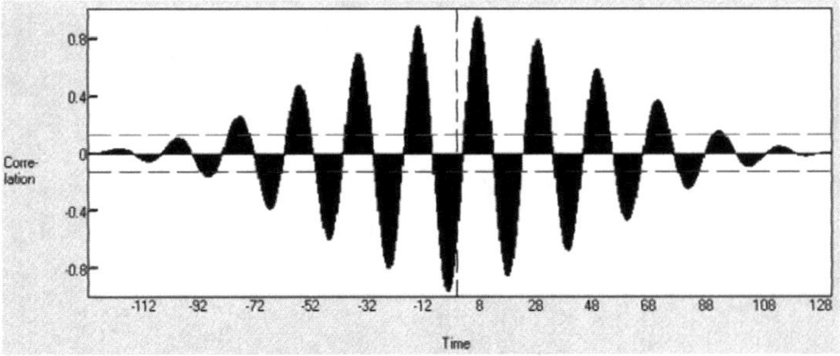

2017 AIChE® SPRING MEETING & 13TH GLOBAL CONGRESS ON PROCESS SAFETY

Process Interaction Map

Intuitive visual rendering of number crunching results

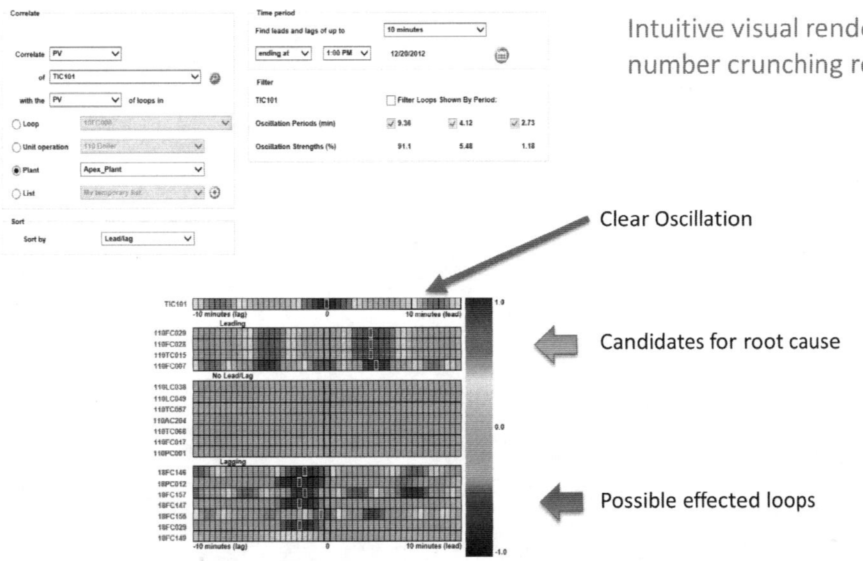

Cause & Effect

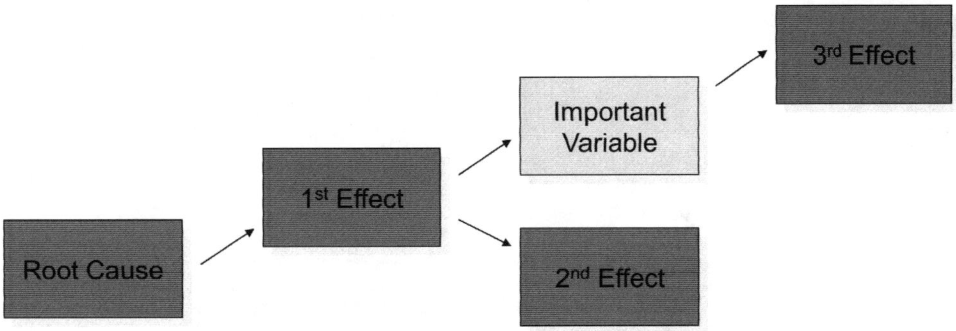

2017 AIChE® SPRING MEETING & 13TH GLOBAL CONGRESS ON PROCESS SAFETY

Even Bigger Data !

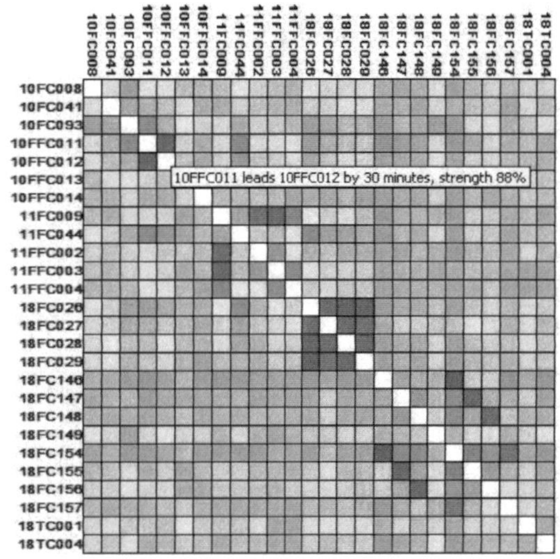

- Days of data; thousands of measurement samples.
- Map out interaction across whole plants / sites even.
- Intuitive visual rendering of the interacting pairs.

17 AIChE Spring Meeting, San Antonio, TX
13TH GLOBAL CONGRESS ON PROCESS SAFETY

Drill Down method

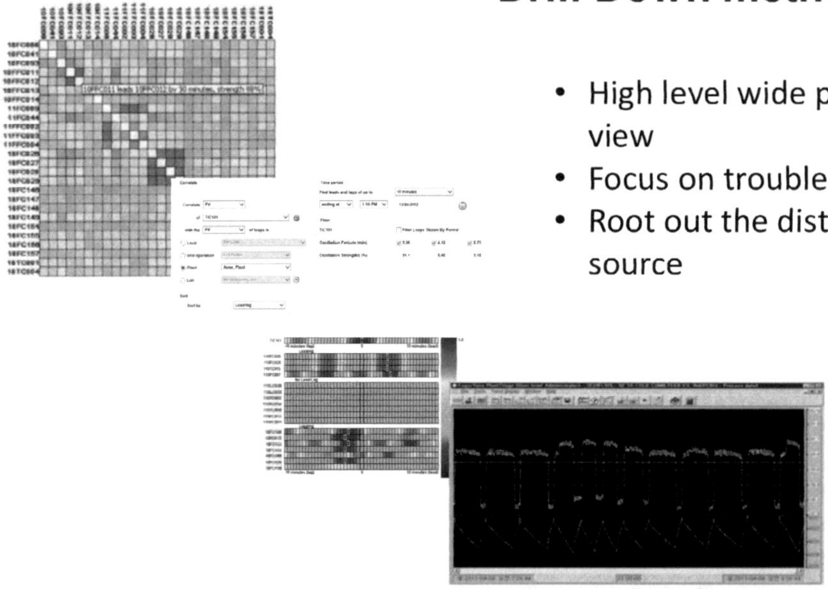

- High level wide perspective view
- Focus on trouble point
- Root out the disturbance source

2017 AIChE® SPRING MEETING & 13TH GLOBAL CONGRESS ON PROCESS SAFETY

17 AIChE Spring Meeting, San Antonio, TX — 13TH GLOBAL CONGRESS ON PROCESS SAFETY

Make Correct Diagnosis

- Oscillation
- Pattern recognition
- Tightly coupled variables

→ Sticking Valve

Measurement

Controller Output

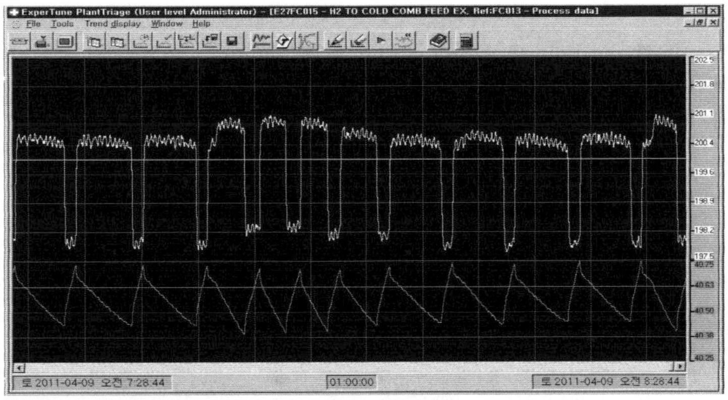

17 AIChE Spring Meeting, San Antonio, TX — 13TH GLOBAL CONGRESS ON PROCESS SAFETY

Conclusions

- Resolving upsets quickly and accurately is a profitable use of resources.
- The data is already in your DCS / Historian.
- Mathematical analysis tools are available.
- Analysis results can present a clear direct path to source of the issue.

17 AIChE Spring Meeting, San Antonio, TX

13TH GLOBAL CONGRESS ON PROCESS SAFETY

Thank You

For more articles and information visit www.ExperTune.com

2017 AIChE® SPRING MEETING & 13TH GLOBAL CONGRESS ON PROCESS SAFETY

Empirical
Process
Solutions

P.O. Box 130581
Houston, Texas 77219
Tel: 713-942-0257
Fax: 713-513-5798
www.EmpiricalProcess.com

Building a Solid Foundation for a Successful APC Application

Keith A. Lehman
Empirical Process Solutions, LLC

17 AIChE
Spring Meeting, San Antonio, TX

Empirical Process Solutions - providing APC services & solutions for the process industries.

Empirical
Process
Solutions

Introduction

- I have been designing and implementing APC/MVC applications for 25+ years, independent since 2004
- APC/MVC technology, design, implementation, maintenance, etc. are all keys to success, but ...
- The importance of the "foundation" is often overlooked and under-appreciated (*most important IMO*)
- Any application is only as good as the weakest link

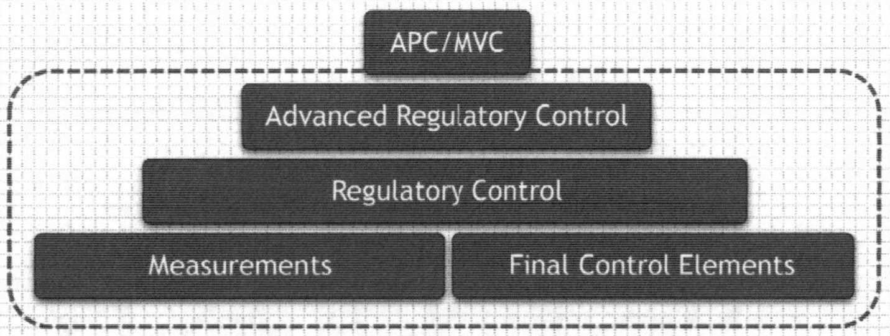

Empirical
Process
Solutions

Inputs & Outputs
Pressure & Temperature Measurements

Pressure Usually measure DP between the process and atmosphere, or two process connections

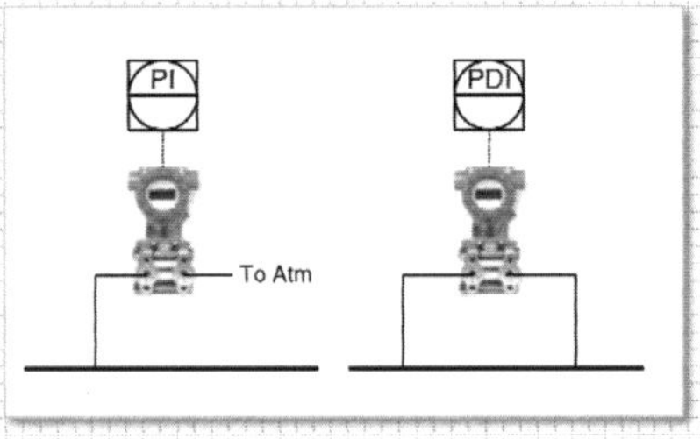

- Plugging, fouling and phase change in the tubing may affect the pressure reading
- Differential pressure can be measured directly or calculated from two individual pressure transmitters

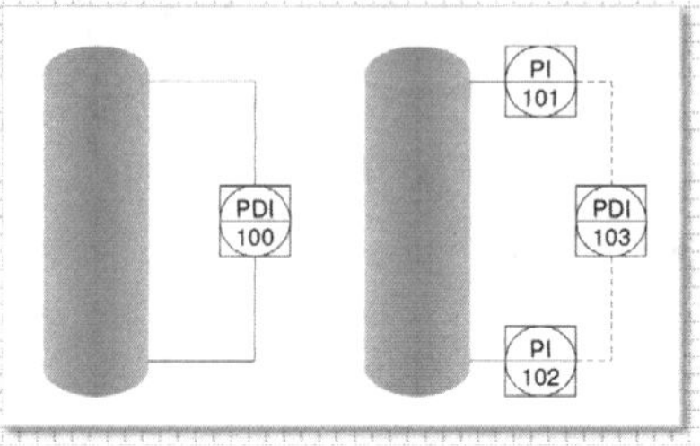

Temperature Thermocouples vs Transmitters

Thermocouples have a large range, and therefore have poor resolution

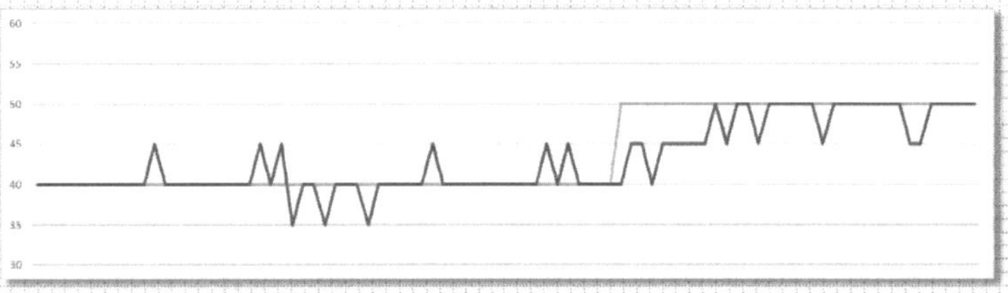

- Try to use temperature transmitters for control, and as inputs to important calculations if possible
- If you need additional temperature measurements for an APC/MVC application, remember to look for existing thermowells!

Empirical
Process
Solutions

Inputs & Outputs
Flow Measurements

Flow Most are based on DP and use design conditions and a constant SG to calculate flow

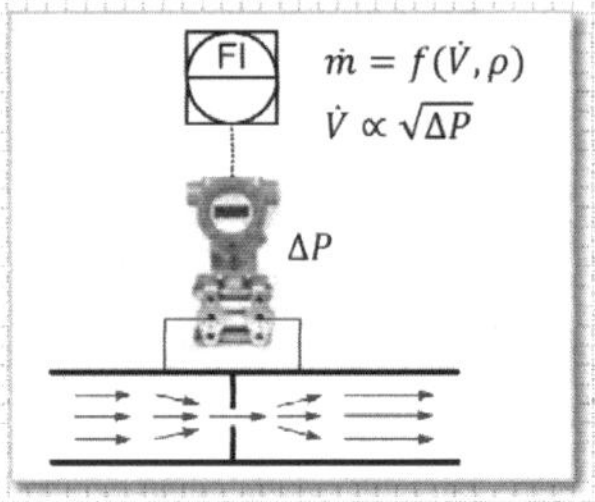

- Flow = f (ΔP, SG and flow element details)
- Changes in process conditions and stream composition can affect accuracy
- Pressure & Temperature compensation can help (P & T comp for gases, T comp for liquids)
- Plugging, fouling and phase change in the tubing may affect the pressure reading

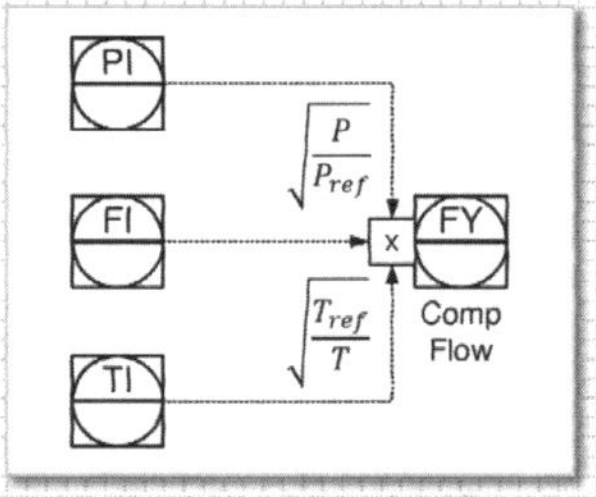

Flow measurements can be extremely noisy or unreliable at the low end, especially if the overall range is very large

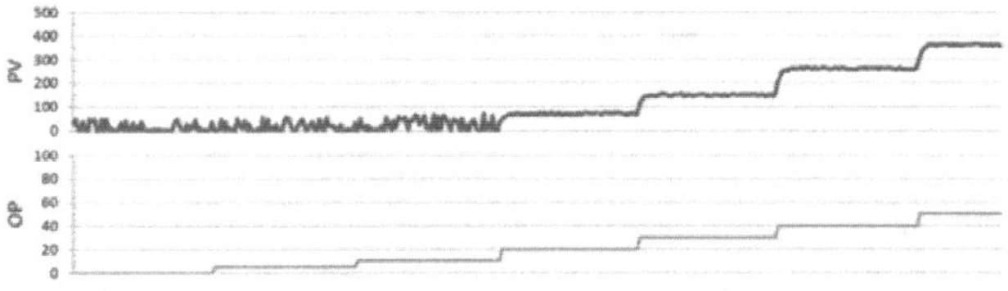

This makes it difficult to control in AUTO (or to use SP as MV)

If necessary, consider using a combination of low and high range flow meters

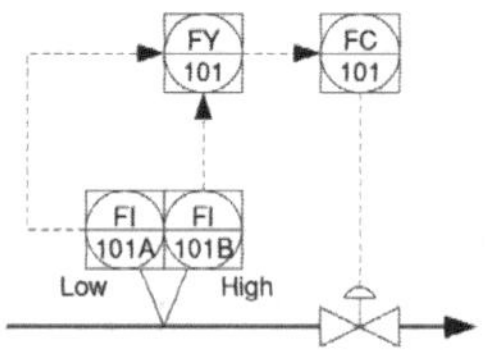

Hint: Remember to square-root the DP (only once)! And hopefully the orifice isn't backwards!

Empirical Process Solutions

Inputs & Outputs
Level Measurements

Level Most common level measurements are based on DP and use a constant SG to infer level

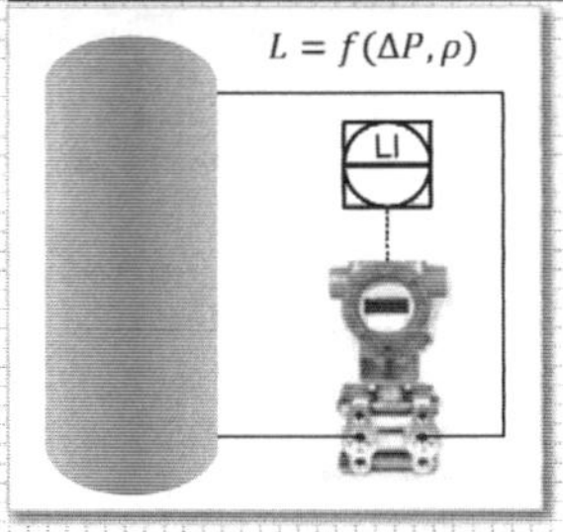

- Level = f (ΔP, SG)
- Plugging, fouling and phase change in the tubing may affect the DP reading
- Horizontal drum levels may need to be linearized, depending on the level span
- Improper calibration, foaming and/or changes in material composition (SG) can result in false level readings
- Errors in min/max level readings can cause dangerous situations, may not be apparent during normal operation (with SP=50%)

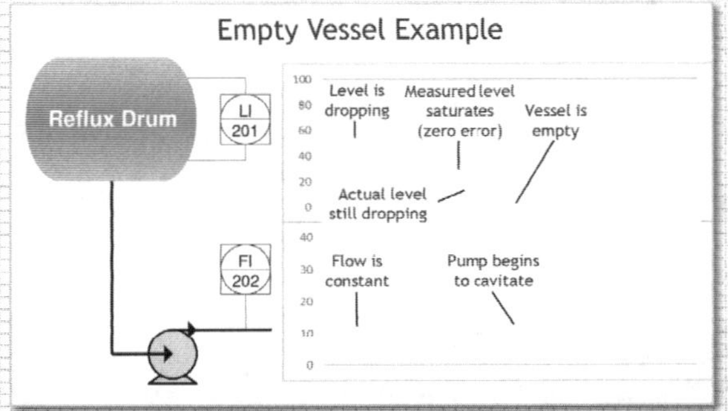

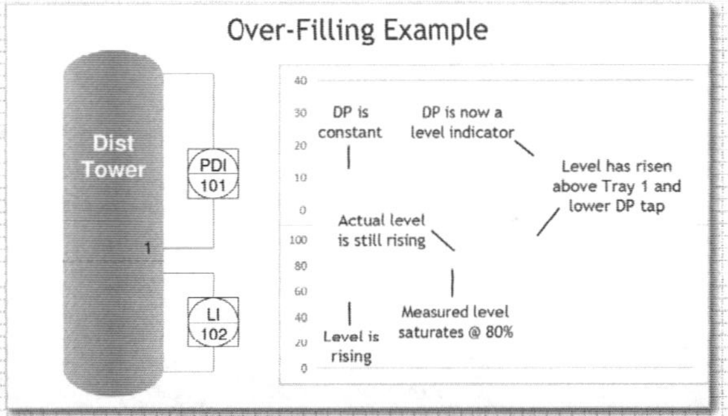

Empirical Process Solutions

Inputs & Outputs
Analyzers & "Black Boxes"

Analyzers Many APC/MVC applications rely on process analyzers

- Calibrate & Validate!
  - May need to improve preventive maintenance program
  - Use signal validation (high/low, jump, freeze)
- Most common problems:
  - Sample system (low/no flow, phase change)
  - Incorrect calibration material
  - Stream switching & purging
- Consider data resolution (analyzer or data link)
- For intermittent analyzers, such as GCs:
  - Avoid (or be aware of) excessive cycle times
  - APC/MVC applications need a "NEWPV" flag

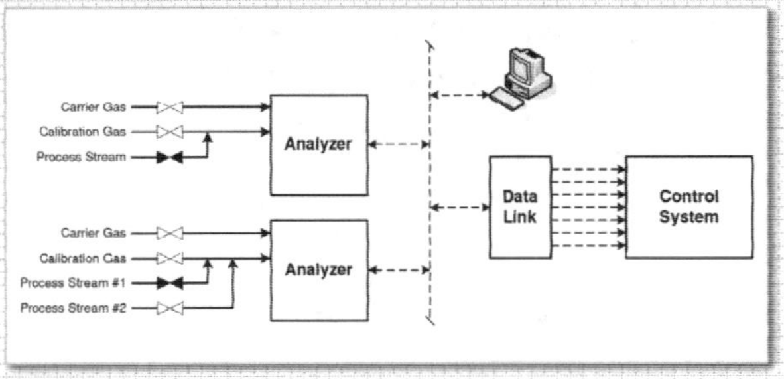

"Black Boxes" Machinery control and other independent systems can create challenges

- "Black box" systems often contain their own controls, overrides, limits, switches, etc.
- Internal settings & parameters may be hard to find, and even harder to change
- Validate inputs to protect against data link issues
- Consider data update frequency & resolution

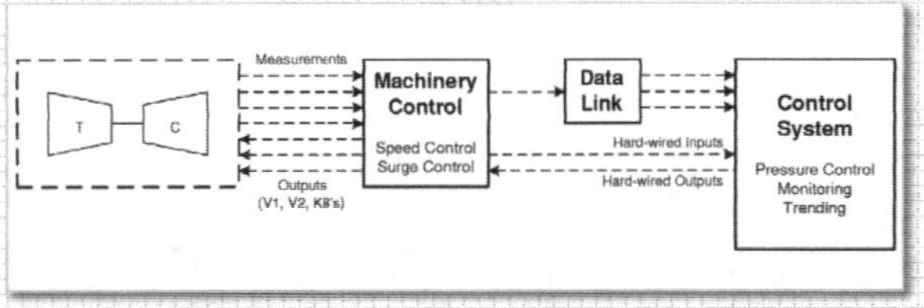

Empirical Process Solutions

Inputs & Outputs
Control Valves & Other "Final" Elements

Control Valves Most common final control element - *many* types, *many* parts, *many* potential problems

- Stiction (usually overly tight packing)
- Hysteresis (mechanical linkage)
- Non-Linear flow characteristic
- Positioner overshoot
- Incorrect valve size or type
- Incorrect valve calibration
- Instrument air problem / leaking diaphragm

Many of these problems can be fixed, or at least improved, by installing a smart positioner.

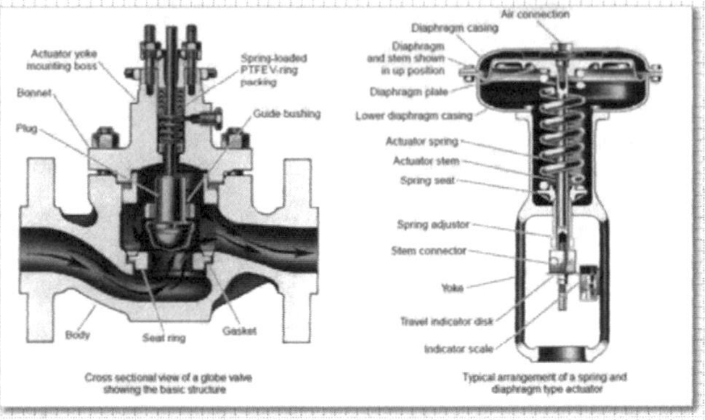

Other Final Elements:

- Damper, Guide Vanes: Stiction, Hysteresis, Calibration
- Variable Speed Drive: Hysteresis / Dead Band
- Discrete Devices: Logic, Timing, Staging (multiple)
- Feeders (solids): Stiction, Hysteresis, Calibration

Before attempting to tune a control loop, try to identify & fix problems with the valve or other final element.

Empirical Process Solutions

Inputs & Outputs
Control Valve Mechanical Issues

Stiction Valve internals get "stuck" and require additional force to get the valve to move

- "Stiction" = Static Friction
- Characterized by "sawtooth" OP pattern and "square-ish" PV pattern
- The pattern changes depending on how close the SP is to an "achievable" PV

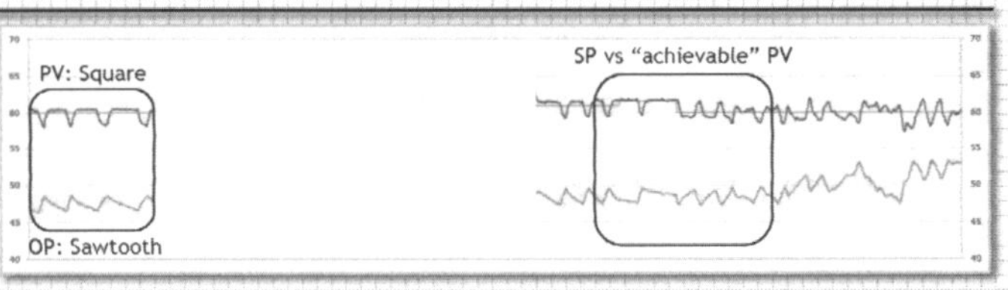

- Tuning can change the behavior or pattern, but doesn't fix it
- OP response is inconsistent and hard to model, especially for small SP changes
- Nearly impossible for random disturbances

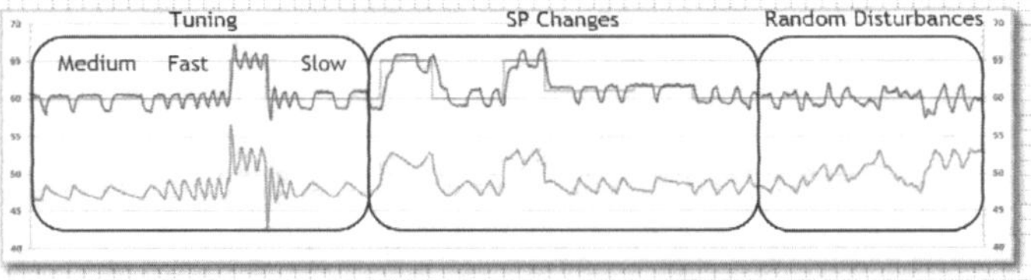

- Making a series of small OP changes can help identify the stiction "threshold"
- Often the result of over-tight packing (*Performance* vs. *Environmental*)
- Consider installing a smart positioner

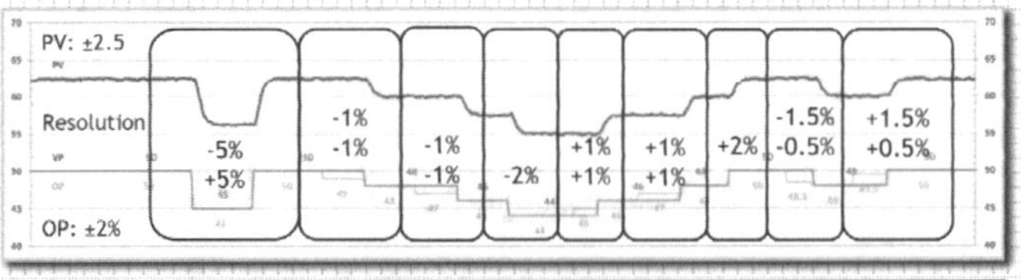

Hint: Instrument techs will "stroke" the valve, and then say this is the best the valve can do.

Empirical Process Solutions

Inputs & Outputs

Control Valve Mechanical Issues

Hysteresis "Slack" between controller output and valve position when the output changes direction

- Hysteresis is also known as Dead Band
- Hysteresis can be difficult to see during normal operation
- Often mistaken for poor tuning or the unavoidable effect of random disturbances

- Hysteresis may also be hard to see if SP changes or disturbances are large
- A planned series of small SP changes or load disturbances can reveal the problem
- Equivalent steps in opposite directions will produce different OP responses

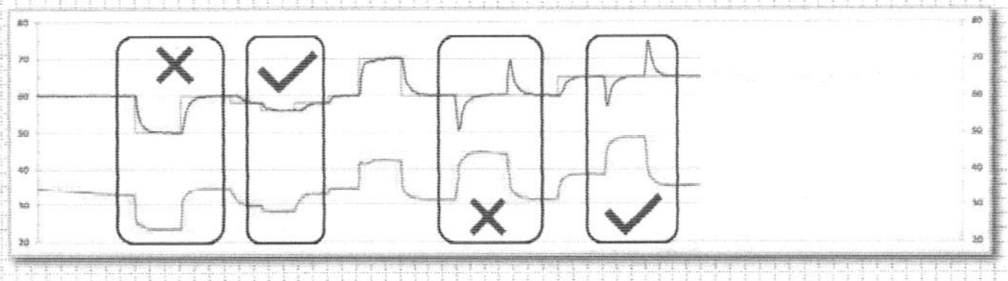

- Making a series of small OP changes can help identify the hysteresis "threshold"
- After passing the threshold, the valve moves normally - until it changes direction

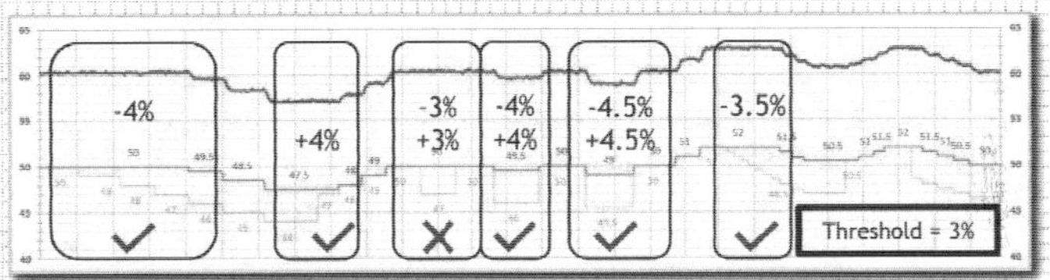

HINT: If the hysteresis behavior is consistent, a custom control algorithm can be used to accumulate moves within the "threshold".

Empirical Process Solutions

Regulatory Control

Configuration & Tuning Objectives

The Ultimate Goal

Robust	• Consider operating range, operating conditions, instrument reliability • Design for usability (operations), maintainability (engineers)
Responsive	• Maintain/repair instrumentation & valves, tune control loops • Use advanced regulatory control techniques where applicable
"Linear"	• Use linearization techniques for control valves & process variables • More consistent response over a wider operating range
"First Order"	• Avoid overly aggressive tuning that might upset the process • Use decoupling techniques to "simplify" dynamics of interactive systems

Configuration:
- Use advanced regulatory controls
- Everything in its proper place...
  - Some controls & calcs should be at the DCS level
  - Other things can be done in APC/MVC, etc.
- Reliability / Usability / Maintainability
- Bumpless Transfer:
  - Initialization / Tracking / Shedding / Wind-Up

Tuning:
- Lots of tuning philosophies, techniques & tools
- My approach is a practical combination of experience, trial-and-error and empirical methods
- Consider control objectives: tight SP control, minimal OP movement
- Tuning software can be helpful, especially for identifying & prioritizing problems

Hint: Think in terms of PVs & OPs becoming CVs in a dynamic model.

Empirical Process Solutions

Regulatory Control
Tuning Rules of Thumb

Prioritize & Classify
So many loops, so little time...

- Focus on the most critical loops (biggest process or operational impact)
- In a complex control structure, tune the fastest loops first
- Apply the right configuration & tuning methods

Process Dynamics
In the world of control, dynamics rule...

- Execution frequency: all tags, calculations, and control loops should execute faster than the one(s) above
- Filtering: use the minimum amount of filtering necessary, the filter becomes part of the process dynamics
- Cascade Control: secondary control loops should be 4-5x faster than their primaries

Process Characteristics
All processes are <u>not</u> the same...

- Self-Regulating Process:
  - PV settles out at a new equilibrium value after an OP change or load disturbance
  - Use PI control with Proportional on *Error*
- Integrating Process:
  - PV continues integrating (or ramping) after an OP change or load disturbance
  - Use PI(D) control with Proportional on *PV* (or SP filtering) - allows larger gain for disturbance rejection
- Pseudo-Integrating Process:
  - Processes with very long dynamics can be treated as integrating processes

Empirical Process Solutions **Building a Solid Foundation for a Successful APC Application 11**

Empirical Process Solutions

Regulatory Control
Linearization

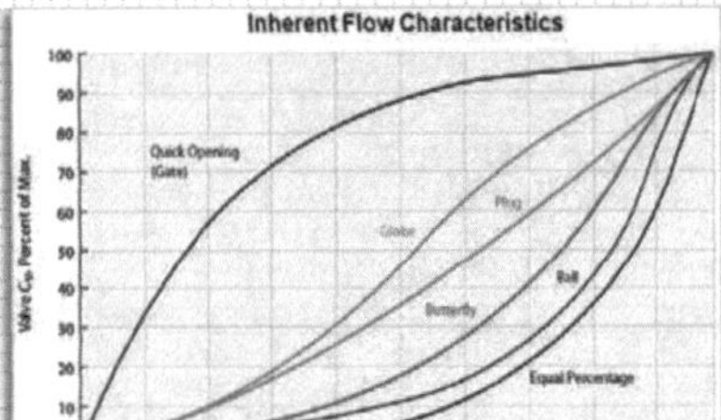

Most (all) control valves have an inherently non-linear OP vs flow characteristic

The installed characteristic of the valve will not match the design (variable DP, operating conditions)

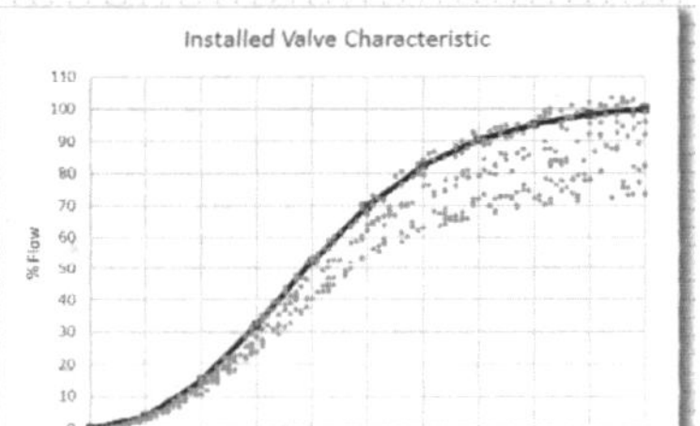

Inconsistent response, poor performance and/or unstable control in certain OP regions

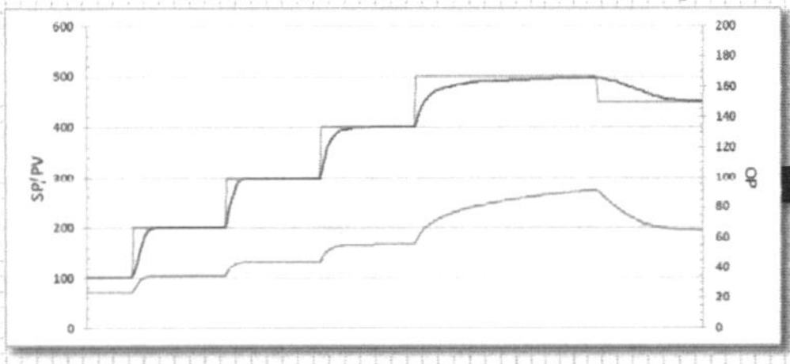

PV vs OP and K vs OP

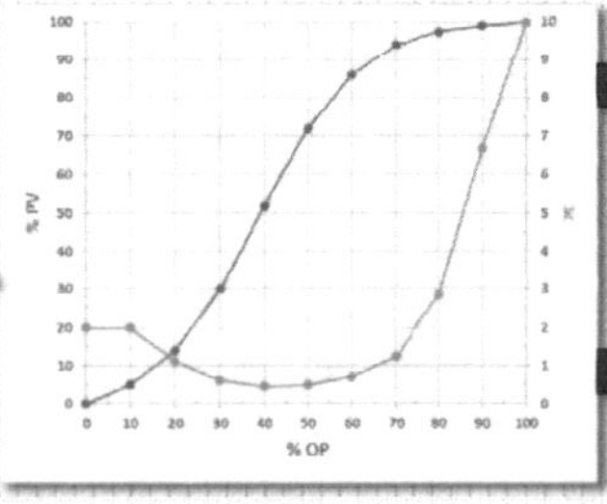

OP Characterization

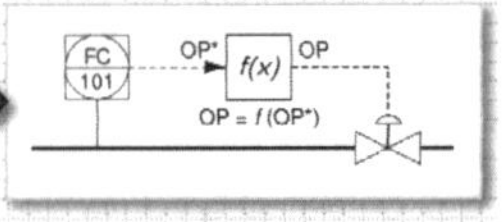

Gain Scheduling

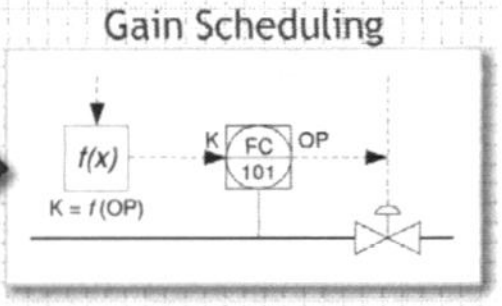

Hint: Linearization techniques can also be used for non-linear process variables such as pH and composition.

Empirical Process Solutions **Building a Solid Foundation for a Successful APC Application 12**

Empirical Process Solutions

Regulatory Control
Multiple Control Valves

Two (or more) control valves or other final elements in a series or parallel arrangement may be used to control a single process variable or multiple process variables.

DMC*ism*: Extra degrees of freedom are not "free".

For multiple control variables:
- Tune one loop fast (more important and/or faster responding) and tune the other loop 4-5x slower
- Or use FF to decouple (fast → slow)

For a single control variable:
- Use split-range control (linearization, overlap)
- Or use one valve to control the process variable and use the other to control its valve position

Two Valves in Parallel (Two Control Variables)

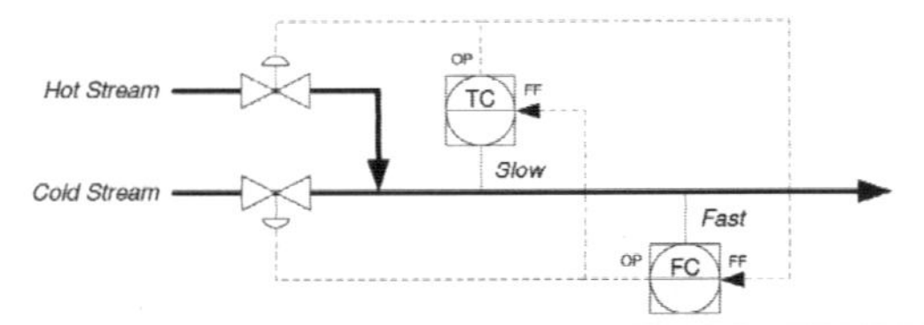

Two Valves in Parallel (Single Control Variable)

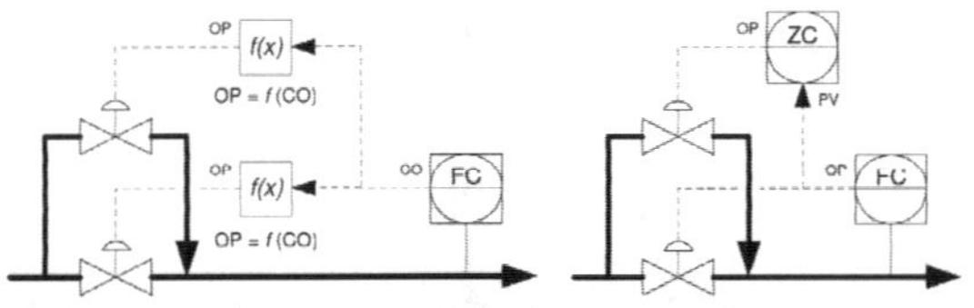

Two Valves in Series (Header Pressure & Flow)

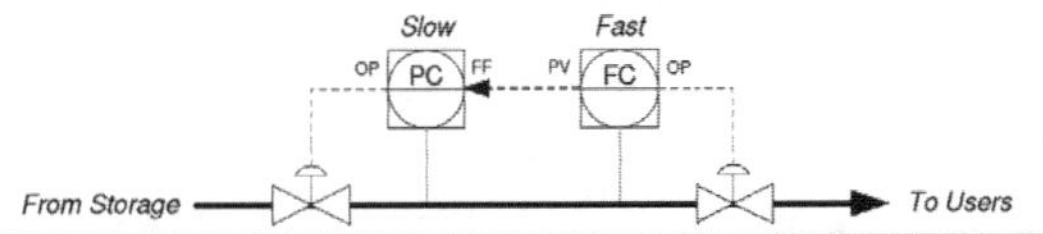

Fun With Fans!

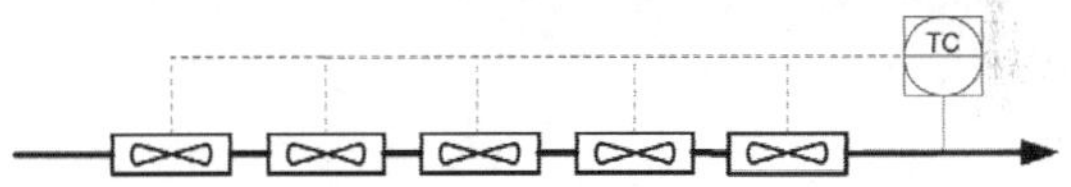

Empirical Process Solutions

Regulatory Control
Overrides, Clamps, Bypasses, etc.

Overrides & Clamps Represent potential unexpected AWS/saturation conditions

- Be aware of all overrides, SP/OP clamps and rate-of-change limits within the control structure
- Local overrides & clamps: Not much we can do
- Include as constraints in APC/MVC & disable if possible
- Be sure to pass the correct AWS status and SP value to APC/MVC

Manual Valves & Bypasses Can significantly change process gains and characteristics

- Be aware of all manually-operated valves, including control valve bypasses, that can impact APC/MVC
- If possible, try to use valve position or status information in the APC/MVC application
- A valve position indicator is the best, but least likely, option
- An operator-entered value can be used - with extreme caution

Other Considerations

- Non-Linear Controllers (Gap Control, "Averaging" Level Control) can cause inconsistent process responses
  - APC/MVC impact (good or bad) depends on their location in the control structure
- Changing Process Flow & Equipment Line-Ups should be considered as part of APC/MVC design & structure
  - Can significantly change process behavior, may even invalidate some cause-effect relationships

Empirical Process Solutions

Advanced Regulatory Control
Overview

We can use Advanced Regulatory Controls, along with known relationships & correlations, to make the process more:

ROBUST **RESPONSIVE** **"LINEAR"** **"FIRST ORDER"**

Cascade Control	Examples: Level → Flow, Temperature → Duty
FeedForward Control	Improves disturbance rejection and helps with decoupling
Ratio Control	Examples: Reflux/Feed, Steam/Feed
Other	Linearization, Decoupling, Model-Based Control, etc.

Empirical Process Solutions

Advanced Regulatory Control
Cascade Control

DMC*ism*: Feedback is your friend.

Cascade Control is the most common type of Advanced Regulatory Control

One control loop (secondary) is used to control another loop (primary) in a cascade configuration

Level to Flow Cascade

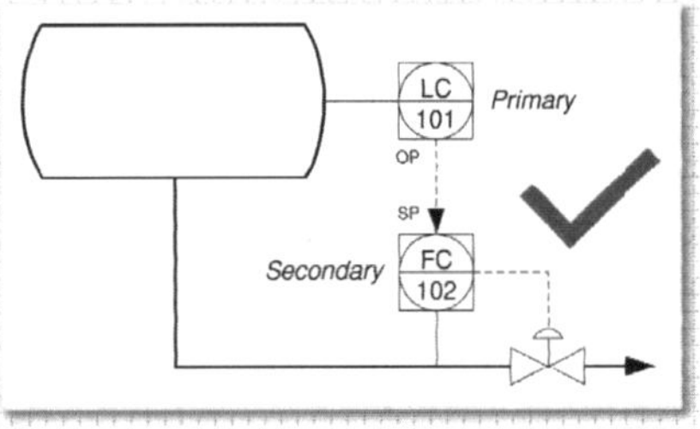

Checklist:
- ✓ The flow certainly has an effect on the level
- ✓ The flow controller will reject pressure disturbances
- ✓ The flow loop dynamics are much faster than the level

Furnace Outlet Temperature

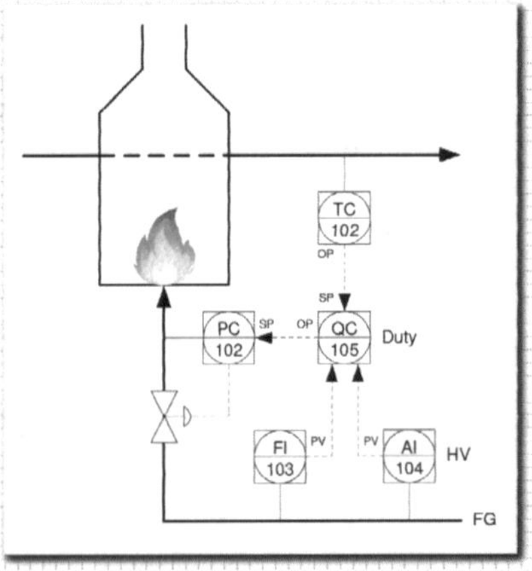

Cascade Control Requirements:
- The secondary loop must directly affect PV of the primary loop
- The secondary loop should help prevent disturbances from affecting the slower primary loop (or it's not necessary!)
- The dynamics of the secondary loop should be 4-5x faster than the primary loop

Hint: Make sure you consider bumpless transfer (*Initialization / Tracking / Shedding / AWS*).

Empirical
Process
Solutions

Advanced Regulatory Control
FeedForward & Ratio Control

FeedForward Control can help improve a control loop's response to disturbances by acting proactively, instead of reacting to a deviation through feedback.

There are two main types of FeedForward Control:

FeedForward (additive) Control

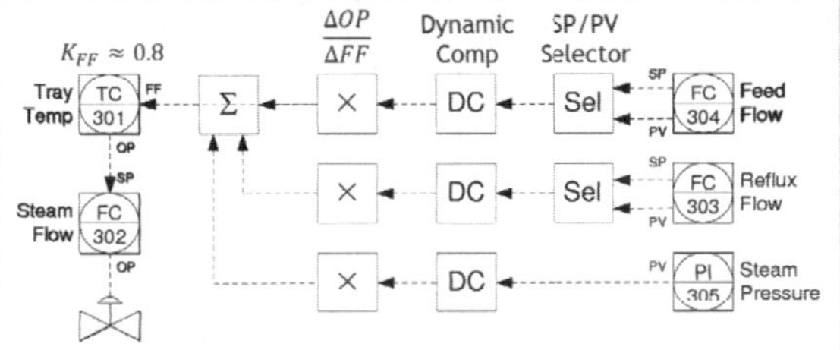

- FF Control should always be combined with feedback trim
- General requirements for a good FF candidate:
  - Must be measurable & reliable
  - Should cause a consistent, predictable process response
  - Provides enough lead time to be better than feedback
- Dynamic compensation can help avoid inverse responses
- Calculate actual FF gain, and then apply $K_{FF} < 1.0$
- Multiple FF signals can be individually scaled & combined

Ratio Control

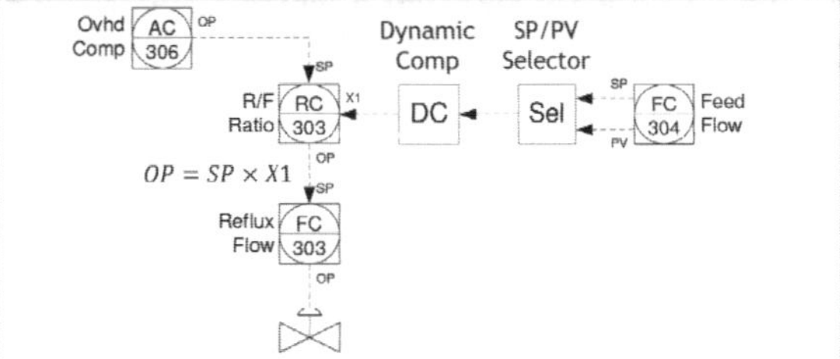

- Ratio Control is useful for process variables that generally need to be adjusted in proportion with another variable (like distillation tower reflux vs feed flow)
- Most commonly used as a secondary in a cascade loop
- Same general requirements as FF control
- Best practice is to use a Ratio Control algorithm (not PID)
  - Avoids adding unnecessary dynamics
  - Handles the inherent non-linearity of a ratio PV better

DMC*ism*: A bad FF is worse than no FF.

Empirical
Process
Solutions

Conclusion

- An operating plant usually has *multiple* constraints
- The *optimum* operating point is at an intersection (or corner) of constraints
- We can shrink (or tighten) the operating range by fixing instrumentation, tuning control loops and using advanced regulatory control techniques
- Then, *and only then*, we can use APC/MVC to effectively and consistently optimize plant operations

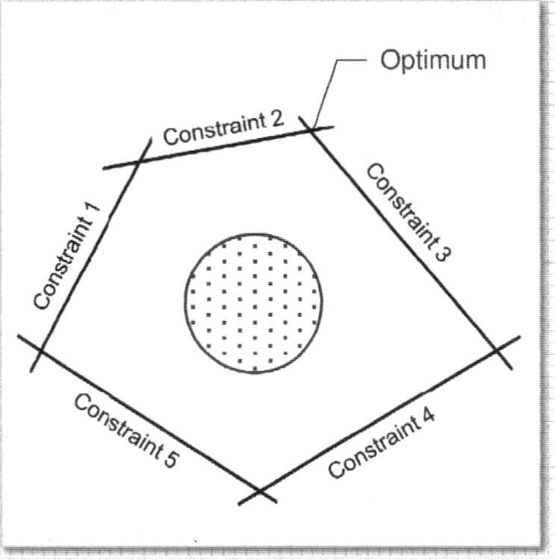

A solid foundation is essential for a successful APC application...

Any application is only as good as the weakest link

481598 Put the Correct Economics into Your Multivariable Controller

Abstract:

Probably the major shortcoming of Multivariable Control Applications are the poor costs put into the economic optimizers in the controllers. The current practice in most applications of the technology is to put cost into the optimizer to make the controller go where Operations feels comfortable. This approach makes people content, but usually leaves significant money on the table. The control engineer goes along with this scenario either because he does not want to fight the politics of the status quo or because he does not know how to put the correct economics into the optimizer. Putting the correct economics in is not as simple as it may appear from the surface. The complexity comes from the need to integrate the Dynamic Controller Model with a Steady State Yield Model. Some of the independent variables in the Yield Model do not match those in the Controller Model. This paper will give the details of the math required to put these models in the right context.

480087 Real-Time Performance Monitoring and Diagnostics of Rod Pump System for Oil Elevation Using Streaming Dynamometer Card Data

Abstract:

Rod pump systems are the most commonly used mechanism for oil lift in the world. The ability to monitor the rod pump operation and diagnose any impending problems with the pump and/or the well is therefore of primary interest to oil field operation. Downhole dynamometer cards (obtained by plotting the measurements of the rod displacement vs. the computed load on the pump, originally developed by Walton E. Gilbert) are viewed as the main information source for performance monitoring and diagnostics of the rod pump and the well today.

Methods to deduce the operating condition of the well and the pump from downhole dynamometer card data are reported both by academia and industry. The use of downhole dynamometer cards are at the heart of many commercial software/hardware products that target performance monitoring and diagnostics of rod pump systems in oil fields around the world.

In this paper, we present an algorithm for automated performance monitoring and diagnostics of the well and pump operating conditions where the streaming card data is processed in real-time. The proposed algorithm uses a hybrid classification approach where both labeled and unlabeled card data would be consumed to generate insight for the operators. The key objective here is to enable the detection of an anomaly in the operation by maximally utilizing prior knowledge while enabling automatic capture of the unique characteristics of a well/rod pump unit solely based on the live operation data.

480732 Challenges in Applying Data Analysis to Industrial Processing: Inverse Problem Viewpoint

Abstract:

Production data analysis is often a regular part of the evaluation and troubleshooting for operations. This is especially true when looking at issues of quality, conversion rates, production rates, or rates of degradation in the process. Ideally, it would be valuable to understand what is actually happening inside the production units. In particular, production problems would be easier to address if it could be determined where and when in the process significant changes are occurring. This can be considered an inverse problem of determining the process configuration that gives the measurements seen.

The underlying physics can provide insights to data analysis and is probably necessary to solve the "Inverse Problem". For example, the physics of a fixed bed reactor will produce relationships about the flows, pressure drop, temperature, heat generated (absorb), and stream composition changes. Applying these physics in an industrial setting will require numerous assumptions about the process. There may be competing assumptions that need to be considered. For the reactor example, these assumptions will be capture in parameters that designate the flow resistance and reaction kinetics. The competing assumptions will generate different model structures and possibly different types of parameters. The analysis of the production data will drive refinements of these parameters.

This presentation examines the challenges and tools for analyzing production data and for determining the appropriate physic driven models that can capture the behavior of the production units. This will include approaches for doing time series analysis, uncertainty quantification, principle component analysis, machine learning, objective function determination and data structure analysis. In particular, we will look at how these tools can provide different insights into the underlying production behavior.

Investigating Polymeric Compositions
via Clustering of Production Analytical Characterization Data

Alix Schmidt, Jennifer Wightman
3/29/17

Agenda

- Silicone Resin Overview
- Introduction to Data
- Clustering
- Interpreting Cluster Results
- Conclusions

2

Silicone Overview

From quartz to highly specialized materials, >7,000 products

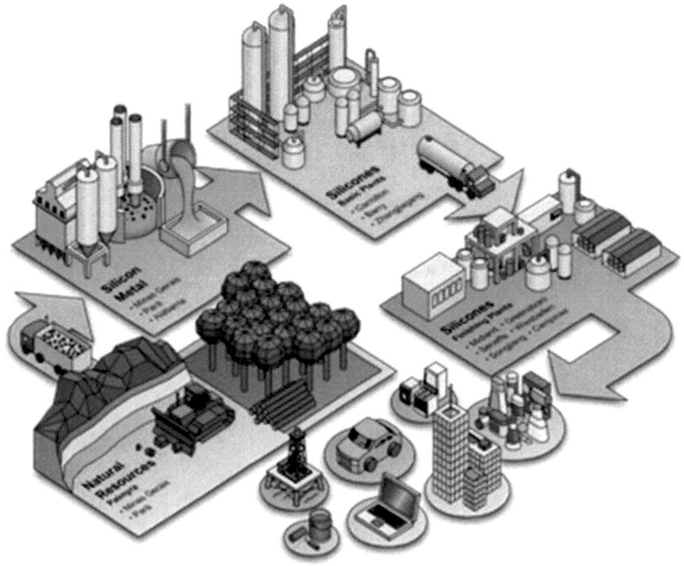

3

Silicone Resin Production

Nomenclature: M = $R_3SiO_{1/2}$, D = $R_2SiO_{2/2}$, T = $RSiO_{3/2}$, and Q = $SiO_{4/2}$

Hydrolysis and Condensation of Si monomers to grow MW:

- Chlorosilanes, Alkoxysilanes, Silicates

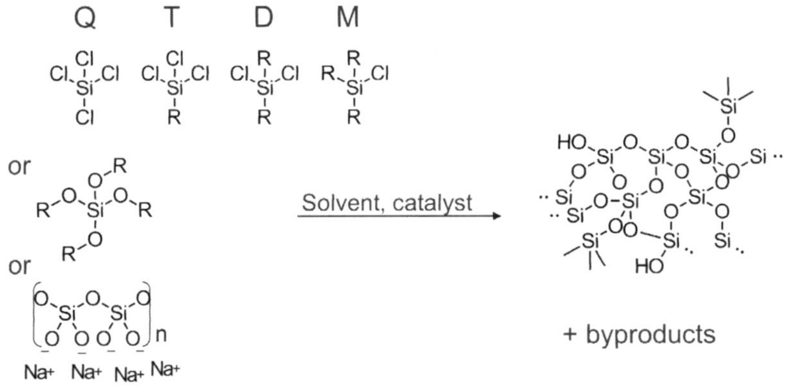

4

Silicone Resin Types

M = $R_3SiO_{1/2}$, D = $R_2SiO_{2/2}$, T = $RSiO_{3/2}$, and Q = $SiO_{4/2}$

MQ

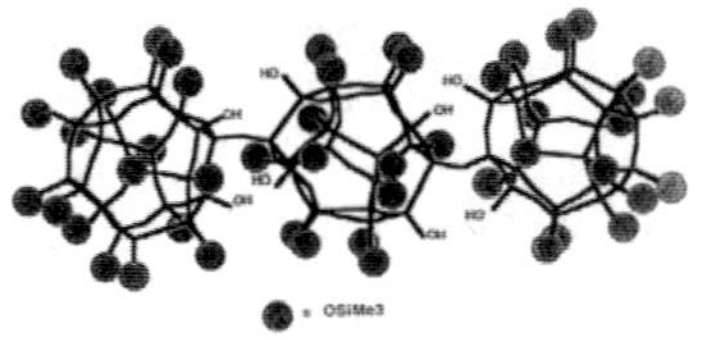

$MT^{Ph}D^{Me}$ [1]

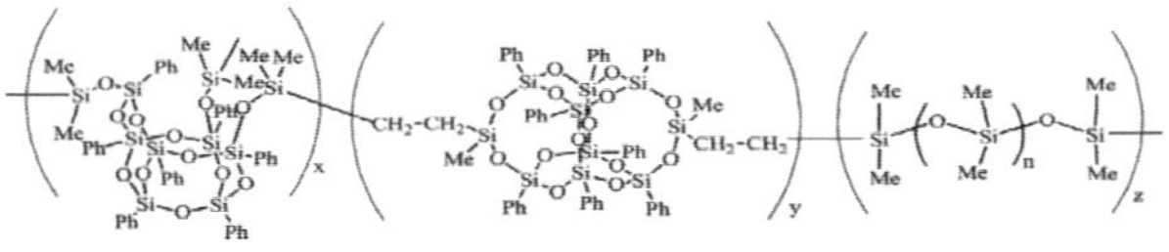

T

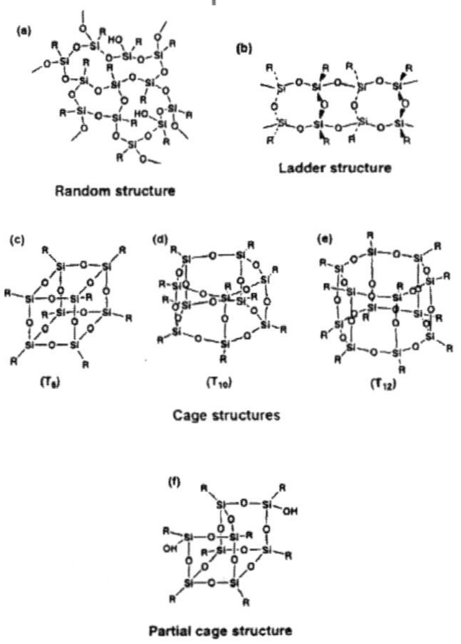

Endless possibilities for structured, amorphous, and hybrid materials....

DOW

[1] M. Aminuzzaman, A. Watanabe and T. Miyashita, *Journal of Materials Chemistry,* vol. 18, no. 42, pp. 5092-5097, 15 August 2008.

5

Resin Characterization Challenges

Molecular characterization of these complex structures is not straightforward

- Gel Permeation Chromatography (GPC) for size
- Viscometry/rheology for size and flow properties
- Gas chromatography for impurities/byproducts
- Titrations for impurities
- Optical properties for structure/impurities
- FTIR for structure

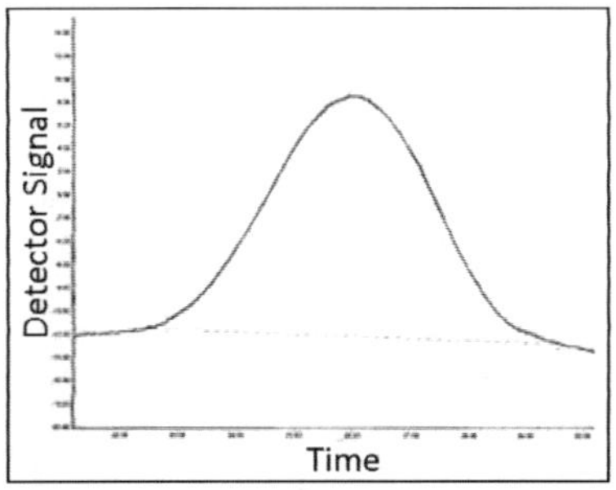

DOW

6

What is meant by “Product Quality”?

Options for quality control:

- Many univariate specifications
- Multivariate control charting
- Advanced modeling of conditions

OR

Can we find a multivariate classifier to define a specific *structure or composition,* rather than the physical properties?

Hypothesis:

Different process conditions produce different resin structures, which inherently have different feature patterns

DOW

7

What can we do with historical data?
▸ Unsupervised Learning

The silicone resin structure exploration is an *unsupervised* learning problem → we have all x’s and no y’s

- Not a predictive model for product quality (yet)
- Mining the data to identify patterns that relate to our hypothesis
- Data analytics as a tool in the scientific method

DOW

8

Introduction to the Data

14 Analytical Characterization Results for 770 Batches

Purity:	Physical properties:
• Non-volatile content (NVC) • Residual Impurity A • Residual Impurity B • Residual chloride	• Viscosity • Refractive index (RI) • "Cut viscosity" at a fixed NVC

Extent of the condensation reaction:
• Silanol (SiOH) concentration
• Molecular size from gel permeation chromatography (GPC):
  • Mn, Mw, Mz, and polydispersity (PD) from GPC Method 1
  • Mw from GPC Method 2
  • A proprietary molecular weight descriptor, "HMS"	

9

Exploratory Data Analysis

Inspect distributions of the 14 variables

- Values range from 10^{-3} to 10^{7} → all were standardized to ~N(0,1)
- All very normally distributed except Impurity A and Mz

Viscosity

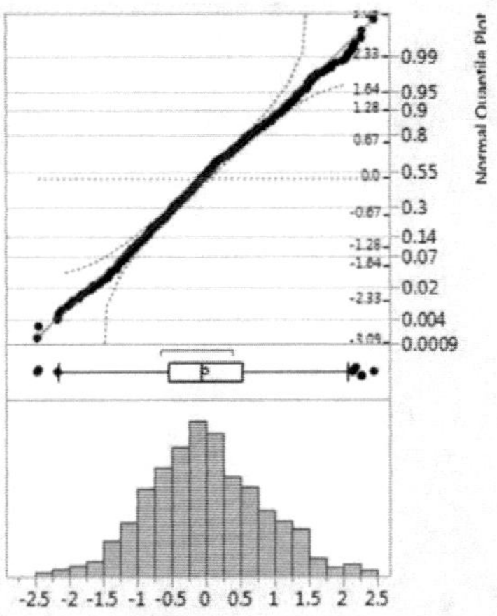

Mz

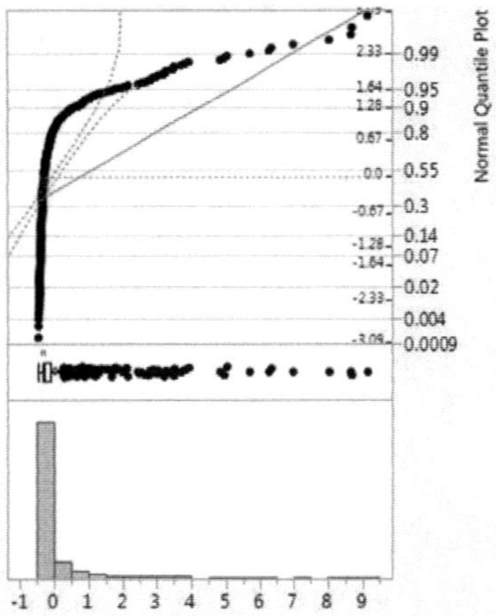

10

Cluster Analysis

Finding the best way to group the observations when you don't have the "right" answer

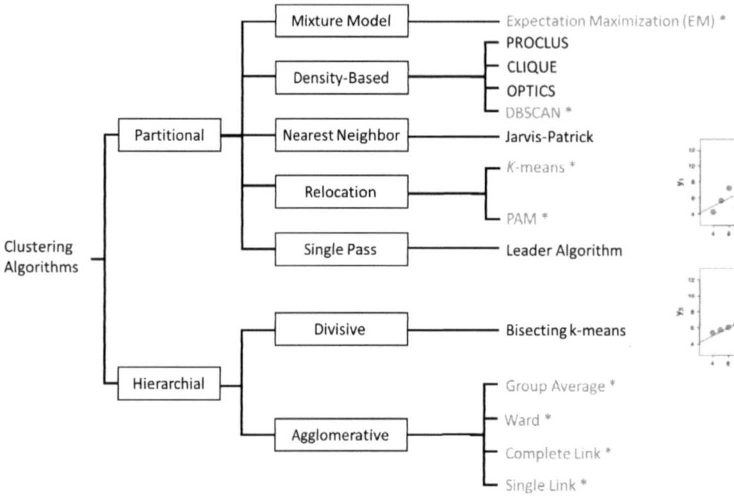

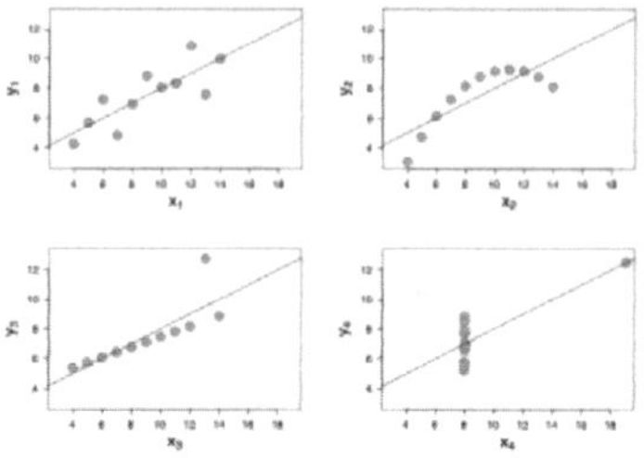

(left) [3] G. Downs and J. Barnard, "Clustering Methods and Their Uses in Computational Chemistry," in *Reviews in Computational Chemistry*, vol. 18, K. Lipkowitz and D. Boyd, Eds., John Wiley and Sons, Inc., 2002, pp. 1-40.
(right) [4] Anscombe, Francis J. (1973) Graphs in statistical analysis. American Statistician, 27, 17–21

13

Visualization of 14-Dimensional Clusters

Principal Components Analysis can help

No obvious cluster separation

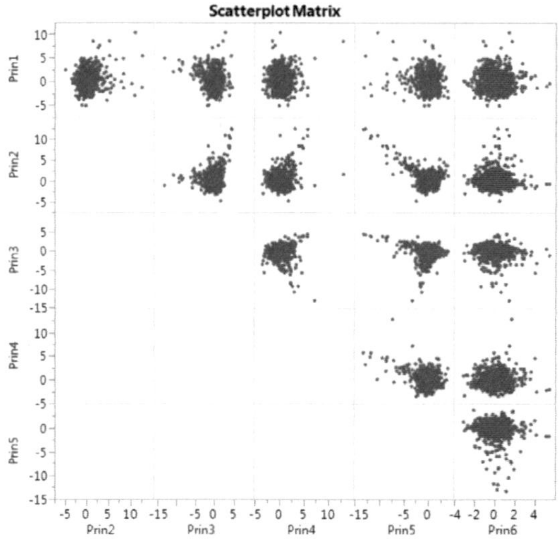

14

K-means and K-Medoids (PAM)

Short story: choose the number of clusters (*k*), assign every point to the closest cluster center, then iterate locating the center and reassigning points until the clusters are stable

Tuning parameters:

- What does cluster "center" mean?
- What does "closest" mean?
- How do you choose *k*?

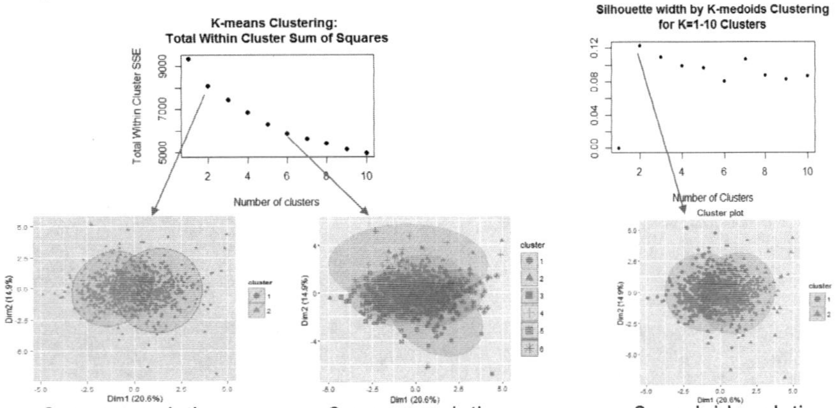

2-means solution 6-means solution 2-medoids solution

15

Density-Based (DBSCAN)

Short story: can identify clusters of complex shapes by detecting areas of low data density between groups

Tuning parameters:

- Minimum # of points to define a cluster vs. "noise"
- The max distance allowed

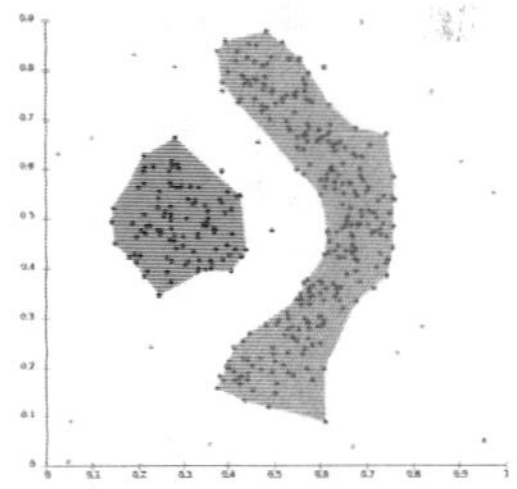

Example of DBSCAN clustering, drawn from Chire [6]

Results: Found 1 cluster with lots of noise

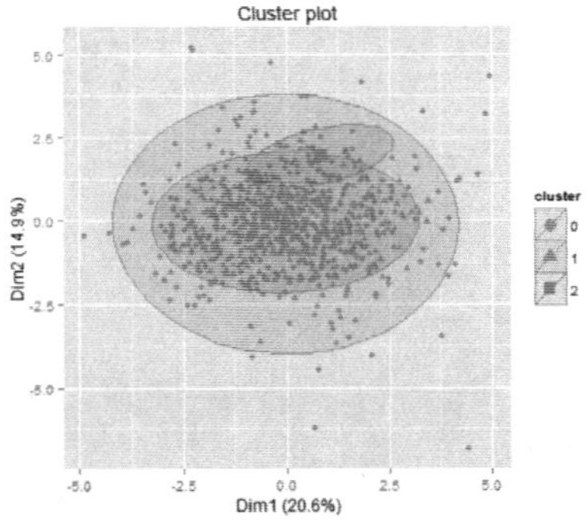

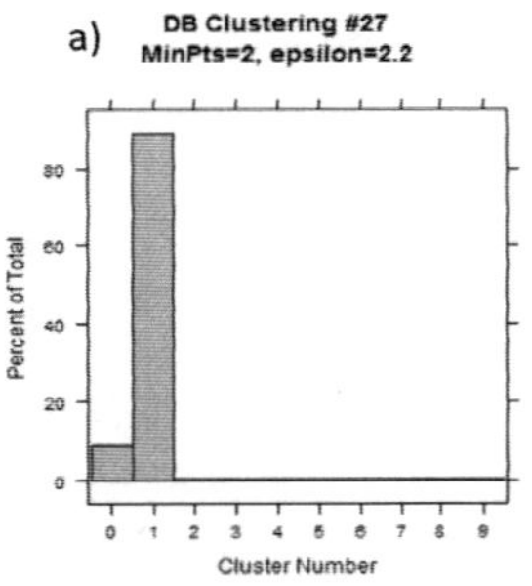

17

[6] Chire, "DBSCAN-density-data.svg", commons.Wikimedia.org, 2012, CC-BY-SA-3.0

Expectation Maximization (EM)

Short story: Estimate parameters for many mixtures of *k* multivariate normal distributions, then select the mixture and *k* which maximizes BIC

Tuning parameters: *k*

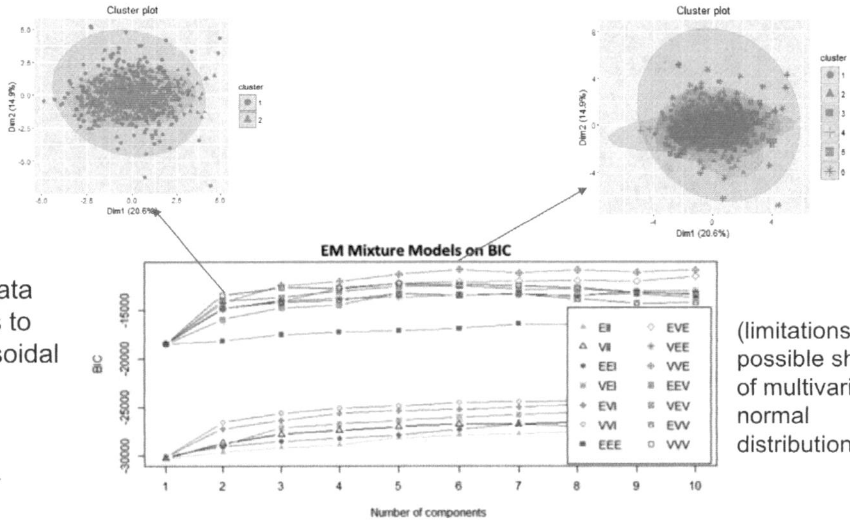

Resin data appears to be ellipsoidal

(limitations on possible shapes of multivariate normal distributions)

18

Hierarchical (Connectivity-Based)

Short story: Find the two closest points/clusters, combine them, then find the next closest, and the next, and the next… from N to 1 clusters

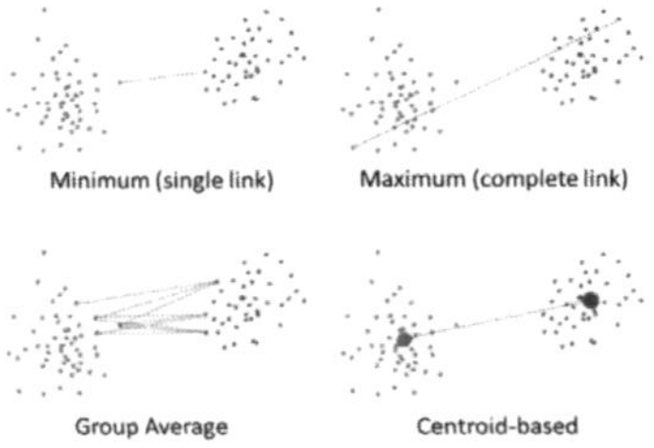

Tuning parameters – what does "closest" mean?

- Connection type
- Distance metric

Results: Wide range of solutions based on connectivity!

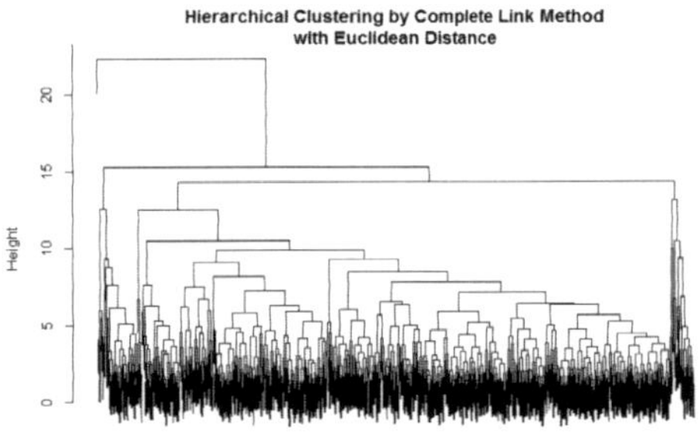

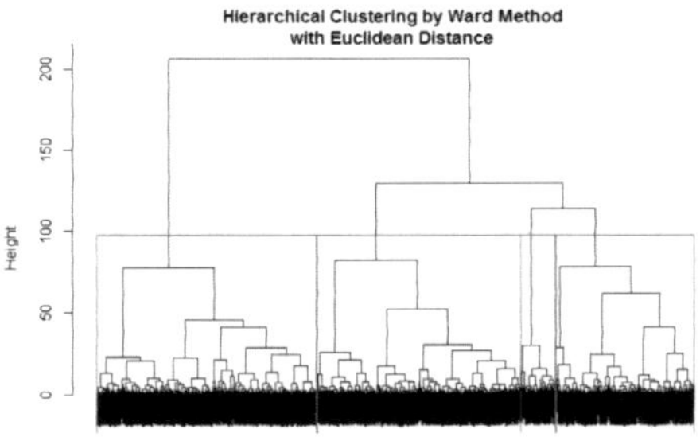

20

Which Results are Best?

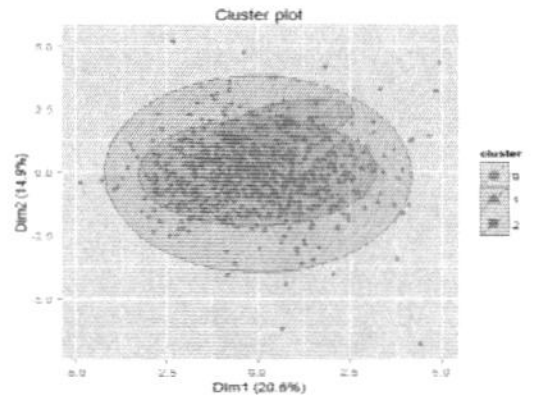

DBSCAN solution

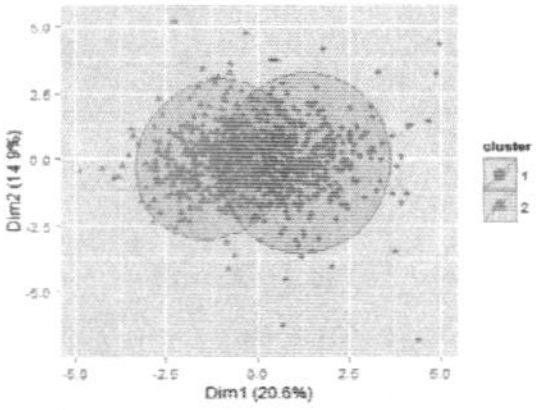

2-means solution

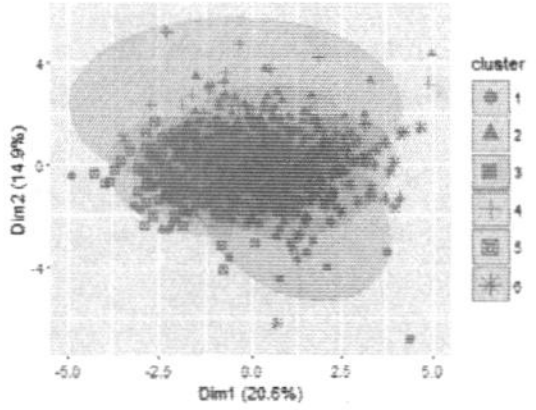

6-means solution

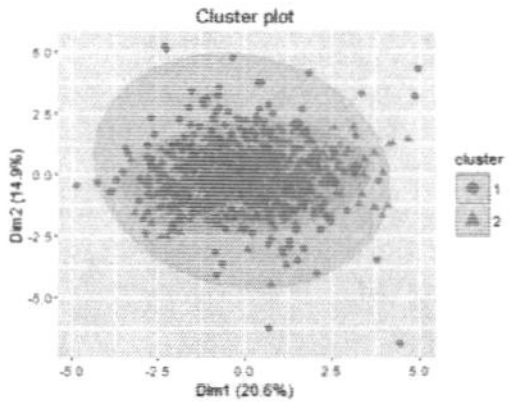

EM 2 solution

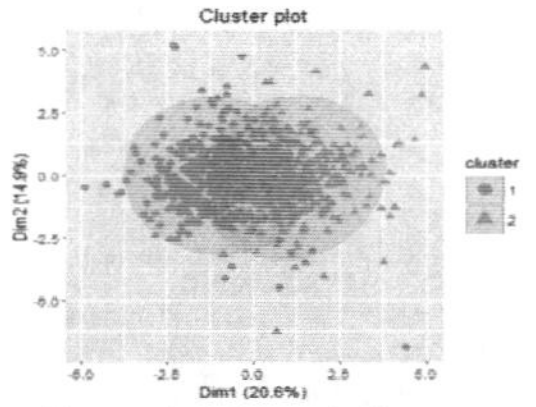

2-medoids solution

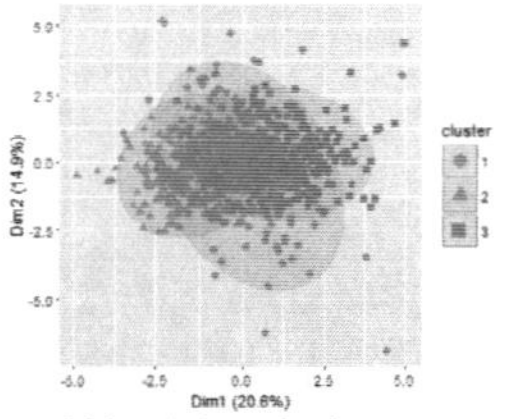

Ward 3 solution

21

Internal Cluster Validation Indices

Cluster validity indices (CVI) compare the within-cluster cohesiveness to the between-cluster separation

Challenges :

- Like R^2, some CVI necessarily improve with more clusters
- CVI ignore small clusters [8]
- Trouble comparing different size or density clusters [8]
- Many CVI penalize ellipsoidal clusters[9]

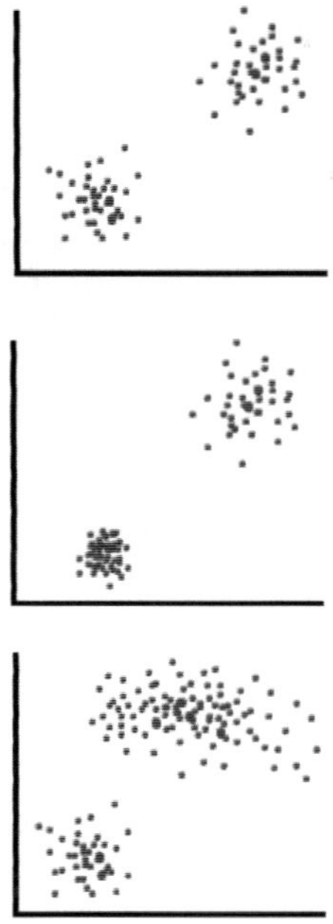

[8] K. Zalik and B. Zalik, *Pattern Recognition Letters,* vol. 32, pp. 221-234, 2011.
[9] O. Arbelaitz, I. Gurrutxaga, J. Muguerza, J. Perez and I. Perona, *Pattern Recognition,* vol. 46, pp. 243-256, 2013.

22

CVI Results

No one solution stands out

Clustering Method	# of clusters	Cohesion		Separation		Ratios, etc.		
		Max Diameter	Average WSSE	Min Separation	Average BSSE	Dunn	C-H	Average Silhouette Width
		↓*	↓*	↑*	↑*	↑*	↑	↑
K-means.2	2	21.9	4.1	0.8	4.8	0.035	126.6	0.14
DBSCAN.2	2	22.3	4.4	0.9	4.5	0.040	17.5	-0.07
PAM.2	2	22.3	4.2	0.8	4.7	0.034	107.3	0.12
EM.2	2	22.3	4.1	0.9	5.0	0.039	37.0	0.14
Ward.2	2	22.3	4.3	1.0	4.6	0.045	82.1	0.08
Ward.3	3	22.3	4.0	1.0	4.7	0.045	73.7	0.09
K-means.4	4	20.8	3.8	0.7	4.7	0.035	97.1	0.12
PAM.4	4	22.1	3.9	0.5	4.6	0.025	79.6	0.10
Ward.4	4	21.7	3.8	1.0	4.7	0.046	78.0	0.10
K-means.6	6	18.9	3.4	0.7	4.7	0.038	95.2	0.13
EM.6	6	22.3	3.9	0.6	4.6	0.025	19.3	-0.02
Ward.6	6	18.1	3.6	1.0	4.7	0.055	70.3	0.09
DBSCAN.8	8	22.3	4.0	2.3	6.6	0.103	6.2	0.04

23

Revisiting the Hypothesis

Hypothesis:
Different process conditions produce different resin structures, which inherently have different feature patterns

Findings:

There is no evidence to suggest distinct structures are formed, and/or the feature patterns of the compositions are not well resolved from process noise

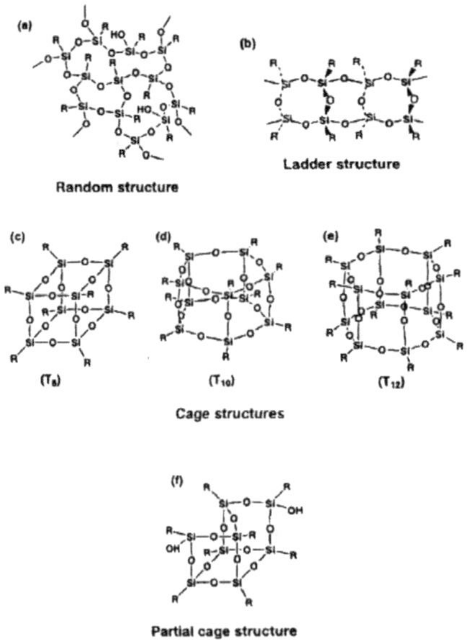

25

Interpreting Cluster Results – Fundamental Perspective

The 2-means solution did have predictive power for downstream processing results

Work with the subject-matter experts to interpret the meaning of the clustering results

Focus on both what is different *and* what is the same

Keep in mind practical significance, not just statistical significance

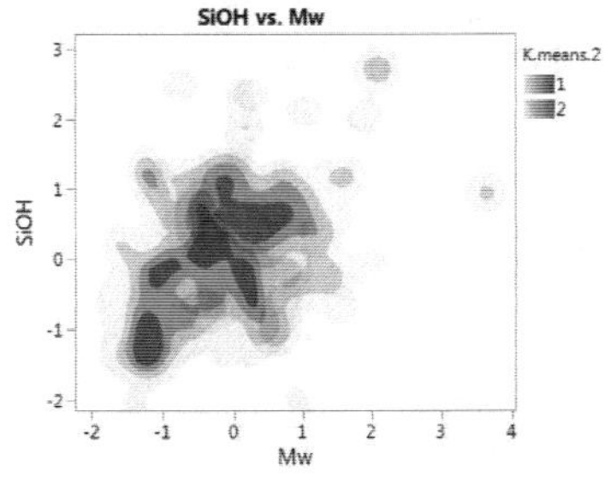

26

Conclusions

We strive to align our process control strategy to fundamental engineering and chemistry… not so easy with polymeric mixtures

Multivariate Analysis can help… but clustering has a lot of caveats

Even poorly separated clusters can still be useful predictors for product performance

30

Acknowledgements

Dr. Jamie Riggs
Dr. Gary Wieber
Thomas Bekemeier
Kirk Opella
Robert Fosdick
Tim Morrison
Dr. James Mentele
Dr. Kurt Brandstadt

References

[1] M. Aminuzzaman, A. Watanabe and T. Miyashita,*Journal of Materials Chemistry,* vol. 18, no. 42, pp. 5092-5097, 15 August 2008.

[2] R. G. Lawson and P. C. Jurs, "New Index for Clustering Tendency and Its Application to Chemical Problems," *Journal of Chemical Information and Computer Science,* vol. 30, no. 1, pp. 36-41, 1990.

[3] G. Downs and J. Barnard, "Clustering Methods and Their Uses in Computational Chemistry," in *Reviews in Computational Chemistry*, vol. 18, K. Lipkowitz and D. Boyd, Eds., John Wiley and Sons, Inc., 2002, pp. 1-40.

[4] Anscombe, Francis J. (1973) Graphs in statistical analysis. American Statistician, 27, 17–21

[5] P.-N. Tan, M. Steinbach and V. Kumar, Introduction to Data Mining, Pearson, 2005

[6] Chire, "DBSCAN-density-data.svg", commons.wikimedia.org, 2012, CC-BY-SA-3.0

[7] T. Hastie, R. Tibshirani and J. Friedman, The Elements of Statistical Learning, 2nd ed., Springer, 2009.

[8] K. Zalik and B. Zalik, *Pattern Recognition Letters,* vol. 32, pp. 221-234, 2011.

[9] O. Arbelaitz, I. Gurrutxaga, J. Muguerza, J. Perez and I. Perona, *Pattern Recognition,* vol. 46, pp. 243-256, 2013.

R Libraries

• amap	• fpc
• analogue	• ggplot2
• caret	• mclust
• cluster	• NbClust
• clustertend	• Pvclust
• dbscan	• vegan

481850 Environmental Impact of Mining

Abstract:

Mining can have adverse effects on surrounding surface and groundwater if protective measures are not taken. The result can be unnaturally high concentrations of some chemicals, such as arsenic, sulfuric acid, and mercury over a significant area of surface or subsurface.[3] Runoff of mere soil or rock debris -although non-toxic- also devastates the surrounding vegetation. The dumping of the runoff in surface waters or in forests is the worst option here. Submarine tailings disposal is regarded as a better option (if the soil is pumped to a great depth).[4] Mere land storage and refilling of the mine after it has been depleted is even better, if no forests need to be cleared for the storage of the debris. There is potential for massive contamination of the area surrounding mines due to the various chemicals used in the mining process as well as the potentially damaging compounds and metals removed from the ground with the ore. Large amounts of water produced from mine drainage, mine cooling, aqueous extraction and other mining processes increases the potential for these chemicals to contaminate ground and surface water. In well-regulated mines, hydrologists and geologists take careful measurements of water and soil to exclude any type of water contamination that could be caused by the mine's operations. The reducing or eliminating of environmental degradation is enforced in modern American mining by federal and state law, by restricting operators to meet standards for protecting surface and ground water from contamination. This is best done through the use of non-toxic extraction processes as bioleaching. If the project site becomes nonetheless polluted, mitigation techniques such as acid mine drainage(AMD) need to be performed.

The five principal technologies used to monitor and control water flow at mine sites are diversion systems, containment ponds, groundwater pumping systems, subsurface drainage systems, and subsurface barriers. In the case of AMD, contaminated water is generally pumped to a treatment facility that neutralizes the contaminants.[5]

A 2006 review of environmental impact statements found that "water quality predictions made after considering the effects of mitigations largely underestimated actual impacts to groundwater, seeps, and surface water".[6]

THE POTENTIAL USE OF SOLAR PHOTOCATALYTIC OXIDATION FOR REMOVAL OF EMERGING PHARMACEUTICALS FROM EFFLUENT OF WASTEWATER TREATMENT PLANT: A PILOT PLANT STUDY

Fares Almomani, Rahul Bhosale, Anand Kumar, Majeda Khreisheh, Ujjal Gosh
Department of Chemical Engineering, Qatar University, P.O box 2713, Doha, Qatar

Abstract

The treatment of effluent wastewater containing pharmaceutical by solar advanced oxidation processes were studied in a pilot plant. The removal efficiency of four groups of pharmaceuticals (antibiotics, estrogens, acidic, and neutral), commonly detected as emerging contaminants, were tested by different solar-driven photochemical processes including solar -photolysis, solar-photocatalytic oxidation using Fe(III) or TiO_2 and solar based photo-Fenton reaction. Results showed that Solar-photolysis is not effective for the complete removal of the selected contaminants, on other hand solar-photocatalytic oxidation using TiO_2, solar based photo-Fenton reaction showed significant removal efficiencies and high mineralization.
Solar photo-catalytic systems produced a final solution with lower toxicity if compared with original solution.

Keywords: photons, emerging contaminates, solar excitation, hydroxyl radicals

Introduction

Incident presence of what is called emerging contaminates(ECs)such as pharmaceuticals, pesticide in water and effluents of wastewater treatment plants have raised a lot of concern regarding the performance of conventional treatment processes in removing and treating these contaminants (Ternes, 2001, Kolpin et al., 2002). It was reported that the main reason for the presence of these ECs is related to the continual release of these contaminates into this eco-system from different point sources (Huber et al., 2004). ECs are considered non-biodegradable, and consequently it is expected that their concentration in the environment will continue to increase unless they are effectively treated and eliminated. Reasonable concentration of these contaminates have been detected in secondary treatment effluents of wastewater treatment plants (WWTPs) (Kolpin et al., 2002). Contamination of groundwater with ECs was reported from different sources(Ternes et al., 2002). Thus, there is a big need for an effective and cheap post-treatment to remove ECs from wastewater before discharge to the environment.

ECs can be effectively removed from water solution using different treatment processes such as activated carbon, nano-filtration and reverse osmosis(Rosenblatt et al., 1967) (Reddersen et al., 2002, Webb et al., 2003). However, the cost associated with these treatment alternatives limits their applicability.

Advanced oxidation processes (AOPs), on other hand, can be employed as an effective and promising technology for the removal of any traces of ECs from water and wastewater. AOPs which based on the production of highly oxidative radicals are considered as destructive processes. These processes have the capacity to mineralize different contaminates from wastewater in a short period of time. With such technologies, problematic compounds can be partially or totally oxidized into smaller by-products that can be easily removed in conventional biological processes. Applying this strategy would potentially reduce the size of both oxidation and biological reactors, and may reduce both capital and operating costs. This synergy has the potential to open up a new niche in the water treatment field for treating contaminants which are conventionally reserved for other treatment technologies (e.g. activated carbon). The high cost associated with the use of AOPs (reagent and UV-light) limit their applicability. However, most recently, solar advanced oxidation processes were effectively used in the treatment of water and wastewater containing non-biodegradable and/or toxic organic pollutant (Almomani et al, 2010, Al

Momani and Jarrah, 2009). Using solar irradiation as a source of photon can reduce the cost of AOPs and make them a suitable treatment options.

The present work aims to investigate the suitability of using solar induced AOTs for removal of ECs from water and wastewater. The study will be carried out in a compound parabolic collector (CPC) prototype pilot plant solar reactor. The CPC technology allows all the received solar radiation to be utilized for oxidation. Thus, oxidation performance increase. The outcome of this project will allow to develop a new technology for the treatment of water and wastewater containing ECs by use of solar energy which is abundant in Qatar and the Arab gulf region.

Materials and Methods

Experimental Set Up

Experiments set up used to carry out pharmaceutical solar oxidation is shown in (Fig. 1). The solar pilot plant is equipped with (1) a wastewater containing pharmaceutical. The tank was immersed in water bath to control the working temperature, (2) compound parabolic collector (CPC) solar reactor . the CPC allows for almost all the solar irradiation received at the reactor aperture area (not only direct, but also diffuse) to be collected being available for oxidation. The solar consists of six quartz tubes with a length of 50 cm, an inner diameter of 3 cm and outer diameter of 3.1 cm. This pilot plant has the capacity to treat up 5 L of wastewater, the solution was circulated through the reactor with a flow rate of 20 L. h^{-1} and the sun's movement was tracked manually.

Analytical methods

The pharmaceutical concentrations were measured with a Hewlett-Packard 1050 series HPLC equipped with an Ultra Aqueous C_{18} column from Restek (3.2100 mm, 5 mm) and a variable wavelength detector. The eluents consisted of 10 mM phosphoric acid (A) and pure acetonitrile (B). The eluent flow rate was 0.5mL.min^{-1} and the detection limit was 10 μg.L^{-1}. The 95% confidence intervals for a single measurement were typically 3–5%.

The dissolved organic carbon (DOC) was measured using a Shimadzu TOC analyzer. The analyzer was calibrated using potassium phthalate standard solution and following the instrument standard method. Each sample was analyzed three times and measurements were averaged. Measurements reproducibility showed a precision in the range of ±0.1 mg.C.L^{-1}.

Water samples

Wastewater (WW) was collected from the effluent of secondary treatment train from a Doha north WWTP. Water samples were filtered through a 0.45 μm Millipore filter, and stored at 4 °C prior to use. Table 1 shows the characteristics of Wastewater treatment plant effluent.

Table 2: Water quality parameters for surface and wastewaters.

Parameter	Wastewater treatment plant effluent Average, (std. dev.)[1],
pH	7.9 (0.4)
DOC (mg.L^{-1})	13(1.2)
Alkalinity (mg$CaCO_3$.L^{-1})	145 (6.0)
Nitrate (mg-N.L^{-1})	31 (1.4)

Nitrite (mg-N.L^{-1})	15 (0.4)
Sulfate (mg.L^{-1})	50 (2.0)
Chloride (mg.L^{-1})	15 (1.2)

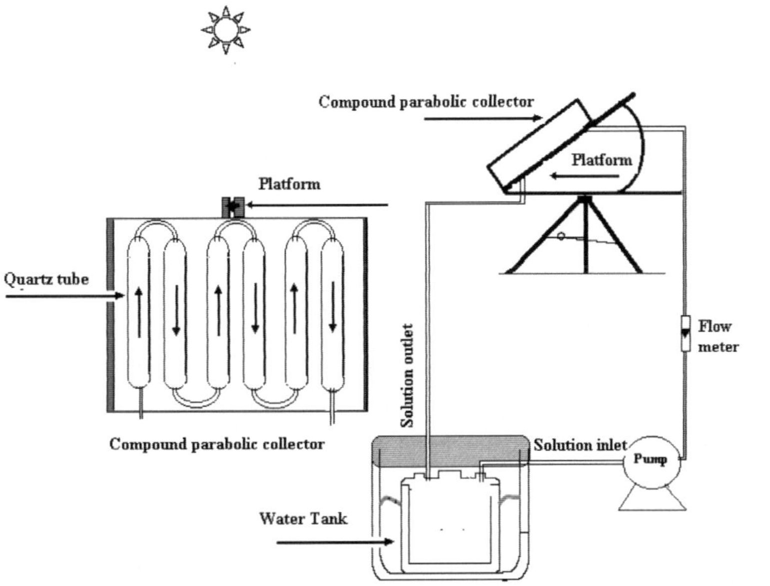

Figure 1: Solar Experimental Setup

Experimental Procedure

Experiments were conducted Doha, Qatar (25.2854° N, 51.5310° E) in a semi-batch mode, under sunlight and at ambient temperature. At the beginning of the reactor was loaded with 5 L of the aqueous solution of pharmaceuticals and the required amount of catalyst was added afterwards. Then, the solution was recirculated through the CPC by a centrifugal pump at a flow rate of 20 L.hr^{-1} so that the flux pattern inside the tubes was laminar (the Reynolds number was 2500). For the first 10 min, the CPC tubes were kept covered to avoid sunlight illumination. This time was enough to reach the adsorption equilibria of pharmaceuticals onto TiO_2 (if used). After that, the CPC was uncovered and the reaction was was started. Also, if needed, hydrogen peroxide was added at this point. Samples were withdrawn at intervals through a sampling port to determine the concentration of pharmaceuticals as function of reaction time, the remaining total organic carbon (TOC), and the concentrations of hydrogen peroxide, ozone and Fe(II). The ozone concentration in the outlet gas stream was also continuously monitored with in-line ozone analyzers.

The initial concentration of pharmaceuticals used in experiment are presented in table 1. The initial concentrations were selected to improve the chemical analysis accuracy and facilitate mineralization kinetic study. The initial Fe(III) and TiO_2 concentrations were 3 and 100 mg L^{-1}, respectively.

Toxicity analysis

Toxicity bioassays were performed to evaluate the toxicity of water containing pharmaceuticals before and after ozonation. Tests were carried out using Microtox® bioassays, which involved the use of the marine bacterium *Photobacterium phosphoreum* as an indication of sample toxicity. The standard Microtox protocol was followed, and data analysis was achieved using the Microtox™ Omni Software (SDI, 2002). Sample toxicity was evaluated at 5 and 15 min from the time of mixing of the sample with marine bacterium at various dilutions. The toxicity was determined by measuring the EC_{50} of the sample. The EC_{50} is the concentration of the test sample at which there is 50% reduction in bacterial light production. This value was related to sample toxicity using the following equation:

$$TU = \frac{100}{EC_{50}} \quad [1]$$

where *TU* is the toxicity unit and EC_{50} is the concentration of the test sample at 50% reduction in bacterial light production.

Results and Discussion

Solar -Photolysis of pharmaceuticals

Figure 2 A and B show the evaluation of pharmaceutical as a function of solar radiation time during solar photolysis process. Direct solar photolysis was not effective in removing or degradation of pharmaceuticals from wastewater. This is an expected fact as the solar radiation with wavelength < 290 nm does not reach Earth's surface as it is mostly retained by atmosphere [32]. In addition, all the target pharmaceuticals do not absorb radiation above 300 nm significantly. The removal efficiencies ranged from 9.1- 12.1%, 13-15.6%, 13.9 -16 and 5.4 and 8.2 for antibiotics, estrogen, acidic and neutral pharmaceutical, respectively.

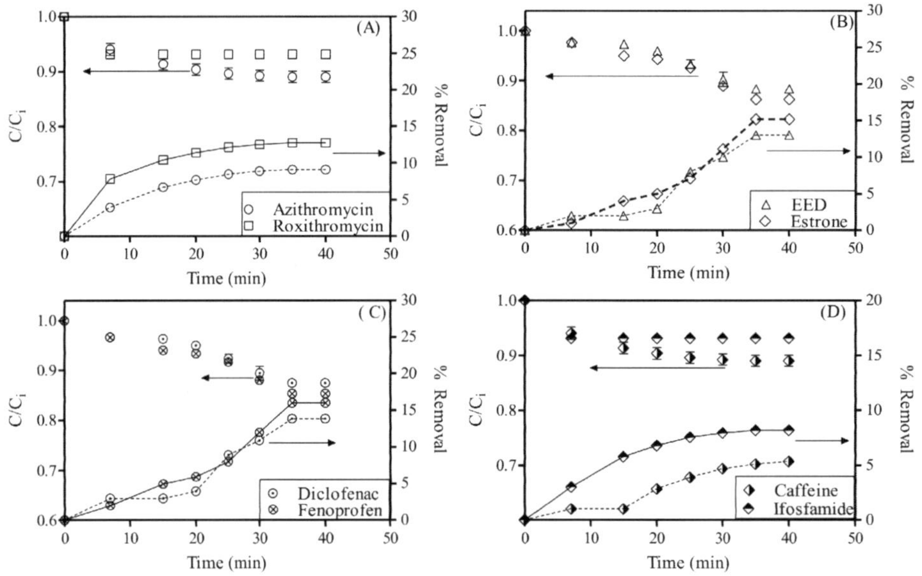

Figure 2: Evolution of the normalized concentration $[C]/[C]_i$ and % removal of antibiotics, Estrogens, Acidic and natural pharmaceuticals during solar photolysis

Solar-photocatalytic oxidation of pharmaceuticals

Iron (Fe(III) and titanium dioxide were used as catalyst for this part of experiments. Figure 2 shows the results for experiments carried out with iron as catalyst in the AOPs including Fe(III)/solar-hv and Fe(III)/solar-hv/H_2O_2. The reactions involved in these oxidation processes are summarized in equations (1 and 2).

$$Fe^{3+} + H_2O \rightleftarrows Fe(OH)^{2+} + H^+ \quad \text{(Reaction 1)}$$

$$Fe(OH)^{2+} \xrightarrow{h\upsilon} Fe^{2+} + {}^{\cdot}OH \quad \text{(Reaction 1)}$$

During the photocatalytic oxidation with Fe(III)/hv Pharmaceuticals oxidation occurs by the formation of aqua complexes, mainly $Fe(OH)_2^+$ (see reaction (1)), which can undergoes photolysis under UV radiation ($\lambda > 300$ nm) to generate hydroxyl radicals as shown reaction 2. As it can be seen in Figure 3, the combination Fe(III)/hv led to reasonable degradation of the pharmaceuticals. The removal efficiencies for antibiotics, estrogen, acidic and neutral pharmaceutical ranged from 74 – 77.8%, 69-71.9%, 70.8-80.9% and 57.6-66.8, % respectively. The results suggest that solar-photocatalytic oxidation can be used to oxidize these pharmaceuticals. However, the reaction time should be increase of there is a need to add reagent to assist in oxidation.

Adding H_2O_2 to Fe(III)/hv change the system a well-known oxidation system called photo-Fenton. In this case, H_2O_2 accelerates the production of hydroxyl radicals as in reactions (3) and (4) and change Fe (III) to Fe(II) (reaction 5) which, in turn, can reduce Fe(III) to Fe(II) as in reaction (6). The net results is more hydroxyl radicals and more pharmaceuticals oxidation.

$$Fe^{2+} + H_2O_2 \rightarrow Fe^{3+} + OH^- + {}^{\cdot}OH \quad \text{(Reaction 3)}$$

$$H_2O_2 \xrightarrow{h\upsilon} 2{}^{\cdot}OH \quad \text{(Reaction 4)}$$

$$H_2O_2 + {}^{\cdot}OH \rightarrow HO_2^{\cdot} + H_2O \quad \text{(Reaction 5)}$$

$$Fe^{3+} + HO_2^{\cdot} \rightarrow O_2 + Fe^{2+} + H^+ \quad \text{(Reaction 6)}$$

Solar photo catalytic oxidation by TiO_2 is well-known and it has been widely used to oxidize non-biodegradable organic matter in water systems.

The change in solution toxicity after different oxidation process were evaluated. For experiments carried out with photolysis, it was observed that oxidized water samples toxicity does not change a suggesting that the low oxidation reaction did not change water toxicity. For water samples with higher oxidation such as photo-Fenton and TiO_2/hv system lower acute toxicity than the initial wastewater was observed. As an example, water containing 3.75 mg/L of Roxithromycin showed a TU value of 17, oxidation photo-Fenton and TiO_2/hv system decreased the TU to 1 and 1.2, Respectively. Similar results were observed for other pharmaceutical. The reduce in toxicity is related to the reaction between radicals and the studied pharmaceuticals at tertiary amines bonds leading to the cleavage of the N–C bonds. This reaction can lead to either the destruction of the chemical structure of pharmaceuticals or the loss of the methyl group from the amino group. Li et al., 2001 have confirmed that the loss of the methyl group of roxithromycin eliminates its apparent activity. Besides, the oxidation reaction can cause cleavage of the phenolic structure in pharmaceuticals and reduce its estrogenicity significantly. On the basis of these considerations, solar photocatalytic oxidation of pharmaceuticals results in significant reduction in pharmacological effects of these pharmaceuticals

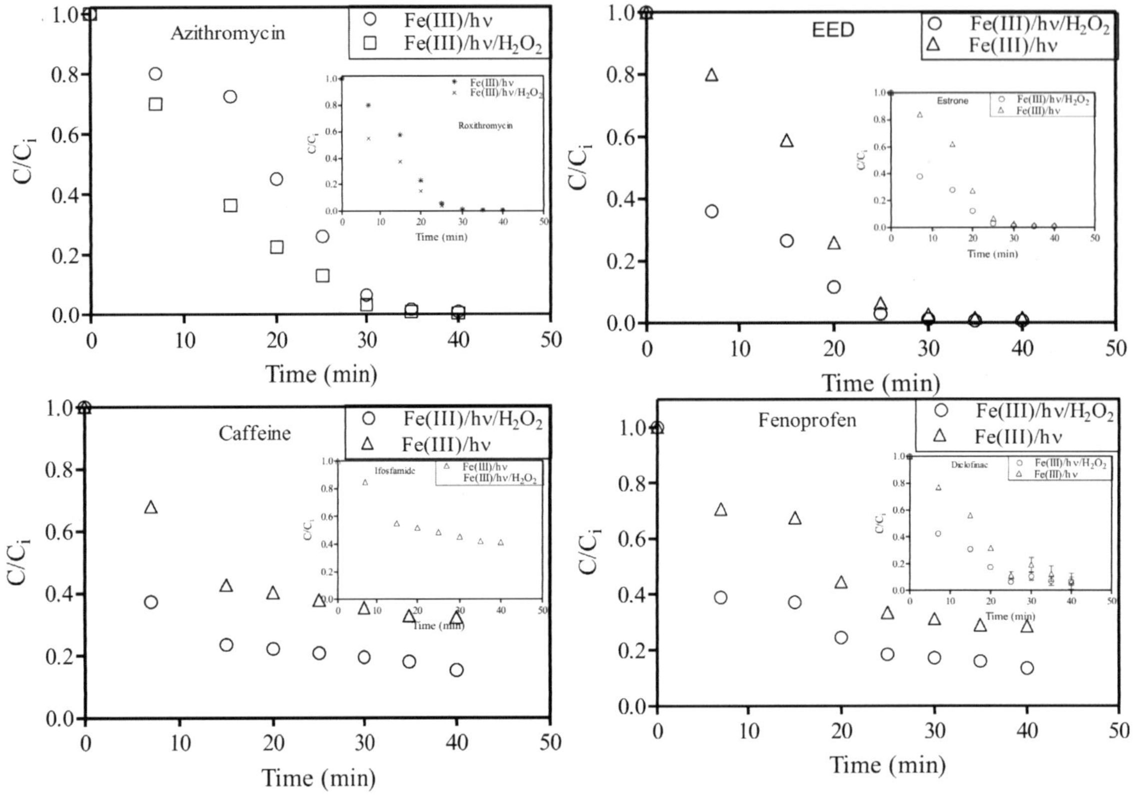

Figures 3 : Evolution of the normalized concentration $[C]/[C]_i$ of antibiotics, Estrogens, Acidic and natural pharmaceuticals during solar-photocatalytic oxidation.

Conclusion:

Oxidation of four groups of pharmaceuticals (antibiotics, estrogens, acidic and neutral) with solar advanced oxiadation was investigated in effluent of WWTP. Solar photocatalytic reaction can be used successfully to eliminates nob-biodegradablke pharamacuticals from WWTP effluets. The degradation rate of pharmaceuticals was strongly accelerated by adding catalys or hydrogen peroxide to the system. 98 % of pharmaceutical degradation was achieved by solar photo-Fenton or photo-catakytic oxidation with TiO_2in 40 min reaction time.

Toxicity tests using Microtox^{TM} bioassay were completed to evaluate the effect of the oxidation reactor on the toxicity of water solution before and after the treatment. Solar oxidation reduced the a cute toxicity of all the pharmaceuticals in the wastewater suggesting that this technology can be coupled with wastewater treanment process to complete the removal of the toxic and non-biodegradable materials

Acknowledgements

This publication was made possible by the NPRP grant (NPRP No.: 6 - 1436 - 2 - 581) from the Qatar National Research Fund (a member of Qatar Foundation). The statements made herein are solely the responsibility of authors. The authors gratefully acknowledge the support provided by the Qatar University to do this work..

REFERENCES

DODD, M. C. & HUANG, C.-H. (2004) Transformation of the Antibacterial Agent Sulfamethoxazole in Reactions with Chlorine: Kinetics, Mechanisms, and Pathways. *Environmental Science & Technology,* 38, 5607-5615.

GAMAL EL-DIN, M., SMITH, D. W., AL MOMANI, F. & WANG, W. (2006) Oxidation of resin and fatty acids by ozone: Kinetics and toxicity study. *Water Research,* 40, 392-400.

HUBER, M. M., KORHONEN, S., TERNES, T. A. & VON GUNTEN, U. (2005) Oxidation of pharmaceuticals during water treatment with chlorine dioxide. *Water Research,* 39, 3607-3617.

HUBER, M. M., TERNES, T. A. & VON GUNTEN, U. (2004) Removal of Estrogenic Activity and Formation of Oxidation Products during Ozonation of 17alpha -Ethinylestradiol. *Environmental Science & Technology,* 38, 5177-5186.

KOLPIN, D. W., FURLONG, E. T., MEYER, M. T., THURMAN, E. M., ZAUGG, S. D., BARBER, L. B. & BUXTON, H. T. (2002) Pharmaceuticals, Hormones, and Other Organic Wastewater Contaminants in U.S. Streams, 1999-2000: A National Reconnaissance. *Environmental Science & Technology,* 36, 1202-1211.

LOVATO, M. A. E., MARTÃ-N, C. A. & CASSANO, A. E. (2009) A reaction kinetic model for ozone decomposition in aqueous media valid for neutral and acidic pH. *Chemical Engineering Journal,* 146, 486-497.

PINKSTON, K. E. & SEDLAK, D. L. (2004) Transformation of Aromatic Ether- and Amine-Containing Pharmaceuticals during Chlorine Disinfection. *Environmental Science & Technology,* 38, 4019-4025.

REDDERSEN, K., HEBERER, T. & DANNBIER, U. (2002) Identification and significance of phenazone drugs and their metabolites in ground- and drinking water. *Chemosphere,* 49, 539-544.

ROSENBLATT, D. H., HULL, L. A., DE LUCA, D. C., DAVIS, G. T., WEGLEIN, R. C. & WILLIAMS, H. K. R. (1967) Oxidations of Amines. II. Substituent Effects in Chlorine Dioxide Oxidations. *Journal of the American Chemical Society,* 89, 1158-1163.

TERNES, T. (2001) Pharmaceuticals and Metabolites as Contaminants of the Aquatic Environment. *Pharmaceuticals and Care Products in the Environment.* American Chemical Society.

TERNES, T. A., MEISENHEIMER, M., MCDOWELL, D., SACHER, F., BRAUCH, H.-J. R., HAIST-GULDE, B., PREUSS, G., WILME, U. & ZULEI-SEIBERT, N. (2002) Removal of Pharmaceuticals during Drinking Water Treatment. *Environmental Science & Technology,* 36, 3855-3863.

WEBB, S., TERNES, T., GIBERT, M. & OLEJNICZAK, K. (2003) Indirect human exposure to pharmaceuticals via drinking water. *Toxicology Letters,* 142, 157-167.

ZHANG, P. & LIU, J. (2004) Photocatalytic degradation of trace hexane in the gas phase with and without ozone addition: kinetic study. *Journal of Photochemistry and Photobiology A: Chemistry,* 167, 87-94.

481851 Oil Pollution, Environmental Degradation and Restoration of Nigeria's Ogoniland

Abstract:

Human activities especially those of oil exploration and exploitation raise a number of issues such as depletion of biodiversity, coastal and riverbank erosion, flooding, oil spillage, gas flaring, noise pollution, sewage and wastewater pollution, land degradation, soil fertility loss and deforestation, which are all major environmental issues.

Ogoniland is characterized by typically deltaic features; uneven terrain, numerous creeks, shallow brackish water bodies and a variety of vegetation types including swamp forest. The land has a groundwater either as localized aquifers or in hydraulically interconnected aquifers. Ogoniland is not an Island. This has two implications. The first is that pollution from Ogoniland has the potential to reach and cross its boundaries as well as entering Ogoniland from extreme sources.

Petroleum hydrocarbons are naturally occurring hydrocarbons substances and depending on the length of the carbon, can occur in gas, liquid and solid form. However, hydrocarbons refer to chemical substances formed exclusively from carbon and hydrogen by the decay of organic substances trapped within sedimentary rocks.

Hydrocarbons substantially alter the physical and chemical properties of the soil once it comes in contact with it. The contact could be from the natural seepage of hydrocarbons to accidental seepage of crude oil on the ground. In the case of water, hydrocarbons can cause both physical and chemical effects by preventing oxygen transfer in the water column, thus affecting aquatic life-support systems. As regards vegetation, where spillage is not immediately attended to, oil spills often lead to fires, causing total or partial destruction of vegetation. Dissolved or emulsified oil in the water column can contaminate the plankton, algae, fish, eggs and the vertebrae larvae. Also, physical contacts with oil destroys the insulation of fur and feathers, and more often, birds ingest oil which have lethal or sub-lethal impacts through, for example, liver and kidney damage.

Residents in these areas, face the risk of dangerous exposure to hydrocarbons when they breathe, bathe, eat fish, drink water or accidentally eat or touch soil or sediment that is contaminated with oil. In addition to the chemical pollution by hydrocarbons, there are other environmental concerns linked with oil industry operations. These range from clearance of land for oilfield facilities, hydrological changes due to construction of roads and pipelines and contamination from chemicals other than hydrocarbons.

There must be maintenance of oilfield facilities, decommissioning of oilfield facilities and prevention of illegal activities, oil spill response and remediation of contaminated sites. Environmental restoration includes the establishment of an Integrated Contaminated Soil Management Centre (ICSMCS), setting up of mini-treatment centers and rehabilitation of mangroves.

Part of the actions is the amendment of the official gazette establishing the Hydrocarbon Pollution Restoration Project (HYPREP), to reflect a new governance framework comprising a Governing Council, a Board of Trustees and Project Management.

There must be an end to the widespread pipeline sabotage, crude oil theft and illegal refining that are the main cause of environmental damage in Ogoniland and the wider Niger Delta.

There should be transfer of oversight of the Environmental Guidelines and Standards for Petroleum Industry in Nigeria (EGSPIN), to ensure constant surveillance. Also, there should be the provision for social and health impact assessment as an integral part of the overall environmental impact assessment (EIA) process for all new oil and gas facilities and upgrades to existing facilities in line with international best practice. This means that the Environmental Impact Assessment (EIA) Act should be amended to accommodate these changes. It also proposes that there should be a review of the provisions of the National Oil Spill Detection and Response Agency NOSDRA (Establishment) Act, 2006 against NOSDRA's current operational responsibilities. Oversight of clean-up should be given to a separate governmental department.

2017 AIChE®

SPRING MEETING&

13TH

GLOBAL CONGRESS

ON PROCESS SAFETY

March 26–30, 2017
Henry B. Gonzalez
Convention Center
San Antonio, TX

17 AIChE
Spring Meeting, San Antonio, TX

13TH GLOBAL CONGRESS ON PROCESS SAFETY

17 AIChE
Spring Meeting, San Antonio, TX

13TH GLOBAL CONGRESS ON PROCESS SAFETY

The role of APC in Industry 4.0

Filippo Trivella
Alpha Process Control

17AIChE Spring Meeting, San Antonio, TX — 13TH GLOBAL CONGRESS ON PROCESS SAFETY

Agenda

- Brief Introduction
- Impact of Industry 4.0 on Process Automation
- Status of Advanced Process Control
- APC requirements for Industry 4.0
- The transformation of APC

17AIChE Spring Meeting, San Antonio, TX — 13TH GLOBAL CONGRESS ON PROCESS SAFETY

ALPHA Process Control

Brief intro

- About the Author: Filippo Trivella
  - Chemical Engineer
  - 25 years of experience in design and implementation of APC applications
  - 180+ implementations of APC in refining and petrochemical processes

- About the Company: Alpha Process Control
  - Independent provider of APC solutions
  - Founded in 2011
  - Customers in Europe, Middle East and Asia
  - Executed more on than 100 APC projects in Refining, Petrochemical and Polymer industries

17 AIChE Spring Meeting, San Antonio, TX | 13TH GLOBAL CONGRESS ON PROCESS SAFETY

What is Industry 4.0?

- Industry 4.0 is defined as the digitization of the manufacturing industry
  - Increased computing power
  - Adoption of Advanced analytics and intelligence
  - Advances in Human Machine interaction
  - Digital to physical conversion

2017 AIChE® SPRING MEETING & 13TH GLOBAL CONGRESS ON PROCESS SAFETY

17 AIChE Spring Meeting, San Antonio, TX | 13TH GLOBAL CONGRESS ON PROCESS SAFETY

Industry 4.0 – new "disruptive" technologies

Digitization of the manufacturing sector – Industry 4.0

Data, computational power, and connectivity

Big data/open data
Significantly reduced costs of computation, storage, and sensors

Internet of Things/M2M
Reduced cost of small-scale hardware and connectivity (e.g., through LPWA networks)

Cloud technology
Centralization of data and virtualization of storage

Analytics and intelligence

Digitization and automation of knowledge work
Breakthrough advances in artificial intelligence and machine learning

Advanced analytics
Improved algorithms and largely improved availability of data

Human-machine interaction

Touch interfaces and next-level GUIs
Quick proliferation via consumer devices

Virtual and augmented reality
Breakthrough of optical head-mounted displays (e.g., Google Glass)

Digital-to-physical conversion

Additive manufacturing (i.e., 3D printing)
Expanding range of materials, rapidly declining prices for printers, increased precision/quality

Advanced robotics (e.g., human-robot collaboration)
Advances in artificial intelligence, machine vision, M2M communication, and cheaper actuators

Energy storage and harvesting
Increasingly cost-effective options for storing energy and innovative ways of harvesting energy

SOURCE: McKinsey

2017 AIChE® SPRING MEETING & 13TH GLOBAL CONGRESS ON PROCESS SAFETY

17 AIChE Spring Meeting, San Antonio, TX — 13TH GLOBAL CONGRESS ON PROCESS SAFETY

Industry 4.0 in process automation

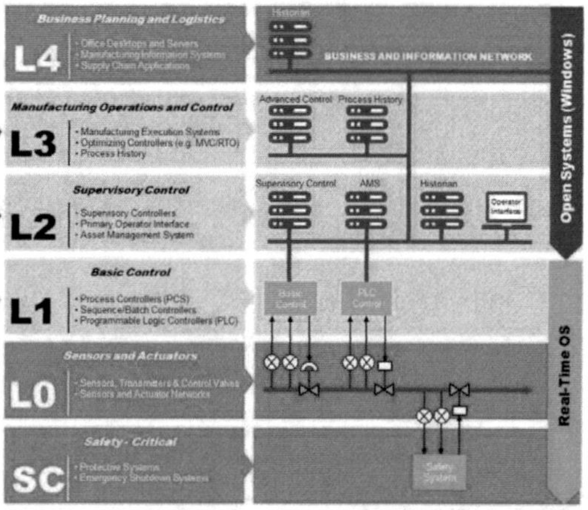

Modified excerpt from "How is Process Control Different from IS?" presentation. Authored by Johan B. Nye, EMRE for SANS SCADA Security Summit 3/06.

Current status

Systems are "closed"

Hierarchical structure

Proprietary HW/SW

17 AIChE Spring Meeting, San Antonio, TX — 13TH GLOBAL CONGRESS ON PROCESS SAFETY

Industry 4.0 in process industry automation

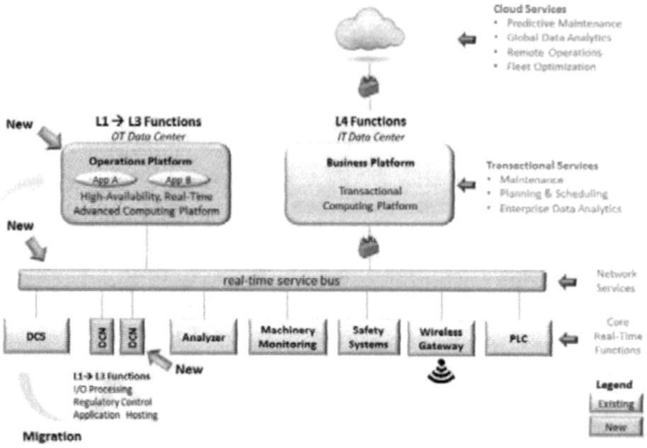

New Components of ExxonMobil Automation Vision
(Sources: ExxonMobil, Lockheed Martin)

The future

Open systems

Cloud based

The impact on process industry

- In the next 3-5 years industry will be significantly impacted:
  - Change of Hardware/Software architectures
  - Review of current operating practices
  - Change/upgrade of operator skills

The Cloud effect

- Main impact will be on architecture of hardware and software

- Effect also on operating practices
  - Access to a wider range of tools and technologies due to "open" architecture

17 AIChE Spring Meeting, San Antonio, TX — 13TH GLOBAL CONGRESS ON PROCESS SAFETY

The Big data effect

- Process industries have huge amount of data
  - Currently the use of these information is very limited
- Sophisticated analytics will bring increased capabilities in extracting knowledge from data
  - Increase use of existing data for APC development
- **Process knowledge will remain the key factor**

17 AIChE Spring Meeting, San Antonio, TX — 13TH GLOBAL CONGRESS ON PROCESS SAFETY

The status of APC

- Widely applied technology in the continuous process industry
  - Probably >10,000 applications in Oil&Gas and Petrochemicals
- The most successful technology is multivariable predictive control (MVC)
  - 3-4 major technology players
- Success factor of MVC is: to safely operate units on their most profitable constraints
  - Adapt to changing operating conditions
- High return on investment
  - 3-6 months ROI are typical

The challenges of APC

- O&G and Petrochem markets are "saturating"
  - Limited adoption in other industries
    - Minerals and mining
    - Food
    - Polymers
- No major developments in APC technology
  - Algorithms are essentialy the same as early days
  - Development of tools aimed at making deployment/maintenance easier (e.g. "adaptive" modeling)
  - Software development concentrated on OS compatibility issues etc.
  - Human Machine Interfaces have not been developed significantly
- APC resource pool shrinking
  - No young generation of APC engineers
  - Loosing experience and knowledge

Current APC and Industry 4.0

- APC technology is not Industry 4.0 ready
  - Software is not cloud based
  - Technology is "closed"
  - HMI interface is very basic
  - Limited data analysis capabilities

17AIChE Spring Meeting, San Antonio, TX
13TH GLOBAL CONGRESS ON PROCESS SAFETY

The future of APC

- Significant changes:
  - Shift from Client/Server to Cloud based
  - Shift from closed to open system
  - Improvements in HMI
  - Integration of existing and new technologies

17AIChE Spring Meeting, San Antonio, TX
13TH GLOBAL CONGRESS ON PROCESS SAFETY

Main developments

- The main developments should be on the HMI technology
  - Shift from “passive” presentation of data to “proactive” data analysis
  - Introduction of ***artificial intelligence*** elements in the HMI

APC and AI

APC technology can represent the cornerstone of the development of an AI system

APC allows the integration of engineering and operations

APC and AI

- APC has unique characteristics that are essential to an AI system
  - Knowledge of the unit (steady state model)
  - Knowledge of economics (objective function)
  - Knowledge of safe operations (equipment and process constraints)
  - Knowledge of Production (product specifications)
  - Knowledge of the dynamics of the process (dynamic model)

The role of operators now

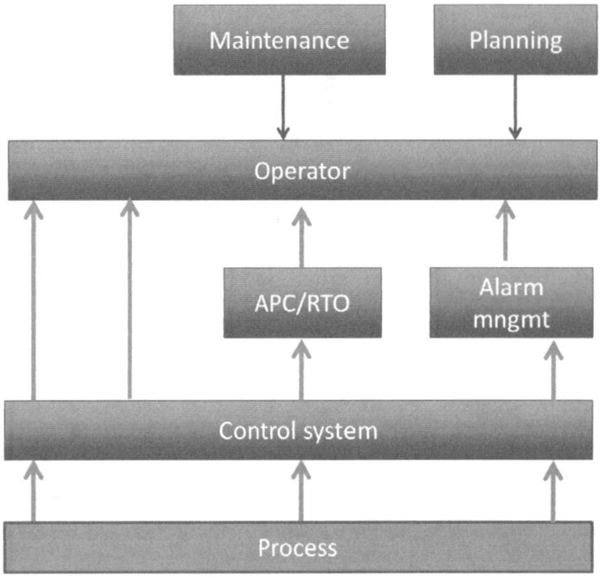

Operator is receiving info from many different sources

- *Control*
- *Alarms*
- *Planning*

All these sources are independent

Operator needs to elaborate a bign umber of data and info

The role of operators in Industry 4.0

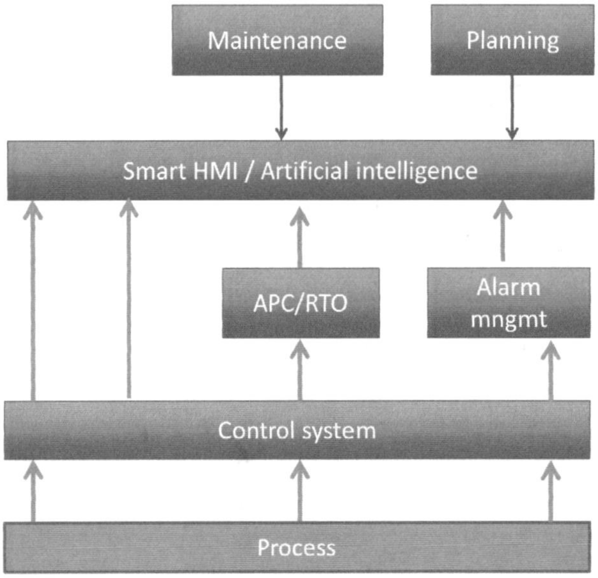

An AI system will process data and give meaningful information to operators

Operators will make high level decisions

New generation of Operators with specific training and skills

Integration of engineering and operations

From Advanced to Smart Process Control

- Introduction of data analytics capabilities are essential
- HMI needs to be redesigned with embedded AI capabilities
- SW architecture needs to open to external applications

17 AIChE Spring Meeting, San Antonio, TX
13TH GLOBAL CONGRESS ON PROCESS SAFETY

MPC in Industry 4.0

- MPC is the most widely applied APC technology

- Base for MPC success are:
  - Use of a dynamic model of the process
  - On-line economic optimization

2017 AIChE® SPRING MEETING & 13TH GLOBAL CONGRESS ON PROCESS SAFETY

Dynamic simulation

- Dynamic process simulation is not used for design and maintenance of APC strategies
  - Integration of rigorous steady state and dynamic simulation into APC
  - Increase of use of models developed for process design

Realtime optimization

- RTO has limited implementation in the O&G and Petrochem industries:
  - Useful for non-linear processes
  - Typically sets targets for MPC
- Integration of RTO and MPC in one package
  - Linear and non-linear processes

What's next?

- Short term (1-3 years):
  - Main focus will be on the HW and SW architecture
- Mid-term (3-5 years)
  - New generation of APC technologies will start replacing current and expanding into different

17 AIChE Spring Meeting, San Antonio, TX
13TH GLOBAL CONGRESS ON PROCESS SAFETY

What's next?

- New generation of APC will require different skills in workforce
  - Smart APC will proactively help operators in making decision
  - New ways of interacting with the automation system
  - Specific training and education will be required for the workforce of the future

2017 AIChE® SPRING MEETING & 13TH GLOBAL CONGRESS ON PROCESS SAFETY

17 AIChE Spring Meeting, San Antonio, TX

13TH GLOBAL CONGRESS ON PROCESS SAFETY

Summary

- Impact of Industry 4.0 on Process Automation
- Status of Advanced Process Control
- APC requirements for Industry 4.0
- The transformation from Advanced to Smart

Syncrude

LEVERAGING ADVANCED PROCESS CONTROL FOR OPERATIONAL EXCELLENCE

Perry Nordh (Honeywell), John MacGowan (Syncrude)

Honeywell

Syncrude

Agenda

1. Presentation Outline
2. What is Advance Process Control
3. Why embark on Advanced Process Control:
   a) Capture of knowledge
   b) Continuity of operation
   c) Improve safety & reliability
   d) Improve operational learning
   e) Sustained return on investment
4. Sustainment Challenges
   a) Operating Philosophy
   b) Operating Envelop
   c) Instrumentation reliability
   d) Operations Acceptance
   e) Operations conservatism
5. Conclusion

Honeywell

Advanced Control – What's It all About?

Syncrude

- Advanced control and optimization is key to running process plants reliably, efficiently and responsively
- Typically projects give payback time of < 1 year
- Wide range of methods available
  - Pick the right one for the job
    - Cost vs benefit

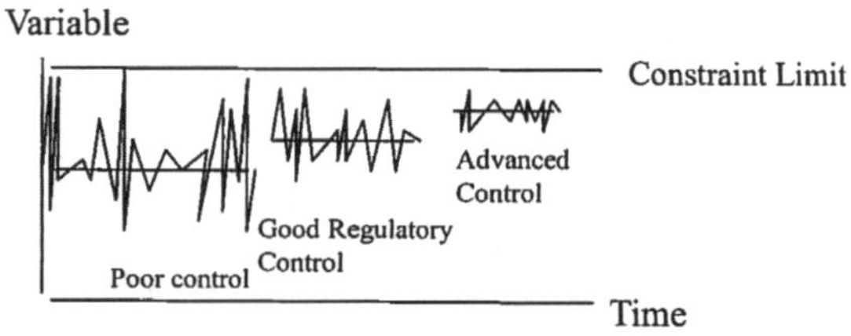

Honeywell

Some Preliminary Observations

Syncrude

- Multivariable Predictive Control (MPC) is a solution to many operating problems
- MPC is field proven, across many industries, with THOUSANDS (5000+) of running applications
- The MPC underlying technology is NOT NEW, introduced almost 30 years ago
- Countless success stories
- The ongoing maintenance of MPC applications has become a completely manageable issue
- These solutions are DCS/PLC agnostic, for the most part
- ROI very attractive
- Deep domain knowledge of process dynamics and interactions is key – UOP collaboration with HPS provides best scenario possible

Honeywell

Syncrude 4

PID control versus MPC

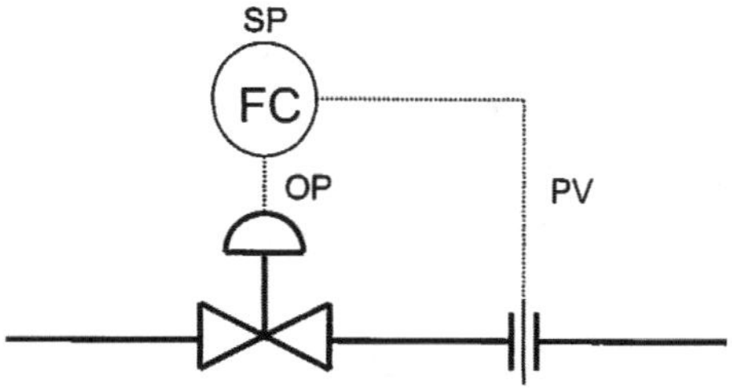

PID Control

- Single Input / Single Output control
- Controls to a setpoint
- Does not handle interactions well
- Little/no optimization
- Customisation needed for complex control
- Requires an error before it responds
- Like driving while looking backwards!

Honeywell

Syncrude 5

PID control versus MPC

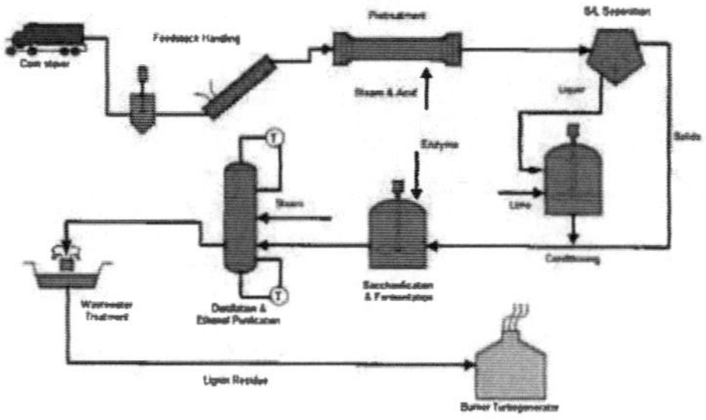

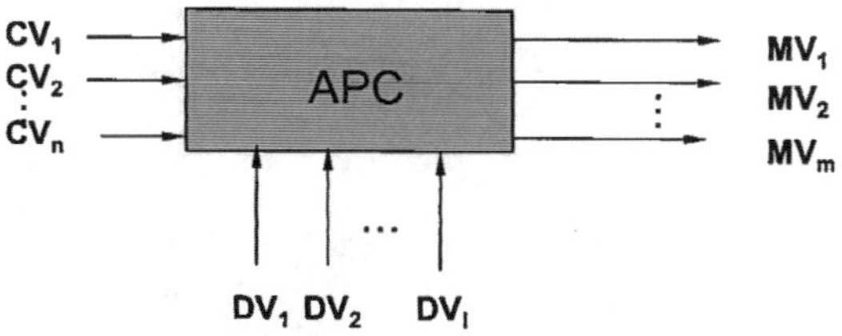

Multivariable Predictive Control

- Application residing internal or external to DCS
- Contains a model of the process
- Controls within a range
- Multi-input / Multi-output control
- Little / no customization needed
- Optimization / constraint handling built-in
- Corrects for future errors

Honeywell

Syncrude

Defining APC

- Advanced Regulatory Control (ARC)
  - Typically control that goes beyond Basic PID control
    - But typically done within the confines of the control system.
  - Single Input / Single Output control
  - Feedforward, Lead/Lag, Decoupling, etc.
  - Coding/Customization at DCS/PLC level
  - Little/no optimization

- Multivariable Process or Model Predictive Control, (MPC)
  - Application residing internal or external to DCS
  - Multi-input / Multi-output control
  - Little / no customization
  - Optimization / constraint handling built-in

Honeywell

Syncrude

Advanced Control Solutions

- Multivariable Predictive Control
- Soft Sensor Calculations
- Tuning and Mai

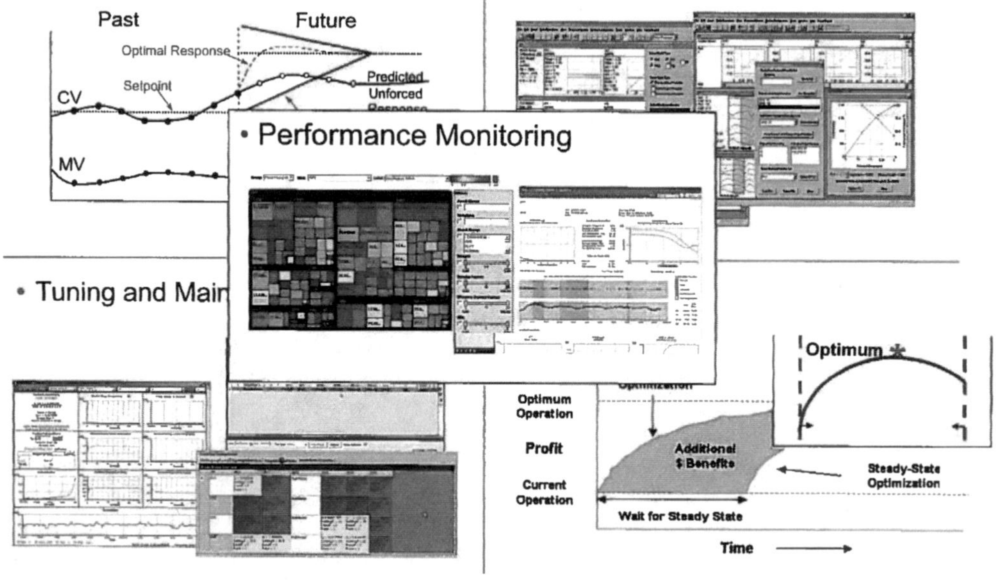

Honeywell

8

Syncrude

Why APC?

Honeywell

9

Syncrude

Why Advanced Process Control

- Capture of knowledge & experience
- Continuity of operation
- Operational learning
- Improve safety & reliability
- Value Generation

Honeywell

Syncrude 10

Capture of Knowledge & Experience

- Best Panel Operator running all consoles 24x7
- Similar operations
- Aging workforce –
- Systematic way to
  - Handles/links
  - Moves sizes
  - Sanity Checks
  - Abnormal operati
  - Operational "land
- Download before they are lost to organization

Honeywell

Syncrude 11

Continuity of Operation

- Process units are "yanked" around from shift to shift (operator to operator)
- Even with the best panel operator – there is inefficiencies during shift turnover
- With APC the plant is operated consistently and optimally – all the time

Honeywell

Syncrude 12

Operational Learning

- Even the best organizations learn something while designing and implementing APC
- Process interactions become more discreet and observable
- Considerations of failed instrumentations supports thought provoking safe guarding
- Requirement for considering and managing “land mines”

Honeywell

Syncrude 13

Improved Safety & Reliability

- You could argue the previous 3 benefits contribute to a safe and reliable operation
- Auto pilot is used after takeoff & before landing – why?
- Human factors studies suggest monotonous work will have people lose focus – that mountain snuck up on me!
- Reduction of key stroke errors
- Automatic monitoring and management of the operating envelope

Honeywell

Eliminate Process Variability

Syncrude 14

- Plants with Best Process Control – Have Fewer Slowdowns and Shutdowns

Correlation Between Uptime and APC Implemented[2]

- Analyze constraints and operate within them
- Reduces operator manipulation by 80%

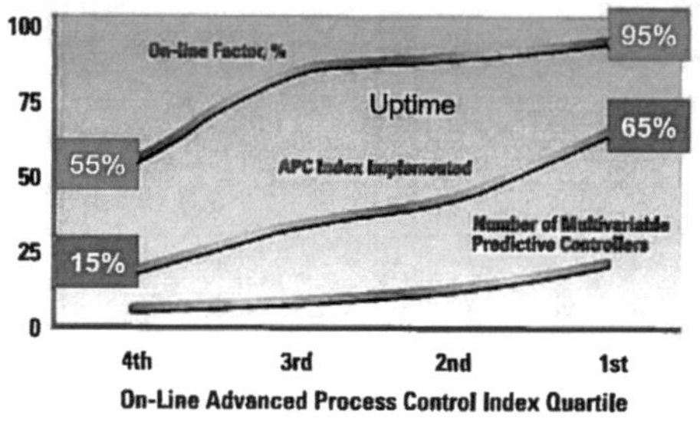

2. Birchfield, G., Solomon Associates – "Olefin Plant Reliability"

Honeywell

Value Generation

Syncrude 15

- Even the best operator can not monitor all operating envelops & change SP each minute to optimize
- Manages disturbances (ie rate or composition changes) proactively
- Highest level of APC can manage changing economics
- Stretch your existing infrastructure
- Consistently operating optimally

Honeywell

Syncrude 16

Sustainment Challenges

Honeywell

Syncrude 17

Syncrude
APC model for sustaining benefits

- Responsible for 21 on-line APC controllers and benefits
- 20 online Inferential property models
- APC controller installed on Hydrogen plants , HGO/LGO Hydrotreaters/ Naphtha stripper/ HGO Fractionator/ Diluents recovery units/ Vacuum units/Naphtha recovery units
- Most of APC controllers are in the upgrading area , two NRU APC Controllers are in Extraction
- Average service factor was 93%
- Total accumulated APC benefits value closing on $1B

Honeywell

Syncrude 18

Sustainment Challenges

- Operator Acceptance
  - Operating Philosophy
  - Operating Envelopes
- Operations conservatism
- Change Management
- Instrument reliability

Honeywell

Syncrude 19

Operator Acceptance

- There must be consensus on:
  - What the process objectives are **AND** how to achieve them
  - What the process operating envelop is **AND** how to maintain & prioritize them
- Often individual panel operators (or teams) have differing views on how to operate their process – see continuity of operation
- Those that are not in agreement with these design details will either turn the controller off or subvert the controller
- Mitigation:
  - High level sign off of agreed to operating philosophy
  - Increase Operator's knowledge & understanding of the controller
  - Agreed to operating envelopes in place and monitored

Honeywell

Syncrude 20

Operator Acceptance
Example of Operating Philosophy

- Operating safely and reliably
- Maintain product quality
- Minimize catalyst deactivation:
  - Minimize overtreating
  - Minimize Weighted Average Bed Temperature (WABT) and peak bed temperatures
  - Maximize front-end conversion
  - Maximize hydrogen partial pressure
  - Coordinate bed deactivation
    - » Minimize the difference between Bed 3 and 4 WABT
    - » Minimize the difference between Reactor A & B WABT

Honeywell

Syncrude 21

Operator Acceptance – Knowledge & Understanding

- Commission:
  - Detailed commissioning procedure
  - Commissioning steps approved by panel operator
  - On only during period when support for first few weeks
  - Daily follow up
  - Service factor turnover criteria (2-3 months)

- Training:
  - All panel operators, full day class training by panel rep
  - One-on-one training as we commissioned
  - Training during regular monitoring and maintenance
  - Refreshers

Honeywell

Syncrude 22

Operator Acceptance – Knowledge & Understanding

- Operators need to be able to find information
  - What is the controller doing?
    - Horizontal Gain maps
    - Move cause tips/information
    - Information on the design of the controller
- How-to Manuals
- Point origin diagrams
- Operating Envelope monitoring tools
  - Limit/constraint handling information
- Approved Operating philosophy

Honeywell

Syncrude 23

Operator Conservatism

- Some operators are more conservative than other operators – controller can be run on the edge:
  - Continual monitoring of key specs and operating envelop
- Lack of trust of the controller resulting in subverting of the controller
- Mitigation:
  - Monitoring & Maintenance
  - 1 on 1 demonstrations – see commissioning & Training
  - Acceptance of constraint priorities – see knowledge & Understanding of controller

Honeywell

Operator Conservatism – Monitoring & Maintenance

- Daily monitoring and follow up
- Monthly operations stewardship
- Quarterly management stewardship
- Annual audits
- Documentation
- Operational KPIs

Honeywell

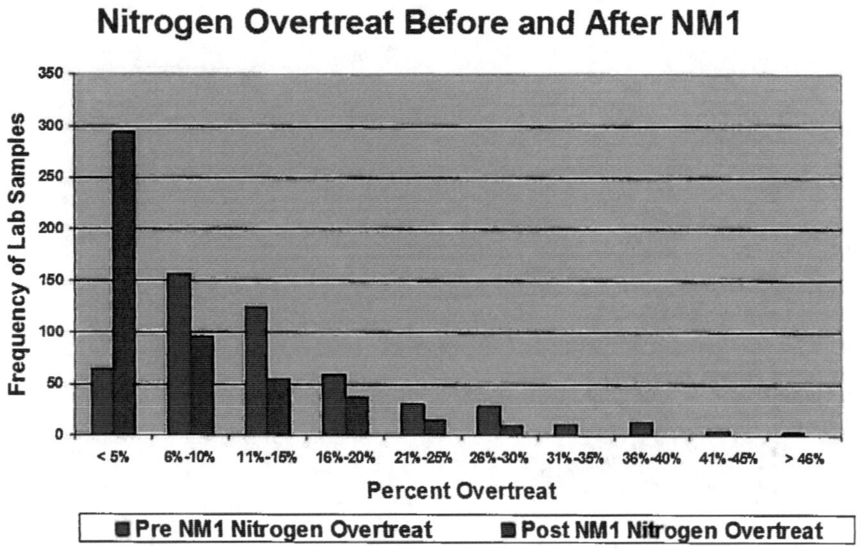

Honeywell

Change Management

- Control Engineers like to please

+

- Panel operators have varied opinions

= tendency for too many changes

- Mitigation:
  - Every month file check tools report created for each controller to check any APC tuning change or MV/CV limit changes.
  - Objective for file check to keep track on tuning changes or any abnormal changes in controller parameter.
  - Any major changes of APC controller tuning follows management of change
  - File check tool track changes in CPT file and engineering changes on LCN

Instrument Reliability

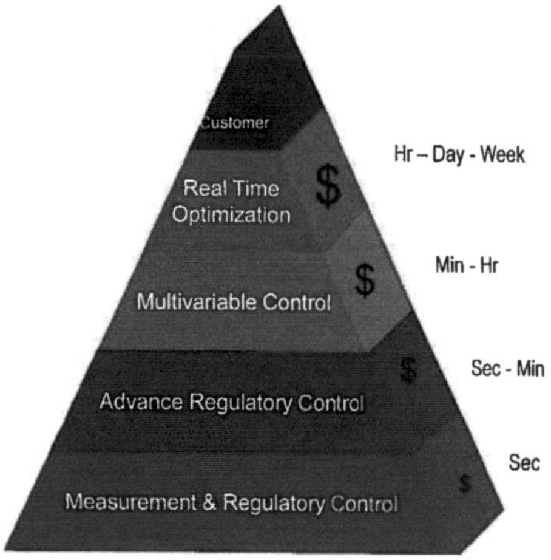

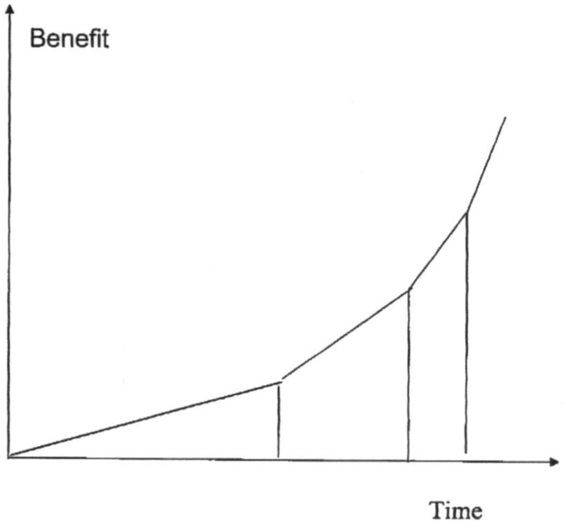

Instrument Reliability

- Base of the pyramid is measurement & regulatory control
- This needs to be good **and** reliable
- Projects should be set aside if this is not there and can not be robust
- This include Inferential models and their inputs

Honeywell

Framework for successful APC at Syncrude

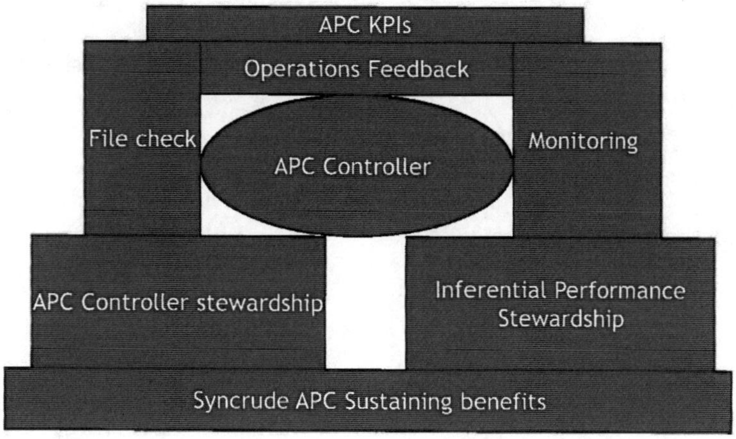

Honeywell

30

Syncrude

Conclusion

Honeywell

31

Syncrude

Conclusion

- APC is a proven technology
  - Decades of experience
- Benefits form a foundation to running your operations safely, reliably and profitably
  - Safety and reliability benefits are often neglected
  - Benefits are often not tracked over time or made visible
- Sustainment can be systematic
  - Requires focus on both control and people

Honeywell

Syncrude 32

Questions?

Honeywell

Syncrude 33

Thank You!

Honeywell

Syncrude 34

Informational Notice and Disclaimer

- This information is not for project planning purposes
- Please note that roadmaps are subject to change
- Please contact your Honeywell Account Manager to discuss project planning

Honeywell

PIL: The experts in process improvement technologies

process integration limited

Recent Developments in the Optimisation of Crude Oil Distillation Systems

29/03/2017

Lluvia Ochoa-Estopier
lluvia.ochoa-estopier@processint.com
www.processint.com

www.processint.com	+44 161 974 0090	Station House, Stamford New Road, Altrincham, Cheshire, CW5 7BH, UK

PIL: The experts in process improvement technologies

Contents

- Introduction
- Methodology
- Case studies
- Conclusion

Slide Number: 2

Knowledge transfer partnership

PIL: The experts in process improvement technologies

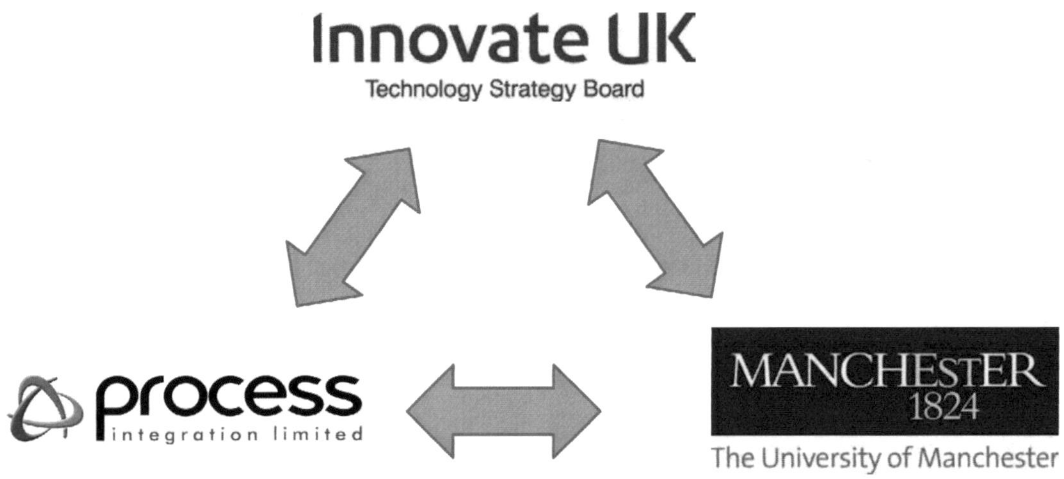

Slide Number: 3

Process Integration Limited (PIL)

PIL: The experts in process improvement technologies

process
integration limited

Consultancy, software and training for oil, gas and petrochemical industries

Provides cutting-edge technology and expertise to its clients to achieve cost-effective, practical solutions

Slide Number: 4

Crude oil refining

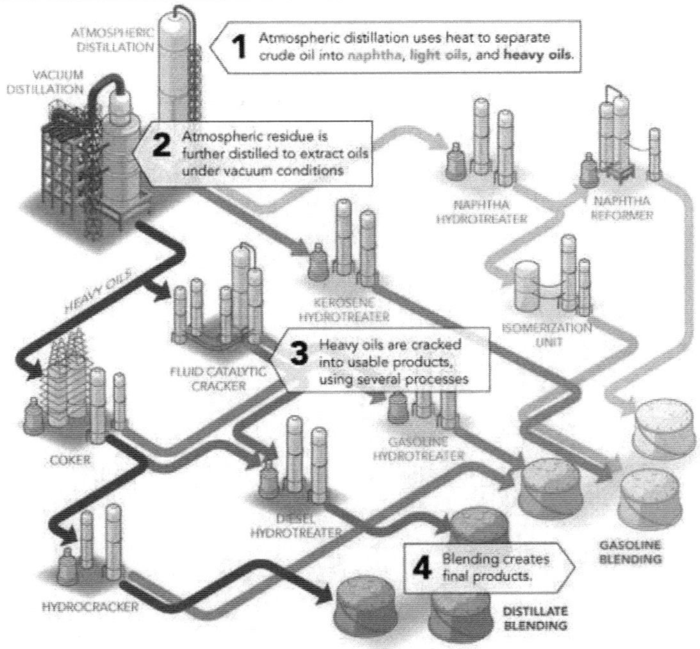

PIL: The experts in process improvement technologies

Slide Number: 5

Energy use in crude oil distillation

Medium size refinery

6 million tonnes of crude per year (110,000 bbl/day)

70 MW furnace duty (ADU+VDU)

~ 15 million Euros per year in furnace operating costs

PIL: The experts in process improvement technologies

Slide Number: 6

PIL: The experts in process improvement technologies

Economics in crude oil distillation

Operating costs

- Fuel gas
- Stripping steam

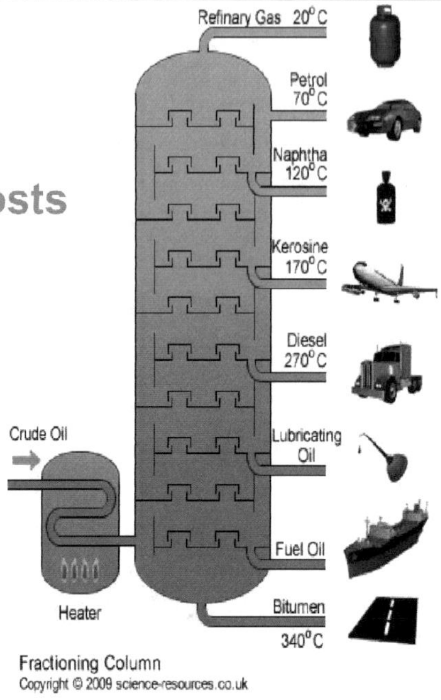

Product revenue

- Petrol
- Naphtha
- Kerosene
- Diesel

Slide Number: 7

PIL: The experts in process improvement technologies

Strategies to reduce energy consumption

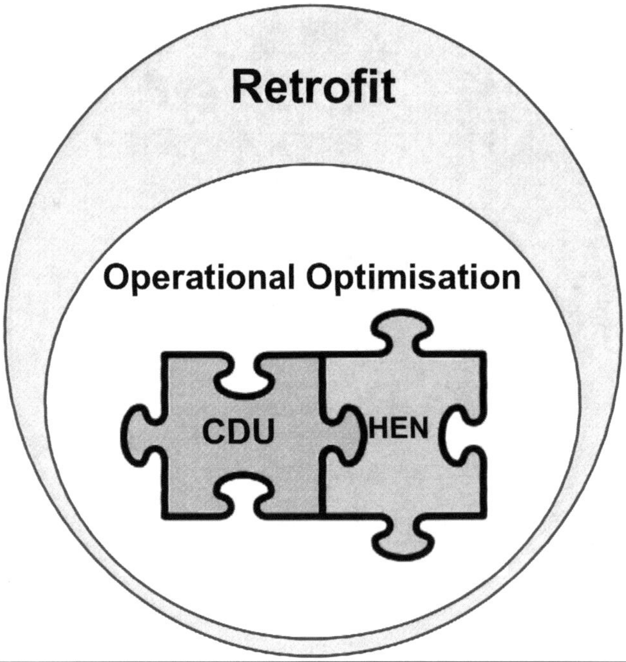

- Capital investment required
- Benefits increase with higher capital investment
- Downtime required for implementation

- No capital investment required
- Limited benefits due to fixed configuration
- Changes can be readily implemented

CDU and HEN should be optimised together

Slide Number: 8

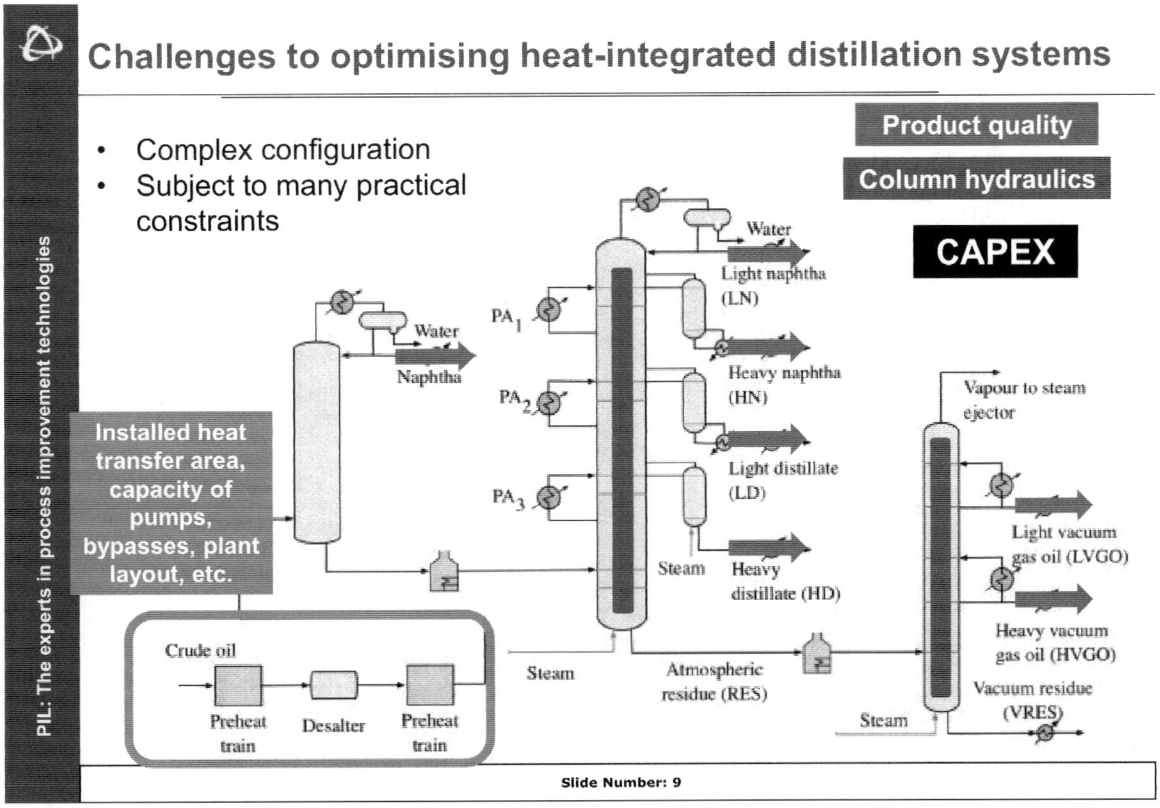

Challenges to optimising heat-integrated distillation systems
PIL: The experts in process improvement technologies
• Complex configuration
• Subject to many practical constraints
Product quality
Column hydraulics
CAPEX
Installed heat transfer area, capacity of pumps, bypasses, plant layout, etc.
Water
Naphtha
PA1
PA2
PA3
Water
Light naphtha (LN)
Heavy naphtha (HN)
Light distillate (LD)
Steam
Heavy distillate (HD)
Vapour to steam ejector
Light vacuum gas oil (LVGO)
Heavy vacuum gas oil (HVGO)
Vacuum residue (VRES)
Crude oil
Preheat train
Desalter
Preheat train
Steam
Atmospheric residue (RES)
Steam
Slide Number: 9

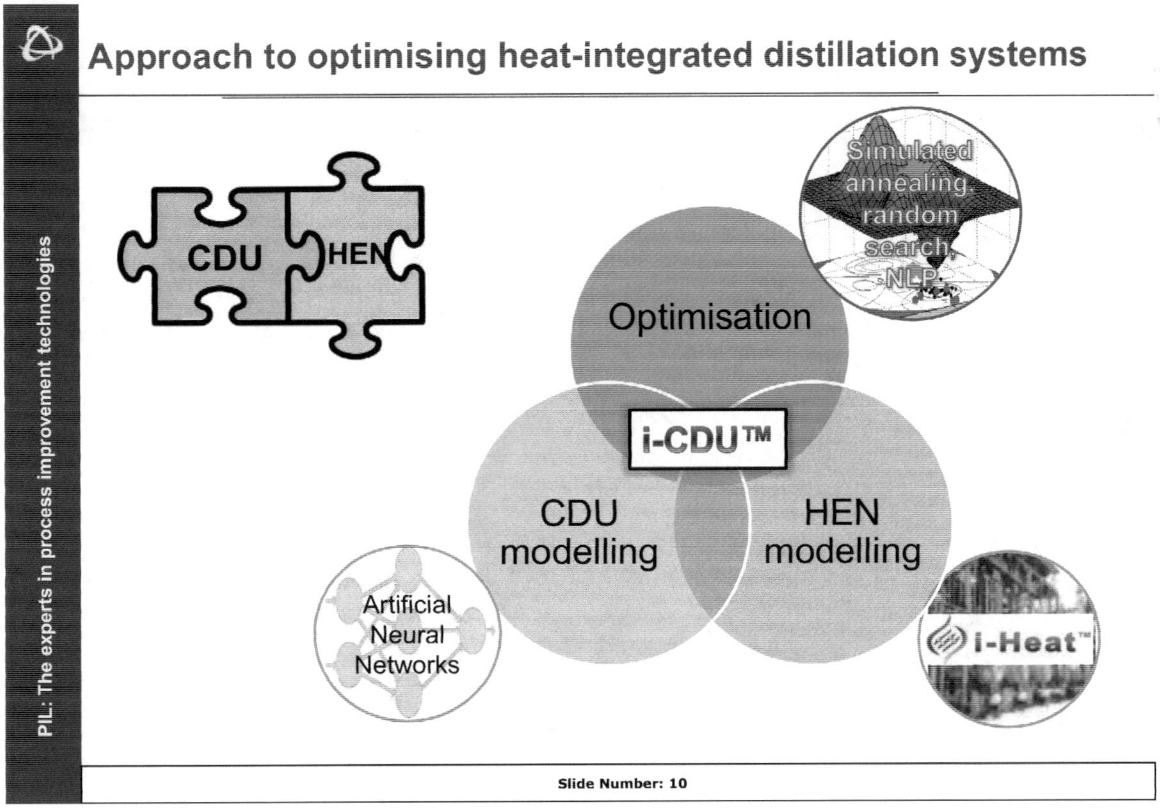

Approach to optimising heat-integrated distillation systems
PIL: The experts in process improvement technologies
CDU
HEN
Simulated annealing, random search, NLP
Optimisation
i-CDU™
CDU modelling
HEN modelling
Artificial Neural Networks
i-Heat™
Slide Number: 10

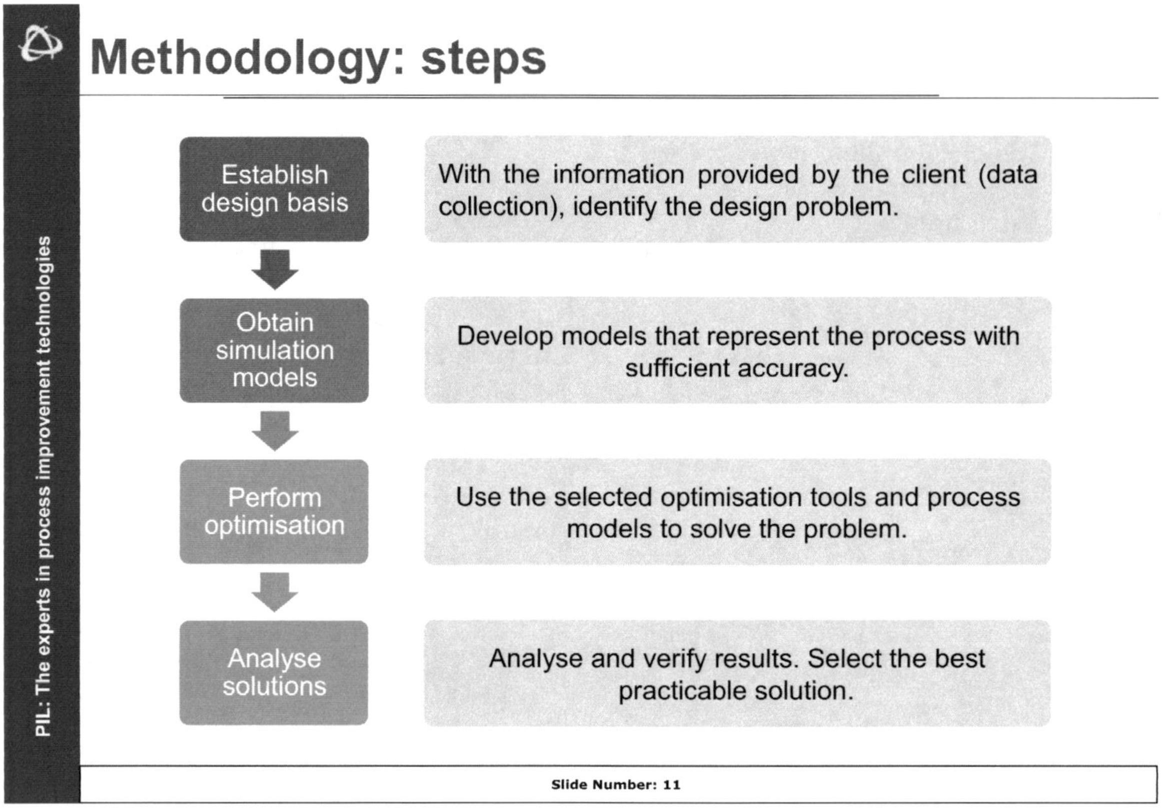

Methodology: steps
Establish design basis
With the information provided by the client (data collection), identify the design problem.
Obtain simulation models
Develop models that represent the process with sufficient accuracy.
Perform optimisation
Use the selected optimisation tools and process models to solve the problem.
Analyse solutions
Analyse and verify results. Select the best practicable solution.
PIL: The experts in process improvement technologies
Slide Number: 11

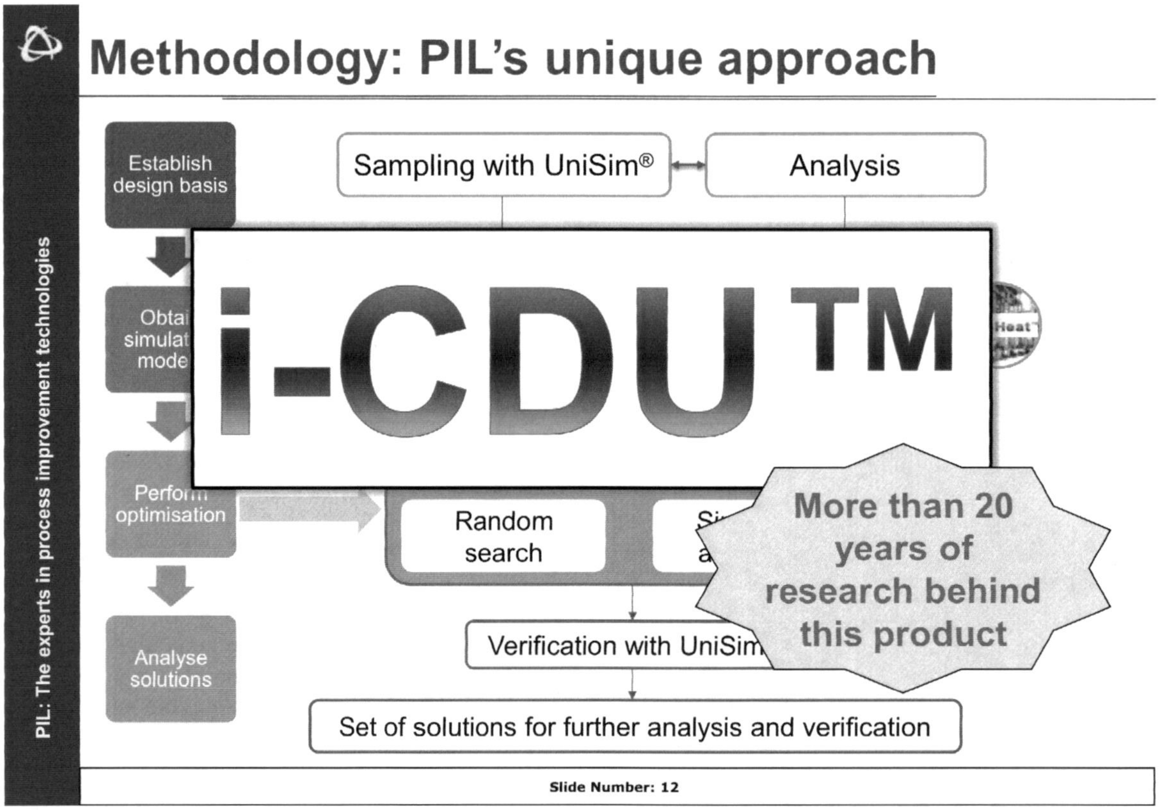

Methodology: PIL's unique approach
Establish design basis
Sampling with UniSim®
Analysis
i-CDU™
Perform optimisation
Analyse solutions
Random search
More than 20 years of research behind this product
Verification with UniSim
Set of solutions for further analysis and verification
PIL: The experts in process improvement technologies
Slide Number: 12

PIL: The experts in process improvement technologies

Methodology

Analysis – new tools allow the identification of the feasible region

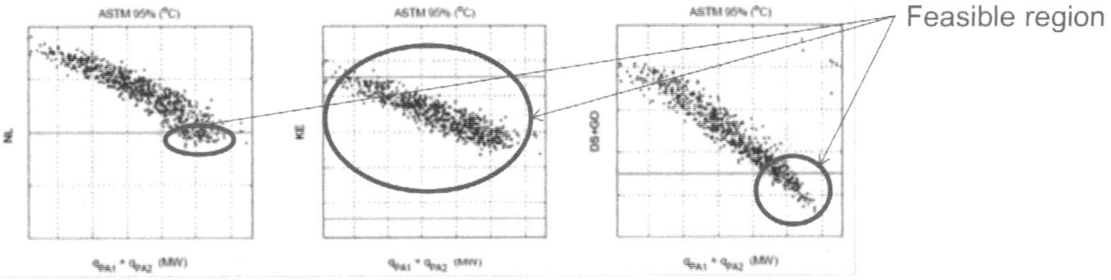

Modelling – Neural networks as accurate as the Unisim simulation, but faster

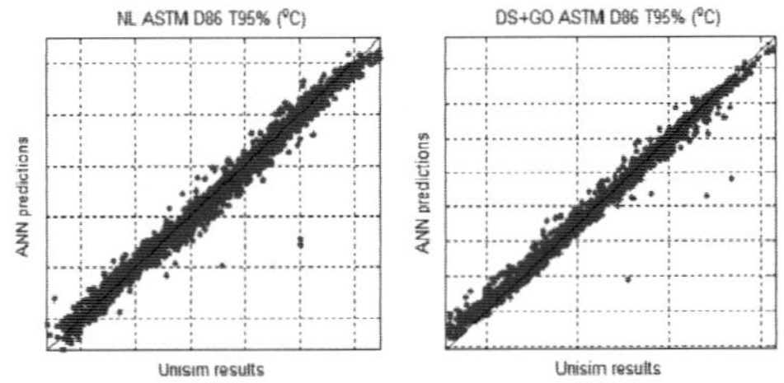

ANN distillation models

- Accurate and robust
- Simulation time of 100 points
  - ❑ ANNs: 0.3 s
  - ❑ Unisim: 20 min

Slide Number: 13

PIL: The experts in process improvement technologies

CASE STUDY

Slide Number: 14

PIL: The experts in process improvement technologies

Industrial Case Study

Project background

- Operational optimisation of a medium-scale European refinery. Two feed stocks. Yield optimisation. ~125,000 bbl/day

Project Objectives

- Improve net profit by operational optimisation

Constraints

- Very tight product quality specifications
- CDU operating over its design capacity
- Column flooding in diesel and MVGO sections
- Condenser bottleneck
- Installed areas, etc.

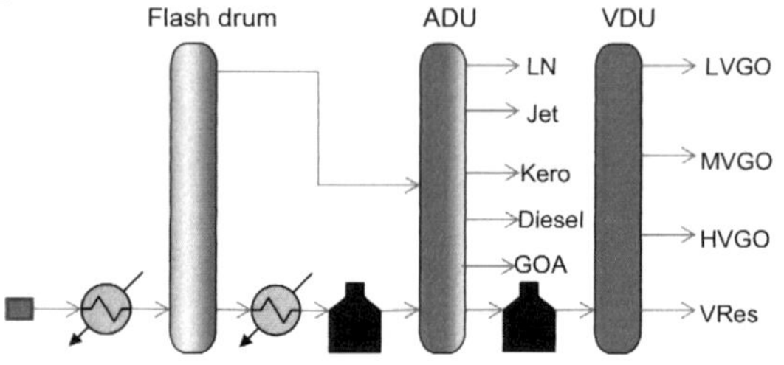

Results implemented on-site

Slide Number: 15

PIL: The experts in process improvement technologies

Industrial Case Study – Operational Optim.

Inputs Optimisation variables	**Outputs** Used to evaluate constraints and objective function
• Product flow rates • Pump-around specifications • Furnace coil outlet temperatures • Stripping steam • Reflux	• Jet flash point • ASTM D86 T5% and T95% temperatures • Hydraulic parameters • HEN stream properties (e.g. temperature, enthalpy, heat capacity)

Slide Number: 16

Industrial Case Study – Operational Optim.

- Estimated economic benefits greater than **5 MM€/y**
- Small increase in operating costs: fuel oil and steam
- Two crude oils considered in study

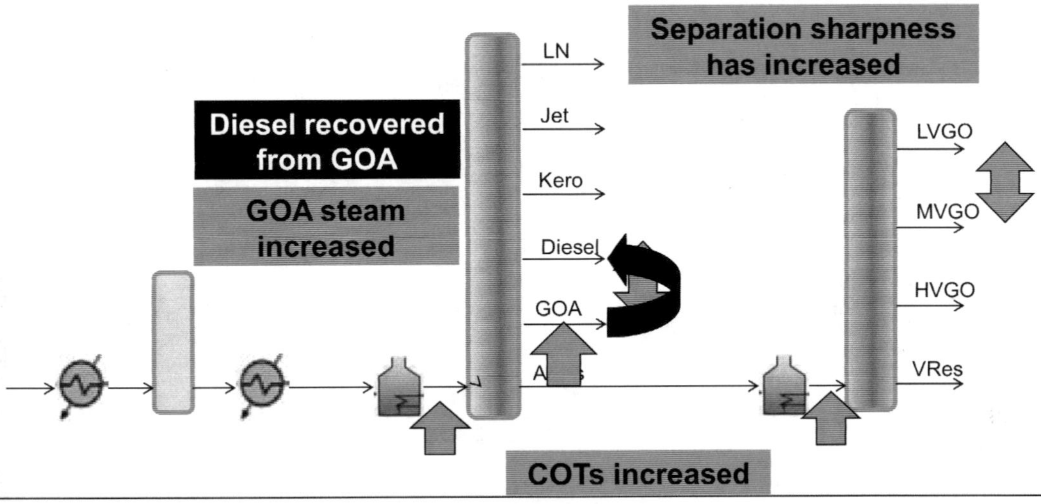

Industrial Case Study – Operational Optim.

PIL: The experts in process improvement technologies

Main operational changes

Description	Change from base case
Furnace outlet temperature increase	< 1°C
Total stripping stream increase	< 1 tonne/h
Yields increased	Kerosene and Diesel

PIL: The experts in process improvement technologies

Industrial Case Study – Implementation Crude 1

DAY 1

Due to the limitations at the furnaces, the COT was increased by reducing the crude oil throughput.

Profit was increased by more than $ 1,000 per hour.

Due to the decrease in crude oil flow rate, the energy costs did not increase

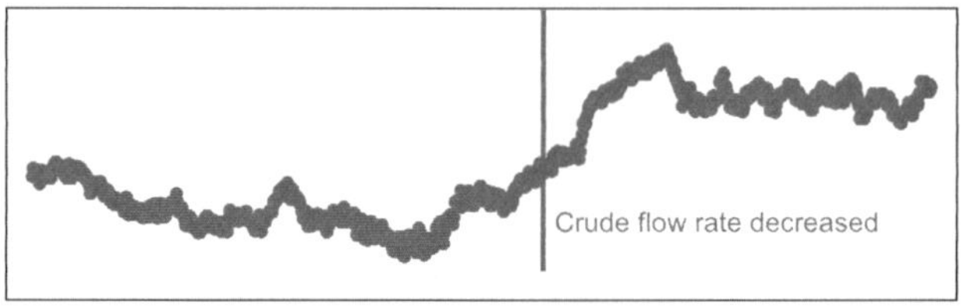

Furnace coil outlet temperature

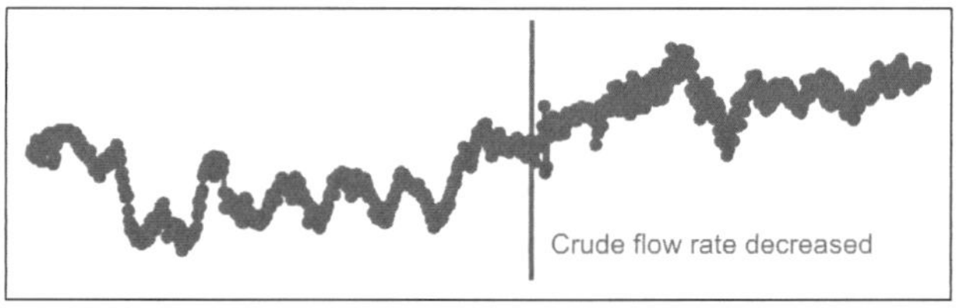

Profit

Slide Number: 19

PIL: The experts in process improvement technologies

Industrial Case Study – Implementation Crude 1

DAY 2

Increasing GOA stripping steam improved the separation between GOA and Diesel

The average net profit further increased after this change

No significant changes to the energy cost compared to the base case.

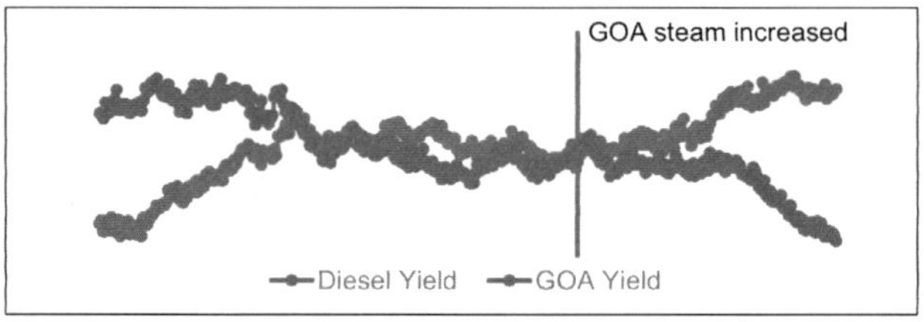

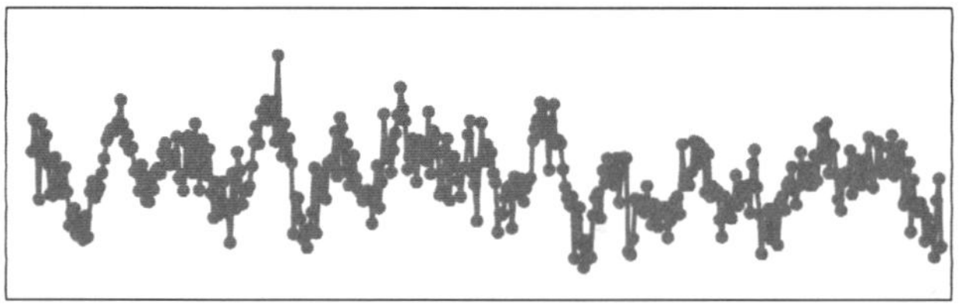

Profit

Slide Number: 20

PIL: The experts in process improvement technologies

Industrial Case Study

- Comparison between estimated and implemented <u>profit improvement</u> for Crude 1
- Assuming Crude 1 is processed 40 % of the year

1.1 MM€/yr **Estimated**

3.5 MM€/yr **Implemented**

Slide Number: 21

PIL: The experts in process improvement technologies

Industrial Case Study – Summary

- The new methodology was able to identify main operational changes to the distillation unit considering two crude oil feed stocks:
  - Increase COT
  - Increase GOA stripping steam
- Significant economic benefits were achieved during implementation from increasing product yields
- Energy requirements from fuel gas and stripping steam were not significantly increased

Slide Number: 22

PIL: The experts in process improvement technologies

CONCLUSIONS

Slide Number: 23

PIL: The experts in process improvement technologies

Conclusions

- Recent developments in the optimisation of crude oil distillation systems include
  - New analysis tools to screen solutions
  - A robust, simple and accurate CDU model
  - Consideration of practical constraints such as product quality, column flooding, etc.
- New methodology is able to find solutions with increased profit even for heavily constrained systems
  - Product quality maintained during optimisation
  - Practicable solutions (e.g. column hydraulics and pump capacities checked)
  - Savings found for operational optimisation, with **no capital investment required**

Slide Number: 24

480109 Plug-in Analytics Device Hardware and Applications for Rockwell Automation's Connected Enterprise and the Internet of Things

Abstract:

The abundance of sensor data, demand for lean and efficient plant operations, and added complexity of supply chains have evolved industrial requirements and expectations for advanced analytics applications, yet many proposed solutions to meet these new challenges require extensive engineering expertise or massive historical data.

Rockwell Automation has developed innovative, plug-in IoT solutions for analytics. These hardware-based applications require minimal engineering expertise by combining Control Logix technology with data-driven process modeling concepts. Further, these applications access data streams on the connected bus to learn and adapt, minimizing the need for any process historian.

We present a three pump application, which utilizes two of Rockwell's core plug-in analytics capabilities: automated stream-based modeling and advanced dispatch scheduling to meet anticipated demands.

481014 Model Reduction for Complex Systems Analysis

Abstract:

We present results from the application of the ALAMO methodology to the development of reduced-order models (surrogate models) for several case studies of interest to the design, analysis, control, and optimization of chemical processes. We begin by considering several simple systems, including flash calculations and steam tables. We then assess the capabilities of ALAMO to build simple and accurate models for an air separation unit. Air separation is an energy intensive process that involves heat integration and material recycles between columns. As a result, this system is often difficult to model. A high-fidelity model for an air separation unit was developed in gPROMS. ALAMO was utilized in order to reduce the complex simulation model to a reduced-order algebraic surrogate model. Extensive computational results are presented to compare the two models in terms of run time and relative accuracy.

481016 Feasibility Mapping and Optimization Using Big Data

Abstract:

With the advent of high-fidelity model-based computer simulations and advanced sensors, more and more data is made available for the determination of feasible operating region and the optimization of complex chemical processes. Specifically, data-driven methods can be advantageous for black-box process optimization, where the analytical forms of the objective function and constraints are either unknown or are difficult to formulate explicitly leading to lack of derivative information. However, the high-volume of data and the high-dimensionality of real industrial problems can introduce significant challenges for data-driven feasibility mapping and optimization. Current data-based approaches are mostly limited to small data sets or samples and perform well only if the problem has a few decision variables. To this end, we propose an efficient method for the approximation of convex or nonconvex, continuous or disjoint feasible regions of constrained black-box problems. Here, we approximate a feasible region as the region described by the convex hull of all feasible samples and their neighboring points while subtracting the infeasible samples and their neighboring points from the convex hull. To address high-dimensional problems, we propose a novel approach that transforms a multi-dimensional problem into a single-dimensional parametric problem over a newly introduced auxiliary variable [1]. In this presentation, we will introduce the convex-hull based feasibility mapping and single-dimensional projection methods using data obtained from high-fidelity process models and discuss the results for the intensification of an integrated carbon capture and conversion process to generate syngas utilizing power plant flue gas and natural gas within a single column.

References:

1. Bajaj, I., Hasan, M.M.F. A Novel Derivative-Free Optimization Method based on Single Dimension Projection. Accepted for publication in the Proceedings of FOCAPO/CPC 2017.

481240 Critical Leadership Moments

Abstract:

When situations at work turn from casual to critical, we instinctively react poorly. Our logic center shuts down and we resort to primitive thinking skills. A critical moment is defined by three components which set it apart: 1) high stakes are involved, 2) there are opposing opinions, and 3) strong emotion is present. If not all three of these elements are recognized, employees always silently "act out" their frustrations in negative ways. Sometimes unknowingly. As managers, when you are taught how to "talk through" high-stake, emotionally charged topics, the outcome is dramatically more successful on the performance of individuals, teams, and entire organizations.

480857 Your Career Path Is like a Distillation Column

Abstract:

A typical distillation column will have its distillate stream at the top and residue stream at the bottom. No matter which stage your feed stream is coming from and what the composition og your feed stream is, all the molecules should select a right steam to go. Your career path is same. Normally there are two kinds of career paths(managerial career path and technical/professional career path) that you can follow in your career development. No matter what your level of KSA(knowledge, skill and ability) is when you graduate from your college, you should pick up a right career path to follow. Different career path will require different KSA development so it is better for you to make a decision as early as possible. This presentation will provide you're the detailed comparison of two career paths and your development approach of required KSA.

482028 Sustaining a Zero Incident Culture

Abstract:

Throughout our evolution from an oil company to a global specialty chemicals company, we have advanced by integrating different approaches to a single whole. This presentation will discuss Ashland's highest priority, sustaining a fundamental focus on safe, compliant and responsible operations.

Speaker biography:

Fernandez-Moreno, 53, joined Ashland in 2012 as president, Ashland Water Technologies. In October 2013, Fernandez-Moreno was elected senior vice president of Ashland and became the president of Ashland Specialty Ingredients. In April 2015, he was promoted to senior vice president of Ashland and president, Chemicals Group. In this expanded role, he continues to lead Specialty Ingredients and has assumed additional management responsibility for Ashland Performance Materials. He also serves as a member of Ashland's Executive Committee.

Ashland Specialty Ingredients offers industry-leading products, technologies and resources for solving formulation and product-performance challenges in key markets including personal care, pharmaceutical, food and beverage, coatings, construction and energy. Using natural, synthetic and semisynthetic polymers derived from plant and seed extract, cellulose ethers and vinyl pyrrolidones, Specialty Ingredients offers comprehensive and innovative solutions for today's demanding consumer and industrial applications.

Since joining Ashland in November 2012, Fernandez-Moreno has been instrumental in revitalizing Ashland Water Technologies. Under his leadership, Water Technologies has simplified the business, focused on key growth opportunities and delivered better service and value for customers.

A 30-year veteran of the global chemical industry, Fernandez-Moreno previously served as executive vice president of Arch Chemicals, Inc. where he was responsible for the wood protection and HTH water products businesses. Prior to that, he served as business group vice president of Dow Coating Materials which was formed after Dow Chemical acquired Rohm & Haas in 2009. He previously spent more than 25 years with Rohm & Haas in a series of leadership roles spanning across Europe, Latin America and the United States.

Fernandez-Moreno earned a Bachelor of Science degree in chemical engineering from Universidad Iberoamericana in Mexico City. He also completed the Wharton Management Program at The Wharton School at the University of Pennsylvania.

480668 Moving the Needle with All Data, Big and Small

Abstract:

Dow combines the power of science and technology to passionately innovate what is essential to human progress. As a result, data science has been a vibrant theme at Dow for decades, in both the R&D and manufacturing organizations. More recently, the Advanced Analytics team was formed to apply these skill sets to answer broader business questions. Join us in this session to learn about Dow's analytics journey, using big and small data technology to drive growth. Through forecasting, data and text mining, and optimization and simulation methods, the Dow Advanced Analytics team is poised to add tremendous value to the company's top line, while becoming an essential partner for decision makers across every level of Dow.

478986 Digital Transformation in Pharmaceutical Product Development - Value, Strategy and Implementation

Abstract:

Pharmaceutical product development is the act of selecting and transforming a molecule to a valued medicine subject to patient, regulatory, manufacturing and business requirements. Organic Chemists, Molecular Biologists, Chemical Engineers, Mechanical Engineers and Analytical Chemists work together to develop the product design and its critical quality attributes, the process design and its various critical process parameters to deliver the right medicine in the right presentation with the right product & process knowledge to the right patient at the right time. The digitization of the underlying data is a monumental task, and requires a carefully conceived strategy. This talk will articulate the business drivers for this digital transformation, the three key elements of the strategy, namely Data Integration, Data Structure, and Analytics, and the progress on the implementation plan within Lilly.

480533 Big Data Analytics for Batch Performance Improvement

Abstract:

Batch processes are ubiquitous in food, chemical, pharmaceutical and semiconductor industries. Big Data Analytics efforts in chemical industry in particular have largely focused on continuous processes up to this point in literature (1–3). In this paper, we present some of our findings in using Big Data Analytics methods to explore batch-based datasets. We present two case studies in using batch datasets to improve batch production by identifying key steps and variables associated with longer cycle time.

In the first case study, we examine a batch system with multiple operating units. The general approach consists of: (1) identify general step sequences, (2) assess the correlation between steps and other common process interruptions, and (3) identify key variables in the steps of interest for improvement. The methodology is helpful to understand the sources of the highest variability that affect the batch quality variable of interest. The proposed methodology is validated on an industrial system where three batch units (with over 100 variables and many recipe steps) are operating in series. After performing the data analysis and identifying critical steps and process variables, the production rate of this batch system has shown consistent improvement.

In the second case study, we combine data-driven models constructed from batch process data with a controller to reduce process variation of a batch reaction system. After analyzing the batch dataset from this process, key recipe step was identified to be the source of variability. However, conventional control strategies cannot be applied easily due to the unique process equipment. As a result, we propose an alternative control strategy based on the model from the batch data analysis, where both normal operating data and design of experiment data were utilized. In addition to minimizing variability, the safety and equipment constraints are also considered in the control action, leading to robust and stable control. The effectiveness of the proposed batch control strategy was demonstrated to show considerable improvement through both simulated case study and plant implementation.

References

1. Qin SJ. 2014. Process data analytics in the era of big data. *AIChE J.* 60(9):3092–3100

2. Colegrove LF, Seasholtz MB, Khare C. 2016. Big data: getting started on the journey. *Chem. Eng. Prog.*, March, pp. 41–45

3. Kadlec P, Gabrys B, Strandt S. 2009. Data-driven soft sensors in the process industry. *Comput. Chem. Eng.* 33(4):795–814

Distillation Tower Inspection for Turnarounds and Revamps

185 Kister Distillation Symposium 2017

Doug Bouck
Technical Consultant
Separation Solutions, Ltd

Safety

- Personal
  - Basic Refinery/Chemical Plant Safety Training
  - Confined Space Safety Training
  - Hole Watch Communication
    - Voice or Radio (Compatibility Issue)
  - Know Your Own Limits
    - Heights
    - Closed In Spaces
    - Darkness
    - Flexibility
    - Conditioning

2

Safety

- Hazards
  - Climbing
    - Outside: Cluttered Platforms and Ladders
    - Inside: Cords, Hoses, Damage
  - Encumbrances & Hang-ups: Harness, Meters, Radio, Tools, Coveralls
  - Falling Objects: Tray Parts & Tools
  - Trays & Other Internals: Loose, Corroded, Displaced, Sharp Edges
  - Partner or Other Workers/Inspectors
  - Tight Spaces – Exit Plan

3

Safety

- Hazards
  - Dust, Scale
  - Old Equipment - Asbestos Gaskets
  - Pyrophoric Dust & Scale Producing SO2/H2SO3 Over 104F
  - Packing Fire Prevention:
    - "Causes and Prevention of Packing Fires", Chemical Engineering, p. 34 (July 2007).

- YOU are Responsible for YOUR Safety

- YOU are Responsible for OTHER'S Safety

4

5

Typical Tower Internal Manways

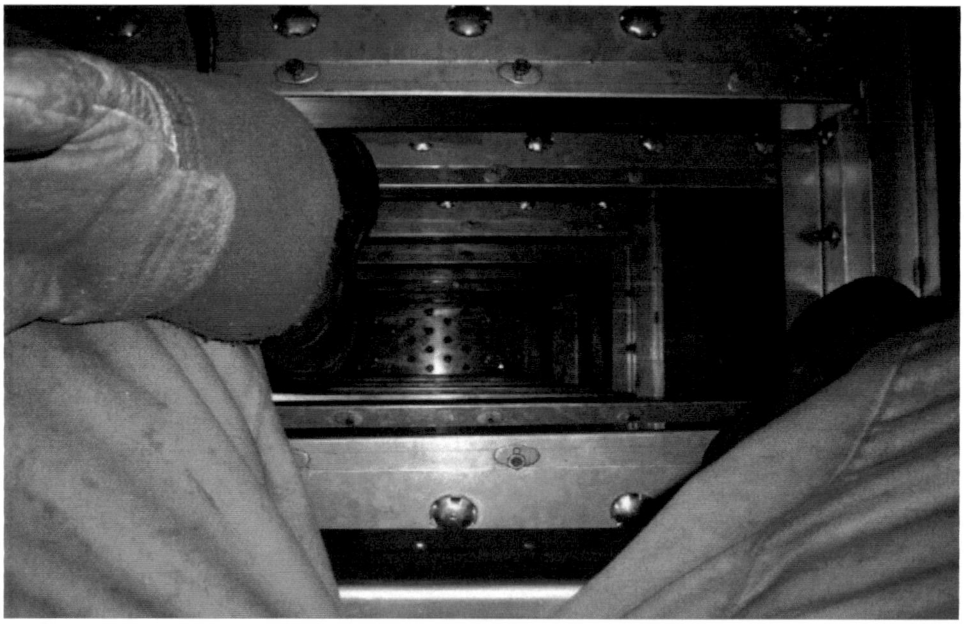

6

Tight Spaces

7

Multiple Personnel

8

Tight with Tray Damage

9

Cluttered Manway & Internals

10

Scaffold Inside Tower

11

Workplan Responsibilities

- Process Inspection
- Maintenance/Mechanical Inspection
- Vendor Reps./Inspectors
- Contractor Inspectors
- Operations Inspection
- Training Inspection Tours
- Punch Lists and Change Orders
- Follow-up Inspections
- Signoff Responsibilities

12

Inspection Equipment

- Typical in Waist Pack
  - Lock-out Lock and Tag
  - Zip Lock Bags & Magnet
  - Soapstone + Yellow Crayon + Black Marker
  - 10-25 ft Tape Measure
  - 9/16 Wrench & Plyers
- Other
  - Shin Guards and/or Knee Pads
  - Note Pad, Drawings & Inspection Plan
  - 2 Light Sources with Fresh Batteries
  - 2+ Pens/Pencils
  - Camera (Hot Work Permit)

13

Waist Pack Contents

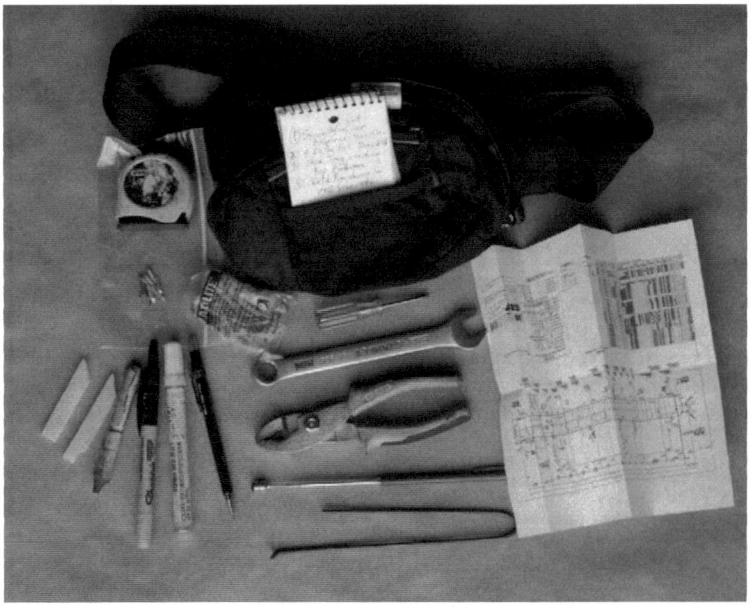

14

Other Equipment

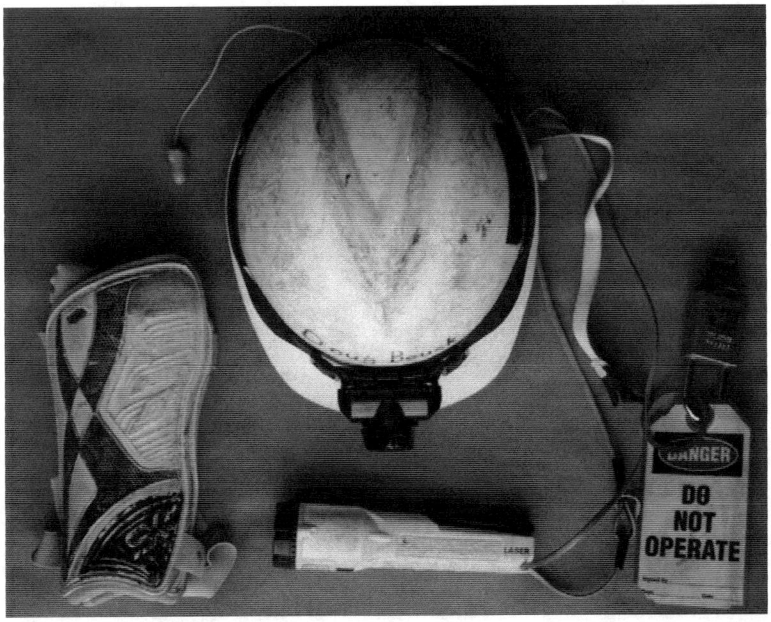

15

Inspection Equipment

- Special
  - Goggles, Dust Mask or Face Mask
  - Weir/Downcomer Dimension Template
  - 6 Inch Steel Rule
  - Dimension & Tolerance Table
  - Detailed Tray/Internals Drawings
  - Punch List and/or Drawings
  - Level Device: Bubble, Water or Laser
    - Device Accuracy Test
  - Distributor Test Equipment
  - Water Tightness Test Equipment

16

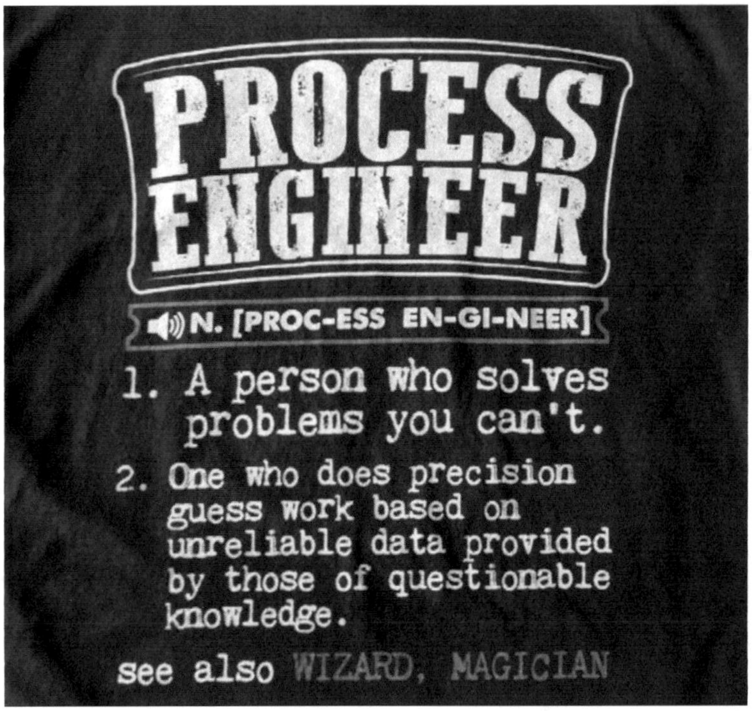

17

Keys to Process Inspection

- Understand How the Process Works
- Understand How the Parts Fit Together
- Understand How the Parts Work Together
- Know Why the Specs are Important
- Know How Deviations Affect the Process
- Be Able to Explain Operating and Economic Effects of Deviations
- Consider the Next Fix Chance May be 5 Years
  - Know When to Hold'em and When to Fold'em

18

Basic Preliminary Tray Inspection

- Number Trays for Reference
- Note Cleaning Issues – Take Photos
  - Does It Need Cleaning Before Full Prelim.?
  - Loose Deposits or Hard Scale?
  - Tray Decks/Downcomers and On/Under Valves
  - Any Special Cleaning Techniques Required?
- Note & Mark Missing/Stuck Valves
- Note & Mark Loose or Missing Hardware
- Note Cracked/Damaged/Corroded Trays
- Note Condition of Internal Manway Hardware
- Note Condition of Feed Pipes & Vortex Breakers
- Prepare Recommendations & Punch List

19

TAR Tray Inspection 1

- Objectives:
  - Return Equip. to Efficient Operation
  - Remember the Next Fix Opportunity May Be Years
- Cleanliness
  - Loose Scale/Deposits on Downcomers & Walls
    - Any Loose Scale Can Move to Downstream Equipment
  - Deposits on Decks
    - Plugging or Restriction of Valves, Holes or Caps
    - Bubble/Tunnel Caps Annular Space Must Be Clear
  - Accumulation in Downcomers, Sealpans & Drawpans

20

TAR Tray Inspection 2

- Missing/Worn/Stuck/Plugged Valves
  - One (1) Percent Acceptable Exception
  - Moving Valves Must Move Full Travel
    - Remove Deposits on Top, Legs or Around Holes of Valves
  - Check for Worn Valves & Tray Decks
  - Replacement Valves: Legs Bent for Retention
  - Replacement Valves with Cages: Crimping of Leg Tabs

21

Valve Cage Installation Problem

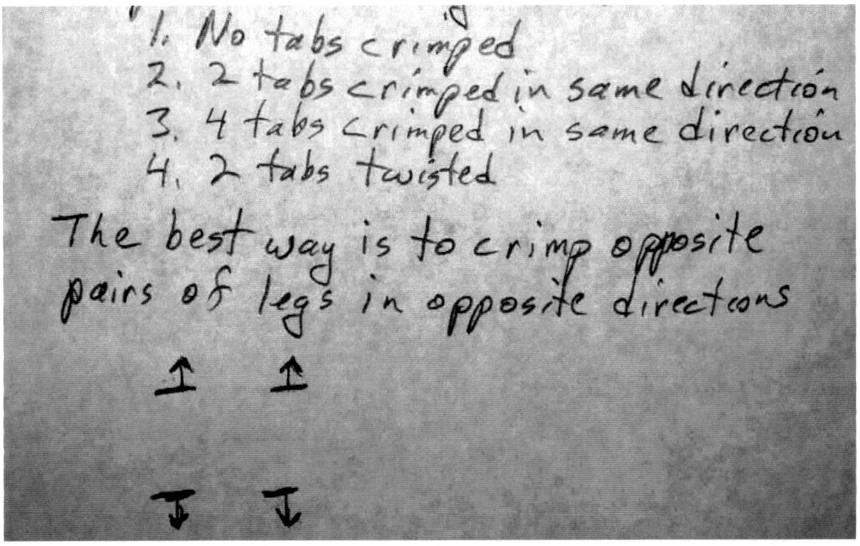

22

TAR Tray Inspection 3

- **Most Frequent Problem:**
  - Frozen/Damaged Internal Manway Hardware
    - Result of Hammer Release and/or Corrosion
    - Frozen Hardware Guarantees Poor Re-Closure
    - Poor Closure Causes Vibration Cracking of Trays
    - Poor Closure Causes Less Security Against Upset
    - Poor Closure May Result In Operating Deficiency
  - **Hardware Problems Must be Corrected Before Approval for Internal Closure**

23

TAR Tray Inspection 4

- Nutter MW Hardware Problems:
  - Subtle Correct Orientation
  - Rectangular Slots and Bolts
  - Boilermaker Hammer Release Damage
  - Same Poor Closure Issues
- Blind Fastening
  - Unusual
  - Difficult
  - Two Different Types of Hardware
  - Two Different Techniques
  - Boilermakers May Need Instruction

24

Nutter Internal Manway Hardware Orientation

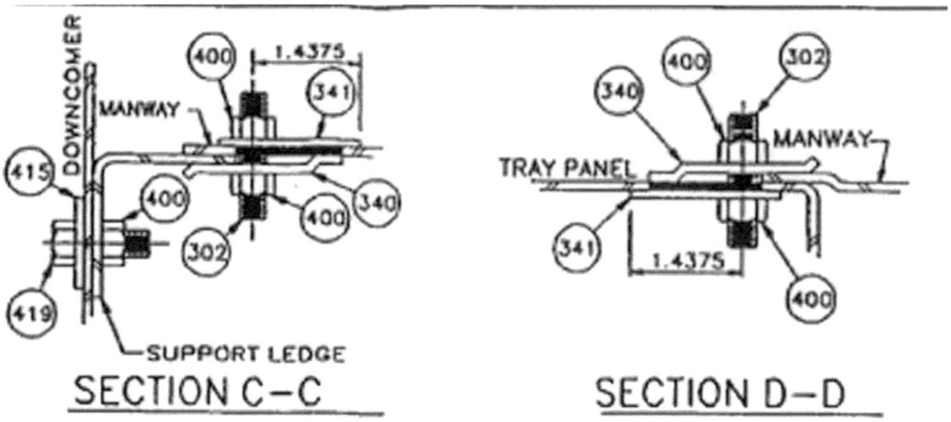

25

Blind Manway Bolting

26

Blind Manway Bolting

27

TAR Tray Inspection 5

- Other MW Hardware Problems
  - Light Wt. Football Washer Bends Causing Poor 2nd Closure
  - 3/16 in. Thick Heavy Duty Washers (Footballs) Recommended

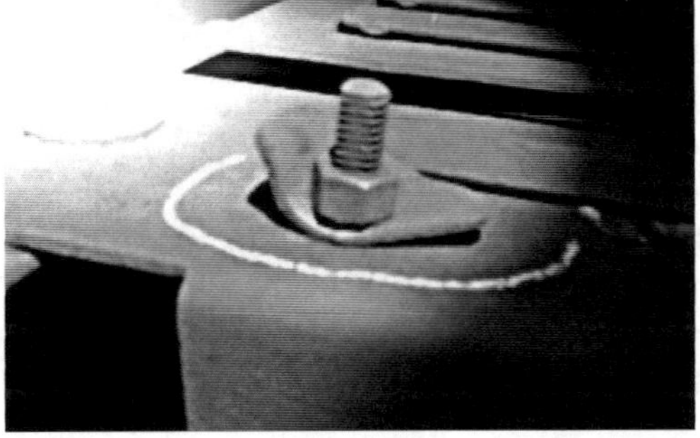

28

Holes in Downcomers

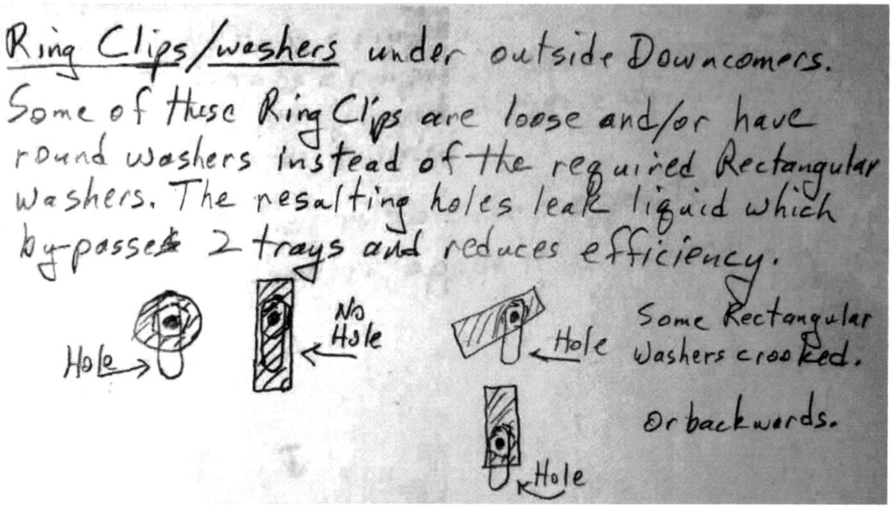

29

Internal Downcomer Loose Scale

30

Downcomer Scale

31

Dirty Round Valve Tray

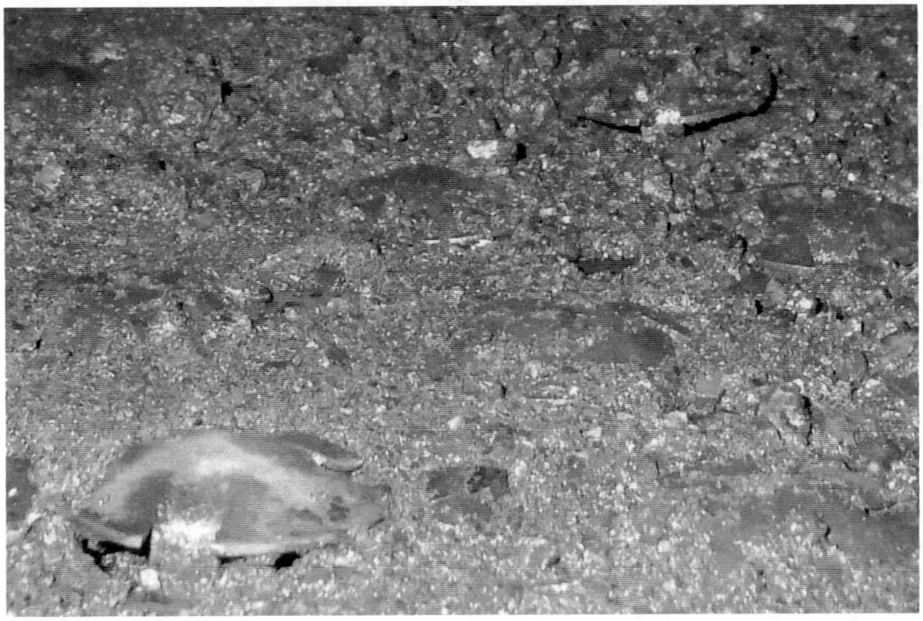

32

1- 1.5 Inch Deposits

33

Corroded and Plugged Bubble Caps

34

Displaced Weir

35

Bent/Loose Downcomer Clip

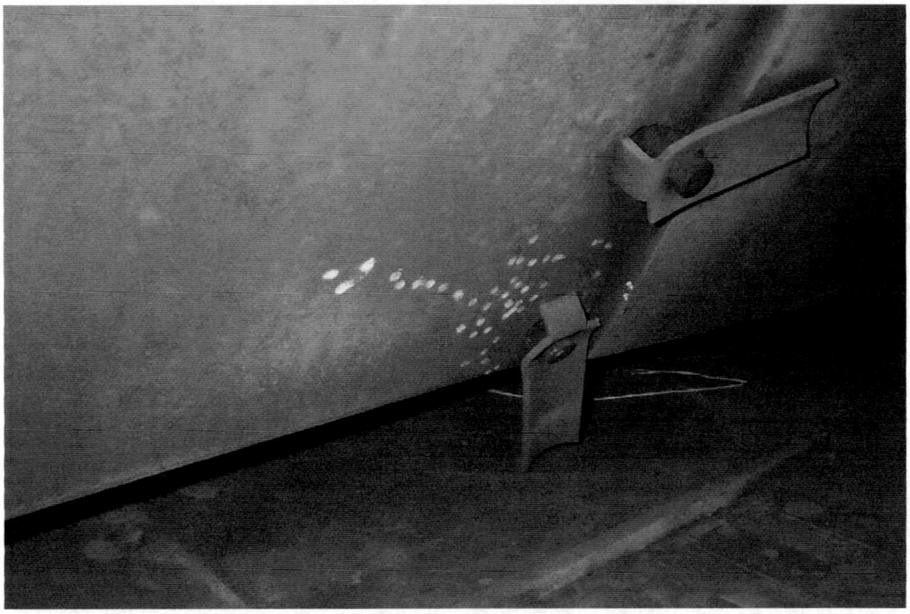

36

Bent Tray Downcomer & Clip

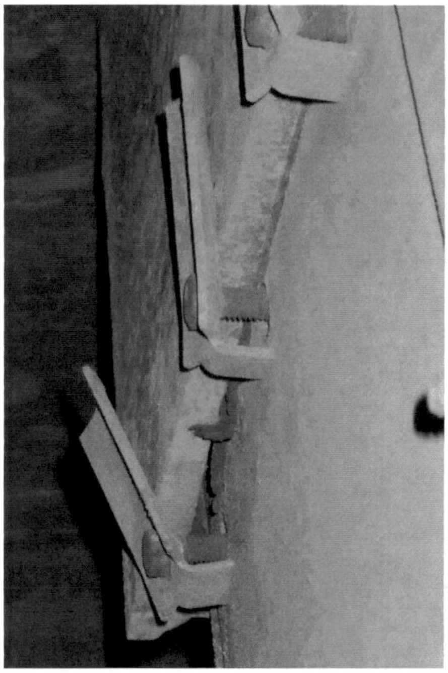

37

Ring Clip Askew

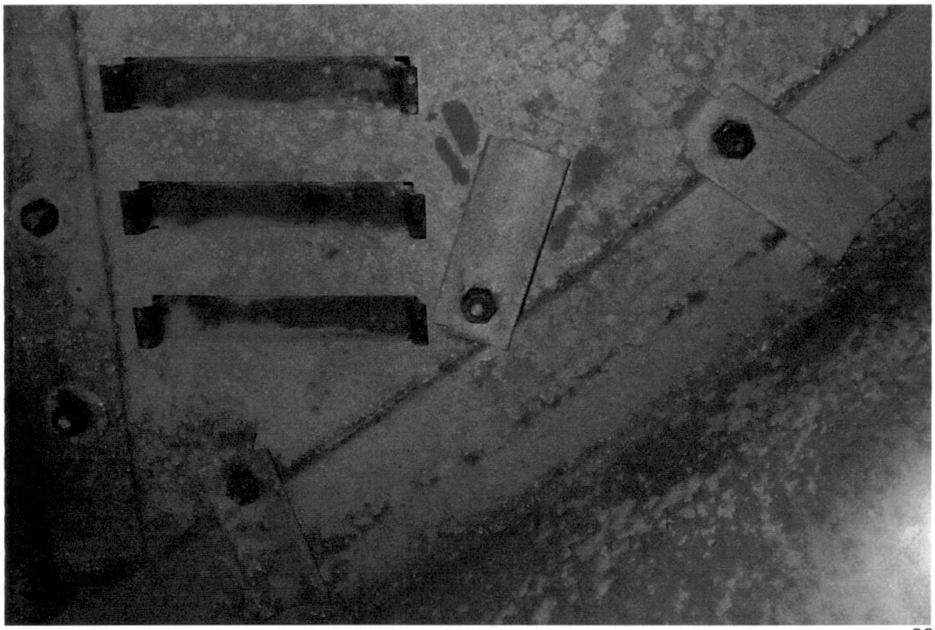

38

Displaced Valves

39

Displaced Valves

- Check the Legs Condition
  - Valve Legs Worn?
  - Improper Installation?
    - Legs not Spread Properly?
- Check the Hole Condition
  - Worn By Erosion-Corrosion?
- Blown Loose by Process Problem?
- Deck Thinning from Corrosion?

40

Missing Bubble Cap

41

Missing Swept Back Weir Piece

42

Missing Deck Bolts

43

Tray Damage Analysis

- Trays Bent or Cracked
  - Fatigued by Age and/or Weight
  - Overtightening
  - Level Loss or Explosion
- Trays Pushed Up
  - High Level + Percolating Vapor
  - Water Explosion Below
- Trays Pushed Down
  - High Level + Rapid Level Loss
  - Water Explosion Above

44

Boss says:

DOUGBERT, THE CEO SAYS THAT THE PROBLEM DIB COLUMN IS OPEN FOR INSPECTION AND YOU ARE THE MAN!

Dougbert says:

WHY ME? THAT TOWER IS OVER 200 FEET TALL!

Boss says:

HE SAYS YOU ARE THE ONLY ONE WHO KNOWS HOW THE INSIDES ARE SUPPOSED TO WORK.

45

Bent Tray Panel at Support Ring

46

Results of Loss of Level Control with Vapor Pushing Liquid Up Through the Trays

47

Uplifted Downcomer Panel

48

Lift Damage

49

Lift Damage

50

Lift Damage

51

Push Down Damage

52

Push Down Damage

53

Brittle Failure

54

Process Revamp Review

- Pre-TAR Revamp Process Review
  - What is Different
  - Objective of Change
  - Impact of Change on the Process
  - Will the Changes Meet the Objectives
  - Are the Changes Consistent and Valid Engineering?
- Pre-TAR Mechanical Review
  - Mechanical Design meet Process Design?
  - Integration with Surrounding Systems: High Capacity Trays + Reboiler + Condenser?

55

Revamp Inspection

- Installed Mechanical & Process Check:
  - Installed per Drawings
  - DC Installation Blocks (Record DC Escapes & Any DC Changes)
  - Alignment of Tray Sections (Centering & Panel to Panel) Out of Round?
  - Correct Hardware Installation
  - Hardware Orientation & Proper Tightening
  - Punch List & Final Process Check

56

Revamp Field Inspection Examples

- Parts Verification Check: Example Valve Lift
- Small Parts Check and Security: Examples Twr. Closure Hold-up & Missing Nozzles
- Field Process Check: Example Missing Downcomers
- Field Dimension Check: Example Stripper Dbl. Flood
- Field Dimension Check: Support Ring Thickness
- Inspection Timing: Example Redesigned Feed Pipe
- Field Layout: Example Chimney Tray Water Draw

57

Example: Chimney Tray Water Draw

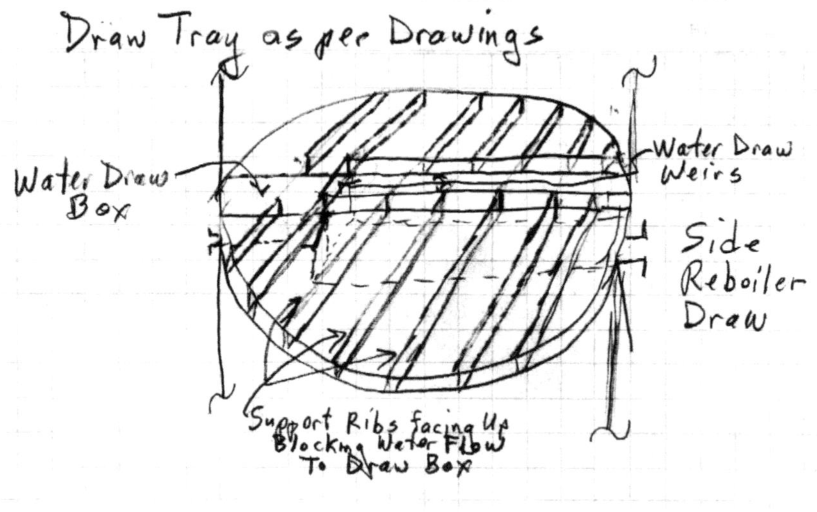

Bouck, Doug, Distillation Revamp Pitfalls to Avoid, Chemical Engineering Progress, p. 32-38, Feb. 2014

58

Special Inspections 1

- Feed and Return Pipe Distributors
  - Holes/Slots/Nozzles
    - Clear?
    - Corroded?
  - Flange Bolts Tight (Wrench Check)
  - Distributor Support Bolting Tight (Dbl. Nutting Use)
- False Downcomers & Distributor Impact Panels
  - Double Check for Loose Hardware
  - Need for extra security from vibration?
  - Vibration initiated cracking?

59

Distributor Hole Corrosion

60

Special Inspections 2

- Material Balance or Draw Problems?
- Flow or Level Control Problems?
- Suspected Leaks?
- Chimney Trays & Draw Boxes
  - Corrosion & Deposits in Draw and/or Piping
  - Thickness Checks in Suspect Corrosion Areas
  - Potential Leak Areas at Welds or Gasketing
  - Potential Leak Areas at Bottom (Under-deposit Corrosion)
  - Loose or Bent Hats, Chimneys or Decks

61

Chimney Draw Tray

62

Draw Box Corroded Through at Bottom

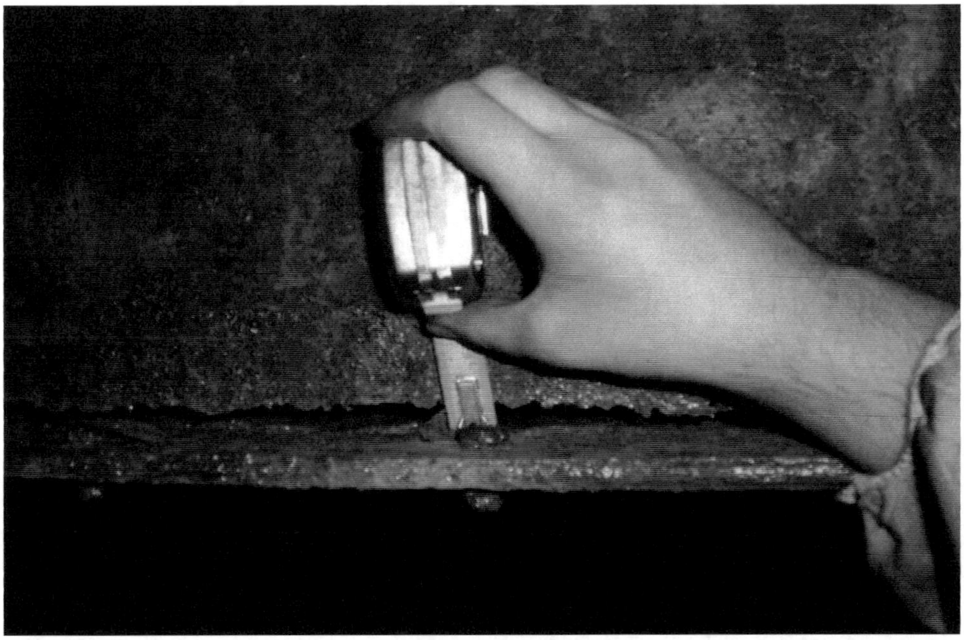

63

Draw Box Corroded Through at Welds

64

Special Inspections 3

- New or Revamped Chimney Trays
  - Gasketed or Welded
  - Leak Testing
    - Water Quality (Chlorides for Certain 300 Series Stainless and/or Solids in Firewater)
    - Quantity - Multiples Greater Than Leak Rate
    - Pressure - Pump Head Requirements (Water)
    - Test Depth - Approx. Operating Level
- Distributor Testing
  - On Site Spray Distributor Example
  - Off Site Packing Distributor Vendor Testing

65

Special Inspections 4

- Final Inspection Before Closure
  - All Internal Repairs & Cleaning Complete
  - Internal Manway Hardware Fixed or Replaced
  - Account For All Internal Manways – Directional?
  - Tools & Misc. Picked Up

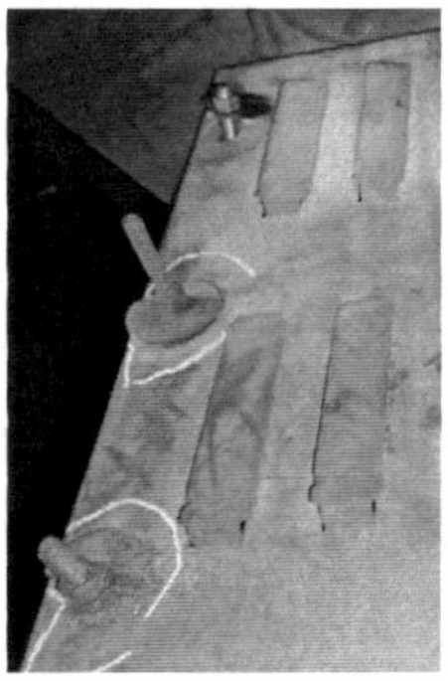

66

New Tower Inspection 1

- Tolerances
- Level Checks
  - Trays – Dependent on ring levelness
  - Weirs (If Adjustable)
- Downcomer Width – Top & Bottom
- Downcomer Escape
- Weir Heights (If Adjustable)
- Alignment/Centering of Tray Sections
- Valve Rise & Valve/Hole Counts

67

New Tower Inspection 2

- Hardware
  - Per Drawing
  - Correct Metallurgy
  - Bolting Snug Tight (~15 ft-lbs for 3/8 inch)
  - NOT Overtight
    - Bend Tray
    - Bend Clamps
    - Bend Washers
    - Gall Bolts

68

Pyrophoric Fire Prevention

- Pyrophoric Deposits
  - Iron Sulfide Can Be Most Anywhere Sulfur Is
  - Conversion Treatments Not Perfect
  - Prefer One-Step Processes Like ZymeFlow
  - Preventive Measures + Monitoring
  - FeS + O2, Dry and 104+ Degrees F. = Exothermic
  - Examples: Amine, Main Fractionators
- Carbon, Coke & Other Organics
- Hot Work
  - Read Chemical Engineering July 2007 Article on "Causes and Prevention of Packing Fires"

69

Packed System Inspection with NO Packing Removal

- No Specific Operating Problems
- Pre-TAR Gamma Scan
- Distributor Piping – Clean & Bolting Tight
- Pre-Distributor - Clean & Functional
- Distributor Pan, Troughs or Spray
  - Clean
  - Holes or Nozzles Clear
- Visual Inspection Top of Packing
- Visual Inspection Bottom of Packing

70

Structured Packing Installation

Feed Inlet Pipe
Predistrubutor
Distributor
Optional Holddown
Top Layer of Packing
90 Deg. Oriented Next Layer

Alternating Orientation of Layers

Bottom Layer of Packing
Packing Support
Liquid Collector

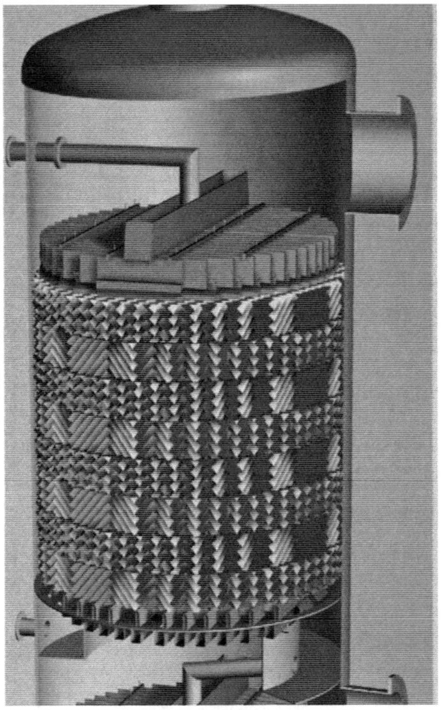

Sulzer Chemtech Brochure

71

Packed System Inspection With Packing Removal

- Before Re-Installation of Packing Check:
  - Draw Integrity – Check for Corrosion & Cleanliness
  - Support Pan/Grid – Clean/Unplugged
  - Check Support Orientation
  - Check Connections and Support Bolting Tightness
- Structured Packing Installation:
  - Verify Packing Size, Type & Metallurgy
  - Starting Orientation is Important
  - 90 Degree Alternate Layers (45 or 60 Degree Rotation for Grid)

72

Protect the Packing

- Walk on Planks NOT the Packing
- WHY for Structured Packing??
  - Structured Packing is Very Thin
  - Boots Smash the Top Crimps
  - Smashed Crimps Reduce Capacity
  - Smashed Crimps Reduce Efficiency
  - Hidden Damage is Untraceable
- Make Sure the Contractor is Complying

73

Protect the Packing

Stand on Planks
not the Packing

74

Protect the Hands/Arms

Koch-Glitsch Brochure

75

Installation

- Personal Safety
  - Gloves for Sharp Edges
- Orientation to Support
- Orientation Layer to Layer
- Shoehorns for Multi-piece Installation
- Distribute Excess Space Between Pieces
- Wall Wipers of Various Designs

76

Structured Packing Layers

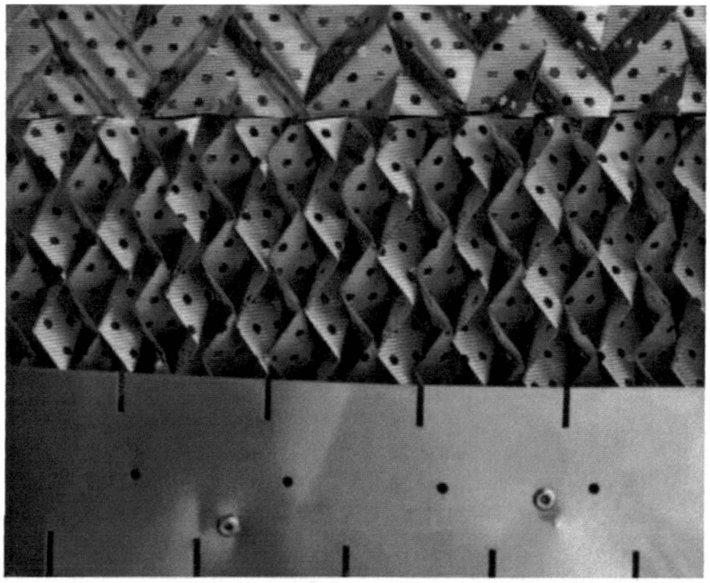

FLEXIPAC® Structured Packing with perforated and textured surface shown with attached wall wiper band

77

Other Wall Wipers

Sulzer Mellapac Structured Packing

78

Holddown Grid – Miss-installed

79

Distributor Inspections

- Holddown – Bolting Tight
- Level Distributors & Pre-Dist. 1/8 in. High to Low
- Installation Hold Point for Level Check/ Adjustments
- All Distributor and Predistributor Bolting Tight
- Distributor Clean
- Predistributor Clean
- Feed Piping Re-installed & Bolted Tight

80

Spray Distributors

- Check all Spray Nozzles and Headers
  - Remove Nozzles (Have Spares on Hand)
  - View Nozzles Internals – Clean or Replace
  - Save Nozzles in Safe Place
  - Check and Clean Headers as Necessary
  - Wrench Check all Nozzles for Tightness
  - Wrench Check all Flange Bolting
  - Double Nut and Check all Header Support Bolting

81

Detached Spray Headers

82

Upset Structured Packing (No Holddown Grid)

83

Redistributor Between Packed Beds

84

Random Packing Installation

Feed or Transfer Pipe
Predistributor
Distributor
Retaining Grid

Random Packing

Support Plate
Support Ring

Vapor Inlet

Liquid Level Collector/Sump

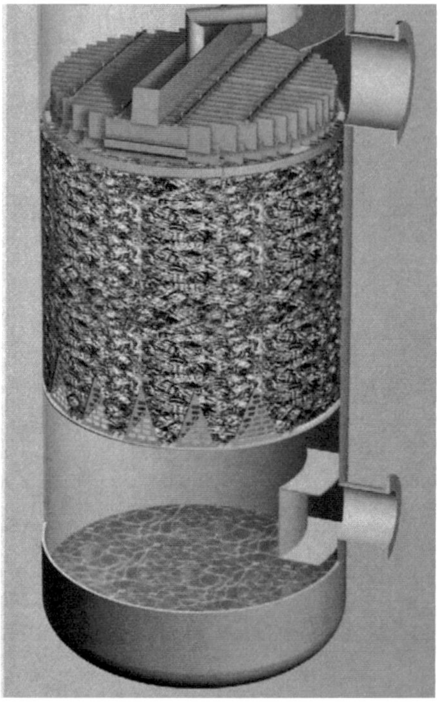

Sulzer Chemtech Brochure

85

Packing - Random

- Removal for Replacement
  - Remove Piping, Distributors & Retaining Grid
    - Safe Storage of All Parts
    - Record Orientations for Re-Installation
    - Clean All Parts
  - Remove Rings by Vacuum
  - Clean Any Collector or Draw System
  - Clean Packing Support
  - Check Support Integrity & Bolting

86

Packing - Random

- Installation
  - Verify Packing Size, Type and Metallurgy
  - Dump New Packing
  - Rake Top
  - **Do Not Overfill!**
  - If Necessary to Enter, Stand on Planks not Packing
  - Re-install Retainer Grid
  - Level Distributors & Pre-Dist. 1/8 in. High to Low
    - Installation Hold Point for Level Check/Adjustments
  - Install Feed Piping
  - Check Tightness of All Bolting

87

Protect the Packing

- Walk on Planks NOT the Packing
- WHY for Random Packing??
  - Random Packing is Thin
  - Boots Crush or Compress the Rings
  - Hidden Damage is Untraceable
  - Crushed Rings Reduce Capacity
  - Crushed Rings Reduce Efficiency
- Make Sure the Contractor is Complying
- Stay Totally Off, If Possible

88

The Process Engineer
(A Bit of Philosophy)

- Little Authority
- Much Responsibility
- Bridge Operations, Maintenance & Management
- Optimization Economics - Beyond Politics
- Relationships Create Influence
- Influence Creates Accepted Recommendations
- Recommendations Create Improvements
- Improvements Create More Profits
- More Profits Create Stable Company and Jobs

89

Take Ownership of the Future. The Role of Process Engineers in Inspection **Is** Important!

2 Peter 1:15

And I will make every effort to see that after my departure you will always be able to remember these things.

YOU ARE the FUTURE!

90

481654 Distillation Tower Inspection for Turnarounds and Revamps: Tutorial and Presentation - Part II

Abstract:

This presentation of "Distillation Tower Inspection for Turnarounds and Revamps" covers the unique safety issues for Distillation Tower inspections plus typical work plan responsibilities and inspection equipment. Both preliminary and follow-up inspection of trays and packed systems are covered. Emphasis is placed on what to look for in typical and unique inspection situations with both trayed and packed towers. Examples will be demonstrated and/or discussed including the unique perspective that the Process Engineer brings to the inspection process especially in replacement and revamp situations.

Value of Process Knowledge and Models in Data Analytics Applications

Jaleel Valappil and David Messersmith

Bechtel Oil, Gas and Chemicals

Andri Rizhakov and Kelly Knight

Bechtel Nuclear, Security and Environmental

AIChE Spring Meeting 2017, San Antonio, TX

Session: Big Data Analytics – Industry Perspective

1.0 Introduction

A typical chemical process facility requires large capital spending to achieve full production. To realize maximum commercial benefit from these capital investments over the lifecycle of the facility, it is important to utilize these existing assets to their full potential. Process facilities have relied on traditional DCS/APC systems and enterprise performance management platforms to maximize their asset utilization. There is a great potential to further enhance the value from the existing assets using new industrial internet and big data analytics technologies.

The industrial internet is a recent development enabled by the convergence of cloud computing, big data, sensor technologies and cheap computing resources. This technology utilizes very large amounts of real-time data from diverse sources to build model based analytics for various applications. The industrial internet combines the physical plant with the digital information to derive new benefits. This is enabled by the decreasing cost of sensors, computing storage/ infrastructure and innovations in data analytics [1]. Recent developments in big data analytics enables wider deployment of this technology [4] in process industries.

2.0 Main Applications and Benefits of Big Data Analytics in Process Industry

2.1 Process, Equipment and Control Monitoring

Monitoring of plant data for abnormal operating modes is an important application of big data analytics. Abnormal operating scenarios result in lost plant productivity. Identifying these conditions and taking corrective actions early before it results in plant downtime is hence very valuable. One of the key applications of industrial internet and big data analytics is real time monitoring of the performance of the process and equipment using the data collected. The data can be used to build either predictive or unsupervised models to monitor the state of the process and equipment and trigger alarms for operator to take corrective actions.

2.2 Asset Management - Predictive Maintenance

Predictive maintenance is the most important application of industrial internet and data analytics that enable identifying equipment problems beforehand and scheduling maintenance at appropriate time. This involves monitoring equipment conditions continuously and predicting when a failure is likely to occur. Traditional way of performing maintenance is preventive, whereby the maintenance is performed at pre-scheduled intervals. This can result in unneeded maintenance at times. In other times, preventive maintenance may be too late if failures result from events or conditions that occur at any time. Predictive maintenance enables maintenance planning based on actual condition of the equipment. This also avoids the unplanned downtime that can be caused by reactive maintenance. Predictive maintenance is not limited to turbomachinery. Other process equipment, including exchangers and valves can benefit from this. Some of the critical valves can be proactively replaced before failure to minimize unplanned downtime. Heat exchangers can be monitored with additional wireless temperature sensors and cleaned before the fouling starts affecting plant production.

2.3 Plant Operational Optimization

This application of big data analytics involves giving operational guidelines to maximize production, energy efficiency or other economic objectives. Currently, most of the process plants in operation are equipped with Advanced Process Control (APC) systems, which maximize the capacity or efficiency. These systems are layered on top of DCS and execute every minute or so. Apart from this, some of the operating plants are also be equipped with RTO (Real Time Optimization), supervisory or advisory systems that aim to enhance the operation of the plant. There is further opportunity to optimize smaller units in the plant. It needs to be noted that these are currently supervisory functions that provide recommendations to the operator (Unlike APC systems, which write back to the DCS on a real time basis).

3.0 Value of Process Data and Knowledge for Big Data Analytics

To realize the maximum return from investments this emerging technology area, LNG industry needs to properly utilize the existing plant information and resources. There are several valuable resources in operating facilities and plants under construction that can be beneficial in this regard.

1. **Process knowledge and information**

   Process design and operations knowledge is an important asset that can enable realizing the maximum benefit from investment in industrial internet. Unlike other domains, process facilities are inherently constrained by thermodynamic relationships. Understanding underlying process relationships can streamline application of this technology.

   There are two areas where process knowledge is important:

   1. This is a key input for the selection of data analytics application and this helps to establish and select the right variables to be included in the data driven models.
   2. This is also critical in identifying the opportunities for optimization applications and selecting the right objectives and scope of the same.

   Process plants recently completed or under construction collect the relevant design data in digital format that is used for engineering. The Smartplant Foundation or similar platforms are used to store 3D models, datasheets and P&ID information with digital links among them. These provide a virtual information repository that can be leveraged to benefit the development and deployment of big data applications.

2. **Fundamental simulation models**

   Process facilities utilize steady state and dynamic simulation models during EPC phase to enhance engineering design and ensure safety. In addition, majority of the plants develop Operator Training Simulator (OTS) to assist in training and startup. These models and OTS can be beneficial in maximizing the value derived from industrial internet. In addition, industrial internet facilitates the deployment of digital twins, which are exact replica of a key plant asset that can be used for monitoring, optimization and troubleshooting.

   The fundamental process models are valuable in various ways in enhancing the data driven models used for industrial internet applications, including:

   a. Selecting the right model structure: Many of the data driven models (Especially for optimization) need causal models with the correct cause and effect. Identifying this from data can be near impossible in certain cases. Process models can identify the correct causal links and also establish the variables that are significant. This feature selection is an important step in data analytics applications.
   b. Boosting the data models: There are instances where sufficient data that cover the operating range may not be available from operational data. In these case, simulation runs can augment the data to provide the required range.

c. Model Based Optimization: The availability of models and the ability to host these remotely opens up new possibilities to optimize the process remotely. These models can be used for steady state optimization to establish the best operating parameters for the plant.

4.0 Case Study – Amine Plant Monitoring

There are several advantages to using data driven analytics for monitoring industrial systems. Two different approached can be used for fault detection in process plants:

a) Model based – Which uses fundamental process models to detect deviations from normal operating conditions.
b) Data driven – Which uses historical process data for fault detection.

The application of graphed anomaly detection techniques in determining abnormal operating conditions in LNG process is presented below. This is applied to amine plant in a currently operating LNG facility. A predictive model consisting of statistical probability distribution (Bayesian distribution modeling) is developed form historical data. The distribution is not limited to normal and can have different shapes. A normalized anomaly score is calculated. The advantages of using the technique presented here include:

a. It reduces plant downtime by identifying operational problems in advance
b. It helps to identify root cause of problems quicker
c. Can be done without historical data, as it learns online
d. Can be set up easily without large investment

One of the key steps in the graphed anomaly detection is the clustering of variables for anomaly detection. Statistical methods like k-means and k-medoids can be used for clustering. If design data is available, that can be used to set up clusters based on equipment. Alternately, the clusters can also be developed utilizing the variables selected based on process knowledge.

4.1 System components and setup

The system components utilized for anomaly detection is shown in Figure 1.

- Splunk is used to index the machine data coming from the LNG process. Splunk is a widely used big data tool and has the ability to index various types of data including streaming data.

- Prelert is used to detect anomalies using machine learning techniques. Prelert is a statistical anomaly detection technique, which develops probability distribution of the variables and updates it as new data comes in. An anomaly score is determined based on the distribution and the incoming real-time data [2].

- Neo4j is the graph database used to store and analyze the design data from Smartplant Instrumentation database and use it for anomaly detection.

4.2 Application to amine system monitoring

Application of above tools for detecting anomalies in amine system used in LNG plant is studied. Amine system removes CO_2 and H_2S from natural gas before it is send for liquefaction. The performance of this unit is very important as any carried over CO_2 from amine unit can freeze in the cold temperatures in liquefaction section causing plant downtime and lost production.

The historical data from an operating LNG plant was used for this analysis. The data used corresponded to a seven-month period of operation after startup. The relevant tags were identified and corresponding data was collected. In addition, design data related to instrumentation and other items were obtained from Smartplant Instrumentation database to aid this project.

The analysis for amine plant data was initially performed using all the data (About 300 tags) for the unit to determine an anomaly score. This was fund to be unsatisfactory as the score was not capturing changes in the plant operation.

The anomaly scores were also calculated using data clustered by key equipment (Amine absorber, amine regenerator etc.). The tags that belong the individual equipment are derived from Smartplant database. A score is calculated for each equipment in this case.

As a third approach, data clustered by groups as determined from operational experience was used for determining the scores. Four different groups related to acid gas extraction, acid gas removal, separation performance and amine inventory were identified for this purpose.

An example of data driven anomaly detection performance is shown in Figure 2. During the operation, the CO_2 composition was found to increase to above 100 ppm while the plant feed gas rate was close to the normal rate. This indicated that there was some upset condition that may be causing this. The inlet feed gas CO_2 composition was found to increase during this period, which affected V-1201 operation (Acid gas removal column). The results shown in Figure 2 demonstrates that the anomaly score specifically for the equipment V-1201 did show an increase well before the CO_2 concentration (red line) increased. This could have been used to alert the operators about the impending problem.

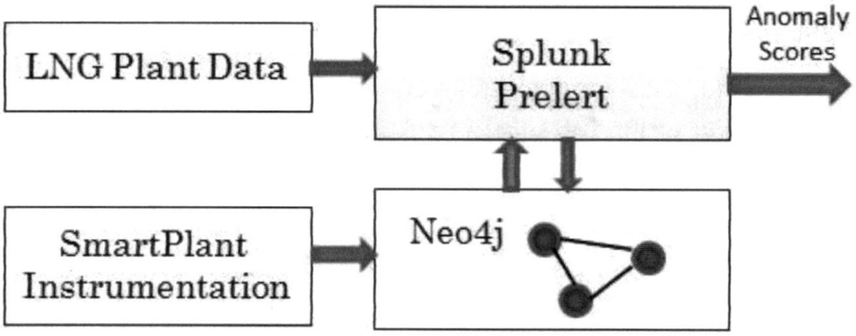

Figure 1: System architecture for anomaly detection

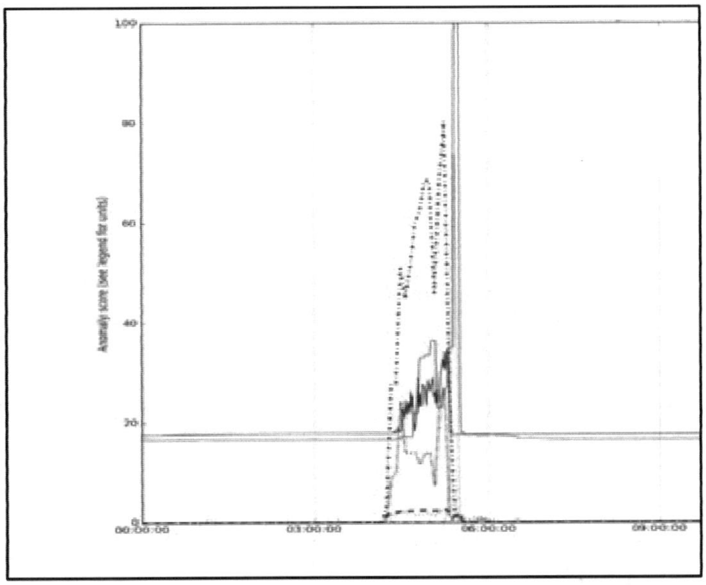

Figure 2: Result of anomaly detection. The dotted blue line shows the score.

5.0 References

1. "*Industrial internet reference architecture*", Published by Industrial Internet Consortium (IIC).
2. Veasy, J. and Dodson, S., "*Anomaly detection in application performance monitoring data*", International Journal of machine learning and computing April 2014, 120-126.
3. Johnson, D., "*IIoT applications enable the process world*", InTech Magazine, Mar-Apr 2016.
4. Perrons, R.K. and Jensen, J.W., *"Data as an asset: What the oil and gas sector can learn from other industries about "Big Data""*, Energy Policy, 81(2015) 117-121.

Connecting the Dots – Gaining Insight from Big Data in Hydrocarbons Manufacturing

Birgit Braun, George Bellos, Stef Luijten, Liwei Li, Mark McAdon, Jason Hamilton

AIChE Spring Meeting 2017

The Dow Chemical Company

"Big Data" and Manufacturing

Big Data = "data sets too large and / or complex for traditional data processing tools and applications" (think Excel)

→ *common and* *<u>not new</u>* *in manufacturing*

- Large data sets
  - Process data, analytical data, meta-data → data bases
  - Global business → about 200 furnaces across 14 plants
- Complex data sets
  - Different routine process scenarios, feeds, technologies
  - Expected and unexpected non-routine operations

New Opportunity

New opportunities for manufacturing data

- ➢ Manage different data repositories
- ➢ Combine / align / refine data
- ➢ Modeling and visualization
- ➢ Improve understanding

3

Graphic: https://inspiredittrainer.com/2013/07/28/what-is-big-data/

New Opportunity

New opportunities for manufacturing data

- ➢ Manage different data repositories
- ➢ Combine / align / refine data
- ➢ Modeling and visualization
- ➢ Improve understanding

- **Multivariate relationships**
- **Better forecasting**
- **More reliable (real-time) monitoring**

4

Graphic: https://inspiredittrainer.com/2013/07/28/what-is-big-data/

Example Applications in Hydrocarbons Manufacturing

- Furnace technology and operation
  - Coil technology differentiation
  - Performance over time
  - Improve process conditions
- Process chemistry
  - Key process / multivariate indicator identification
  - More reliable fouling monitoring

DOW

5

Innovative Approaches and Solutions

Internally created platform

- Interfaces automatically with updating databases
- Integrated code for date alignment and pre-processing
- Streamlined data modeling
- Visualization of results

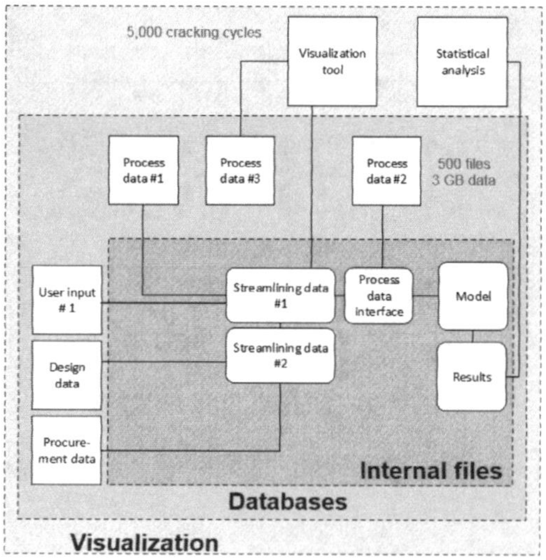

DOW

6

Example: Furnace Runlength vs Different Furnaces

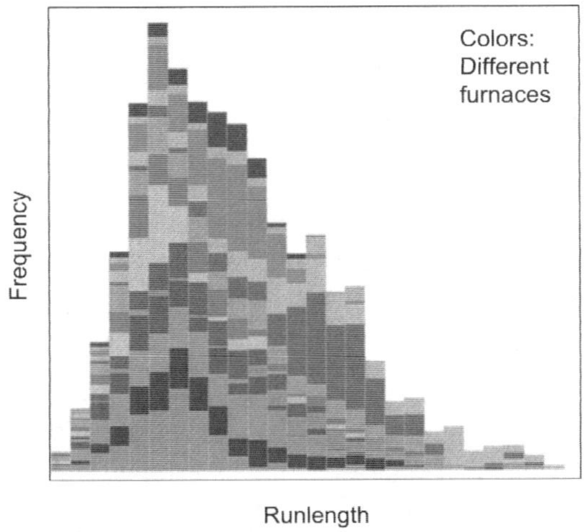

- Multiple years of plant performance data color coded for different furnaces
- Not all plants perform equal
- Identification of biggest improvement opportunities

DOW

7

Example: Furnace Runlength vs Different Technologies

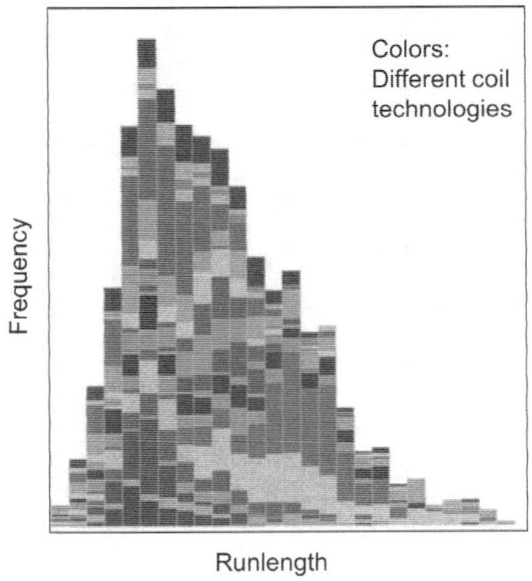

- Multiple years of plant performance data color coded for different coil technologies
- Run length is different for different coil technologies
- Initiation of possible root cause investigation and improvement initiatives

DOW

8

Example: Technology Impact on Coil Age

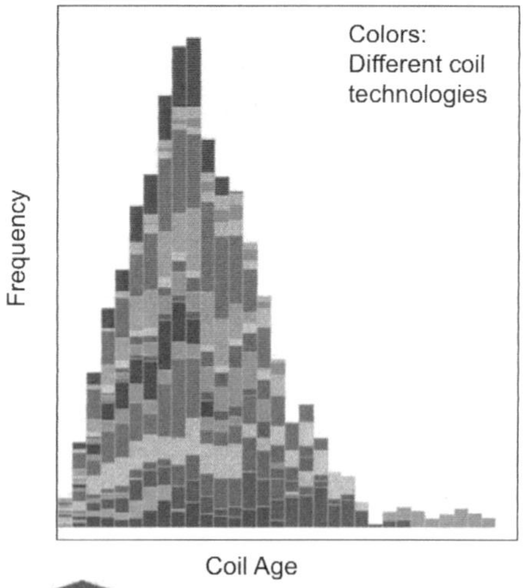

- Multiple years of plant performance data color coded for coil technology
- Shows differences in typical coil age for various technologies
- Identification of advantaged coil technologies

Optimize Practices

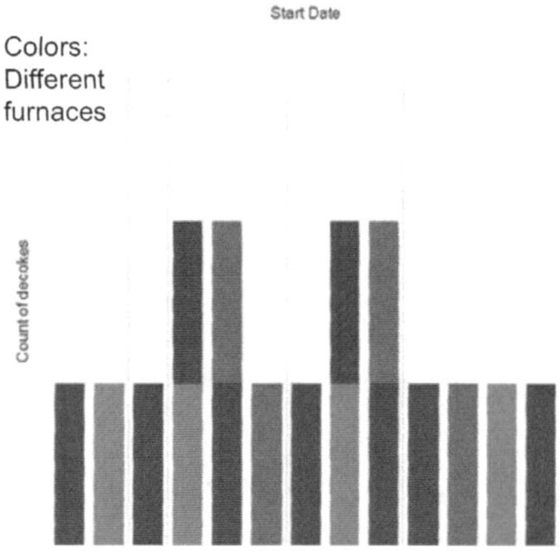

- Identifying furnaces going through simultaneous decoke cycles
- Challenge operating practices

Example: Variation of Process Conditions per Plant

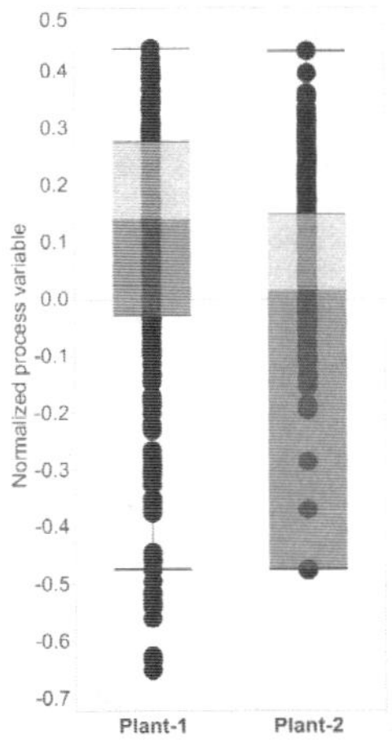

- Identification of different operating practices
- Identify units running outside recommended range
- Contribute to reliability improvement

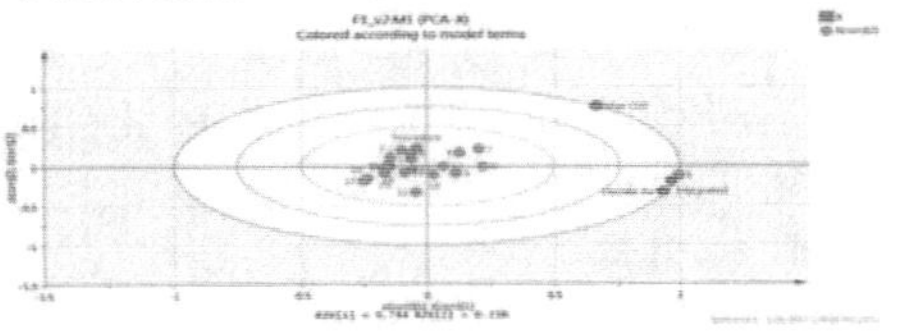

DOW

Information at Your Fingertips → Interactive Dashboard

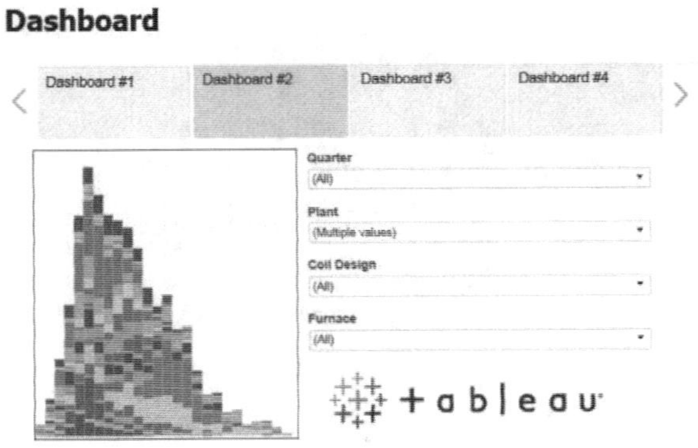

- Share information with a group
- Tailor made data selection
- Access to 5Gb+ data in few seconds

DOW

Identification of Differentiating Patterns

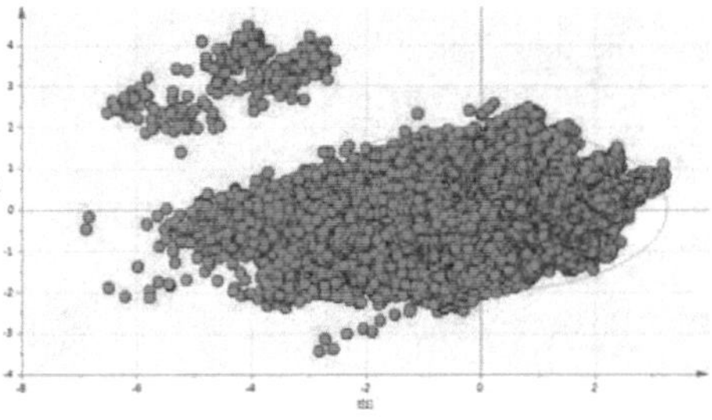

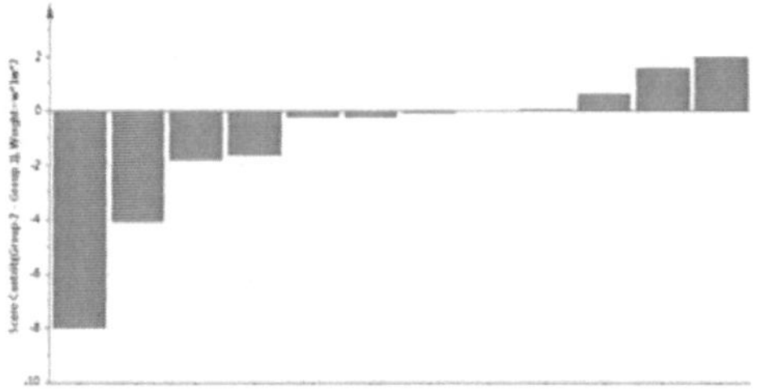

- Pre-processed data sets for efficient PCA analysis
- Identification of different operating conditions and technologies on plant performance

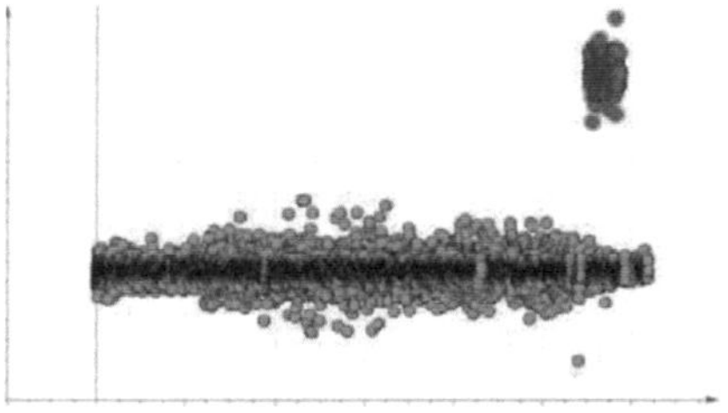

DOW

13

Information Diluted in Noise

Following Up Performace of x Radiant Coils

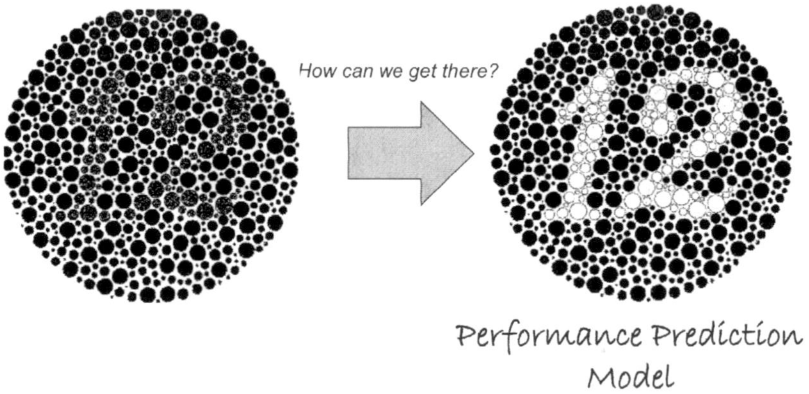

Pictures are taken from https://en.wikipedia.org/wiki/Ishihara_Test#

DOW

14

Performance Prediction Model

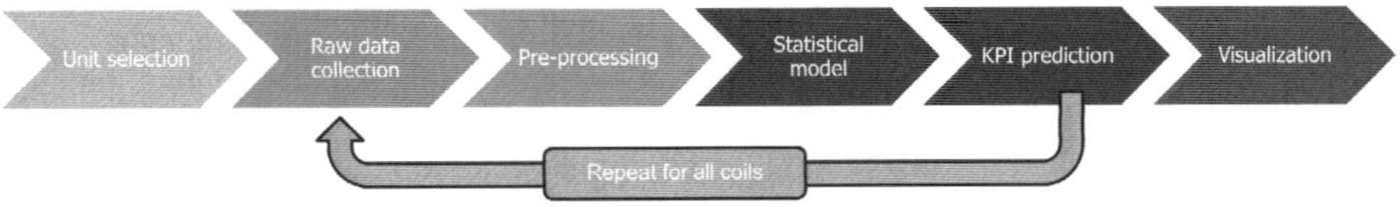

- Unit performance evaluation
  - Different runs
  - Selectable time during run
- Improve process knowledge

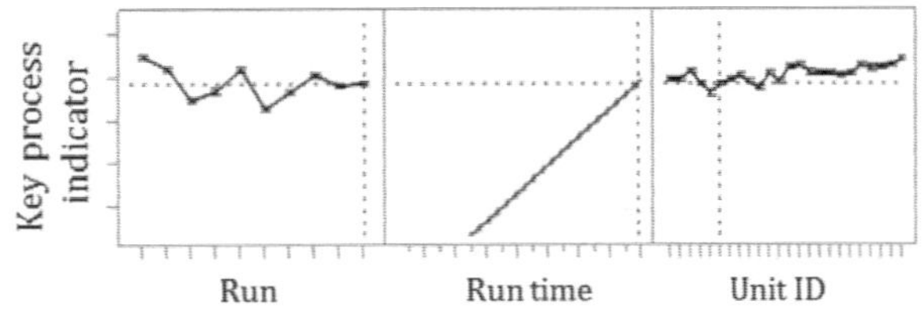

jmp Statistical Discovery.™ From SAS.

DOW

Identification of Relevant Key Process Indicators

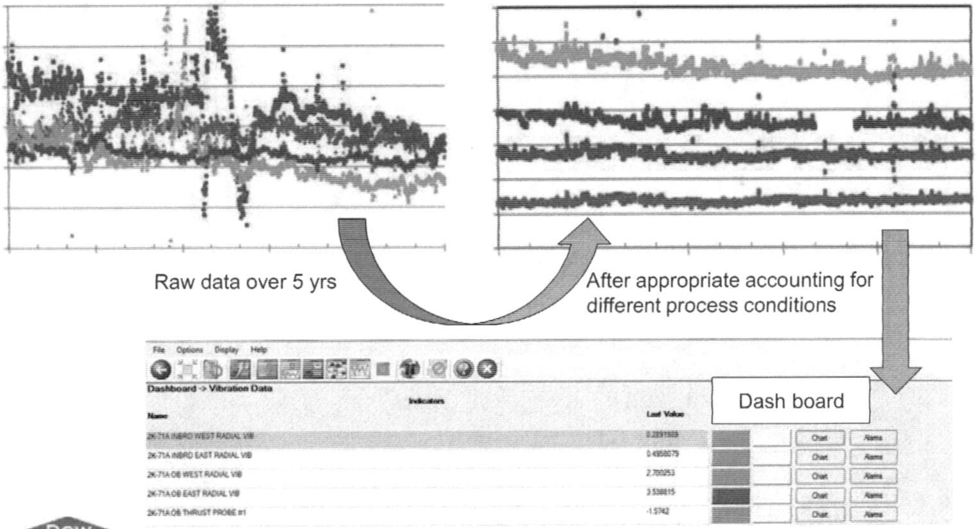

DOW

Conclusions

- Connecting the dots / data from different data sources enables
  - Technology evaluations
  - More detailed plant performance analysis
  - Operation improvements
  - Better performance monitoring
- Critical for success
  - Team work
  - Appropriate preprocessing
  - Automatic data update

DOW

17

DOW

Thank You

478954 Real-Time Batch Process Monitoring in Biopharmaceutical Manufacturing: Big Data Analytics for Small Data

Abstract:

In recent years, the biotechnology-based products have been gaining increasing visibility and success in treating chronic diseases such as arthritis, diabetes and cancer. As per the report published in 2005 [1], approximately 27% of the new medicines in active development are in fact biotechnology-based products.

To allow for an efficient monitoring and control of biopharmaceutical processes, multivariate statistical techniques are routinely deployed. Despite over two decades of research and advancements in industrial batch process monitoring and control, biopharmaceutical process monitoring faces a unique challenge due to scarcity of historical data. In fact, it is common in biopharmaceutical manufacturing to have only a few engineering batches for a product with no commercial production history at a manufacturing facility before the actual good manufacturing practice (GMP) campaign. While limiting the number of engineering runs improves speed to market and reduces manufacturing costs, it also limits the amount of representative data available for establishing a monitoring baseline, which presents several problems. First, with small data sets it is nontrivial to capture the inherent variability under the normal operating conditions (NOC) in their entirety. Second, most of the multivariate monitoring methods, such as principal component analysis (PCA) and partial-least squares (PLS) are big-data technologies that are effective only when there are large historical data sets.

In this talk we will discuss several machine-learning methods to deal with the data scarcity problem in biopharmaceutical process monitoring. We propose a non-parametric modeling approach to generate virtual campaign data under normal operations on-demand. Using virtual campaign data, standard batch monitoring techniques are used to effectively monitor a biopharmaceutical manufacturing process. Encouraging results are reported from practice.

References:

[1]. Hamilton G. The Biotechnology Market Outlook: Growth Opportunities and Effective Strategies for Licensing and Collaboration. Dublin: Research and Markets. 2005.

478993 Quantification of the Effect of Viscosity on Point Efficiency of Sieve Trays in Distillation Columns

Abstract:

Though many physical properties govern the design and performance of distillation columns, the effect of viscosity is critical as it influences both mass transfer and hydrodynamic aspects of distillation columns. While there has been research to analyze the effect of viscosity on distillation, the majority of the research was conducted with liquid viscosities less than 1.0 cP. The distillation efficiency correlations based on low viscosity data may not be reliable for systems that exhibit liquid viscosities more than 1.0 cP at operating conditions. Point efficiency data collected using Oldershaw columns are proven to be effective in representing the commercial point efficiency data. Therefore, an Oldershaw column was recently commissioned at OSU to collect efficiency data for liquids with elevated viscosities. Column validation is done using standard test systems for which commercial efficiency data are readily available. The efficiency data at elevated viscosities (up to 4.0 cP) are collected by Oldershaw efficiency experiments at total reflux. The collected data are crucial in order to evaluate the validity of point efficiency correlations at high viscosities. Also, the elevated viscosity data are critical for extending existing correlations and developing new correlations for viscous systems.

479217 Investigating the Impact of Feed-Air Saturation in Air-Water Spray Entrainment

Abstract:

Entrainment characterization is critical for a variety of separation processes in the chemical process industry, including distillation and absorption. Liquid entrainment is a key aspect of these applications that is not well understood. In order to support further research into this field, it is important to determine the correct methods by which to analyze entrainment in laboratory systems. This work investigates the impact of saturating the feed air to an air-water entrainment system, particularly on the droplet size and velocity distributions as measured by Phase Doppler Interferometry. Entrainment measurements were made at the same experimental conditions with and without pre-saturating the feed, and the results were compared to determine the impact on those measurements. These method evaluations are essential as they guide the larger research effort in correlating entrainment with efficiency and capacity.

Effect of Liquid Viscosity on Mass Transfer for Packings

Di Song[a,b], *A. Frank Seibert*[b], *Gary T. Rochelle*[a,*]

[a]*McKetta Department of Chemical Engineering, The University of Texas at Austin, 200 E Dean Keeton St., C0400, Austin, TX 78712, USA*

[b]*Separations Research Program, Pickle Research Center, The University of Texas at Austin, 10100 Burnet Rd., Austin, TX 78758, USA*

Abstract

It is important to understand the effect of liquid viscosity (μ_L) on both the effective mass transfer area (a_e) and liquid film mass transfer coefficient (k_L) for packings to correctly predict the separation rate and properly design the column. Existing correlations are mostly based on water system and the application of them to viscous media could be problematic.

In this work, a_e and k_L were measured in a semi-works scale packed column (16.8 in. I.D.) with 10 ft packings installed as the internal. Seven structured packings (M 125Y, M 250Y, M 250X, GTO 250Y, B1 250MN, GTP 350Y, GTP 500Y), one random packing (RSR 1.5), and one hybrid (RSP 250Y) were tested. Liquid viscosity was varied from 0.8 to 70 cP by adding glycerol to water. The a_e was determined by reactive absorption of ambient CO_2 into dilute NaOH, and the k_L was determined by air stripping of toluene from the liquid. Liquid viscosity showed insignificant effect on a_e over the range of μ_L investigated. The total dependence of k_L on μ_L is -0.72, of which -0.35 is from the indirect influence through diffusivity, and -0.37 is from the direct influence of μ_L on k_L through liquid turbulence.

Models of a_e (Eq. 1) and k_L (Eq. 2) were developed based on the nine measured packings together with other 18 packings in the database of the Separations Research Program (SRP). The effect of secondary mass transfer (wall and end effect) has been included in the models.

$$\frac{a_{e,packing}}{a_P} = 1.26 \cdot \eta \cdot \left(We \cdot Fr^{-\frac{1}{3}}\right)^{0.117} = 1.26 \cdot \eta \cdot \left[\left(\frac{\rho_L}{\sigma}\right) g^{1/3} \left(\frac{u_L}{a_P}\right)^{4/3}\right]^{0.117} \quad (1.a)$$

$$a_{e,total} = a_{e,packing} + a_{secondary} \approx a_{e,packing} + a_{wall} \quad (1.b)$$

$$Sh = 0.124 \cdot Sc^{0.5} \cdot \left(Re \cdot Ca^{1/5} \cdot Fr^{-1/3}\right)^{1.08} \quad (2.a)$$

$$k_L = 0.124 \cdot u_L^{0.576} \cdot \mu_L^{-0.364} \cdot D^{0.5} \cdot a_P^{-0.44} \cdot g^{0.36} \cdot \rho^{0.58} \cdot \sigma^{-0.216} \quad (2.b)$$

With a relatively large equipment size and packing database, this work has the greatly expanded the data of mass transfer in viscous liquids available in the open literature.

Introduction

Many industrial applications encounter mass transfer in viscous liquid, such as electrolyte solutions, organic/polymer solutions, crude oil, and ionic liquids, etc. The increase in μ_L can significantly decrease k_L. Take post-combustion CO_2 capture with aqueous amine for example, the high solvent viscosity can cause slower diffusion of CO_2 to the bulk liquid, slower diffusion of free amine from the bulk liquid to the gas-liquid interface (surface depletion of alkalinity), slower diffusion of loaded amine back to the bulk liquid (which affects equilibrium partial pressure of CO_2 at the interface), and less liquid turbulence on the packing surface. Most of the existing k_L (k_La) correlations investigate only water and thus provide unsatisfactory predictions for the influence of μ_L [1]. Therefore, the application of them to amine solutions could lead to either inefficient or insufficient column design.

The SRP database for the air-water column contains valuable hydraulic and mass transfer data for various structured, random, and hybrid packings. Based on the database, models of a_e and k_L have been developed previously by Tsai [2] and Wang [3], respectively. The Tsai area model was developed seven years ago, and new data with totally different packing geometry have been generated continuously. For the k_L model developed by Wang, though it is relatively new, a large amount of new data with water and aqueous glycerol was generated during the past two years. Therefore, it is necessary to update both models with the expanded database.

Methods and Equipment

A systematic method for packing characterization with water has been developed at the Separation Research Program in the University of Texas at Austin. The a_e has been measured with chemical absorption of ambient CO_2 into aqueous NaOH because of the well-established kinetics [4]. The k_L has been measured with air stripping of toluene from water.

Glycerol was chosen as the liquid viscosity enhancer for its complete and easy dissolution in water, relatively cheap price, and the Newtonian behavior of its aqueous solution. Liquid viscosities of 65, 15, 4, and 1 cP (approx.) were obtained by adding 85, 65, 44, and 0 wt % glycerol to water. The reaction kinetics for CO_2/NaOH/H_2O/glycerol was calculated from a model developed based on the wetted-wall column data [5]. The glycerol will not change the controlling mechanism for mass transfer in the existing water systems. For toluene air stripping, though the infinite dilution activity coefficient of toluene decreases rapidly with increasing glycerol, the value will still be well above 100 at the highest proposed glycerol, which ensures the liquid film domination. For the reactive system, the effect of physical liquid film mass transfer (i.e. surface depletion of alkalinity) has been corrected in the kinetic model and is believed to be not an issue for the packed column experiments considering the relatively low CO_2 concentration and flux.

Packed column

The PVC packed column used in this work is sketched in Fig. 1. The inside diameter of the column was 16.8 in. The maximum packing height was 10 ft. The gas and liquid had countercurrent contact inside the column with air supplied by a 30-kW blower with variable speed drive from a duct (8 in outside diameter) below the packing support. The liquid was discharged by a centrifugal pump regulated with a variable speed drive with a maximum capacity of 150 gpm from a 350-gallon storage tank. The liquid could either flow to the column top and distributed by an F40 distributor (40 drip points/ft^2) or flow back to the column sump and storage tank via a bag filter for the pump to operate at a more favorable drive-speed region. Above the distributor was a Trutna collector to knock out any liquid reaching the column exhaust. Typical liquid inventory size was 250-300 gallon. Typical gas and liquid rates were 150-750 acfm and 1-30 gpm/ft^2 respectively.

Approximate 10 ft packing was installed for a_e measurement and k_L measurement with aqueous glycerol. A shorter 6 ft bed was used for k_L measurement with water because of the difficulty of GC analysis for the low outlet toluene concentration caused by high mass transfer efficiency. Each element was installed with 90° rotation with adjacent elements to facilitate liquid distribution as a routine practice.

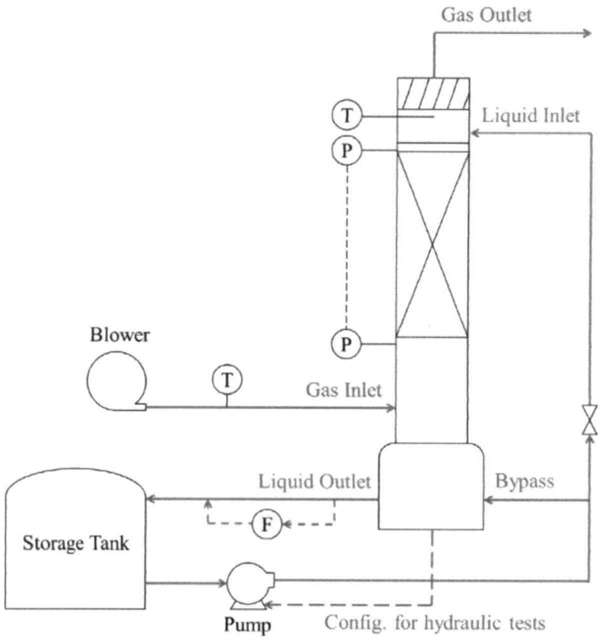

Fig. 1: Packed column experimental system.

Auxiliary Equipment

The CO_2 concentration in both the inlet and outlet gas was measured online by a XStream® CO_2 analyzer (Rosemount). Inlet CO_2 was ambient 370-400 ppmv, and outlet CO_2 was 100-300 ppmv depending on the packing and gas flow rate. Toluene concentrations of the liquid samples were determined by gas chromatograph (Hewlett Packard 6890) analysis with 1-bromo-4-fluorobenzene used as the internal standard. Typical inlet toluene was 100-200 ppm$_w$, and outlet being 0.5-50 ppm$_w$ depending on the liquid flow rate and viscosity. A viscometer (TA Instrument AR-2000ex) was used for liquid viscosity measurement. A densitometer (Mettler Toledo DE40) was used for density measurement to determine glycerol concentration for reaction kinetic calculations.

Calculations

The total mass transfer area in the column, $a_{e,total}$, was determined experimentally by Equation 3. The k_g' for aqueous NaOH was calculated using the model of Pohorecki and Moniuk [4]. The k_g' for caustic glycerol solutions was calculated from a model developed based on the wetted-wall column data [5]. The k_La_e was determined experimentally by Equation 4.

$$a_{e,total} = \frac{K_{OG}a_e}{K_{OG}} \approx \frac{K_{OG}a_e}{k'_g} = \frac{u_L}{k'_g Z}\ln\left(\frac{C_{CO_2,\ in}}{C_{CO_2,\ out}}\right) \tag{3}$$

$$k_L a_{e,total} = \frac{u_L}{Z}\ln\left(\frac{C_{toluene,\ in}}{C_{toluene,\ out}}\right) \tag{4}$$

Physical properties (density, viscosity, surface tension, diffusivity, Henry's constant of CO_2) of the aqueous glycerol solutions were calculated using models developed previously [1].

The secondary effect (end and wall effect) on mass transfer area ($a_{secondary}$) has been corrected to give effective mass transfer area of packing surface ($a_{e,\mathrm{packing}}$). To calculate the wall area, it has been assumed that the column wall was totally wetted not only in the packing section, but also in the auxiliary section (for this case with only one packing section, it's only the section between packing bottom and liquid sump):

$$a_{wall} = \pi \cdot D_{column} \cdot (Z_{packing} + Z_{auxiliary}) \tag{5}$$

The top area includes area generated by the liquid streams from distributor to packing top (both stream peripheral and impact area). Similarly, the bottom area includes area generated by the liquid streams from packing bottom to sump. It has been demonstrated in later section that wall area is always dominant in $a_{secondary}$.

Model of Mass Transfer Area (a_e)

The $a_{secondary}$ has been deducted from $a_{e,total}$ to give $a_{e,\mathrm{packing}}$ (Eq. 5). With 16.8 in. I.D. column, the secondary area is not significant for normal to fine packings, but it starts to play an important role for course packings (Fig. 2). Since the wall area is dominant in the secondary area for the whole spectrum of packing a_P, top and bottom area have been omitted in the future correction of secondary effect for simplification. The column diameter will greatly affect the $a_{secondary}$ as expected (Fig. 3). Therefore, this correction will enable wider application of the area model to column with different sizes. Different regimes of packing area for MG 64Y have been shown in Fig. 4 for illustration.

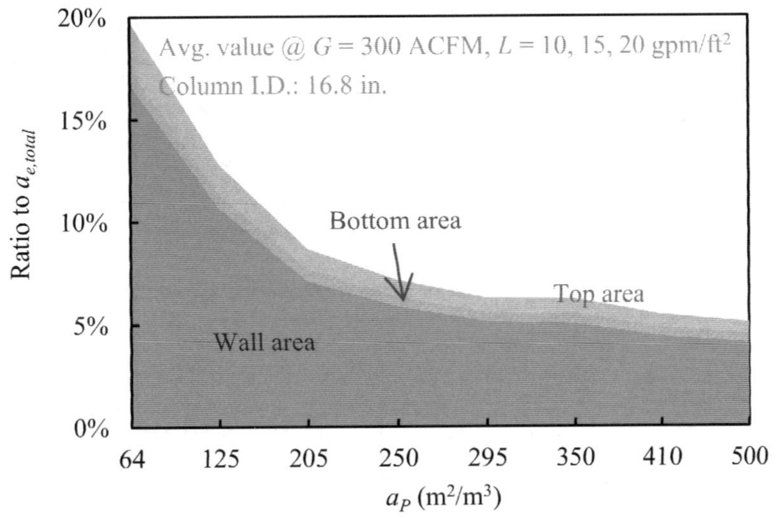

Fig. 2: Effect of a_P on $a_{secondary}$

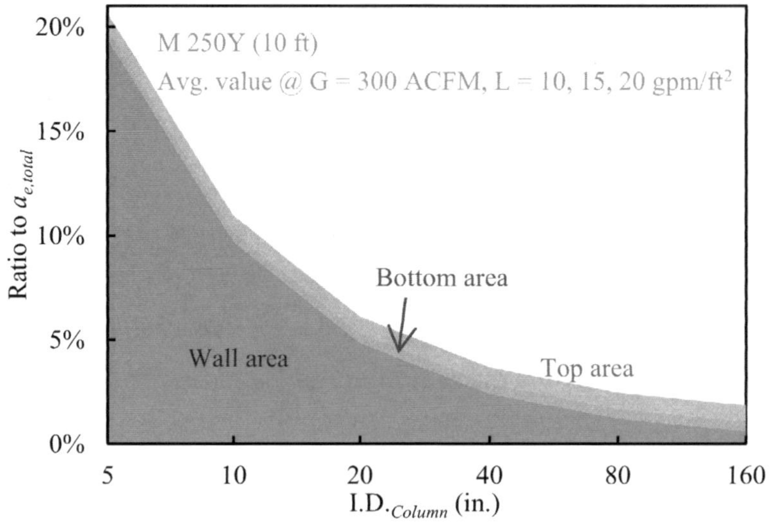

Fig. 3: Effect of column size on $a_{secondary}$

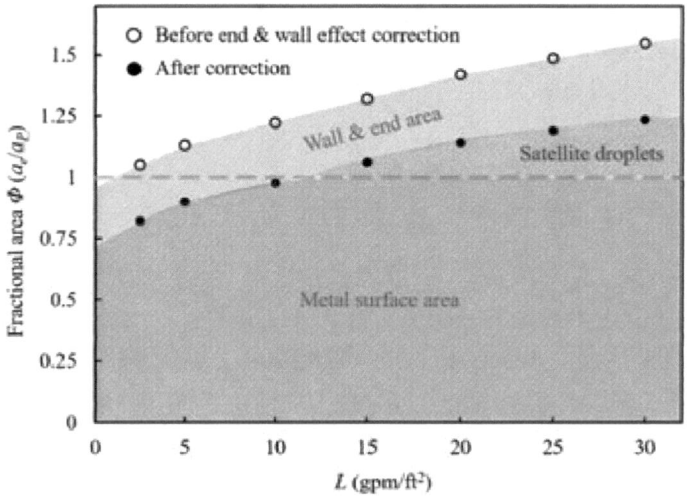

Fig. 4: Packing area regimes of MG 64Y

Figure 5 compares the measured and calculated area with water. An empirical correction factor has been included in the model to account for deviation of random/hybrid packings and plastic packings (Eq. 6 and Tab. 1). For 27 packings with drastically different geometry, the model shows an average absolute deviation (AAD) of only 6.3 percent (Fig. 6).

$$\eta = \eta_{type} \cdot \eta_{material} \tag{6}$$

Tab. 1: Packing correction factors for area model

η_{type}	Structured	Random/Hybrid
	1.0	$-2\times10^{-5}a_P^2+7\times10^{-3}a_P+0.64$
$\eta_{material}$	Stainless Steel	Plastic
	1.0	0.7

Per the model, surface tension affects $a_{e,packing}$ but not viscosity. This is confirmed by Fig. 7 & 8 where no bias of error was observed as a function of the two physical properties. For viscosity, both solutions of polyethylene glycol (PEG) and glycerol have shown no difference in the $a_{e,packing}$ compared to water.

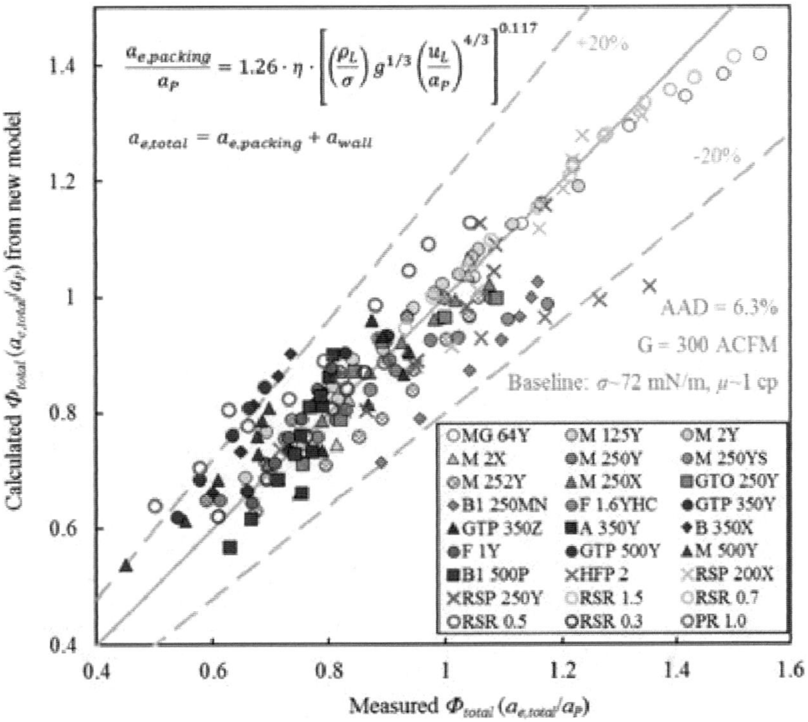

Fig. 5: Comparison of measured and calculated area from Eq. 1 for water

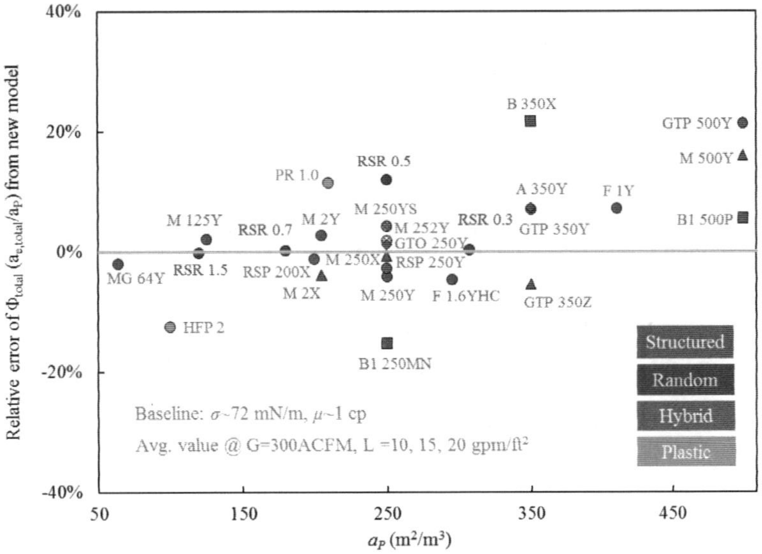

Fig. 6: Relative error of Eq. 1 for different packings

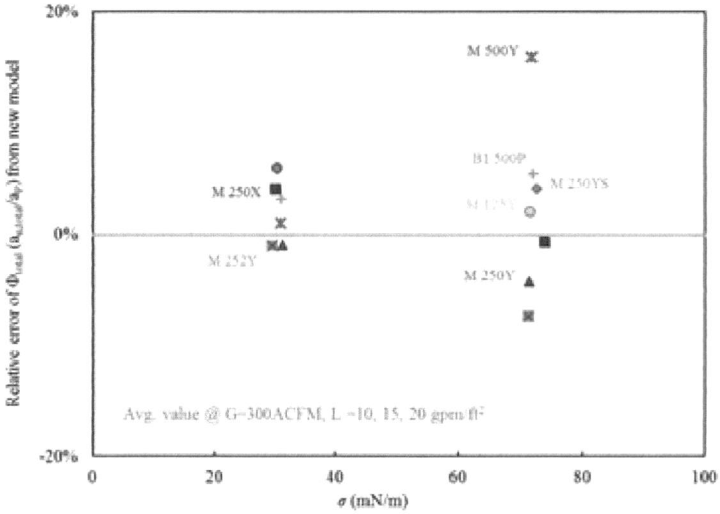

Fig. 7: Relative error of Eq. 1 for different surface tension

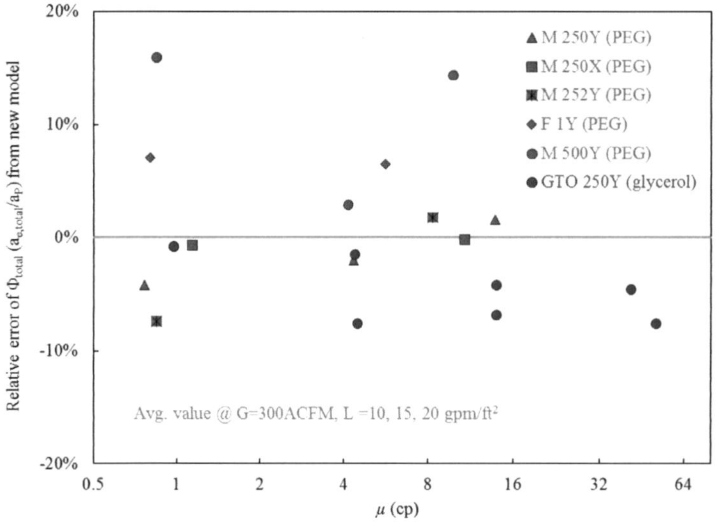

Fig. 8: Relative error of Eq. 1 for different viscosity

Model of Liquid Film Mass Transfer Coefficient (k_L)

The experimental protocol for k_L measurement of SRP has been changed over time. A total of 10 10 ft of packing was initially used. The height was then reduced to 6 ft due to the low outlet toluene concentration that is close to the detection limit of the gas chromatograph (GC). Theoretically, k_L should not be affected by packing height when there is no issue with liquid distribution and GC detection. However, greater k_L has been observed with the 6 ft bed than the 10 ft bed, regardless of packing type and experiment operators (Fig. 9).

The difference of k_L with different bed height may be caused by both the poorer liquid distribution and by issue with the GC detection limit at greater packing bed height. In the development of k_L model, k_L measured with 10 ft bed will be multiplied by 1.32 to be able to compare with those measured with 6 ft bed.

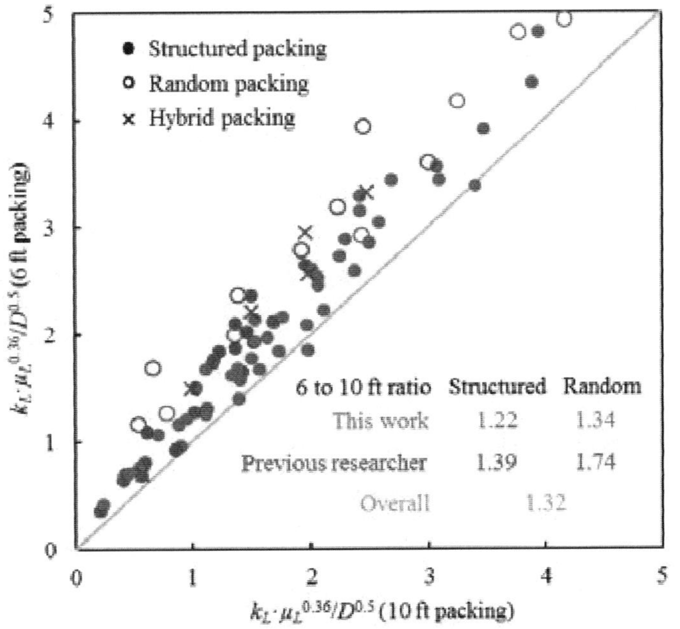

Figure 9: Comparison of k_L measured with 10 ft and 6 ft packing

The average k_L of structured, random, and hybrid packings at 300 ACFM gas rate and 10, 15, and 20 gpm/ft^2 liquid load have been plotted against a_P (Fig. 10). For structured packings, there is a counter-intuitive trend that the k_L decreases with increasing a_P. This might be caused by the longer flow path (and thus greater turbulence) of the liquid on the surface of course packings. The corrugation angle does not have a clear effect on k_L.

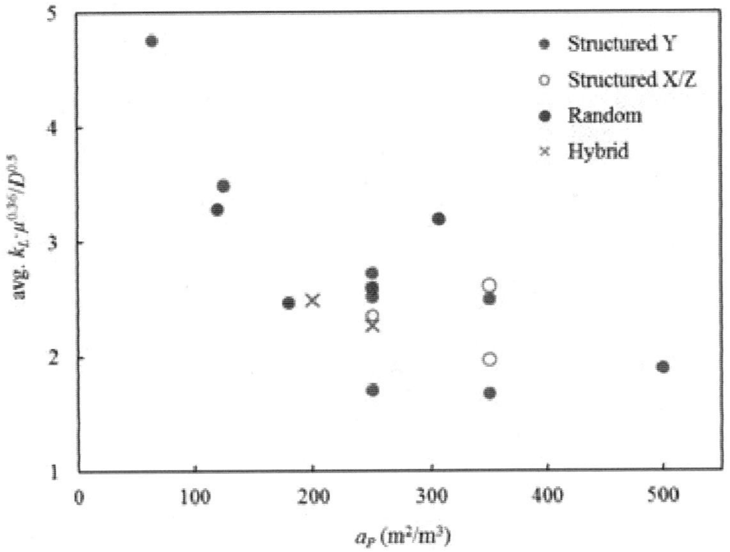

Fig. 10: Average k_L as a function of a_P

Figure 11 compares the measured k_L and calculated k_L from Eq. 2 for 21 packing. Glycerol and water tests have been performed on nine of the packings, and water tests only on the rest of 12 packings. The new model shows an AAD of 20% for all 21 packings. Unlike the area model, the k_L model does not require a correction for random or hybrid packings.

Figure 12 shows the total dependence on μ_L of k_L. The slope (-0.72) is the same for structured, random, and hybrid packings, which means the impeding effect of μ_L on k_L is the same for all three types of packings. Of the total -0.72 dependence, -0.35 is from the indirect influence of μ_L through diffusivity, and -0.37 is from the direct influence of μ_L on k_L through liquid turbulence. The indirect part is believed to be system dependent, and requires knowledge on the D-μ relationship. The direct part is universally applicable to all Newtonian liquids.

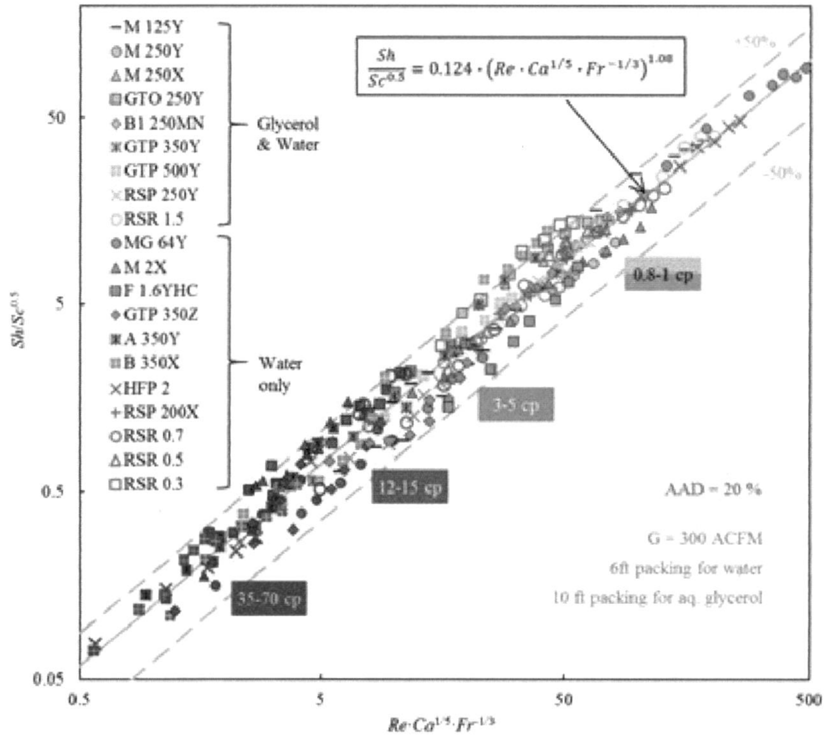

Fig. 11: Comparison of measured and calculated k_L from Eq. 2

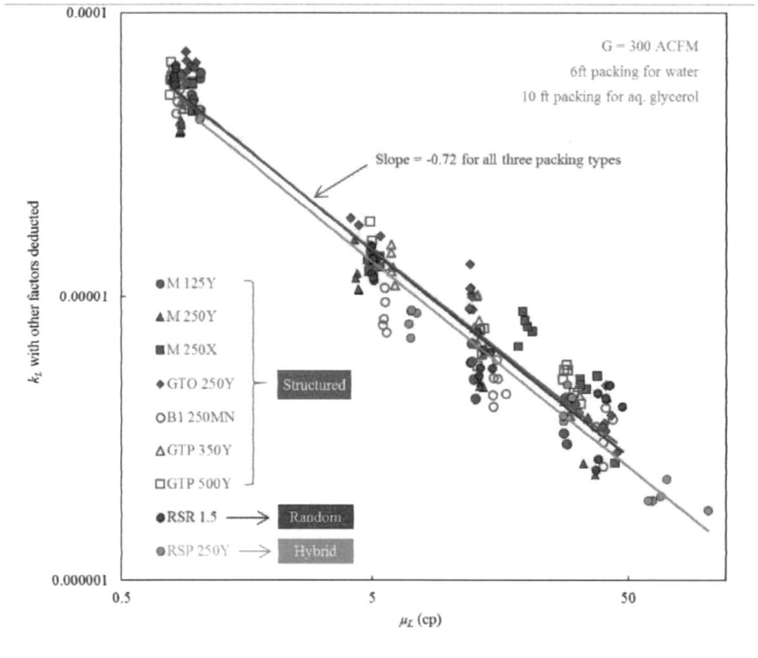

Fig. 12: Total dependence on μ_L of k_L

Figures 13 & 14 shows the relative error of Eq. 2 as function of a_P, and μ_L, respectively. No bias of error was observed as a function of a_P for water in Fig. 13. The model underpredicts k_L for water slightly. In Fig. 14, no bias of error was observed as a function of μ_L.

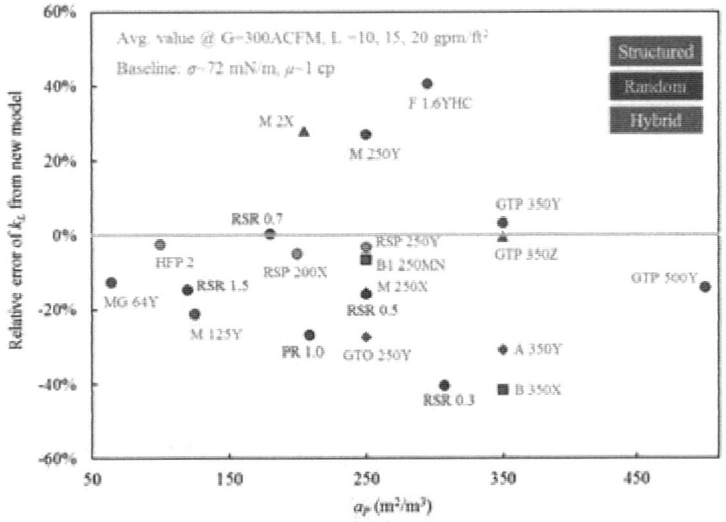

Fig. 13: Relative error of Eq. 2 for different a_P

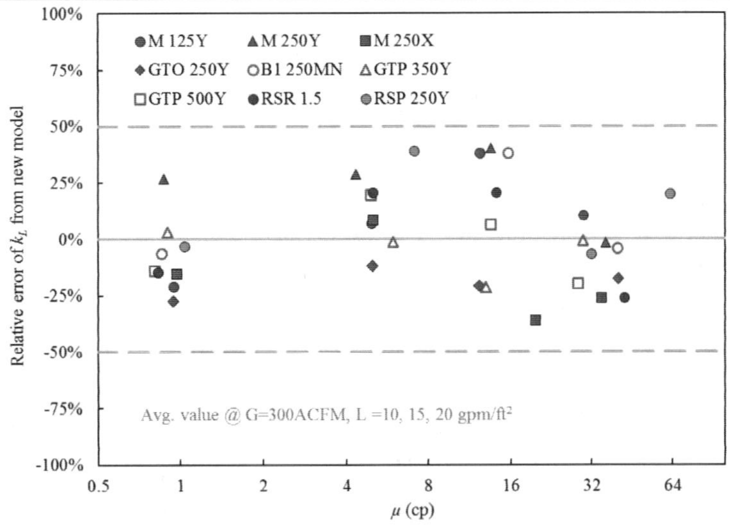

Fig. 14: Relative error of Eq. 2 for different μ_L

Conclusions

Models of a_e and k_L (Eq.s 1 & 2) have been developed based on the SRP air-water column database with 27 packings of various type, a_P, and materials. Liquid viscosity has been varied from 0.8 to 70 cP by the addition of glycerol. The μ_L does not affect $a_{e,packing}$, but it affects k_L with a total dependence of -0.72, of which -0.35 is from the indirect influence through diffusivity, and -0.37 is from the direct influence of μ_L on k_L through liquid turbulence.

The wall effect is dominant in the secondary effect on a_e, and it has been included in the a_e model. The k_L decreases with increasing a_P for structured packings, and shorter packing bed consistently shows greater k_L than taller bed with the same packing internal. With a relatively large equipment size and packing database, this work has the greatly expanded the data of mass transfer in viscous liquids available in the open literature.

Nomenclature

a_e – effective area of mass transfer per volume of column, m^2/m^3
a_P – specific area of packing per volume of column, m^2/m^3
C – concentration, ppmv or ppmw
D – diffusivity, m^2/s
d – characteristic length in dimensionless groups (liquid film thickness in a_e model, and $1/a_P$ in k_L model), m
g – gravitational acceleration, m/s^2
k_g' – liquid film mass transfer coefficient with chemical reactions expressed in gas unit, $mol/m^2\cdot Pa\cdot s$
k_L – liquid film mass transfer coefficient, m/s
K_{OG} – overall mass transfer coefficient expressed in gas unit, $mol/m^2\cdot Pa\cdot s$
u – characteristic velocity in dimensionless groups (liquid film velocity in a_e model, and liquid superficial velocity in k_L model), m/s
Z – packing height

Greek letters

σ – surface tension, N/m
μ – viscosity, cP
ρ – density, kg/m^3
η – correction factor for area model

Dimensionless groups

Ca – Capillary number, $\mu\cdot u/\sigma$
Fr – Froude number, $u^2/g\cdot d$
Re – Reynolds number, $d\cdot u\cdot \rho/ \mu$
Sc – Schmidt number, $\mu/\rho\cdot D$
Sh – Sherwood number, $k_L\cdot d/D$
We – Webber number, $\rho\cdot u^2\cdot d/\sigma$

Packing names

B1 – MONTZ-Pak Type B1TM
F – Flexipac®
GTO – GT-OPTIMPAKTM
GTP – GT-PAKTM
HFP – Hiflow® Plus
M – MellaPakTM
MG – MellaGridTM
PR – Pall Ring

RSP – Raschig Super-Pak®
RSR – Raschig Super-Ring®

References

1. Song, D., A.F. Seibert, and G.T. Rochelle, *Effect of Liquid Viscosity on the Liquid Phase Mass Transfer Coefficient of Packing.* Energy Procedia, 2014. **63**(0): p. 1268-1286.

2. Tsai, R.E., *Mass Transfer Area of Structured Packing*, in *Department of Chemical Engineering*. 2010, The University of Texas at Austin: USA.

3. Wang, C., *Mass Transfer Coefficients and Effective Area of Packing*, in *Department of Chemical Engineering*. 2015, The University of Texas at Austin: USA.

4. Pohorecki, R. and W. Moniuk, *Kinetics of reaction between carbon dioxide and hydroxyl ions in aqueous electrolyte solutions.* chemical Engineering Science, 1988. **43**(7): p. 1677–1684.

5. Song, D. and G.T. Rochelle, *Reaction kinetics of carbon dioxide and hydroxide in aqueous glycerol.* Chemical Engineering Science, 2017. **161**: p. 151-158.

480985 Novel Solvent Exchange Distillation Process- a Comparison Study with Batch Systems

Abstract:

A novel solvent exchange distillation process for both zeotropic and azeotropic equilibrium, which has been developed for replacement of one solvent with a second solvent while keeping the concentration of dissolved nonvolatile solute constant, is described. Detailed simulations for methanol or ethanol replaced by water show that solvent to feed ratios near 1.0 are often sufficient. Comparison of the results with detailed batch distillation simulations show that the novel solvent exchange method gives the same purity with lower solvent and energy consumption. When the less volatile solvent has to be replaced with a more volatile solvent, batch distillation, which requires very large amounts of added solvent and of energy, would normally not be used. Thus, the new solvent exchange distillation process represents a totally new approach to this exchange problem.

AUTHOR INDEX

AUTHOR INDEX

AUTHOR INDEX